McFarlin Library
WITHDRAWN

COMPUTER APPLICATIONS IN RESOURCE ESTIMATION

Prediction and Assessment
for Metals and Petroleum

COMPUTERS and GEOLOGY

a series edited by Daniel F. Merriam

1976 *Quantitative Techniques for the Analysis of Sediments*
1978 *Recent Advances in Geomathematics*
1979 *Geomathematical and Petrophysical Studies in Sedimentology*
     (edited by D. Gill & D. F. Merriam)
1981 *Predictive Geology: with Emphasis on Nuclear-Waste Disposal*
     (edited by G. de Marsily & D. F. Merriam)
1986 *Microcomputer Applications in Geology*
     (edited by J. T. Hanley & D. F. Merriam)
1990 *Microcomputer Applications in Geology, II*
     (edited by J. T. Hanley & D. F. Merriam)

Professor Merriam also is the Editor-in-Chief of *Computers & Geosciences*—an international journal devoted to the rapid publication of computer programs in widely used languages and their applications.

Other Related Pergamon Publications

Books

HOLLAND
Microcomputers and Their Interfacing
HOLLAND
Illustrated Dictionary of Microcomputers & Microelectronics
MARSAL
Statistics for Geoscientists
NORRIE & TURNER
Automation for Mineral Resource Development

Journals

Acta Geologica Sinica
Acta Seismologica Sinica
Applied Geochemistry
Automatica
Computers & Geosciences
Computer Languages
Information Processing & Management
Information Systems
International Journal of Rock Mechanics and Mining
Sciences & Geomechanics Abstracts
Nuclear Geophysics

Full details of all Pergamon publications/free specimen copy of any Pergamon journal available on request from your nearest Pergamon office.

COMPUTER APPLICATIONS IN RESOURCE ESTIMATION

Prediction and Assessment for Metals and Petroleum

Edited by

GABOR GAÁL

*Chief Geologist, Geological Survey of Finland,
and Chairman, COGEODATA*

and

DANIEL F. MERRIAM

*Endowment Association Distinguished Professor of the Natural Sciences,
Wichita State University, Wichita, Kansas*

PERGAMON PRESS
Member of Maxwell Macmillan Pergamon Publishing Corporation

OXFORD · NEW YORK · BEIJING · FRANKFURT
SÃO PAULO · SYDNEY · TOKYO · TORONTO

U.K.	Pergamon Press plc, Headington Hill Hall, Oxford OX3 0BW, England
U.S.A.	Pergamon Press, Inc., Maxwell House, Fairview Park, Elmsford, New York 10523, U.S.A.
PEOPLE'S REPUBLIC OF CHINA	Pergamon Press, Room 4037, Qianmen Hotel, Beijing, People's Republic of China
FEDERAL REPUBLIC OF GERMANY	Pergamon Press GmbH, Hammerweg 6, D-6242 Kronberg, Federal Republic of Germany
BRAZIL	Pergamon Editora Ltda, Rua Eça de Queiros, 346, CEP 04011, Paraiso, São Paulo, Brazil
AUSTRALIA	Pergamon Press Australia Pty Ltd., P.O. Box 544, Potts Point, N.S.W. 2011, Australia
JAPAN	Pergamon Press, 5th Floor, Matsuoka Central Building, 1-7-1 Nishishinjuku, Shinjuku-ku, Tokyo 160, Japan
CANADA	Pergamon Press Canada Ltd., Suite No. 271, 253 College Street, Toronto, Ontario, Canada M5T 1R5

Copyright © 1990 Pergamon Press plc

All Rights Reserved. No part of this publication may be reproduced, stored in a retrieval system or transmitted in any form or by any means: electronic, electrostatic, magnetic tape, mechanical, photocopying, recording or otherwise, without permission in writing from the publisher.

First edition 1990

Library of Congress Cataloging-in-Publication Data
Computer applications in resource estimation : prediction and assessment for metals and petroleum / edited by Gabor Gaál and Daniel F. Merriam.
p. cm. -- (Computers and geology)
Based on a symposium cosponsored by COGEODATA and the International Association for Mathematical Geology, held in Helsinki, Finland, July 21-23, 1988.
Includes index.
1. Prospecting — Data processing — Congresses. 2. Petroleum — Prospecting — Data processing — Congresses. I. Gaál, Gabor. II. Merriam, Daniel Francis. III. COGEODATA. IV. International Association for Mathematical Geology. V. Series: Computers & geology.
TN270.A1C635 1990   622'. 1--dc20   90-7896

British Library Cataloguing in Publication Data
Computer applications in resource estimation : prediction and assessment for metals and petroleum.
1, Mineral deposits. Prospecting. Applications of geophysics. Applications of computer systems
I. Gaál, Gabor II. Merriam, Daniel F. (Daniel Francis)
622.150285
ISBN 0-08-037245-7

Printed in Great Britain by BPCC Wheatons Ltd, Exeter

CONTENTS

List of Contributors ix

Preface, by G. Gaál and D.F. Merriam xiii

DATA INTEGRATION IN MINERAL EXPLORATION BY STATISTICAL AND MULTIVARIATE TECHNIQUES

Statistical pattern integration for mineral exploration,
    by F.P. Agterberg, G.F. Bonham-Carter, and D.F. Wright 1

Statistical classification of regional geochemical samples using
    local characteristic models and data of the geochemical
    atlas of Finland and from the Nordkalott Project,
    by N. Gustavsson and M. Kontio 23

A map-comparison technique utilizing weighted input parameters,
    by U.C. Herzfeld and D.F. Merriam 43

Intrinsic sample methodology,
    by D. Harris and Guocheng Pan 53

Evaluation of the gold potential of the Bohemian Massif,
    by J. Janatka and P. Morávek 75

Comparison of subjective and objective methods in quantitative
    exploration: case studies,
    by C. Kliem and Th. Petropulos 83

Analysis and integration of reconnaissance data in a mineral resource-
    assessment program of Austria,
    by H. Kürzl 97

REGION-SCANDING - mineral forecasting computer system,
    by E.A. Nemirovsky 119

Man-machine analysis of geological maps,
by V.V. Marchenko and E.A. Nemirovsky 125

GEONIX - an UNIX-based automatic data-processing system applied
to geoscience information,
by E. Sauzay, H. Teil, M. Vannier, and L. Zanone 131

Methods and techniques of the prediction of metallic and nonmetallic
raw materials using microcomputers in Czechoslovakia,
by C. Schejbal and J. Hruska 147

Use of characteristic analysis coupled with other quantitative techniques
in mineral-resources appraisal of Precambrian areas in São Paulo - Brazil,
by S.B. Suslick and B.R. Figueredo 155

INTERCRAST - the technology for prognosis and quantitative assessment
of mineralization in regions of intrusive magmatism based on
numerical modeling,
by V.G. Zolotarev 185

DATA INTEGRATION IN MINERAL EXPLORATION BY IMAGE PROCESSING AND OTHER TECHNIQUES

Use of image processing and integrated analysis in exploration by
Outokumpu Oy, Finland,
by J. Aarnisalo 195

Mappable data integration techniques in mineral exploration,
by D. Bonnefoy and A.L. Guillen 213

The use of digital elevation models computed from SPOT stereopairs
for uranium exploration,
by P. Leymarie, J. Dardel, and L. Renouard 225

APPLICATIONS IN PETROLEUM EXPLORATION

Conditional simulation in oil exploration,
by H. Burger, M. Eder, A. Mannaa, and W. Skala 239

Computer-assisted estimation of discovery and production of crude oil
from undiscovered accumulations,
by D.J. Forman and A.L. Hinde 253

Geostatistical characterization of selected oil shale and phosphate
    deposits in Israel,
    by D. Gill 273

Pore geometry evaluation by petrographic image analysis,
    by S.M. Habesch 301

Space modeling and multivatiate techniques for prognosis of hydrocarbons,
    by J. Harff, J. Springer, B. Lewerenz, and W. Eiserbeck 321

Petroleum prospect size estimation by numerical methods,
    by T. Jasko 339

The use of expert systems in identification of siliciclastic depositional systems
    for hydrocarbon reservoir assessment,
    by P.G. Sutterlin and G.S. Visher 347

INVENTORIES

Geological comparison of Brazil and China by state,
    by J.C. Griffiths, H. Hu, and H.C. Chou 367

Application of Q-analysis to the GLOBAL databank: a geological
    comparison of the U.S.S.R. and the U.S.A.,
    by D.N. Pilant, J.C. Griffiths, and C.M. Smith 395

Explorational databases at the Geological Survey of Finland,
    by B. Saltikoff and T. Tarvainen 409

RELATED STATISTICAL TECHNIQUES

Regression analysis of geochemical data with observations below
    detection limit,
    by Chang-Jo F. Chung 421

Trend analysis on a personal computer: problems and solutions,
    by J.E. Robinson 435

Index 449

List of Contributors

J. Aarnisalo, Outokumpu Oy Exploration, Outokumpu, Finland

F.P. Agterberg, Geological Survey of Canada, 601 Booth Street, Ottawa, Ontario K1A OES, Canada

G.F. Bonham-Carter, Geological Survey of Canada, 601 Booth Street, Ottawa, Ontario K1A OES, Canada

D. Bonnefoy, Bureau de Recherches Geologiques et Minieres (BRGM), BP 6009, 45060 Orleans Cedex 2, France

H. Burger, Institut fur Geologie, Mathematische Geologie, Freie, Universitat Berlin, Malteserstrasse 74-100, D-1000 Berlin 46, West Germany

H.C. Chou, Department of Geoscience, Pennsylvania State University, University Park, PA 16802, USA

C.F. Chung, Geological Survey of Canada, 601 Booth Street, Ontario K1A OES, Canada

J. Dardel, CEA-DAMN, 31-33 rue de las Federation, 75752, Paris Cedex 15, France

M. Eder, Bundesanstalt fur Geowissenschafter und Rohstoffe (BGR), Hannover, West Germany

W. Eiserbeck, VEB Kombinat Erdol-Erdgas, Stammbetrieb Gommern, GDR

B.R. Figueredo, Instituto de Geochiencias, Unicamp, Brazil

D.J. Forman, Bureau of Mineral Resources, Canberra, Australia 2601

G. Gaál, Geological Survey of Finland, Betonimiehenkuja 4, SF-02150 Espoo Finland

D. Gill, Geological Survey of Israel, 30 Malkhie Israel Street, Jerusalem 95501, Israel

J.C. Griffiths, Department of Geoscience, Pennyslvania State University, University Park, PA 16802 USA

A. Guillen, Bureau de Recherches Geologiques et Minieres (BRGM), BP 6009, 45060 Orleans Cedex 2, France

N. Gustavsson, Geological Survey of Finland, Betonimiehenkuja 4, SF-02150 Espoo, Finland

S.M. Habesch, Poroperm-Geochem Ltd., The Geochem Group, Chester Street, Chester CH14, SRD, UK

J. Harff, Akademie der Wissenschaftern der DDR, Zentral-institut fur Physic det Erde, Telegrafenberg A1F, Potsdam, DDR-1561

D.P. Harris, Mineral Economics Program & Department of Mining and Geological Engineering, University of Arizona, Tucson, AZ 85721, USA

U.C. Herzfeld, Scripps Institution of Oceanography, Geologic Research Division, University of California/San Diego, La Jolla, CA 92093, USA

A.L. Hinde, Bureau of Mineral Resources, Canberra, Australia 2601

J. Hruska, Intergeo, Olbrachtova 3, 14600, Praha 4, Czechoslovakia

H. Hu, China University of Geology, Wuhan, China

J. Janatka, Geoindustia Praha, Komunardu 6, 170 04 Praha 7, Czechoslovakia

T. Jasko, Quartz Scientific Computing Ltd., 16 Melrose Place, Watford WD1 3LN England, UK

C. Kliem, Institut fur Geologie/Mathematische Geologie, Freie Universitat Berlin, Malteserstrasse 74-100, D-1000 Berlin 46, West Germany

M. Kontio, Geological Survey of Finland, P. O. Box 77, SF-96101 Rovaniemi, Finland

H. Kürzl, Logistik-Management-Service, Gesellschaft mbh, Franz-Josef Strasse 6, Postfach 070, A-8700 Leoben, Austria

B. Lewerenz, Akademie der Wissenschaftern der DDR, Zentral-institut fur Physik det Erde, Telegrafenberg A1F, Potsdam, DDR-1561

P. Leymarie, CNRS, URA Geodynamique, Nice - INRIA, route des Lucioles, Sophia Antipolis, 06560 Valbornne, France

A. Mannaa, Institut fur Geologie, Mathematische Geologie, Freie Universitat Berlin, Matteserstrasse 74-100, D-1000 Berlin 46, West Germany

V.V. Marchenko, International Research Institute for Management Sciencees, Oktyabria 9, 117312 Moscow, USSR

D.F. Merriam, Stratigraphic Studies Group, Box 153, Wichita State University, Wichita, KS 67208, USA

P. Morávek, Geoindustia Praha, Komunardu 6, 170 04, Praha 7, Czechoslovakia

E.A. Nemirovsky, International Research Institute for Management Sciences, Oktyabria 9, 117312 Moscow, USSR

Guocheng Pan, Mineral Economics Program & Department of Mining and Geological Engineering, University of Arizona, Tucson, AZ 85721, USA

Th. Petropulos, Institut fur Geologie, Mathematische Geologie, Freie Universitat Berlin, Malteserstrasse 74-100, D-1000 Berlin 46, West Germany

D.N. Pilant, Department of Geoscience, Pennsylvaia State University, University Park, PA 16802, USA

L. Renouard, ISTAR, Les Algorithmes, 2000 route des Lucioles, Sopia Antipolis, 06560, Valbonne, France

J.E. Robinson, Department of Geology, Syracuse University, Syracuse, NY 13244, USA

B. Saltikoff, Geological Survey of Finland, Betonimiehenkuja 4, SF-02150 Espoo, Finland

E. Sauzay, Informatique Gitologique et Miniere (ICM), Centre de Geologie Generale et Miniere, Ecole National Superieure des Mines de Paris, 35, rue St. Honore, 77305 Fontainebleau Cedex, France

C. Schejbal, University of Mining and Metallurgy, Ostrava, Czechoslovakia

W. Skala, Institute fur Geologie, Mathematische Geologie, Freie Universitat Berlin, Malteserstrasse 74-100, D-1000 Berlin 46, West Germany

C.M. Smith, Department of Computer Science, Pennsylvania State University, University Park, PA 16802, USA

J. Springer, Akademie der Wissenschaftern der DDR, Zentral-institut fur Physic det Erde, Telegrafenberg A1F, Potsdam, DDR-1561

S.B. Suslick, Instituo de Geochiencias, Unicamp, Brazil

P.G. Sutterlin, Department of Geology, Wichita State University, Wichita, KS 67208, USA

T. Tarvainen, Geological Survey of Finland, Betonimiehenkuja 4, SF-02150 Espoo, Finland

H. Teil, Informatique Gitologique et Miniere (ICM), Centre de Geologie Generale et Miniere, Ecole National Superieure des Mines de Paris, 35, rue St. Honore, 77305 Fontainebleau Cedex, France

M. Vannier, Informatique Gitologique et Miniere (ICM), Centre de Geologie Generale et Miniere, Ecole National Superieure des Mines de Paris, 35, rue St. Honore, 77305 Fontainebleau Cedex, France

G.S. Visher, Geological Services & Ventures, Inc., 2920 E. 73rd Street, Tulsa, OK 74136, USA

D.F. Wright, Geological Survey of Canada, 601 Booth Street, Ottawa, Ontario K1A OES, Canada

L. Zanone, Informatique Gitologique et Miniere (ICM), Centre de Geologie Generale et Miniere, Ecole National Superieure des Mines de Paris, 35, rue St. Honore, 77305 Fontainebleau Cedex, France

V.G. Zolotarev, VNII Zarubezhgeologia, USSR Ministry of Geology, Moscow, USSR

Preface

"Computer Applications in Resource Exploration" was the subject of a symposium cosponsored by COGEODATA and the International Association for Mathematical Geology (IAMG) in Helsinki on 21-23 July 1988. The aim of the symposium, convened by Gabor Gaal and Daniel F. Merriam, was to review modern methods of resource prediction and resource assessment in exploration. The meeting was attended by 75 scientists from 16 countries and served as an important discussion forum to international experts. Thirty eight papers were presented along with several practical computer demonstrations on such subjects as trend analysis, classification, correlation, expert systems, image analysis, remote sensing, geostatistics, computer mapping, and databases. Applications were on all aspects of resource exploration, assessment, characterization, and exploitation including gold, base metals, phosphates, petroleum, oil shale, and coal.

Applications and the practical aspects of resource exploration prevailed with a strong emphasis on the statistical approach. The techniques ranged from regression and trend analysis through association and characteristic analysis to probability methods. Several papers reported on image analysis and remote-sensing techniques involving satellite images from the French satellite SPOT and the British-developed sensor GLORIA. Other papers were concerned with databases and database-management systems including the French GEONIX system. Many papers reported results of regional or national mineral-resource assessments, usually based on geochemical surveys, for all or parts of Finland, Czechoslovakia, Israel, West Germany, Hungary, Canada, Austria, Brazil, and the USSR; case studies were given in other papers. About one-quarter of the papers were petroleum oriented and about 40 percent concerned with minerals. However, many of the techniques which were reported could be used for either the exploration or exploitation of hydrocarbons or minerals.

It was obvious from the presentations that most of the computing was done on microcomputers, which now are widespread in academic, governmental, and industrial organizations. Although no new techniques were revealed, a different approach to problem solving using true-and-tested methods, were stressed. The papers demonstrated that predictive techniques and those that optimize are important. It also was shown that modeling is popular and effective and that before all, the systems approach and data integration are necessary in todays environment.

The contributions in the present volume highlight the spirit and the results of the meeting. For sake of systematization the topics of the twenty-two papers submitted have been subdivided into five categories which are:

- data integration in mineral exploration by statistical and multivariate techniques,
- data integration in mineral exploration by image processing and other techniques,
- applications in petroleum exploration,
- inventories, and
- related statistical techniques.

DATA INTEGRATION IN MINERAL EXPLORATION BY STATISTICAL AND MULTIVARIATE TECHNIQUES. In their paper on statistical pattern integration *F.P. Agterberg, G.F. Bonham-Carter,* and *D.F. Wright* offer new and original solutions to three important problems (1) how to optimize binary pattern for linear features associated with mineral deposits, (2) integration of linear structures with geochemical and lithological features, and (3) quantifying uncertainty because of missing information. *N. Gustavsson* and *M. Kontio* assign ore potential to various geochemical anomaly patterns using statistical classification techniques on widely spaced samples in the Precambrian terrain of northern Europe. *D.P. Harris* and *Goucheng Pan* abandon the concept of gridding spatial data and introduce a new intrinsic sampling methodology which allows the use of genetic relations to integrate geodata and the optimal discretization of various geofields and deposit-model areas. *U.C. Herzfeld* and *D.F. Merriam* integrate different anomaly patterns of various geoscientific maps by applying weighted parameters. *J. Janatka* and *P. Morávek* combine empirical methods, deposit inventory, and multivariate techniques (cluster analysis and multiple linear regression) in evaluating gold potential of the Bohemain Massif. *C. Kliem* and *Th. Petropulos* integrate the exploration knowledge in the assessment of the mineral resources of parts of Greece and Turkey by various statistical methods. *H. Kurzl* compiled geological, geochemical, geophysical, and deposit data in a Geographic Information System and assesses mineral resources by multivariate techniques. *E.A. Nemirovsky* gives a short description of SCANDING, developed out of the REGION package, a fully integrated resource-assessment software used in Eastern European countries. *E.A. Nemirovsky* and *V.V. Marchenko* present the principles underlying the use of geological map in quantified resource assessment. *E. Sauzay, H. Teil, M. Vannier,* and *L. Zanone* work on the interface between bibliographic and factual data presenting the GEONIX package for data integration by various statistical techniques. *C. Schejbal* and *J. Hruska* supply a general overlook on mineral-resource prediction techniques applied in Czechoslovakia. *S.B. Suslick* and *B.R. Figueredo* apply characteristic analysis to regional geochemical data in part of Brazil for indicating favorable areas for base metals and bauxite. *V.G. Zolotarev* formulizes an approach of genetic knowledge of granitoid-associated hydrothermal ore deposits to estimate precious and base-metal potential.

PREFACE

DATA INTEGRATION IN MINERAL EXPLORATION BY IMAGE PROCESSING AND OTHER TECHNIQUES. In a pragmatic approach, *J. Aarnisalo* presents the use of image-processing techniques by Outokumpu Oy, Finland, in integrating satellite, geophysical, and to some extent, geological data for pinpointing exploration targets. *D. Bonnefoy* and *A.L. Guillen* demonstrate a Geographic Information System MARICA developed by BRGM, France, which is an integrated and interactive package uniting image-processing techniques, statistical treatment, and expert-system techniques. *P. Leymarie, J. Dardel,* and *L. Renouard* introduce new techniques of computation for digital elevation models from SPOT stereopairs used through structural analysis for uranium exploration.

APPLICATIONS IN PETROLEUM EXPLORATION. *H. Burger, M. Eder, A. Mannaa,* and *W. Skala* apply conditional simulation techniques in correlating geological structures with known oil occurrences. With the SEAPUP computer program, *D.J. Forman* and *A.L. Hinde* simulate drilling and discovery of onshore petroleum traps in Australia and assess undiscovered crude-oil resources. *D. Gill* uses geostatistical methods for modeling the major characteristics and the necessary drilling density in exploration of oil shale and phosphate deposits. *S.M. Habesch* uses image-analysis techniques to measure the geometrical parameters of porosity networks in thin sections of oil-reservoir rocks. Mathematical modeling of sedimentary structures and subsidence history is applied to multivariate data in microcomputer to predict hydrocarbon resources in part of East Germany by *J. Harff, J. Springer, B. Lewerenz,* and *W. Eiserbeck. T. Jasko* estimates probable sizes of undiscovered petroleum resources by the Monte Carlo and Latin Squares methods on a microcomputer. *P.G. Sutterlin and G.S. Visher* demonstrate the usefulness of expert systems in identifying not directly measurable geological features which play an important role in the discovery of hydrocarbons.

INVENTORIES. After compiling an inventory of mineral resources and lithological variations of Brazil and China, *J.C. Griffiths, H. Hu,* and *H.C. Chou* compare these to countries on basis of geological diversity and draw conclusions on mineral-resource development of China and Brazil. *D.N. Pilant, J.C. Griffiths ,and M. Smith* apply the GLOBAL data bank for a similar geological comparison of the U.S.S.R. and U.S.A. using Atkin's Q-analysis. *B. Saltikoff* and *T. Tarvainen* demonstrate contents, structure, and use of explorational databases in the Geological Survey of Finland.

RELATED STATISTICAL TECHNIQUES. *Chang-Jo F. Chung* presents a new regression technique based upon the maximum likelihood method for utilization of geochemical data sets with observations below detection limits. *J.E. Robinson* points out the pitfalls of trend-analysis techniques as applied in microcomputer softwares and demonstrates simple tests to avoid them. The contributions in this volume present a representative cross section on world-wide applications of computerized mathematical techniques in resource exploration. It is hoped that the papers and references therein will increase our knowledge in this field of practical mathematical application and will facilitate world-wide contacts among experts and users. The editors would like to express their sincere gratitude to the authors for their efficient work which has enabled the appearance of this volume in such a short time after the closure of the symposium and thus making possible to disseminate fresh information.

Gabor Gaál
Chairman of COGEODATA
Geological Survey of Finland
SF-02150 ESPOO
Finland

Daniel F. Merriam
Stratigraphic Studies Group
Box 153
Wichita State University
Wichita, Kansas 67208
U.S.A.

DATA INTEGRATION IN MINERAL EXPLORATION BY STATISTICAL AND MULTIVARIATE TECHNIQUES

Statistical Pattern Integration for Mineral Exploration*

F.P. Agterberg,
G.F. Bonham-Carter, and
D.F. Wright
Geological Survey of Canada, Ottawa

ABSTRACT

The method of statistical pattern integration used in this paper consists of reducing each set of mineral deposit indicator features on a map to a pattern of relatively few discrete states. In its simplest form the pattern for a feature is binary representing its presence or absence within a small unit cell; for example, with area of 1 km² on a 1:250,000 map. The feature of interest need not occur within the unit cell; its "presence" may indicate that the unit cell occurs within a given distance from a linear or curvilinear feature on a geoscience map. By using Bayes' rule, two probabilities can be computed that the unit cell contains a deposit. The log odds of the unit cell's posterior probability is obtained by adding weights W⁺ or W⁻ for presence or absence of the feature to the log odds of the prior probability. If a binary pattern is positively correlated with deposits, W⁺ is positive and the contrast C=W⁺–W⁻ provides a measure of the strength of this correlation. Weights for patterns with more than two states also can be computed and special consideration can be given to unknown data. Addition of weights from several patterns results in an integrated pattern of posterior probabilities. This final map subdivides the study region into areas of unit cells with different probabilities of containing a mineral deposit. In this paper, statistical pattern integration is applied to occurrence of gold mineralization in Meguma Terrane, eastern mainland Nova Scotia, Canada.

INTRODUCTION

Geoscience maps of different types are to be integrated for target selection in mineral exploration. The geologist stacks these maps and looks for combinations of indicators favorable for occurrence of deposits of different types. The calculations required for mathematical analysis of digitized patterns for points, lines, and areas have been greatly aided by the development of microcomputer based geographic information

---

* Geological Survey of Canada Contribution No. 24088

systems for the treatment of map data (Bonham-Carter, Agterberg, and Wright, 1988). This has led us to develop further a new method for statistical pattern integration simulating the practice of exploration geologists to combine maps for delineating favorable areas. This method was proposed initially by Agterberg (in press) for combining geophysical survey data with prior probabilities of occurrence of massive sulfide deposits in the Abitibi area of the Canadian Shield originally obtained in 1971. The prior probability for a massive sulfide deposit being in a small unit cell was assumed to be constant within a larger cell. The frequency of massive sulfide deposits had been estimated by regression analysis from lithological and other variables systematically coded for such larger cells. Other geoscience data for the same area (Bouguer anomalies, aeromagnetic anomalies, and boundaries between tertiary drainage basins) had been quantified later as patterns of two or more mutually exclusive states by Assad and Favini (1980). The patterns for proximity to aeromagnetic anomalies and boundaries between drainage basins were binary, needing only two colors (black and white) for representation. It was possible to compute weights W_i^+ and W_i^- representing the states of presence and absence in the unit cell for each binary pattern i.

In Assad and Favini (1980), the pattern for the Bouguer anomaly had five distinct states with different colors. In Agterberg (in press), a weight W_j^+ was computed for each color j of this pattern with more than two states. At any point within the study area, the weights for the geophysical variables were added to the log odds of the prior probability. This gave the log odds of the posterior probability. Because the patterns combined with one another all consisted of polygons, the final product was also polygonal with different colors for classes defined for the posterior probability per unit cell. The addition of weights W^+ or W^- is permitted only if the patterns being integrated are conditionally independent of occurrence of deposits. In the Abitibi study it was shown that this condition is satisfied approximately for the geophysical variables. In this paper, the method will be applied to gold deposits in Meguma Terrane, Nova Scotia.

An advantage of the statistical pattern integration method with respect to most existing methods in the field of regional resource evaluation (e.g., logistic regression) is that a pattern need be available only for parts of the study region. However, if one or more patterns are missing at a given place, the estimated posterior probability has less certainty than those based on more or all patterns. This type of uncertainty, the result of one or more missing patterns, will be studied later in the paper.

Finally, special attention should be given to verification of the theoretical assumption of conditional independence. The simple addition of weights for different features is permitted only if this assumption is satisfied. In general, the possibility of occurrence of conditional dependence increases with an increasing number of patterns. Failure of the method in this respect would lead to discrepancies between frequencies as predicted by the posterior probability map and the corresponding observed frequencies. If the assumption of conditional independence is not satisfied, the theoretical frequencies would exceed the observed frequencies in the most favorable parts of the region. At the end of the paper, we provide a statistical test for comparing the theoretical and observed frequencies with one another. First, the method of statistical pattern integration will be explained by using a simple artificial example.

METHOD OF STATISTICAL PATTERN INTEGRATION

Figure 1 illustrates the concept of combining two binary patterns for which it may be assumed that they are related to occurrence of mineral deposits of a given type. Figure 1A shows locations of six deposits, the outcrop pattern of a rock type (B) with which several of the deposits may be associated (see Fig. 1B), and two lineaments which have been dilatated in Figure 1C to provide corridors (C). Within the corridors the likelihood of locating deposits may be greater than elsewhere in the study region. Points situated both on the rock type and within the lineament corridors may have the largest probability of containing deposits (see Fig. 1D). In Figures 1B to 1D, the deposits are surrounded by a small unit area. This allows us to estimate the unconditional probability p(d) that a unit area contains one deposit if it is located randomly within the study area, and the conditional probabilities p(d|b), p(d|c), and p(d|bc) for occurrences on rock type, corridors and overlap of rock type and corridors, respectively. These probabilities are estimated by counting how many deposits occur within the areas occupied by the polygons of their patterns. The relationships between the two patterns B and C, and the deposits, D, can be represented by Venn diagrams as shown schematically in Figure 2.

For the rock type (B) and the corridors (C), the relative areas assigned to the sets and their overlap (BC) in the Venn diagrams are equal to the corresponding relative areas in the study region. The set for deposits (D) is shown as a broken line in Figure 2 to indicate that its relative area depends on size of the unit cell. In Figure 2C, D is divided into four subsets which can be written as BCD, $\bar{B}CD$, $B\bar{C}D$, and $\bar{B}\bar{C}D$ where each bar indicates complement or "absence" of B or C. The relative areas of the subsets is equal to the relative proportions of total number of deposits belonging to the subsets. Suppose that relative area is written as Mes (for measure). Then B and C are conditionally independent of D if

$$p(bc|d) = \frac{Mes(BCD)}{Mes(D)} = \frac{Mes(BD)}{Mes(D)} \times \frac{Mes(CD)}{Mes(D)} = p(b|d) \times p(c|d) \qquad (1)$$

This is equivalent to assuming either

$$p(b|cd) = \frac{Mes(BCD)}{Mes(CD)} = \frac{Mes(BC)}{Mes(D)} = \frac{p(b|d)}{p(d)} \quad \text{or} \quad p(c|bd) = \frac{Mes(BCD)}{Mes(BD)} = \frac{Mes(CD)}{Mes(D)} = \frac{p(c|d)}{p(d)} \qquad (2)$$

The latter two expressions can be readily visualized by comparing Figure 2A to Figure 2C and Figure 2B to Figure 2C, respectively. Obviously, B and C are not necessarily conditionally independent of D. However, this assumption is considerably weaker than assuming that two patterns are statistically independent. For example, if B were independent of C, we would have exactly:

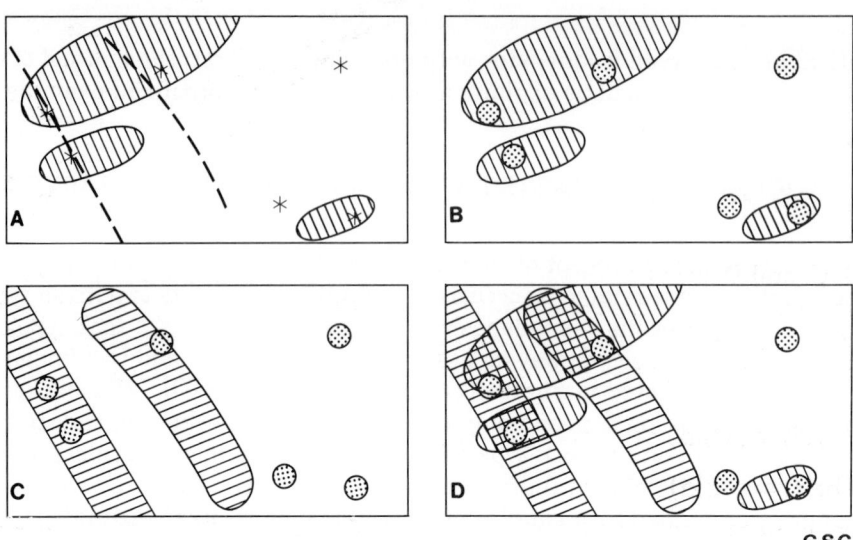

Figure 1. Artificial example to illustrate concept of combining two binary patterns related to occurrence of mineral deposits; (A) outcrop pattern of rock type, lineaments, and mineral deposits; (B) rock type and deposits dilatated to unit cells; (C) lineaments dilatated to corridors; (D) superposition of three patterns.

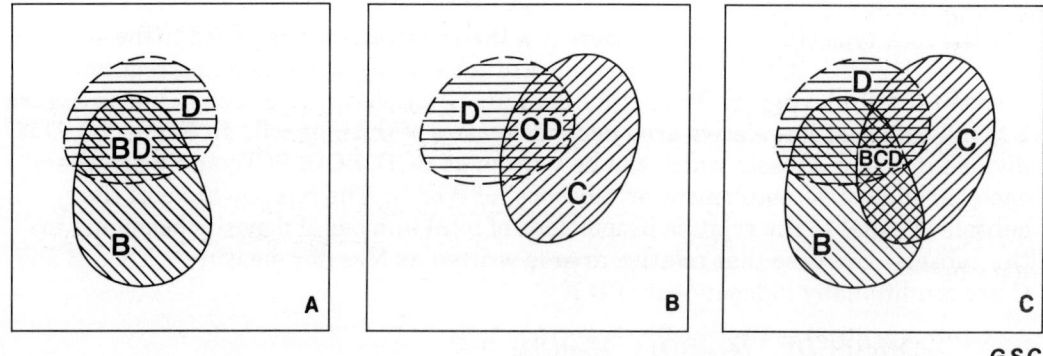

Figure 2. Venn diagrams corresponding to areas of binary patterns in Figure 1; (A) is for Figure 1B; (B) is for Figure 1C; (C) is for Figure 1D.

$$p(bc) = \text{Mes}(BC) = \text{Mes}(B) \times \text{Mes}(C) = p(b) \times p(c) \qquad (3)$$

In our type of application, Mes(BC) is greater than Mes(B) × Mes(C) because both B and C are positively correlated with D.

The relationship between B, C, and D can be expressed by the following (2×2×2) table of probabilities:

	$D = d$			$D = \bar{d}$	
	c	$\bar{c}$		c	$\bar{c}$
b	$p(bcd)$	$p(b\bar{c}d)$	b	$p(bc\bar{d})$	$p(b\bar{c}\bar{d})$
$\bar{b}$	$p(\bar{b}cd)$	$p(\bar{b}\bar{c}d)$	$\bar{b}$	$p(\bar{b}c\bar{d})$	$p(\bar{b}\bar{c}\bar{d})$

Here B, C, and D are regarded as random variables which are either present or absent in a unit cell. Absence is indicated by a bar. The eight probabilities in this table add up to one. If the assumption of conditional independence of B and C holds true, the eight probabilities in the table also are mutually related by:

$$p(bcd) = p(b|d)\, p(c|d)\, p(d) \quad \text{and}$$
$$p(bc\bar{d}) = p(b|\bar{d})\, p(c|\bar{d})\, p(\bar{d}) \tag{4}$$

This result follows from combining the first part of Equation (2) with the identities:

$$p(bcd) = p(b|cd)\, p(c|d)\, p(d) \quad \text{and}$$
$$p(bc\bar{d}) = p(b|c\bar{d})\, p(c|\bar{d})\, p(\bar{d}). \tag{5}$$

Equation (4) implies that all eight probabilities in the table can be determined from only five individual probabilities or functions of probabilities. In our approach we will use for these five constants, the prior probability $p(d)$ and the weights W_b^+, W_b^-, W_c^+ and W_c^- defined as:

$$W/\, p(b|\bar{d})\}; \quad W_b^- = \log_e \{ p(\bar{b}|d) / p(\bar{b}|\bar{d}) \};$$
$$W_c^+ = \log_e \{ p(c|d) / p(c|\bar{d}) \}; \quad W_c^- = \log_e \{ p(\bar{c}|d) / p(\bar{c}|\bar{d}) \}. \tag{6}$$

Weights of evidence W^+ and W^- were previously used by Spiegelhalter (1986).

Two binary patterns, B and C, give four posterior probabilities for $D=d$. These are $p(d|bc)$, $p(d|\bar{b}c)$, $p(d|b\bar{c})$ and $p(d|\bar{b}\bar{c})$. It is convenient to work with odds (O) instead of probabilities with $O = p/(1-p)$ and $p = O/(1+O)$. Then:

$$\log_e O(d|bc) = W_b^+ + W_c^+ + \log_e O(d),$$
$$\log_e O(d|\bar{b}c) = W_b^- + W_c^+ + \log_e O(d),$$
$$\log_e O(d|b\bar{c}) = W_b^+ + W_c^- + \log_e O(d), \text{ and}$$
$$\log_e O(d|\bar{b}\bar{c}) = W_b^- + W_c^- + \log_e O(d), \tag{7}$$

This is the extension of Bayes' rule which holds true only if B and C are conditionally independent with

$$p(bc|d_i) = p(b|d_i) \times p(c|d_i) \tag{8}$$

with $d_i = d$ (for i=1) or $d_i = \bar{d}$ (for i=0).

Previous applications of the assumption of conditional independence in mineral exploration include those by Duda and others (1977) and Singer and Kouda (1988). Even if this assumption is not satisfied, we always have:

$$\log_e O(d|b) = W_b^+ + \log_e O(d) \qquad (9)$$

and equivalent expressions for O(d | b), O(d | c), and O(d | c). The latter are formulations of Bayes' rule which has had many previous geological applications (cf. Harbaugh, Doveton, and Davis, 1977) Extensions of Equation (7) to more than two patterns are readily made. For example, if A is conditionally independent of B and C, then:

$$\log_e O(d|abc) = W_a^+ + W_b^+ + W_c^+ + \log_e O(d) \qquad (10)$$

with seven equivalent expressions.

Part of the usefulness of this approach for mineral exploration results from the fact that it can be assumed that weights such as W_b^+ are independent of the prior probability p(d). For example, if there would be as many undiscovered deposits in the region as there are known deposits, then the prior probability p(d) becomes twice as large. However, weights such as W_b^+ =loge p(b | d)/p(b | d) remain the same even if p(d) is changed provided that the proportion of new deposits associated with B=b would not change during exploration in future.

APPLICATION TO GOLD DEPOSITS IN MEGUMA TERRANE

Wright, Bonham-Carter, and Rogers, (1988) have used regression analysis to determine the multielement lake-sediment geochemical signature that best predicts the catchment basins containing gold occurrences in Meguma Terrane, eastern mainland Nova Scotia (see Fig. 3). Their geochemical signature was reduced to a ternary pattern (Fig. 4) for this study. Bonham-Carter and others (1988) have coregistered and analyzed a variety of regional geoscience data sets for this same study area using a geographic information system. A number of these data sets also are used in this paper.

Bonham-Carter and others (1988) have pointed out that the mechanism of gold mineralization in the study area is not well understood. Various authors have proposed different genetic models, emphasizing stratigraphic control, structural control, or importance of the intrusive granites as a source of mineral-rich hydrothermal fluids. Different processes have played a rôle in the formation of some or all of the gold deposits. By the method given in this paper, the spatial relationships to gold mineralization of patterns based on different genetic models can be compared and integrated with one another.

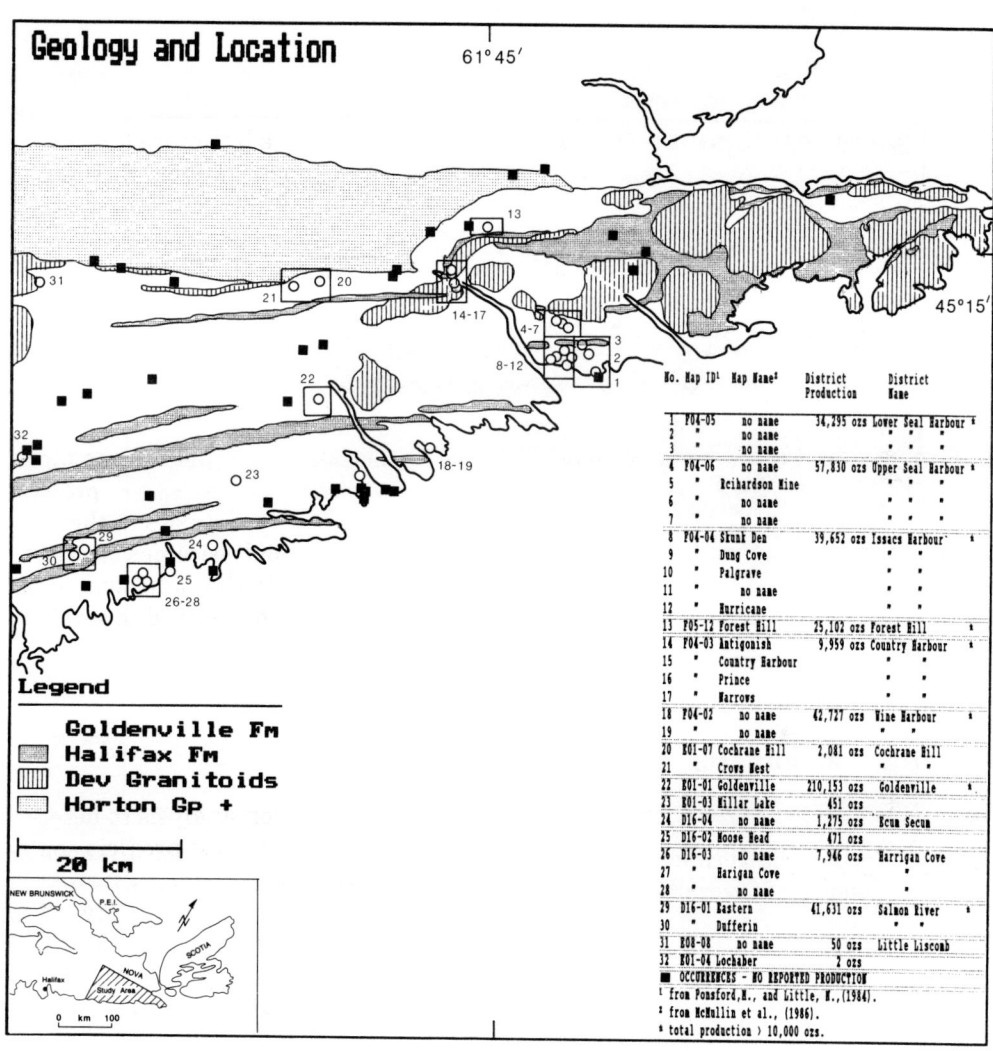

Figure 3. Location of study area with gold deposits in Meguma Terrane, eastern mainland Nova Scotia.

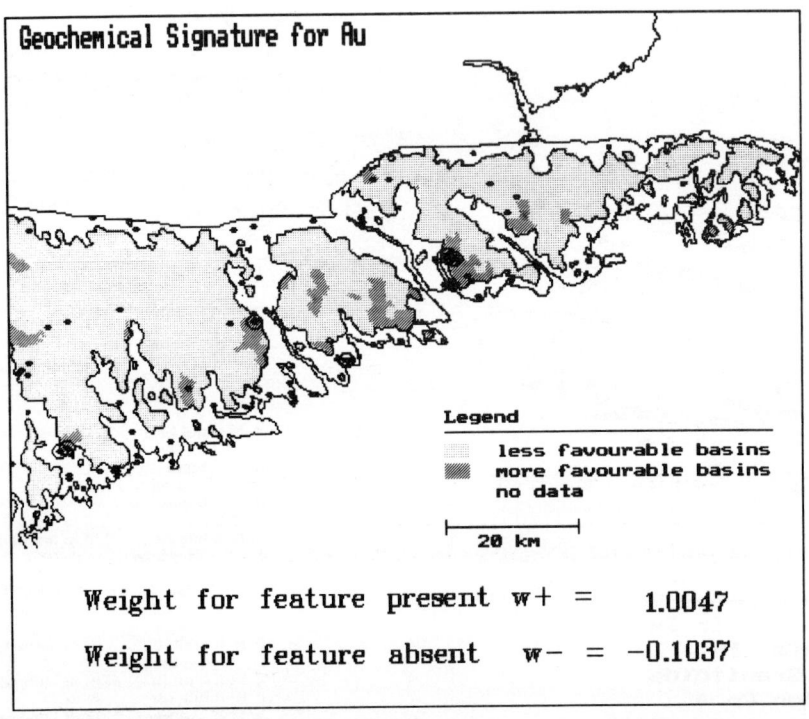

Figure 4. Ternary pattern for geochemical signature (binary pattern for presence or absence of favorable geochemistry plus binary pattern unknown).

The patterns to be combined with one another are: (1) Drainage basins classified according to favorability index derived from lake sediment geochemistry (Fig. 4); (2) Bedrock geology (see Fig. 3); (3) Proximity to axial traces of Acadian anticlines (see Fig. 5); (4) Proximity to NW-trending lineaments; (5) Proximity to Devonian granites (see Fig. 6); and (6) Proximity (within the Goldenville Formation) to the contact between Goldenville and Halifax Formations. The weights estimated for these six patterns are shown in Table 1. The final map (Fig. 7) obtained by adding the computed weights to the log prior odds delineates subareas where most or all favorable conditions exist and can be used in gold exploration.

Four of the six patterns integrated with one another are for proximity to linear or curvilinear features. Binary patterns (e.g., Figs. 5 and 6) were selected in each of these situations after studying how size of neighborhood influences the contrast $C = W^+ - W^-$ which provides a measure of the strength of correlation between a point

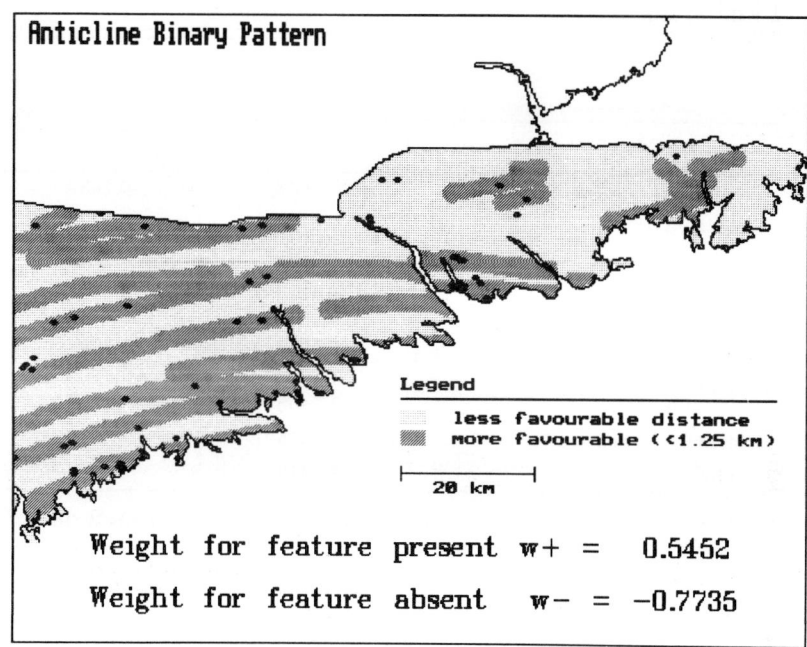

Figure 5. Binary pattern for proximity to axial traces of Acadian anticlines.

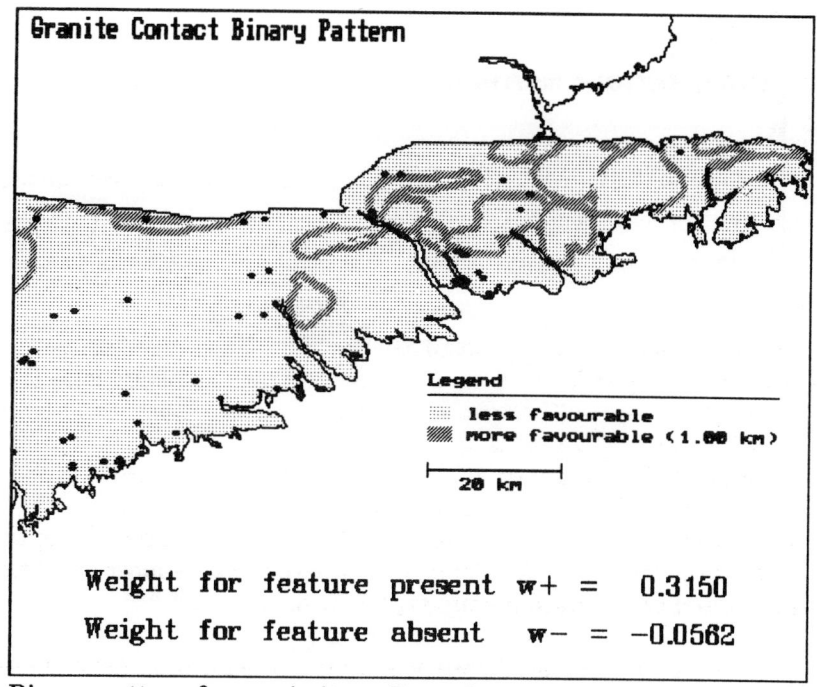

Figure 6. Binary pattern for proximity to Devonian granite contact within Goldenville Formation.

Table 1. Weights for modeling posterior probability of a gold deposit occuring in a 1 km² area.

Map Pattern	W⁺	W⁻
Geochemical Signature	1.0047	-0.1037
Anticline Axes	0.5452	-0.7735
N.W. Lineaments	-0.0185	0.0062
Granite Contact	0.3150	-0.0562
Goldenville-Halifax Contact	0.3682	-0.2685
Bedrock Geology* Halifax Formation	-1.2406	
Goldenville Formation	0.3085	
Granite	-1.7360	

* A ternary pattern where units are mutually exclusive, and weights W⁻ for absence are not used.

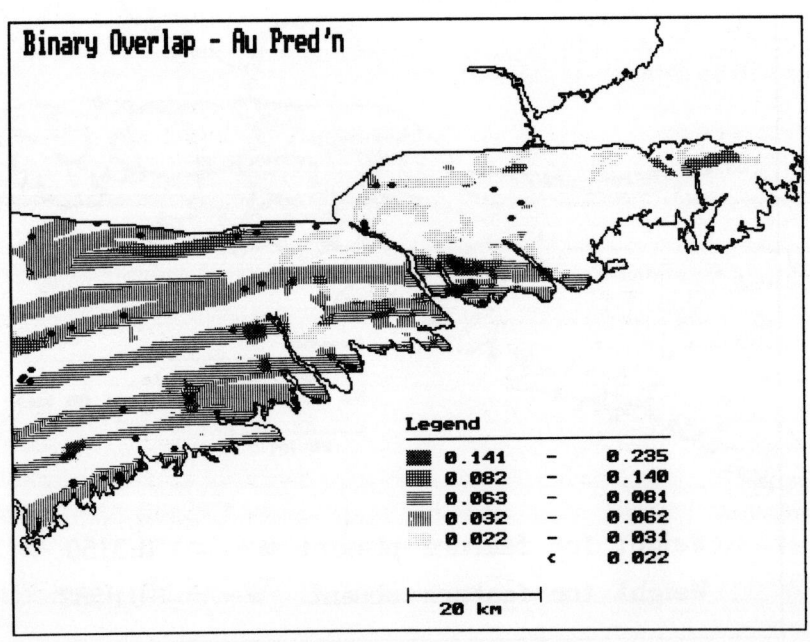

Figure 7. Final map of posterior probabilities using all weights shown in Table 1.

pattern and a binary pattern. The expected value of C is zero if the deposits are randomly distributed with respect to the pattern. The properties of C will be studied later in this paper. The bedrock geology pattern (see Fig. 3) has three states (Goldenville Formation, Halifax Formation, and Devonian granite) and W_j^+ values were computed for each of these states. When the feature is unknown in parts of the study region, no weight is added or subtracted for the unit cells. The feature then has a ternary pattern with discrete states for presence, absence and unknown, respectively. The geochemical favorability index for lake drainage basins was quantified as a ternary pattern (see Figure 4) with W^+ for most favorable signature basins, W^- for less favorable basins, and zero weight ($W^\circ=0$) for parts of the region without lake drainage basins.

Weights for presences or absences of features obtained from different map patterns can be added if the theoretical assumption of conditional independence is satisfied. Although it may not be possible to verify this assumption for all pairs of patterns combined with one another, statistical tests can be used to compare theoretical frequencies of deposits with their corresponding observed frequencies for subareas with the same posterior probabilities on the integrated pattern.

PRACTICAL EXAMPLE OF ESTIMATION OF WEIGHTS

The weights for individual patterns in Table 1 used to obtain the integrated pattern of Figure 7 were obtained from the data which are shown in Table 2. The widths of the corridors for linear features were selected by studying contrasts for different widths as will be explained in the next section.

Table 2. Data used to compute weights W^+ and W^- of Table 1 and their standard deviations $s(W^+)$ and $s(W^-)$.

Map pattern	Corridor width	Area (in km²)	Gold occ.	W^+	$s(W^+)$	W^-	$s(W^-)$
Geochemical Signature		164.9	10	1.0047	0.3263	−0.1037	0.1327
Anticline Axes	2.5 km	1276.4	50	0.5452	0.1443	−0.7735	0.2370
N.W. Lineaments	1.0 km	749.7	17	−0.0185	0.2453	0.0062	0.1417
Granite Contact	1.0 km	382.5	12	0.3150	0.2932	−0.0563	0.1351
Goldenville/Halifax	2.0 km	1029.4	34	0.3682	0.1744	−0.2685	0.1730
Halifax Formation		441.9	3	−1.2406	0.5793	0.1204	0.1257
Goldenville Formation		2020.9	63	0.3085	0.1280	−1.4690	0.4484
Devonian Granite		482.2	2	−1.7360	0.7086	0.1528	0.1248

An example of calculation of the positive weight W^+ and the negative weight of one of the features W^- is as follows: Fifty gold occurrences are situated on the corridors of the anticline axes. The combined area of these corridors is 1276.4 km². The total study area contains n(d)=68 gold occurrences and measures 2945.0 km². The total number of unit cells can be set equal to n=2945. It follows that n($\bar{d}$)= n-n(d)=2877. Our calculations may be based on frequencies (= p×n) instead of on probabilities p. Then:

$$W^+ = \log_e \frac{n(bd)}{n(d)} \bigg/ \frac{n(b\bar{d})}{n(\bar{d})} \; ; \quad W^- = \log_e \frac{n(\bar{b}d)}{n(d)} \bigg/ \frac{n(\bar{b}\bar{d})}{n(\bar{d})} \quad (11)$$

From n(bd) = 50 and n($\bar{b}$d) = 18, it follows that n(b$\bar{d}$) = 1226 and n($\bar{b}\bar{d}$) = 1651. Consequently,

$$W^+ = \log_e \frac{50}{68} \bigg/ \frac{1226}{2877} = 0.5455;$$

$$W^- = \log_e \frac{18}{68} \bigg/ \frac{1651}{2877} = -0.7738$$

The weights reported in Table 2 differ slightly from these numbers, because they were based on slightly more precise estimates of areas.

Table 2 also shows estimates of standard deviations of W^+ and W^-. These were obtained from the variances:

$$s^2(W^+) = \frac{1}{n(bd)} + \frac{1}{n(b\bar{d})} \; ;$$

$$s^2(W^-) = \frac{1}{n(\bar{b}d)} + \frac{1}{n(\bar{b}\bar{d})} \quad (12)$$

These formulae are consistent with the asymptotic expression for the contrast to the discussed in the next section. Spiegelhalter and Knill-Jones (1984) have used similar formula to obtain standard errors of the weights. The only difference between their formulae and ours is that Spiegelhalter and Knill-Jones (1984) applied a correction based on the theory of binary data analysis to help remove bias from their estimated weights as well as from the corresponding variances.

Eight of the sixteen weights in Table 2 are more than twice as large, in absolute value, as their standard deviation. These eight weights probably are different from zero, because the 95 percent confidence interval for hypothetical zero weight is approximately equal to ± 2s.

We have used asymptotic maximum likelihood expressions (cf. Bishop, Fienberg, and Holland, 1975, chapter 14) for s. Such expressions are valid only if a number of conditions are satisfied including the condition that the probabilities in the (2×2) table are neither large (= close to one) nor small (= close to zero). The latter condition may have been violated during estimation of the relatively large standard deviations of negative weights for rock types in the lower part of Table 2, because these are based on relatively few deposits. For example, only two gold occurrences on Devonian granite contribute 0.5 to the variance of their weight (= -1.7360), and therefore, account for most of the value of s(W⁺) = 0.7086 (bottom line of Table 2) which is probably too large.

The standard deviation of a posterior probability can be estimated as follows. The variance $s^2(p)$ of the prior probability p satisfies approximately p/n. For p = 68/2945 = 0.0231, this yields the standard deviation s(p) = 0.0028. The corresponding standard deviation of $\log_e \{p/(1-p)\}$ = -3.7450 is approximately equal to s/p = 0.1213. This follows from the approximate identity for any variable x with mean x:

$$\frac{s(\log_e x)}{s(x)} \simeq \frac{d(\log_e x)}{dx}\bigg|_{x=\bar{x}} = \frac{1}{\bar{x}} \tag{13}$$

Suppose that a unit cell has the following features: Its geochemical signature is unknown; it occurs in the Goldenville Formation not near a granite contact, and in the proximities of an anticline axis, NW lineament, and Goldenville/Halifax contact. Then the log posterior odds is -2.598 as can be seen when the appropriate weights are added. The variance of the log posterior odds is derived by adding variances of weights to the variance of the log prior odds. It follows that the standard deviation of the log posterior odds amounts to 0.401. The posterior probability of the unit cell containing a deposit becomes 0.069 with approximate standard deviation equal to 0.069×0.401 = 0.028. In this way, a standard deviation can be estimated for each of the posterior probabilities on a final integrated pattern. However, it will be shown later that if one or more patterns are missing, the standard deviation of the posterior probability should be increased due to the lack of knowledge. Because no information on geochemical signature is available for the unit cell in preceding example, the final standard deviation becomes 0.042 instead of 0.028 (see later). Although this final value (=0.042) is greater than the standard deviation (=0.028) computed from the uncertainties associated with the prior probability and the weights of Table 2, it is less than the standard deviation (=0.087) of the posterior probability (=0.169) arising when the unit cell considered for example in this section would have favorable geochemical signature.

CORRELATION BETWEEN PATTERN AND DEPOSITS

The contrast $C_b = W_b^+ - W_b^-$ for a pattern B provides a convenient measure of the strength of correlation between B and the pattern of deposits. The (2×2) table of probabilities with marginal totals for B and D is:

	d	d̄	
b	p(bd)	p(bd̄)	p(b)
b̄	p(b̄d)	p(b̄d̄)	p(b̄)
	p(d)	p(d̄)	1.00

If the deposits are randomly distributed within a study region, without preference for b or b̄, this table becomes

	d	d̄
b	p(b)×p(d)	p(b)×p(d̄)
b̄	p(b̄)×p(d)	p(b̄)×p(d̄)

By using the previous definitions of W_b^+ and W_b^-, it then is readily shown that $W_b^+ = W_b^-$ and $C = W_b^+ - W_b^- = 0$.

Table 3 for proximity of gold occurrences to anticline axes in Meguma Terrane, Nova Scotia, shows the contrast C(x) as a function of distance x by which these linear features were dilated (in both directions) to define the binary pattern previously shown as Figure 5. Thus x is equal to one-half the width of the corridors. Inspection of C(x) as a function of x provides a useful tool for deciding on a good value of x. It should be kept in mind, that C(x) will be less precise for smaller values of x. This is because the number of deposits n(bd) from which p(bd) is estimated then may be small and subject to considerable uncertainty. If, as before, total number of unit cells is written as n, we have p(bd)=n(bd)/n with equivalent expressions for the other elements of the (2x2) table.

Table 3. Weights and contrast for anticline binary patterns as function of one-half-width of corridor. Total area sampled = 2945 km²; total number of gold occurrences = 68; * denotes maximum contrast.

CORRIDOR HALF-WIDTH (in km)	CORRIDOR AREA (in km²)	GOLD OCC. ON CORRIDOR	W⁺	W⁻	CONTRAST C=W⁺−W⁻	STANDARD DEV. OF C
0.25	257	16	1.033	-0.181	1.213	0.294
0.50	614	31	0.811	-0.382	1.193	0.248
0.75	809	37	0.707	-0.473	1.180	0.247
1.00	995	43	0.648	-0.599	1.246	0.255
1.25	1276	50	0.545	-0.774	1.319*	0.278
1.50	1488	51	0.408	-0.694	1.101	0.283
1.75	1641	54	0.364	-0.778	1.142	0.302
2.00	1838	57	0.303	-0.857	1.160	0.332
2.25	2007	59	0.248	-0.892	1.140	0.360
2.50	2128	60	0.205	-0.872	1.077	0.379
2.75	2226	61	0.176	-0.878	1.053	0.401
3.00	2341	61	0.124	-0.701	0.824	0.402

Writing $\hat{\alpha} = e^c$, the following asymptotic result for large n (see Bishop, Fienberg, and Holland, 1975, p. 377) can be used:

$$\hat{\sigma}^2_\infty(\hat{\alpha}) = \frac{\hat{\alpha}^2}{n}\left\{\frac{1}{p(bd)} + \frac{1}{p(b\overline{d})} + \frac{1}{p(\overline{b}d)} + \frac{1}{p(\overline{b}\,\overline{d})}\right\} \qquad (14)$$

If $\hat{\sigma}^2_\infty(\hat{\alpha})$ is small compared to $\hat{\alpha}$, it follows from Equation (13) that the standard deviation of C is approximately equal to

$$\hat{\sigma}_\infty(C) \simeq \left\{\frac{1}{n(bd)} + \frac{1}{n(b\overline{d})} + \frac{1}{n(\overline{b}d)} + \frac{1}{n(\overline{b}\,\overline{d})}\right\}^{\frac{1}{2}} \qquad (15)$$

In the last column of Table 3, it is shown how this asymptotic standard deviation initially decreases as a function of distance. Once the one-half-width exceeds 0.75 km, the standard deviation continually increases. An approximate 95 percent confidence interval for C is provided by $\pm 2\hat{\sigma}_\infty(C)$. From this it may be concluded that the values of C shown in Table 3 are significantly greater than zero. Table 4 provides another example of C(x) as a function of x. Both positive and negative values of C occur in Table 4 which is for proximity to Devonian granites. The standard deviation of C now continues to decrease for wider corridors and it is likely that none of the values of C are significantly different from zero. The maximum value of C corre-

Table 4. Weights and contrast for granite contact corridors as function of corridor width. Total area samples = 2945 km²; total number of gold occurrences = 68; * denotes maximum contrast.

CORRIDOR WIDTH (in km)	CORRIDOR AREA (in km²)	GOLD OCC. ON CORRIDOR	W⁺	W⁻	CONTRAST C=W⁺−W⁻	STANDARD DEV. OF C
0.25	121	3	0.074	-0.003	0.077	0.598
0.50	247	6	0.052	-0.005	0.056	0.433
0.75	319	7	-0.051	0.006	-0.057	0.404
1.00	383	12	0.315	-0.056	0.371*	0.323
1.25	478	13	0.167	-0.036	0.203	0.313
1.50	528	13	0.065	-0.015	0.080	0.312
1.75	582	14	0.043	-0.012	0.054	0.304
2.00	670	14	-0.102	0.028	-0.130	0.303
2.25	715	14	-0.168	0.049	-0.217	0.303
2.50	756	14	-0.226	0.068	-0.299	0.303
2.75	799	15	-0.211	0.069	-0.280	0.296
3.00	865	17	-0.170	0.061	-0.226	0.283

sponds to a proximity of 1 km and this binary pattern was selected for use (cf., Tables 1 and 2). The corresponding weights for proximity to Devonian granite (cf. Fig. 6) are relatively small and had relatively little effect on the final map (Fig. 7).

In mathematical statistics, various functions of α have been proposed to express correlation between two binary variables. Yule's "measure of association" $Q=(\alpha-1)/(\alpha+1)$ (see Bishop, Fienberg, and Holland, 1976, p. 378) is comparable to the ordinary product-moment correlation coefficient for two continuous variables in that it is confined to the interval [-1, 1] with $E(Q)=0$ for uncorrelated binary patterns. It is readily shown that

$$\frac{dQ}{d\alpha} = 2/(\alpha+1)^2 \tag{16}$$

which is always positive. Consequently, $\hat{Q}(x)$ as a function of x would reach its maximum at the same value of x as $C(x)$ (cf. Table 3). It may be concluded that the contrast $C=W^+ - W^-$ provides a convenient measure of strength of correlation between the patterns B and D.

UNCERTAINTY BECAUSE OF ONE OR MORE MISSING PATTERNS

In the Introduction, it was pointed out that posterior probabilities do not all have the same precision if some of them are based on fewer patterns than others. This situation arises when data for a pattern are missing in parts of the study region. For example, the geochemical signature based on lake drainage basins is only available for parts of our study area (Meguma Terrane, Nova Scotia). Spiegelhalter (1986, p. 37) has proposed to regard any prior probability $p(d)$ as the expectation of the possible final probabilities $p(d \mid x)$ that may be obtained on observing data x

In general,

$$p(d) = E_X[p(d|X)] = \int p(d|x)p(x)dx \tag{17}$$

For the relationship between B, C, and D:

$$p(d) = \sum_{ij} p(d|b_i c_j) p(b_i c_j) = p(d|bc)p(bc) + p(d|b\bar{c})p(b\bar{c}) + p(d|\bar{b}c)p(\bar{b}c) + p(d|\bar{bc})p(\bar{bc}) \tag{18}$$

The corresponding variance is:

$$\sigma_2^2[p(d)] = \sum_{ij} \left\{ p(d|b_i c_j) - p(d) \right\}^2 p(b_i c_j) \tag{19}$$

If only B is unknown, the information on C can be added to the prior probability in order to obtain updated prior probabilities $p_b(d)$ with variance:

STATISTICAL PATTERN INTEGRATION FOR MINERAL EXPLORATION 17

$$\sigma_1^2[p_b(d)] = \left\{p(d|b) - p(d)\right\}^2 p(b) + \left\{p(d|\bar{b}) - p(d)\right\}^2 p(\bar{b})$$
(20)

This follows from:

$$\sum_j p(d|b_i c_j) p(b_i c_j) = p(d|b_i) p(b_i)$$
(21)

The expressions for the variances $\sigma 1^2$ (one pattern missing) and $\sigma 2^2$ (two patterns missing) are independent of any other patterns for which data were available and used to change the prior probability. Extensions to situations with three or more missing patterns are readily made. In our example, only one pattern is incomplete: Geochemical signature for gold deposits in Meguma Terrane. The ternary pattern representing geochemical signature (Fig. 4) shows those parts of the area where this feature could not be determined. In these places, the probability $p_b(d)$ on the final map (Fig. 7) has partial uncertainty that can be expressed by the standard deviation $\hat{\sigma}1[p_b(d)]$. This uncertainty is partial because it becomes zero in places where all patterns including the geochemical signature are available, although the posterior probabilities in these places have their own uncertainties which can be estimated by using the standard deviations of the weights (see before). The latter type of uncertainty of the posterior probability increases when the pattern for geochemical signature is added. Of course, the uncertainty because of a missing pattern decreases when information on the pattern is added.

The weights $W^+=1.0047$ and $W^-=-0.1037$ for the geochemical signature (cf., Table 1) were determined from likelihood ratios for the entire area. For example, $W^+=\log_e p(b|d)/p(b|\bar{d})=1.0047$ was based on (1) $p(b|d)=p(bd)/p(d)=n(bd)/n(d)$ with $n(bd)=10$ and $n(d)=68$; and (2) $p(b|\bar{d})=p(b\bar{d})/p(\bar{d})=n(b\bar{d})/n(\bar{d})$ with $n(b\bar{d})=164.9 - 10=154.9$ and $n(\bar{d})=2945.0-68=2877.0$. As discussed before, the weight W^+ can be regarded as independent of the prior probability. For this example, approximately the same value of W^+ is obtained when (1) the calculation is based on the subarea (=1765.8 km^2) with known geochemistry; and (2) the prior probability within the area with known geochemistry is equal to that for the total study area (=2945.0 km^2). The second condition implies that there would be about 41 deposits within the area with known data.

In reality, this subarea contains only 24 gold occurrences. A revised weight based on the subarea only would amount to 1.5444 which is greater than $W^+=1.0047$, because the subarea contains a larger proportion (=10/24) of the deposits. The lesser weight ($W^+=1.0047$) was used in Figure 7 and now will be employed for estimating $\sigma 1[p_b(d)]$.

For example, the modified prior probability $p(d)$, which is based on all patterns except geochemical signature, will be set equal to 0.05 and 0.10 within the area without definable lake drainage basins. The log odds of these values are -2.9444 and -2.1972, respectively. Addition of W^+ and W^- provides the required estimates of $p(d|b)$ and $p(d|\bar{b})$. For $p(d) = 0.05$, these conditional probabilities are equal to 0.1257 and 0.0453, respectively. For p(b) which also is needed to determine σ1 the ratio of favorable area (=164.9 km²) to known area (=1765.8 km²) can be used. This gives p(b)=0.0934 and p(b̄)=1-p(b)=0.9066. Consequently, σ1 (0.05)=0.024. By the same method, it follows that σ1 (0.10)=0.042. Previously, it was pointed out that if a unit cell in the Goldenville Formation with unknown geochemical signature is in the proximity of all linear features except granite contact, then its posterior probability is 0.070 with standard deviation equal to 0.028. Addition of the uncertainty because of the missing pattern results in the larger standard deviation of 0.042.

For both revised prior probabilities p(d)(=0.05 and 0.10), the standard deviation expressing uncertainty is the result of missing information is about one-half of p(d), or σ1[p(d)]≈0.5 p(d). This indicates that outside the lake drainage basins where geochemical information is not available, the posterior probabilities on the final map (Fig. 5) are less precise than would follow from the uncertainties associated with the prior probability and the weights (Table 2). It is convenient to express uncertainty due to ignorance by a single statistical parameter (standard deviation σ_1 in this section). It should be kept in mind, however, that this parameter is estimated from a discrete probability distribution approximating an unknown continuous frequency distribution.

TEST FOR GOODNESS-OF-FIT

As pointed out in the Introduction, the final posterior probability map (Fig. 7) provides expected frequencies that can be compared to observed frequencies for the known occurrences. Suppose that p_i represents the posterior probability after classification. For example, p_i may be set equal to the midpoints of the classes of probabilities used for constructing the map on which Figure 7 is based. Suppose that, in total, there are n deposits (n=68 in Fig. 7). For each p_i, the expected frequency amounts to

$$f_{ei} = \frac{A_i p_i n}{\Sigma A_i p_i} \qquad (22)$$

where A_i is the joint area of all polygons with posterior probability pi. The corresponding observed frequency foi is obtained by counting how many deposits actually occur in the polygons with posterior probability p_i. Table 5 shows that expected and observed frequencies are nearly equal to one another for the pattern of Figure 7. It is possible to apply the chi-square test with

$$\hat{\chi}^2 = \sum_i \frac{(f_{oi} - f_{ei})^2}{f_{ei}} = 9.786 \qquad (23)$$

The number of degrees of freedom for the corresponding theoretical $\chi^2(v)$ is not known. Setting ν equal to number of classes -1 would give $\chi^2 2$ 0.05(5)=11.1 for level of significance α=0.05. The estimated value $\chi^2 2$ (=9.8) is less than 11.1 suggesting a good fit of the model.

In this type of application, the theoretical frequencies were determined by assuming conditional independence of all patterns. The test for goodness-of-fit used in this section would suggest that this hypothesis is approximately satisfied. Care has to be taken, however, in interpreting these results, because an upper bound for the number of degrees of freedom (v) was used. Comparison of observed and expected frequencies in Table 5 suggests that observed values tend to exceed expected values in the upper part of the table where pi is relatively large and that the reverse holds that in the lower part of Table 5. This might indicate a minor violation of the assumption

Table 5. Comparison of observed and theoretical frequencies for final integrated pattern of Figure 7.

Class No.	Classes of posterior probabilities	Observed frequency (O)	Expected frequency (E)	$\frac{(O-E)^2}{E}$
1	0.171-0.235	4 ⎫ 7	1.1 ⎫ 5.4	0.474
2	0.141-0.171	3 ⎭	4.4 ⎭	
3	0.101-0.190	1 ⎫ 2	2.7 ⎫ 7.0	3.571
4	0.082-0.100	1 ⎭	4.3 ⎭	
5	0.063-0.081	17	23.9	1.992
6	0.032-0.062	23	16.7	2.377
7	0.022-0.031	5	3.3	0.875
8	0.000-0.021	14	11.6	0.497
			Sum	= 9.786

of conditional independence. If two or more patterns are conditionally dependent with positive "partial association" (cf. Bishop, Fienberg, and Holland, 1975, p. 32), the expected frequencies would exceed the observed frequencies when p_i is relatively large, whereas they would be smaller when p_i is small.

CONCLUDING REMARKS

The application of statistical pattern integration to gold exploration in Nova Scotia was performed using SPANS-a quadtree-based GIS. SPANS runs on IBM PC compatibles under DOS. The work described here was carried out on an 80386 machine with 70 mb hard drive. SPANS accepts a variety of inputs of vector and raster data and permits the user to move readily in and out of DOS, so that other DOS compatible software can be executed on mutually shared data files.

This paper is concerned primarily with three problems: (1) Construction of optimum binary patterns for linear features in order to represent the relationship between these features and occurrence of mineral deposits; (2) Statistical integration of patterns for linear features and polygon patterns for areal features representing geochemistry and rock types; and (3) Development of a measure of uncertainty which is the result of missing information.

In order to resolve the first problem (1), a sequence of increasingly wide corridors around the linear features was constructed using SPANS. The choice of optimum width was made on the basis of the contrast C which measures correlation between a binary pattern and a point pattern. An asymptotic formula was used to estimate the standard deviation of C. Statistical pattern integration (2) was carried out by the addition of weights W^+ or W^- representing presence or absence of features. The addition of weights is based on the assumption of conditional independence of the map patterns with respect to the mineral deposits. This assumption was tested by comparing the posterior probabilities shown on the final integrated map pattern with observed frequencies of gold deposits. Uncertainty resulting from one or more missing patterns (3) was evaluated by considering that no weights for presence of absence of a feature can be added if it is unknown. A measure of uncertainty was based on differences between posterior probabilities computed without the feature, and posterior probabilities computed using the possible outcomes for the feature if its presence or absence would be known. Contrary to the propagation of uncertainty associated with the weights which increases when more patterns are added, the uncertainty resulting from missing information decreases when patterns are added.

ACKNOWLEDGMENTS

This work was supported by the Geological Survey of Canada under the Canada-Nova Scotia Mineral Development Agreement, a subsidiary to the Economic Regional Development Agreement. We acknowledge the contributions of several Nova Scotia Mines and Energy geologists, particularly Peter Rogers. Thanks to Alec Desbarats of the Geological Survey of Canada for critical reading of the manuscript.

REFERENCES

Agterberg, F.P., in press, Systematic approach to dealing with uncertainty of geoscience information in mineral exploration: Proc. 21st APCOM Symposium (Application of Computers in the Mineral Industry) held in Las Vegas, March 1989; Soc. Mining Engineers, New York.

Assad, R., and Favini, G., 1980, Prévisions de minerai cupro-zincifère dans le nord-ouest Québécois: Ministère de l'Energie et des Ressources, Quebec, DPV-670, 59 p.

Bishop, M.M., Fienberg, S.E., and Holland, P.W., 1975, Discrete multivariate analysis: theory and practice: MIT Press, Cambridge, Massachusetts, 587 p.

Bonham-Carter, G.F., Agterberg, F.P., and Wright, D.F., 1988, Integration of mineral exploration datasets using SPANS-A quadtree-based GIS: Application to gold exploration in Nova Scotia; GIS Issue of Photogrammetric Engineering & Remote Sensing, in press.

Bonham-Carter, G.F., Rencz, A.N., George, H., Wright, D.F., Watson, F.G., Dunn, C.E., and Sangster, A.L., 1988, Demonstration of a microcomputer-based spatial data integration system with examples from Nova Scotia, New Brunswick and Saskatchewan: Geol. Survey of Canada, Current Activities Forum Abstract.

Duda, R.O, Hart, P.E., Nilson, N.J., Reboh, R., Slocum, J., and Sutherland, G.L., 1977, Development of a computer-based consultant for mineral exploration: Stanford Research Inst. International, Annual Rept. SRI Projects 5821 and 6915, Menlo Park, California, 202 p.

Harbaugh, J.W., Doveton, J.H., and Davis, J.C., 1977, Probability methods in oil exploration: Wiley Interscience Publ., New York, 269 p.

McMullin, J., Richardson, G., and Goodwin, T., 1986, Gold compilation of the Meguma Terrane in Nova Scotia; Nova Scotia Department of Mines and Energy, Open File 86-055, 056, maps.

Ponsford, M., and Lyttle, N., 1984, Metallic mineral occurrences and map data compilation, central Nova Scotia-map sheets IID, IIE; Nova Scotia Department of Mines Open File Report 599, 12 p.

Singer, D.A., and Kouda, R., 1988, Integrating spatial and frequency information in the search for Kuroko deposits of the Hokuroku District, Japan: Econ. Geology, v. 83, no. 1, p. 18-29.

Spiegelhalter, D.J., 1986, Uncertainty in expert systems, *in* W.A. Gale, ed.: Artificial intelligence and statistics: Addison-Wesley, Reading, Massachusetts, p. 17-55.

Spiegelhalter, D.J., and Knill-Jones, R.P., 1984, Statistical and knowledge-based approaches to clinical decision-support systems, with an application in gastroenterology: Jour. Royal Statist. Soc. A, v. 147, pt. 1, p. 35-77.

Wright, D.F., Bonham-Carter, G.F., and Rogers, P.J., 1988, Spatial data integration of lake-sediment geochemistry, geology and gold occurrences, Meguma Terrane, eastern Nova Scotia: Proc. Can. Inst. Mining Metall. Meeting on "Prospecting in Areas of Glaciated Terrain", held in Halifax, Nova Scotia, September 1988.

Statistical Classification of Regional Geochemical Samples Using Local Characteristic Models and Data of the Geochemical Atlas of Finland and from the Nordkalott Project

N. Gustavsson and M. Kontio
Geological Survey of Finland, Espoo

ABSTRACT

The objective of this study is to detect and identify regional geochemical patterns resembling given models defined by local training areas. The data consist of ICP-analyses on till samples (fine fraction) collected for the Geochemical Atlas of Finland and the Nordkalott project. The samples are composited with the resultant sampling density being only 1 sample/300 km^2. Total and partial leaching are employed to yield two sets of 34 variables.

Supervised learning with nonparametric class-conditional probability functions and the Bayes decision rule is applied to classify the samples using the characteristic features of the training data. The variables are selected according to analytical quality. Outliers are screened out to avoid spurious correlations. The total information of the two sets of variables is compressed into 10 new factors, by factor analysis, and 2 variables, Mahalanobis distances, indicating the rarity of each sample.

The 25 training areas are located at known ore occurrences in Finland and the Nordkalott area of Norway and Sweden. The classification results are presented as a single map showing the maximal probability of all models. Some expected patterns are shown on the maps as well as zones not revealed on single element maps.

INTRODUCTION

This study represents an attempt to recognize and identify regional patterns favorable for ores in geochemical data covering the whole of Finland and the area of the Nordkalott project (Finland, Norway, and Sweden north of 66° N latitude). The data set is integrated from data produced for the Geochemical Atlas of Finland at the

Geological Survey of Finland and from the geochemistry subproject of the Nordkalott project (Bölviken and others, 1986). The Nordkalott project was a joint venture of the Geological Surveys of Finland, Norway, and Sweden.

The objects to be classified are composite samples of fine fraction in till sampled at a density of 1 site/300 km^2. These multivariate data in connection with the preparation of the Geochemical Atlas of Finland and with the Nordkalott project have not yet been utilized exhaustively for prospecting purposes. The present work shows how sparse regional data can be manipulated and interpreted to reveal structures and patterns not apparent in the maps of single elements.

This recognition problem is considered as a supervised classification problem, where each class is defined by representative training samples characterized through chemical measurements. The classification method employed usually is referred to as empirical discriminant analysis (Howarth, 1973). Although not a recent method in geochemistry it has been used at least since 1973 at Imperial College, London. It has been applied locally, possibly not more because of the heavy and high precision computing required. Now that computers are drastically more powerful it becomes highly useful. (New data do not necessarily require the development of new methods. A familiar old method may be more reliable and powerful than a new one.)

One of the major problems in this type of statistical classification of multivariate data is to determine the relevant features that effectively differentiate the given classes. Too many variables would confuse the classes and cause numerical problems in the computations, whereas too few variables would not differentiate all classes. In the present situation the problem is solved by compressing the data by factor analysis and taking the major factors and the pointwise Mahalanobis distance as new statistically significant variables. The Mahalanobis distance is a measure of rarity in the data set and preserves a part of the information lost because of the simplified factor model. Although this procedure certainly will not result in an optimal set of variables for the classification, it does reduce the set of variables in a statistically meaningful way.

In 1986, in connection with the Nordkalott project, a resource-assessment map was produced using empirical discriminant analysis and characteristic analysis to visualize regional and local favorabilities for different types of ores. The classification was based on data integrated from geochemistry, geophysics, and geology. The training data were concentrated in the area of the Nordkalott project and did not involve the same models as in this study. Yet, some interesting similarities nevertheless can be seen between that assessment map and the result of this study.

In the following, a brief description of the data compression and classification procedures we used is given from the statistical point of view. The selection of training areas and evaluation of the results are described from the user's or geologist's point of view. Minimal assumptions were made about the data and geological setting of the studied area. In this way the power of the method as a tool of interpretation of unknown geochemical data is assured.

THE DATA SET

Till was sampled in the Nordkalott project at a density of 1 site/30 km² and for the Geochemical Atlas of Finland at a density of 1 site/300 km². In both situations the samples were composited from locally collected subsamples and sieved for the fine fraction (˙62_m). The compositing reduced the sampling error and increased the statistical stability of the training data. The composite samples of the Nordkalott project were grouped and the grouped samples were analyzed to achieve the same sampling density and the same analytical procedure for both sets of data. ICAP analyses were done for both partial leach (aqua regia) and total dissolution (hydrofluorine and boric acids). The elements included in the classification were for the partial leach Ba, Ca, Cr, Fe, Mg, Mn, Na, P, Ti, V, and Y and for the total dissolution Co, Cu, Fe, La, Li, Mn, Ni, P, Sc, Th, Ti, V, Y, and Zn. Thus the initial number of variables was 24. The total number of composited samples covering the entire area was 1402.

DATA COMPRESSION BY FACTOR ANALYSIS

The number of variables for the classification was reduced from 24 to 16. Factor analysis was employed separately for each data set (the partial leach and total dissolution) to reduce the number of variables.

For both data sets a factor model was fitted to the raw data and the two most significant factors and the point-wise Mahalanobis distance were selected to express the multivariate variation and structure of the data set. The Mahalanobis distance was a useful complement to the two major factors, because it caught a part of the information lost in the compression. The final set of variables consisted of 6 factors and the Mahalanobis distance for the partial leach and 4 factors and the Mahalanobis distance for the total dissolution.

Outliers causing spurious correlations and factor loadings were peeled off on the basis of an initial factor model and the corresponding Mahalanobis distances. Points exceeding the 95% quantile of the distribution of the Mahalanobis distance were excluded before the final model was computed but were included afterwards. The factor scores and the Mahalanobis distances were computed for each sample.

CLASSIFICATION BY EMPIRICAL DISCRIMINANT ANALYSIS

Empirical discriminant analysis is a statistical multivariate method for supervised classification of objects into classes defined by a representative sample of model objects. The variables observed on the model objects and the unknown objects must have a measurement scale that allows a measure of distance between points or samples. Consequently the method is not applicable to nominal variables such as rock type. The model objects — the training data that supervise the classification —

need not be "naturally" clustered in the variable space. This makes the method powerful than many other classification methods in situations where shapes of classes are complicated (Howarth,1973).

The classification procedure encompasses the following steps:

1. Learning of classes from a sample of model objects from each class, the training set, and estimation of class-conditional frequency distribution functions.

2. Definition of the current decision rule with subjective apriori probabilities and a loss (cost on misclassification) function.

3. Testing the quality of the learning by classifying a known test set of objects not included in the training set; if the test result is unsatisfactory, then the training set must be adjusted and the quality of the variables reconsidered.

4. Classification of the unknown objects into the learned classes and display of the obtained results.

The estimation of the class-conditional frequency distribution functions is nonparametric and no assumptions concerning the distribution law of the variables need be made. These functions were estimated using Parzen's window method in which the error of measurement can be tolerated and taken into account. The nonparametric feature of the method makes it safe and flexible in situations where little or nothing is known about the behavior of the variables.

The classification rule was based on Bayes' decision rule, which operates on class-conditional frequencies, apriori probabilities, and a value of loss for each class measuring the cost of misclassification. This rule is optimal in the sense that it minimizes the average overall loss. Here the rule was simplified with equal apriori probabilities and equal values of loss for all classes. A threshold was employed to screen out those samples that did not significantly belong to any of the given classes and therefore were considered as outliers or unknowns. The classification results consisted of the empirical probabilities of the classes and the index of the selected class for each object (Gustavsson, 1983).

The term favorability for a model and a single sample is defined here as the empirical probability for the respective class. The overall favorability is the maximum empirical probability for each sample.

The results of the classification are shown as shaded maps presenting the overall favorability (Fig. 1) and the favorability for single ore types. The favorabilities were smoothed slightly and interpolated to improve the appearance of the maps.

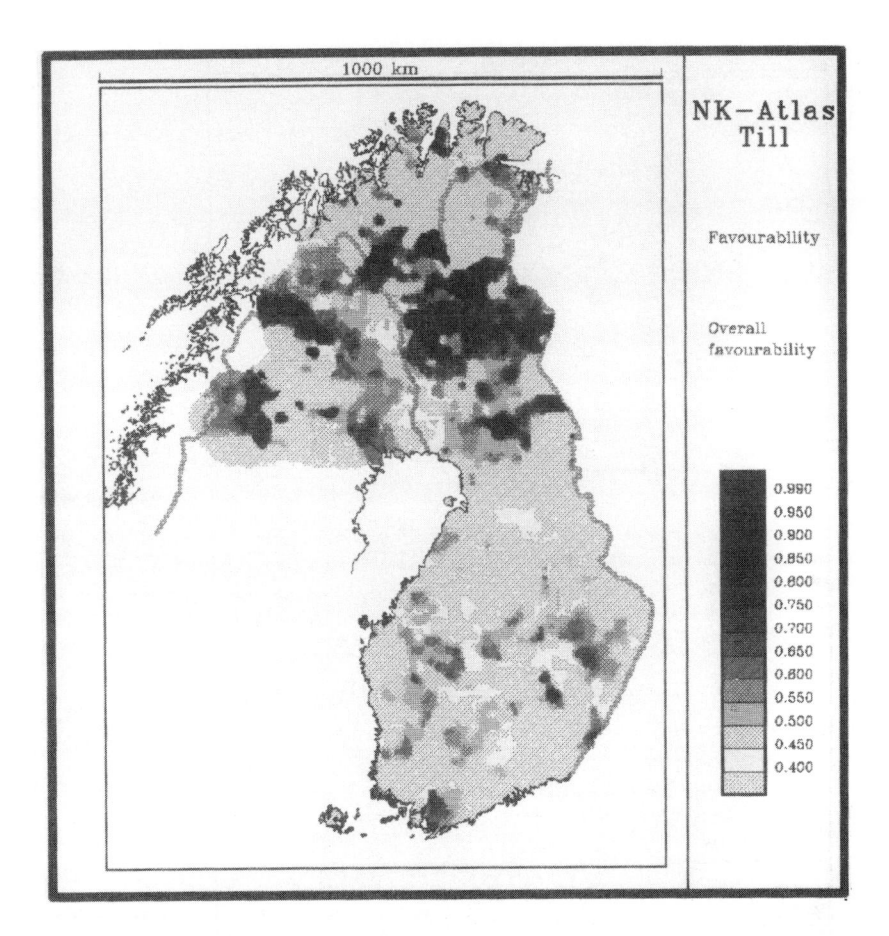

Figure 1. Overall favorability for all models.

SELECTED MODELS FOR ORES

A total of 25 training areas were selected, which contain known formations of ores or showings (Fig. 2). These objects were selected using the Metallogenic Map of Finland compiled by Kahma (1973) and the Metallogenetic Map of Northern Fennoscandia compiled for the Nordkalott project (Frietsch and others, 1986). The demarcation of the training areas was based on the Geochemical Atlas of Finland and the geochemical material of the Nordkalott project. The areas were outlined so as to be as small as possible and in such a way that the geochemical composition of the fine fraction of till within the area distinctly differed from that in the surrounding area. Each area contained at least three sampling sites.

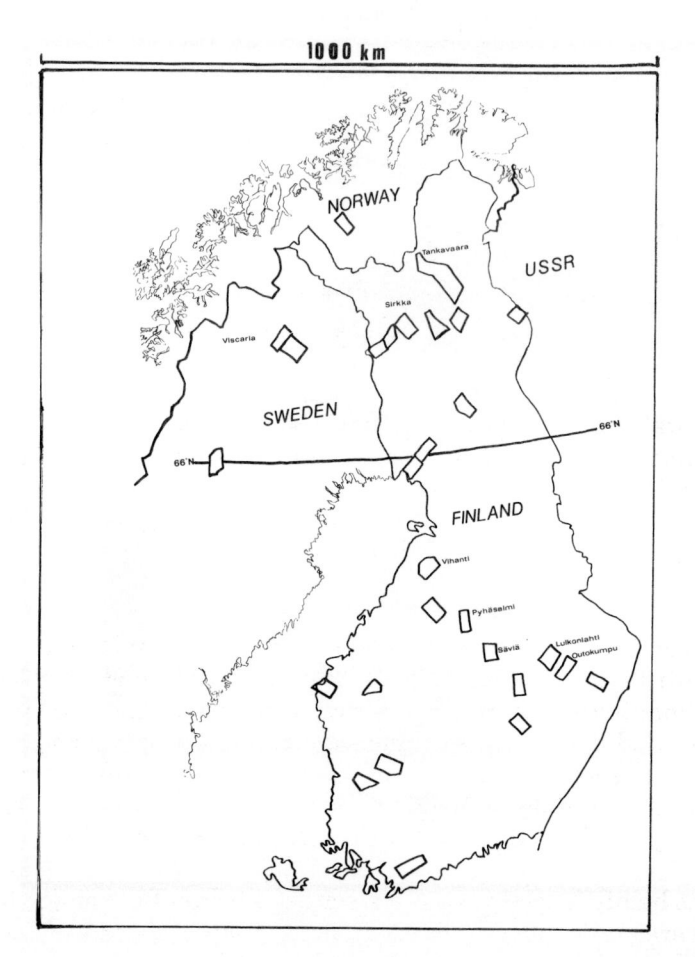

Figure 2. Demarcated training areas corresponding to each model.

THE MEANING OF THE MODELS

The models represent the chemical "fingerprints" of the ore-bearing bedrock areas as reflected in the fine fraction of till. They describe the chemistry of the most easily weathered minerals of the mineralizations and surrounding associated rock types. Areas with similar chemical composition are recognized through comparison with these models.

PERFORMANCE OF THE CLASSIFICATION

Eight (Fig. 2) of the twenty-five models were investigated: Outokumpu, Luikonlahti, Pyhäsalmi, Säviä, Vihanti, Sirkka, Tankavaara, and Viscaria. The Tankavaara model was selected because of showings of placer gold and the other models because of sulfides in the bedrock. The result of the classification with respect to these models was compared against the Metallogenic Map of Finland, the Metallogenetic Map of Northern Fennoscandia, and maps presented in the publication "Tectonic Setting of Proterozoic Volcanism and Ore Deposits", edited by G.Gaál and R.Gorbatschev (1988) (Figs. 3 and 4).

The Outokumpu, Luikonlahti, and Pyhäsalmi models occur as a linear sequence complementing each other on the map. The sequence of anomalies extends from Hammaslahti to the coast of Bothnian Bay south of Vihanti. These models also are recognized in the vicinity of Hämeenlinna and Korsnäs-Petolahti. The models for Luikonlahti and Pyhäsalmi are recognized in the valley of Tornionjoki (Figs. 5,6,7).

The anomalous favorabilities for the Säviä model seem as a sequence to the south of the Outokumpu-Luikonlahti-Pyhäsalmi sequence (Fig. 8). The Säviä model also is recognized in the Sulitjelma region and on both the Swedish and Norwegian sides of the border. Anomalous peaks occur to the west of Boden and at Arjeplog south of Allebuouda. This model is recognized widely near the western border and to the west of the Archean crust.

The anomaly peaks of the favorability for Vihanti occur on both sides of the Outokumpu-Luikonlahti-Pyhäsalmi zone (Fig. 9). The pattern of the Säviä model is recognized in the Vihanti area. In addition, the Säviä model is recognized in the regions of Sulitjelma and Boden, and at Hämeenlinna and Korsnäs-Petolahti and it coincides with the models for Outokumpu, Luikonlahti, and Pyhäsalmi. Further it is seen at Tampere and in the valley of the River Tornionjoki. The area of recognition is bounded in the east by the fault through Outokumpu, Sotkamo, and Misi.

The Sirkka model is highly favorable on the greenstone belt in Finnish Lapland (Fig. 10). This favorable area is bounded by the edge of the granite complex of Central Lapland in the south, the Kittilä granite up to Pulju and Tepasto in the west, and the area of mafic and schistose rocks from Luosto to Tuntsa via Jauratsi in the east. Favorable areas occur at Korvatunturi, Kessi, Savettijärvi, Kirkkoniemi, and Markkina-Naimakka. The model is not favorable on the granulite belt.

The most pronounced part of the anomaly of the Tankavaara model lies within the granulite (Fig. 11) dividing it into different parts. Weaker anomalies occur in the region of Lemmenjoki-Angeli, along the coast from Lyngen to Kvaenangen, and at Tanabron and Sevettijärvi.

The favorable region of the Viscaria-model occurs as a wide band from Lannavaara via Abisko through the Caledonides towards Narvik and Bruvann (Fig. 12). It is bounded by Kiruna and Svappavaara in the south.

EVALUATION OF THE RESULTS

In general the results support existing knowledge. The "arctic bramble" zone (the Ladoga-Bothnian Bay zone), which contains ore formations of various ages and types, is shown distinctly as a belt (the Outokumpu, Luikonlahti, and Pyhäsalmi models), widest in its western part (where it is bounded by the Säviä and Vihanti models).

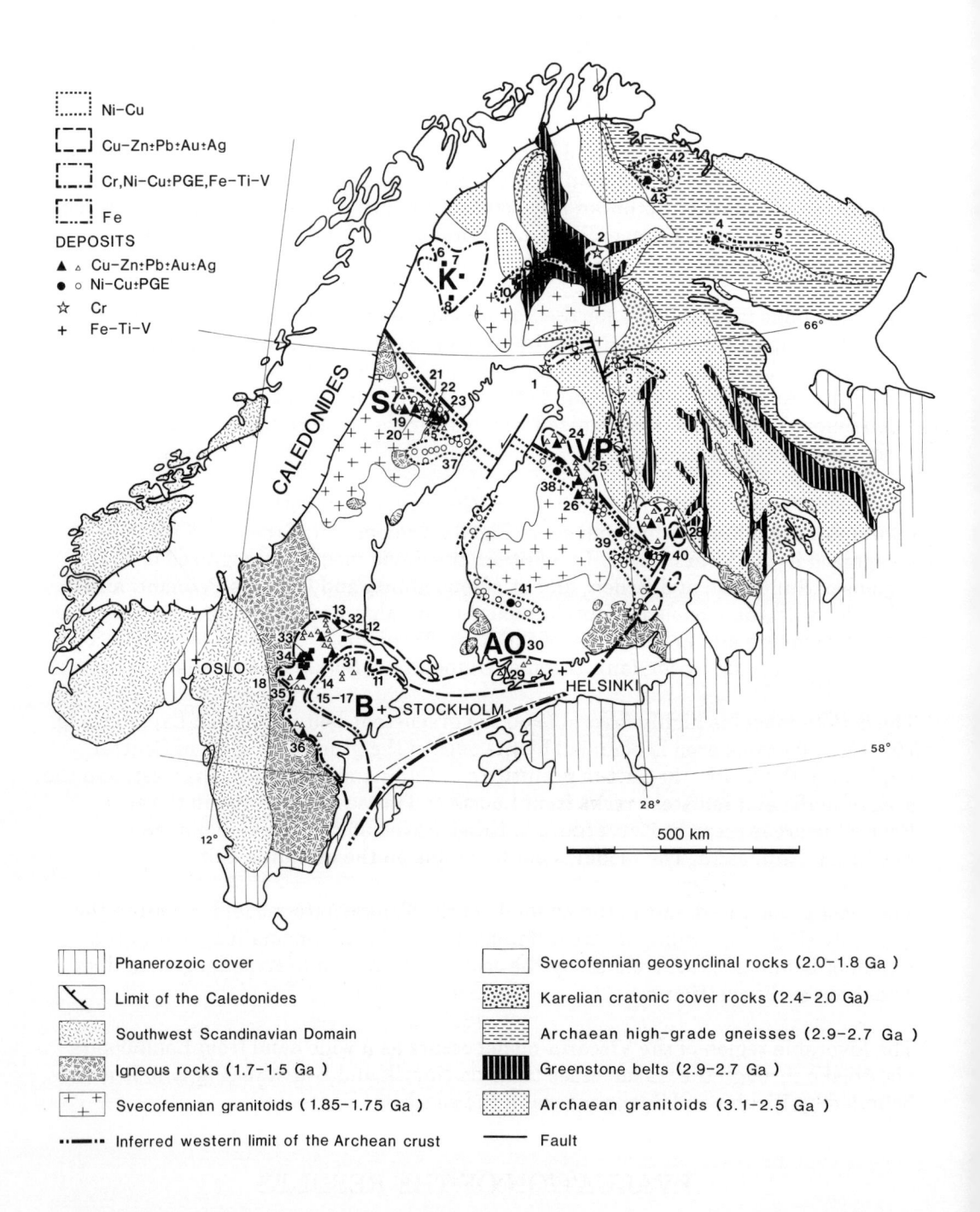

Figure 3. Early Proterozoic metallogenetic provinces and major tectonic subdivisions of Baltic Shield (Gaál and Gorbatschev, 1988).

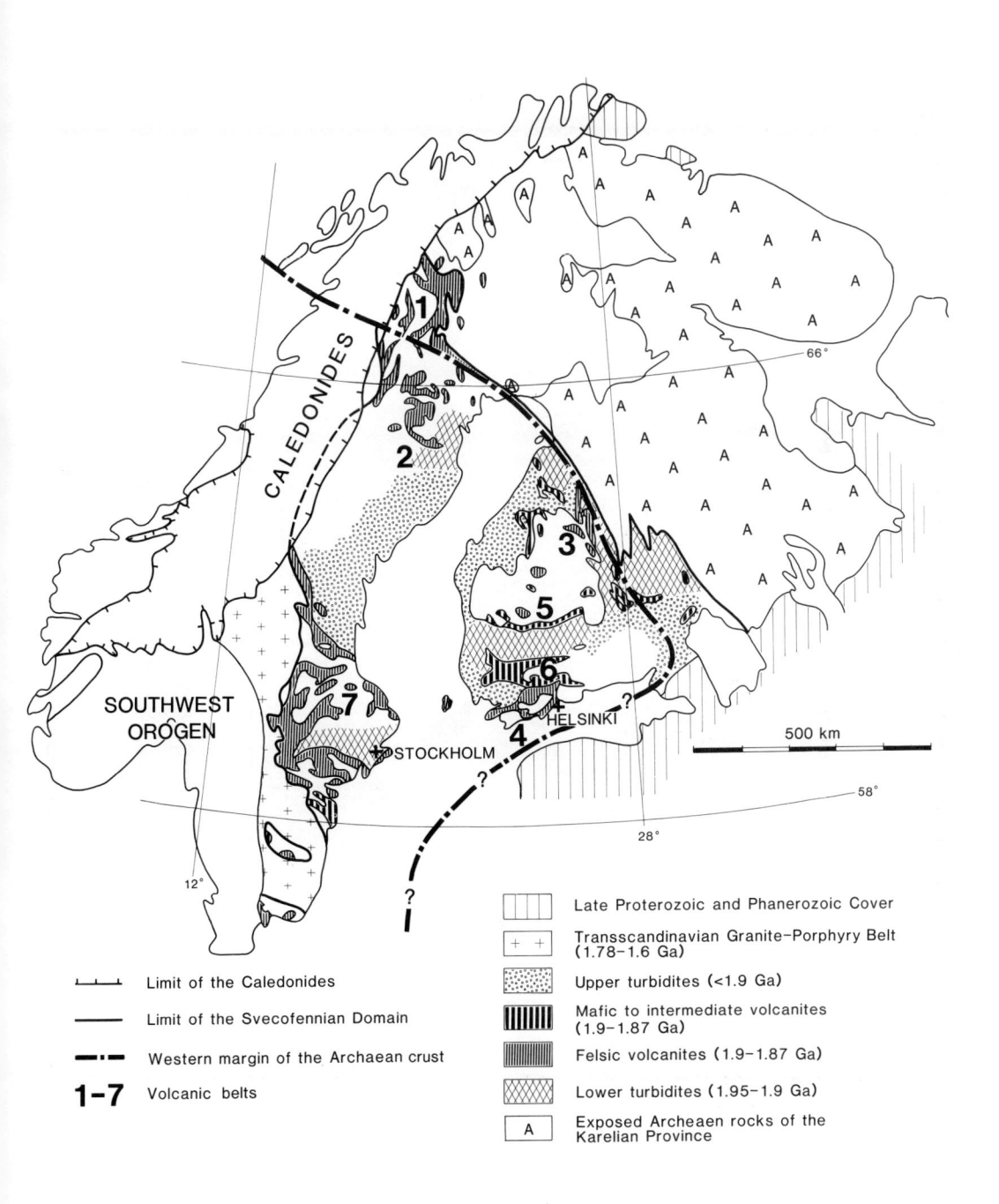

Figure 4. Tectonic setting of volcanic belts of Fennoscandian (Baltic) Shield. (Gaál and Gorbatschev, 1988).

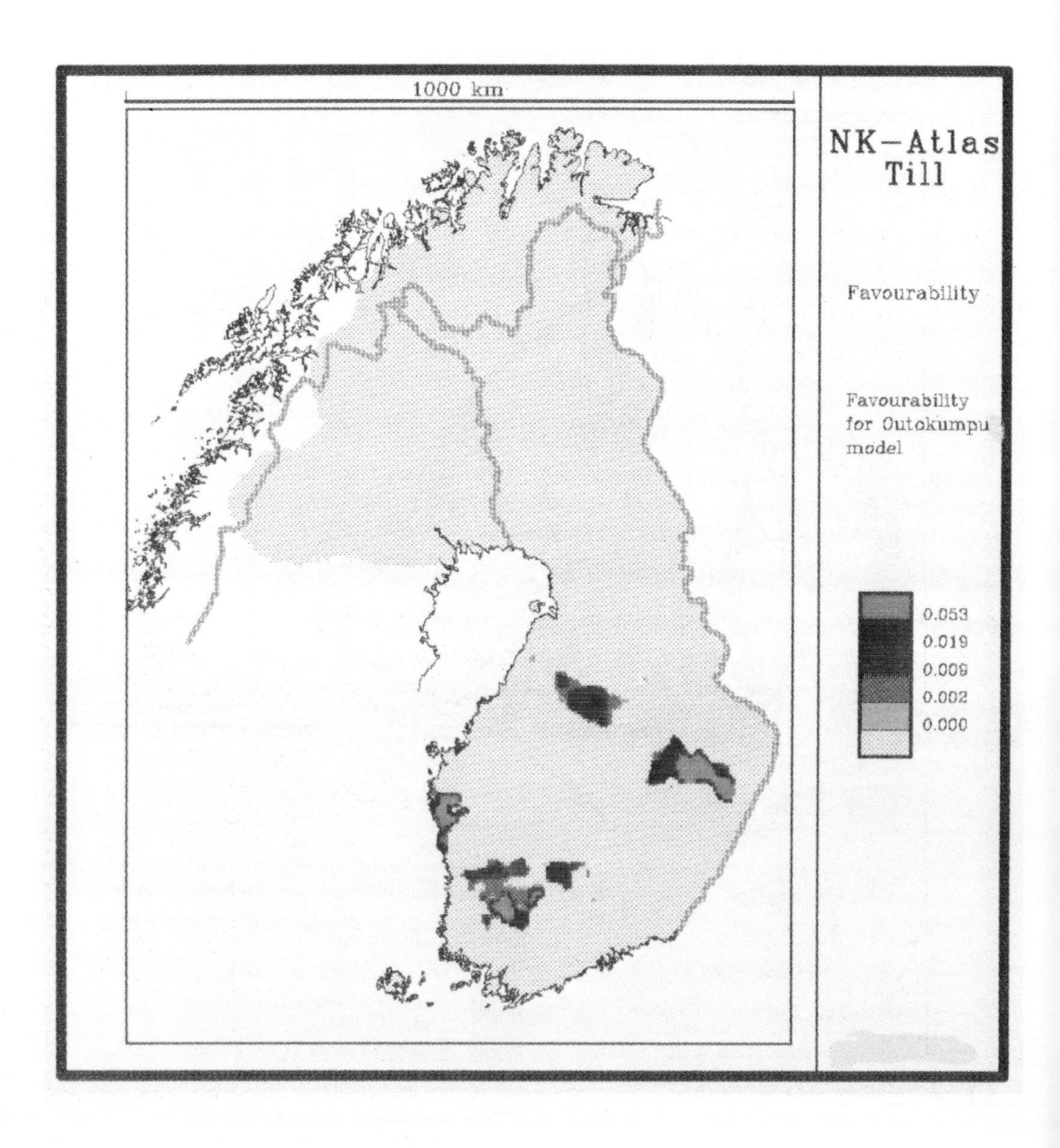

Figure 5. Favorability for Outokumpu model.

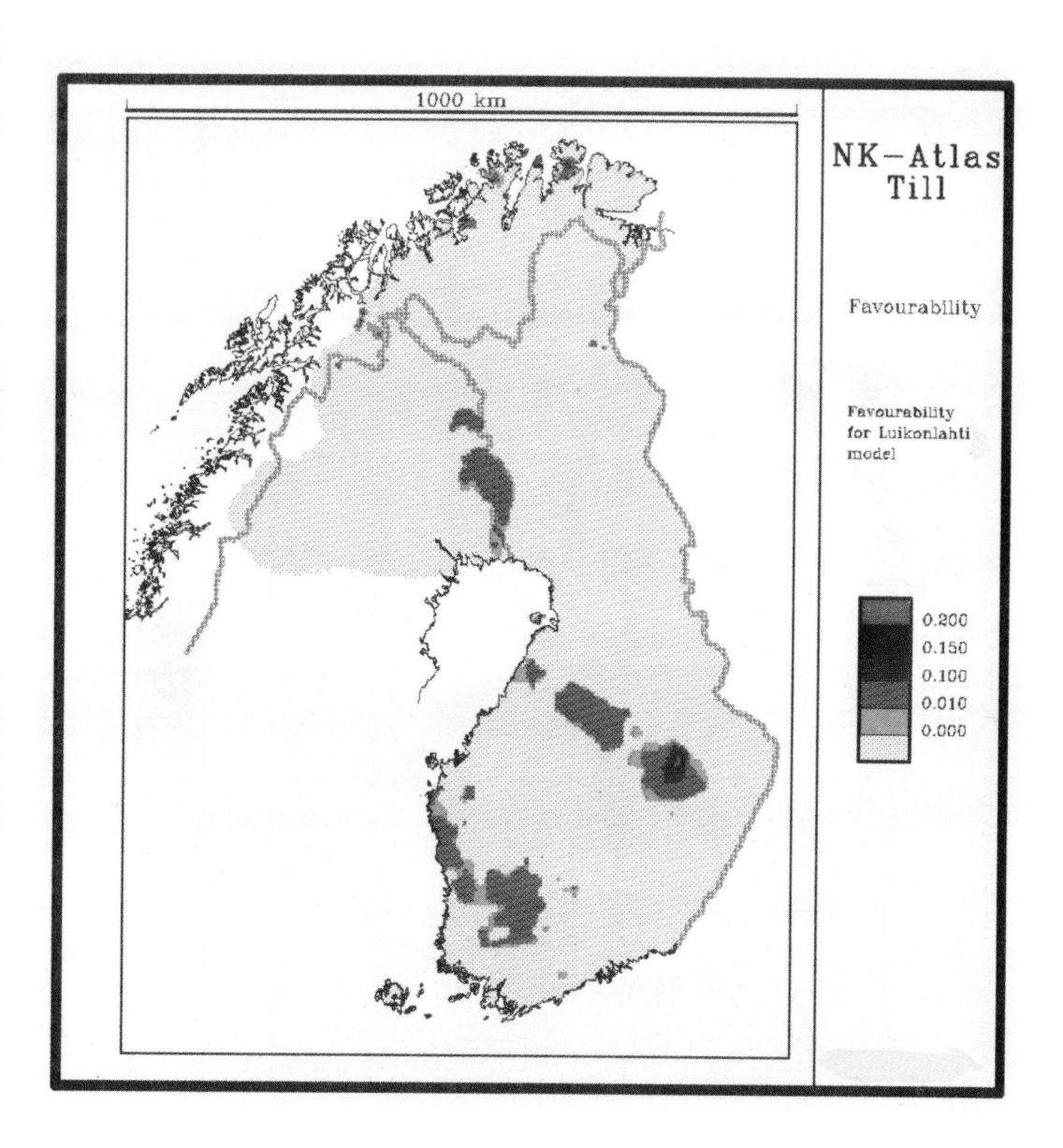

Figure 6. Favorability for Luikonlahti model.

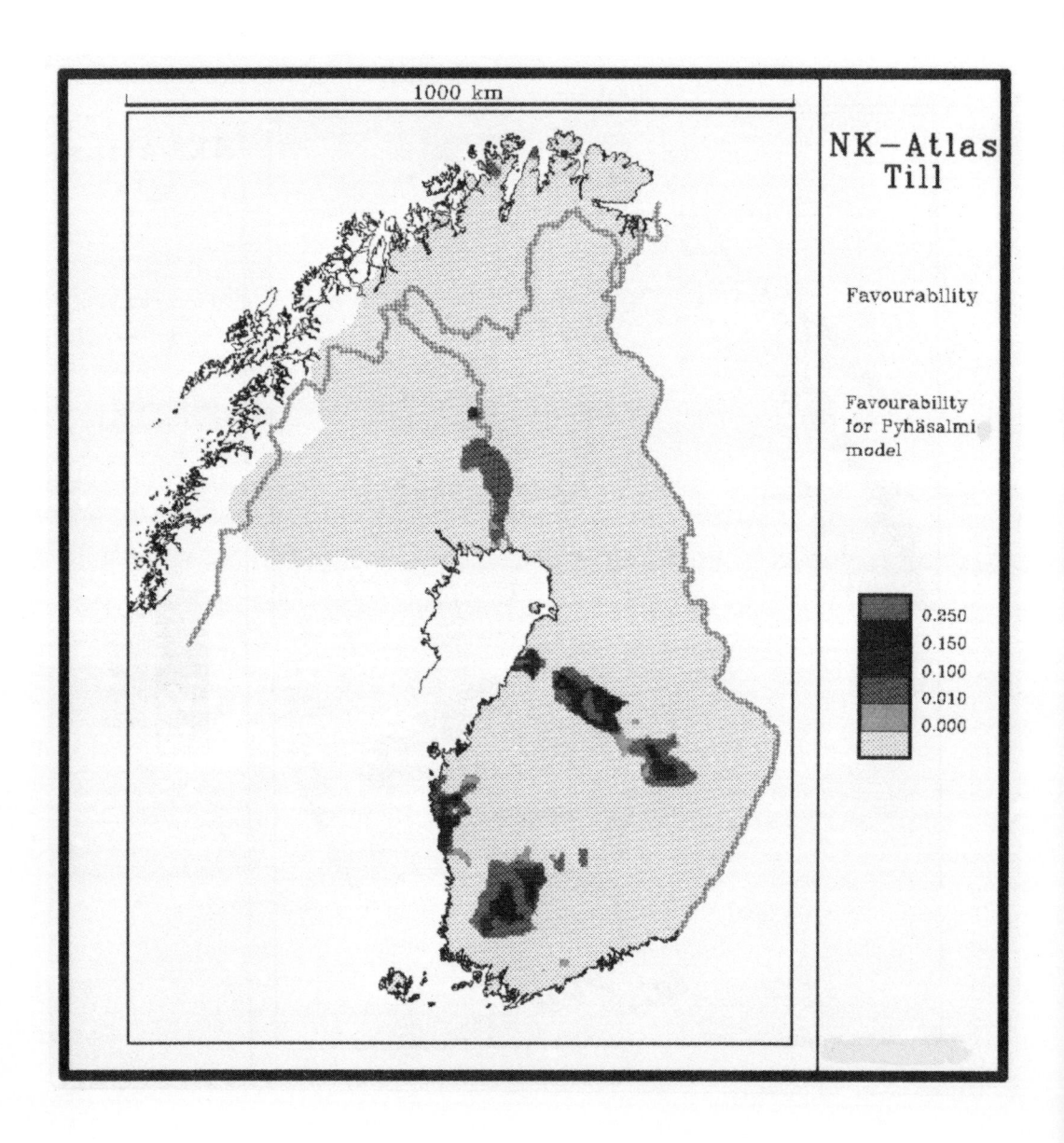

Figure 7. Favorability for Pyhäsalmi model.

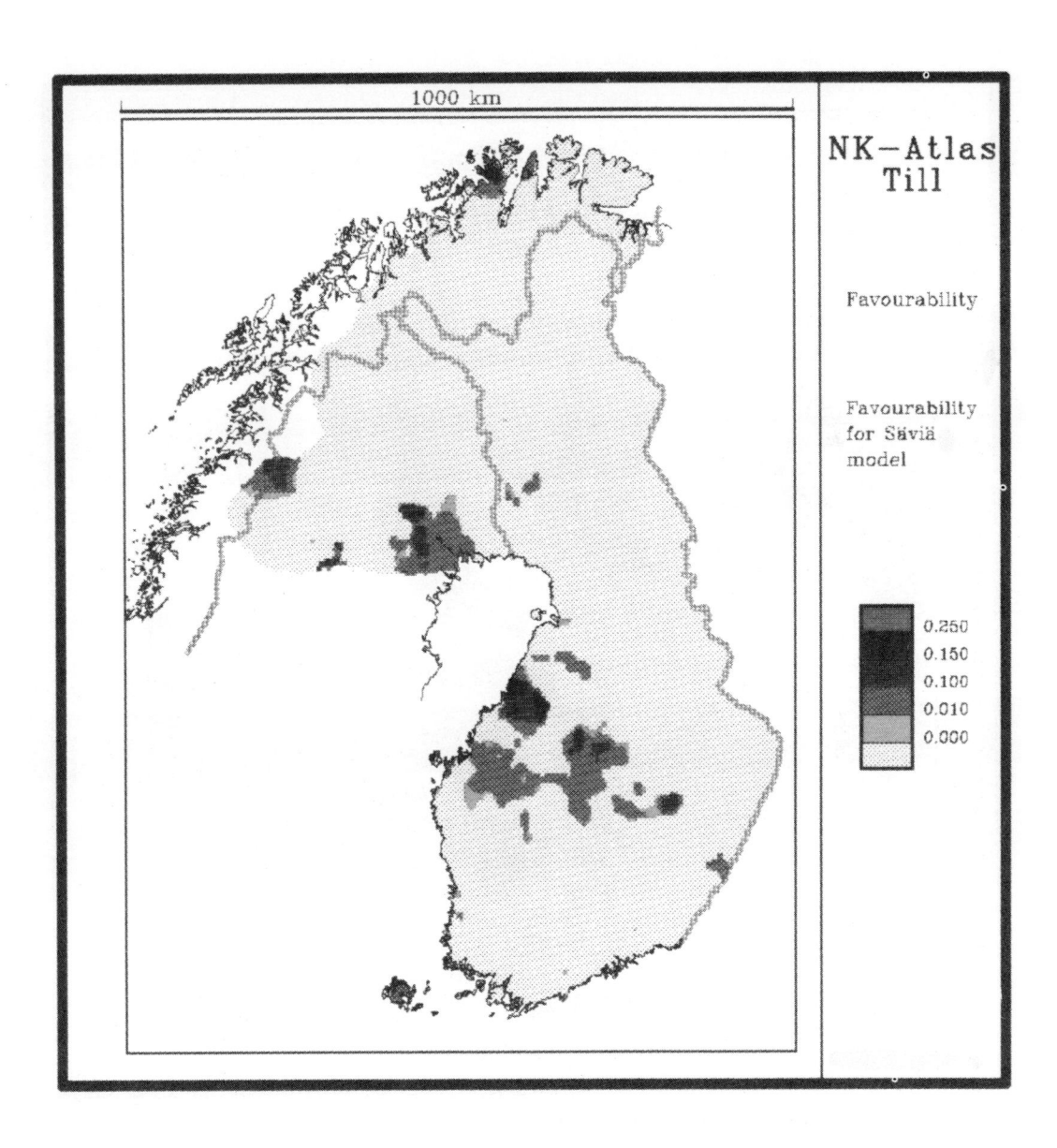

CLASSIFICATION OF SAMPLES FROM NORDKALOTT PROJECT

Figure 8. Favorability for Säviä model.

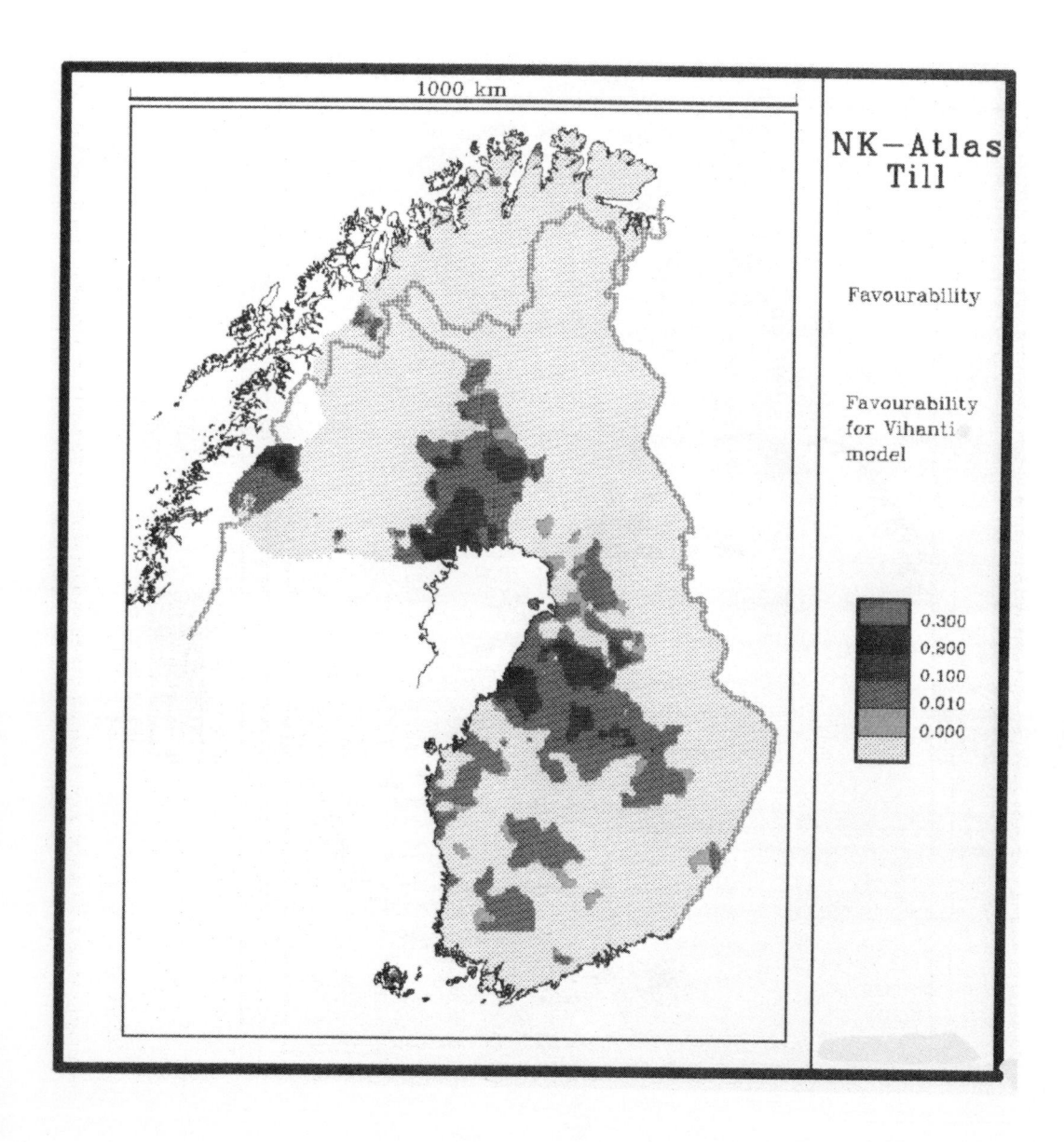

Figure 9. Favorability for Vihanti model.

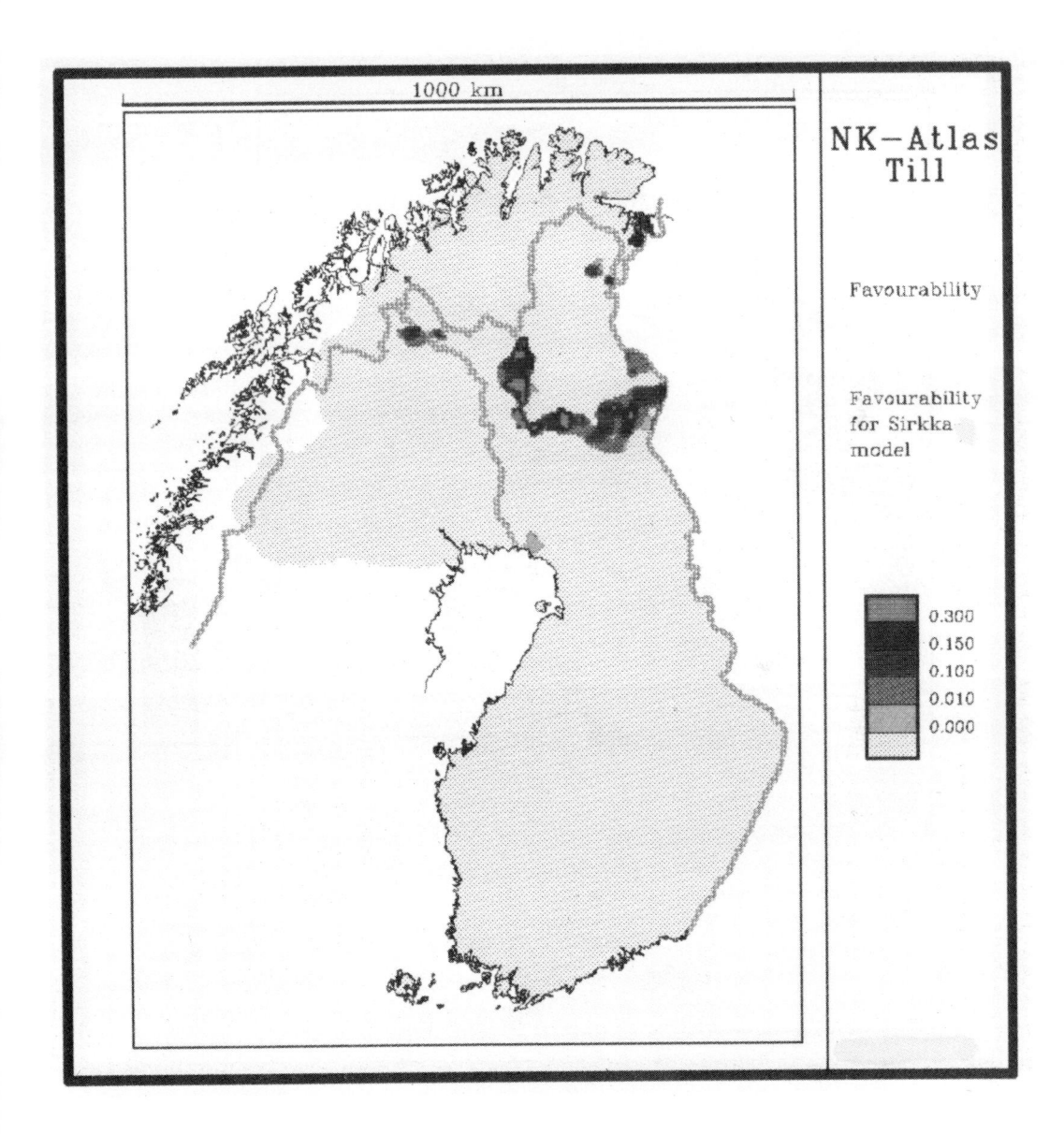

CLASSIFICATION OF SAMPLES FROM NORDKALOTT PROJECT 37

Figure 10. Favorability for Sirkka mode.

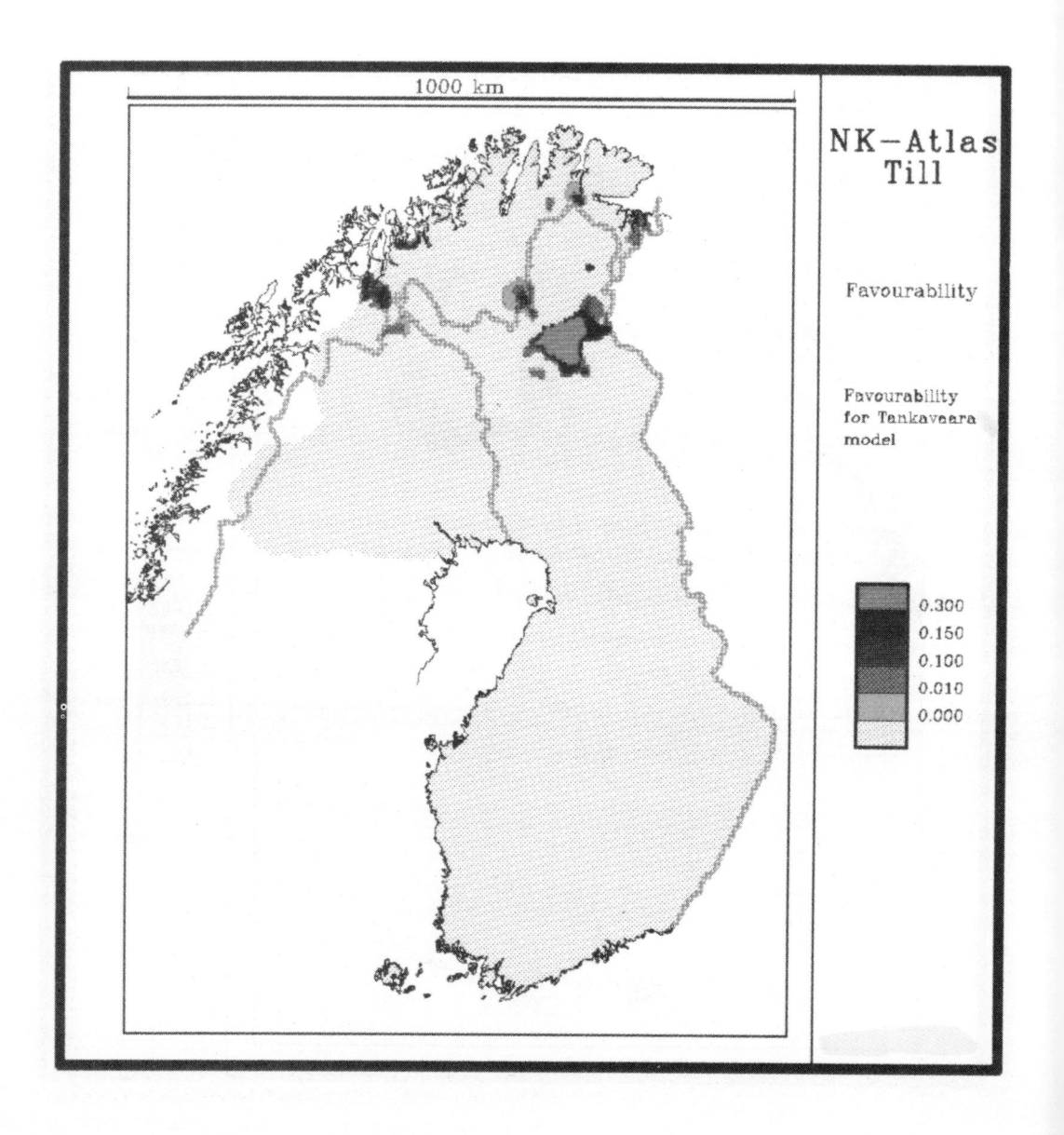

Figure 11. Favorability for Tankavaara model.

Figure 12. Favorability for Viscaria model.

widest in its western part (where it is bounded by the Säviä and Vihanti models). Two other interesting areas are the ore critical regions of Hämeenlinna and Korsnäs-Petolahti (the Outokumpu and Pyhäsalmi models and, weakly, the Luikonlahti and Vihanti models). The Tampere area is shown as a weak anomaly (Vihanti model).

The results strongly correspond to tectonics, age, and facies. The western edge of the Archean crust (Säviä model), the Outokumpu-Sotkamo-Misi-fault (Vihanti-model), the granulite (Tankavaara and Sirkka models) and the greenstone belt (Sirkka-model) seem to control the favorabilities.

The difference between the Vihanti model and the zone of the Outokumpu-, Luikonlahti-, and Pyhäsalmi models is interesting, as is the bounding of the Säviä model along the southern edge of this zone. The favorability of the Vihanti and the Säviä models at Arjeplog and especially at Boden and Sulitjelma in Sweden was unexpected. It is noteworthy that the Viscaria model cuts through the Caledonides. The Sirkka model coincides with a part of the edge of the greenstone belt, but not with the whole belt, which may indicate some unknown geological factor influencing the favorability of this model.

REQUIREMENTS FOR THE DATA

The regional sampling grid should be as regular as possible and have good coverage the analytical results should be comparable from one sample to the next. Thus the data should be free from disturbing effects such as temporary biases resulting from sampling or chemical analysis, otherwise the results are spurious.

Only relevant variables should be incorporated in the factor analysis, and the selected factor model should have a high degree of explanation. This is restrictive, because the least significant factors associated with the least reliable variables are interesting for the interpretation of geological processes.

DISCUSSION

The method is easy to use because models are created by demarcating the desired training areas and the sampling material of the models is of the same type as the sampling material to be classified. The material must be selected to reflect the characteristic features of the areas, however, otherwise the models become too general and the resultant areas of favorability extend too widely.

One limitation of the method is that only areas analogous to the given models of ores or showings are revealed. The Finnish part of the Lapland Granulite Belt serves as a good illustration of this problem: the models need to be taken from the Soviet side because the Finnish Lapland Granulite Belt contains no known ore deposits.

REFERENCES

Bölviken, B., and others, 1986, Geochemical atlas of northern Fennoscandia, The Nordkalott project.

Frietsch, R. and others, 1986, The metallogenetic map of northern Fennoscandia.

Gaál, G., and Gorbatschev, R., 1988, Tectonic setting of Proterozoic volcanism and associated ore deposits: IGCP Field Conference in Finland and Sweden, Geol. Survey Finland, Espoo, Guide 22.

Gustavsson, N., 1983, Use of pattern recognition methods in till geochemistry, *in* Howarth R.J., ed., Statistics and data analysis in geochemical prospecting: Elsevier, New York, p. 303-309.

Howarth, R.J., 1973, FORTRAN 4 programs for empirical discriminant classification of spatial data: Geocom Bull. 6, p. 1-31.

Kahma, A., 1973, The metallogenic map of Finland.

A Map-Comparison Technique Utilizing Weighted Input Parameters

U. C. Herzfeld *
Scripps Institution of Oceanography

D. F. Merriam
Wichita State University

ABSTRACT

A method is presented for comparing thematic spatial data by weighting input parameters according to a preconceived understanding as to their importance. A difference map is built from standardized weighted input data for every grid node and projected to a comparison value. Moving over the map area, a comparison matrix is obtained that can be contoured to yield a resultant map. The algorithm has the capability of utilizing different data forms and accommodating missing data. An example of an application of the technique using geological, geophysical, stratigraphical, and topographical data in an area of south-central Kansas (USA) is given.

INTRODUCTION

Map comparison techniques have been of interest to geologists as an aspect of spatial analysis for a long time. A variety of techniques have been utilized—each practical for different conditions—ranging from simple quick visual comparisons to sophisticated computer-oriented automated techniques. These methods mostly are based on a point-by-point comparison to obtain a coefficient of resemblance (Merriam and Robinson, 1981), or a resultant map (Escher, Robinson, and Merriam, 1979) to give a spatial relation of the parameters being compared. The comparisons usually are made pairwise between maps for a measure of similarity, then a matrix of similarity is constructed (which can be represented as a dendrogram (Merriam and Sneath, 1966), or the parameters are combined so that the similarity is represented as a series of resultant maps. Algorithms giving comparison values per grid node may have edge effects, which are the result of some moving window techniques (cf. Merriam and Sondergard, 1988). Standardization is necessary prior to analysis if parameters are measured in different units, for example geological, geophysical, and geochemical properties.

---

* Part of this work was done while UCH enjoyed a visit to the Wichita State University/Kansas in April 1987.

The technique proposed here is a simple, but effective, one that can be adjusted by the user as to perceived differences in importance of input into the combinatorial map results. The different input parameters first are standardized, then combined on a point-by-point basis at each of the original data distribution points or grid-node values of all maps being considered. The resultant value at each position is plotted and the map values are contoured. The isolines define areas where the combined parameters are reenforced and supportive, that is similar, and areas where they are not—giving a visual spatial representation of results. Because the algorithm employs a functional defined in each point of the map area, it does not produce edge effects. The algorithm is realized in the FORTRAN program MAPCOMP (cf. Herzfeld and Sondergard, 1988).

A subset of geographically distributed data was taken from an area in south-central Kansas for a test situation. The set contains geological, geophysical, and topographical data for a 17 X 17 grid interpreted from published material on a 6-mile interval.

OUTLINE OF METHODS

We aim to compare n maps of the same area. For simplicity, we assume that each map shows one parameter (e.g. geophysical, geological, topographical), and that data are given in terms of a digital terrain model (DTM), that is as grid values. (Otherwise, maps have to be digitized, the parameters separated, and a DTM built; see the worked example).

Standardization

To allow comparison, a standardization of the parameters is necessary. This has to be performed in a geologically sensitive way, depending on the actual problem. All measured values are fitted into the interval (0,1), using ratio standardization defined via

$$z(x) = \frac{x - a}{b - a} \qquad (1)$$

where x denotes the measured value, a and b minimal and maximal values, respectively, of the considered parameter, and z the standardized value. Additionally, the type of scale of the measured parameter must be taken into account, for instance, if it is logarithmic, exponential, or negative-trending. In the situation of a decreasing scale (among increasing other parameter scales, e. g. subsurface topography vs. thickness), the (inverse) ratio standardization formula reads

$$\tilde{z}(x) = \frac{b - x}{b - a} \qquad (2)$$

with notation as in 1, and z the transformed value.

We note that statistical standardization (so-called z-score standardization) is inappropriate as data stemming from geological structures are not random with normal distribution but highly deterministic. In fact, in a test run, z-score standardization gave worse results than ratio standardization.

Algebraic concept

Let M denote the map area under consideration, and $M_1,...,M_n$ the individual maps containing standardized values; then for a grid point $x \in M$ we define a difference matrix $D(x) \in R^{n \times n}$ by

$$D(x) = (d_{st}(x))_{s,t=1}^{n} \tag{3}$$

where $d_{st}(x) = m_s(x) - m_t(x)$ is the difference between values $m_s(x)$ and $m_t(x)$ in maps number s and t respectively, $s, t \in \{1,...,n\}$. As x varies over M, now D(x) becomes a matrix function $D: M \to R^{n \times n}$. If we define a norm on the vector space $R^{n \times n}$, we get a composed function $F: M \to R^{n \times n} \to R$, $x \to F(x) = ||D(x)||$. Let

$$F(x) = \frac{1}{k} \sum_{s<t, t=1}^{n} |d_{st}(x)| \tag{4}$$

with $k = n(n-1)/2$, the number of comparisons. Note that 2F(x) is the arithmetic mean of the absolute values of entries in D(x), and that the right-hand side of Equation 4 defines a norm.

Weights

If we use Formula (4), we suppose that all input maps are of equal importance. However, in a practical situation such as exploration, we can imagine that one might want to make a decision relying mainly on a geological map, say, and a little on both a gravity map and a geochemical map. To meet similar requirements, we assign a weight $w_i \in R_o^+$ to the map M_i for all $i \in \{1,...,n\}$, that refers to the importance of map number i. (Note that all weights must be nonnegative and at least one positive.) The definition of the weights is open to the user of the program, as it is a question of geological responsibility to select among the maps. The formula for F including weighted input parameters is given by

$$F(x) = \frac{\sum_{s<t, t=1}^{n} w_s w_t |m_s(x) - m_t(x)|}{\sum_{s<t, t=1}^{n} w_s w_t} \tag{5}$$

(Note that if a zero weight is used, the right-hand side of (5) actually gives only a seminorm.)

Calculating F(x) over the whole area, that is for all $x \in M$, we build up a DTM for the comparison map F. This DTM can be visualized as a map, printing the grid values or contouring them. The resultant map shows relative values, giving the quality of coincidence of the input maps.

ALTERATIONS

Comparing subsets of n maps

A resultant value on the comparison map F can stem from the fact that all maps are slightly different, or the fact that all but one are similar and one map differs from all the others. In that situation, a visual test might help to spot the outlier. Using the program, it is easy to compare only some of the n maps, and thus determine significant connections. This is done by giving a weight zero to the maps that will be left out in the next run of the comparison.

Missing data

There might be missing data at some of the grid nodes of the map area M, and this could happen at different locations of the individual input maps. The program MAPCOMP offers a choice of two possibilities to handle such situations:

(i) F-algorithm: At location $x \in M$, compare only the maps that have data at x.

(ii) G-algorithm: At location $x \in M$, only do the comparison if all n maps have values at location x.

In subareas where we have values on all maps, F and G coincide. If only one map shows a value at the point x, no comparison is done. In places where 2 to (n-1) maps carry data, we get an F-map but not a G-map. The advantage of the F-map is that comparison is carried through in a larger part of the considered area. On the other hand, G-maps have a higher reliability. For example, moving on an F-map from a part where we compare three maps to a part based on two maps, the information has to be interpreted with caution, in particular if weights are used. MAPCOMP prints a warning flag.

WORKED EXAMPLE

To give a practical test of the suggested method, we use a subset of geographically distributed data from a Kansas area.

The data

Data for the project were hand-digitized from published maps. All of the maps were partitioned by the normal land grid of townships and ranges. The grid node values were interpolated from the township/range grid which are on a 6-mile interval. The area is located in south-central Kansas and is between T. 19-34 S. and R. 4W. and R. 12 E. The 9200-square-mile area contains all or parts of 11 counties—Butler, Chase, Chautauqua, Elk, Greenwood, Harvey, Lyon, Marion, McPherson, Sedgwick, and Summer.

Although the grid is coarse, it is deemed usable for the intent here.

The following maps are used: Precambrian (Figure 1A, digitized from Cole, 1962), Base of Kansas City (Figure 1B, from Watney, 1978), Lansing (Figure 1C, from Merriam, Winchell, and Atkinson, 1958), Arbuckle (Figure 1D, from Merriam and Smith, 1961), Topography (Figure 1E, from U.S. Geological Survey, 1963), total thickness of sediments between the Precambrian surface and the present-day topography (Figure 1F), Aeromagnetics (Figure 1G, from Yargar and others, 1981), and Bouguer gravity anomalies (Figure 1H, from Yargar and others, 1985).

A cross-table of correlation based on z-scored standardized data was employed to decide between ratio and inverse ratio standardization. We used ratio for all geological maps and inverse ratio for the others. As only one map had missing data in a small section (in the south), all comparisons were carried throughout by the F-algorithm. The following code is used to give the weight of the input maps:

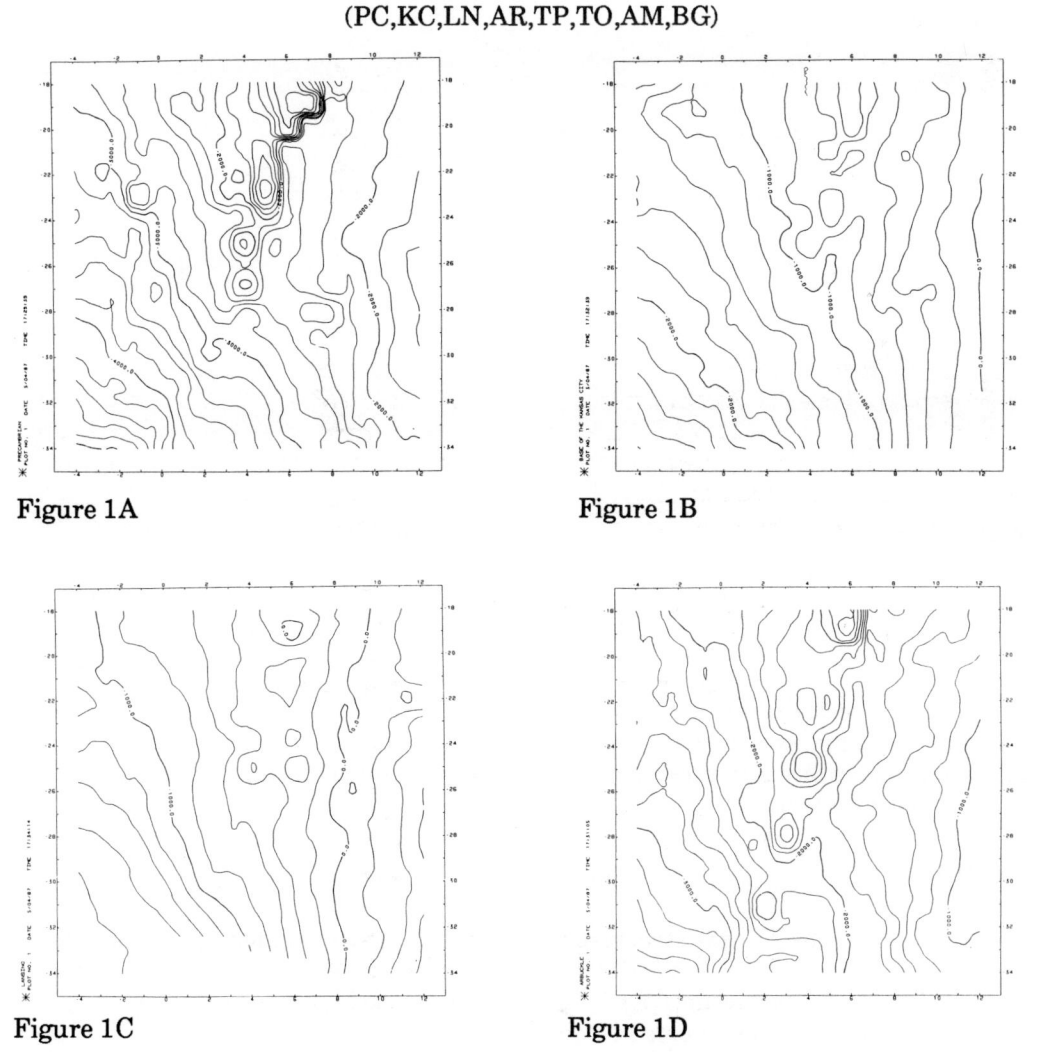

(PC,KC,LN,AR,TP,TO,AM,BG)

Figure 1A

Figure 1B

Figure 1C

Figure 1D

Figure 1. Input maps for thematical comparison; A, Precambrian; B, Kansas City; C, Lansing; D, Arbuckle.

Figure 1E

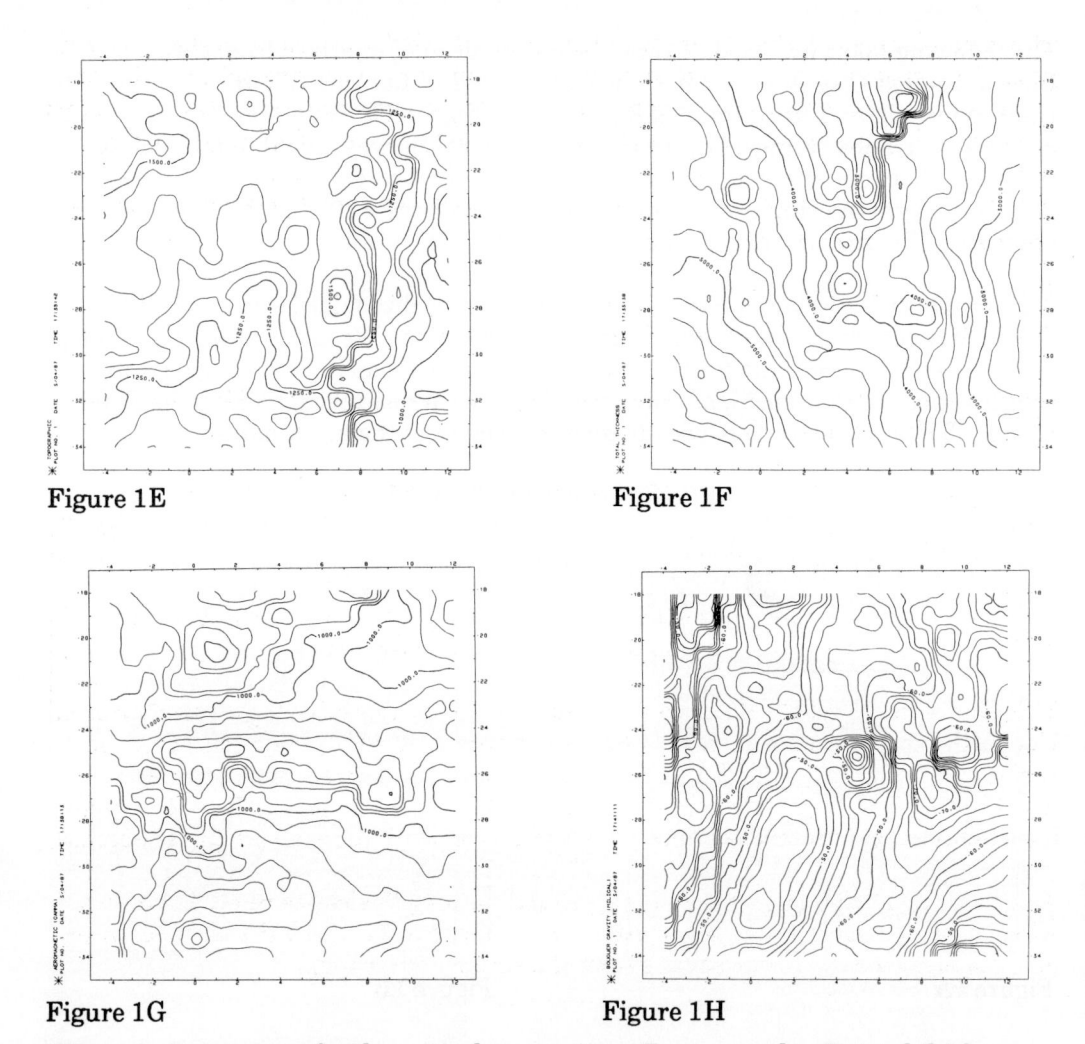

Figure 1F

Figure 1G

Figure 1H

Figure 1. Input maps for thematical comparison; E, topography; F, total thickness; G, aeromagnetics; H, Bouguer gravity anomalies (for references and gridding see text).

is the weight vector where the entries are the weights of Precambrian, base of Kansas City, Lansing, Arbuckle, topography, total thickness of sediments between the Precambrian surface and present-day topography, aeromagnetics, and Bouguer gravity maps respectively.

Results

It is known that the Kansas structure maps show N-S trending features, whereas geophysical maps have E-W trends preferably. In Figure 2 an unweighted comparison of all maps is given [code(1,1,1,1,1,1,1,1)]. We expect a grade of dissimilarity varying according to the scheme in Figure 3.

This can be traced on the comparison map in Figure 2; for example, in the area of the geological structure in the eastern part of the map, there is no geological variation, and high dissimilarity. In the area of the main aeromagnetical anomaly (in the upper center), and no geological features, the comparison map shows high dissimilarity, too.

A demonstration of the weighting input parameter technique is given by the following sequence of comparison maps.

Figure 4 shows a comparison of only the structure maps, N-S trending features dominate, that resemble the usual structure pattern.

Figure 5, giving high importance to the structure maps, some to topography, and little to geophysics, is close to the output in Figure 4.

In the comparison underlying Figure 6, more weight is put on the geophysical information, and consequently the map is more similar in structure to the map in Figure 7, where only geophysical items are represented.

In Figure 7, we weighted aeromagnetics higher then Bouguer gravity, as the actual E-W trending pattern is shown best on the aeromagnetic map.

This shows that MAPCOMP in fact is based on an appropriate weighting algorithm. Hence, the comparison map cannot be viewed only as a such, giving contoured local variation of similarity. Furthermore, it can be considered as a combination map, which resembles the weighted results of a couple of expert's maps. The latter might have application in interdisciplinary studies on one area, for example in assessment.

As a geoscientific result, we note that geology and geophysics are related only somewhat in our test area, whereas all structure maps show basically the same features, and geophysical significance stems from the aeromagnetic map.

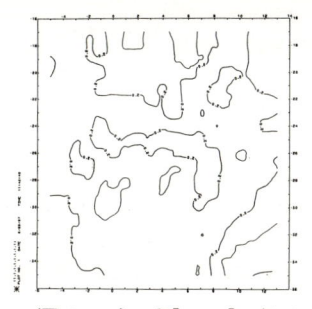

Figure 2. Comparison map (F-type) with code (1,1,1,1,1,1,1,1).

dissimilarity of comparison map		geological variation	
		low	high
geo-physical varia-tion	low	low	high
	high	high	low

Figure 3. Scheme of dissimilarity.

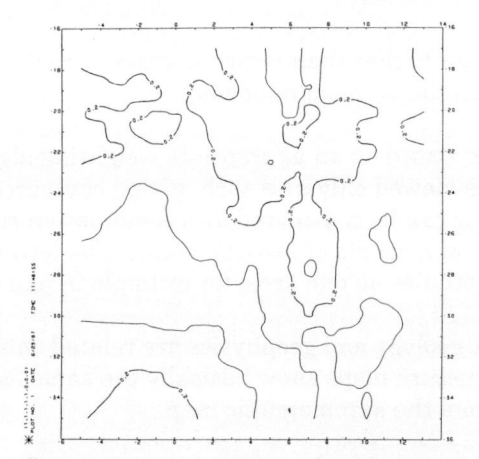

Figure 4. Comparison map (F-type) with code (1,1,1,1,1,0,0,0).

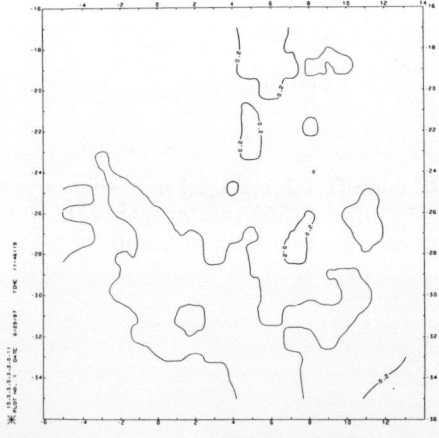

Figure 5. Comparison map (F-type) with code (5,5,5,5,3,3,0,1).

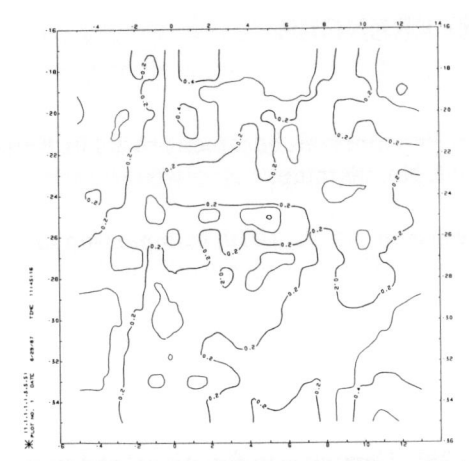

Figure 6. Comparison map (F-type) with code (1,1,1,1,1,3,5,5).

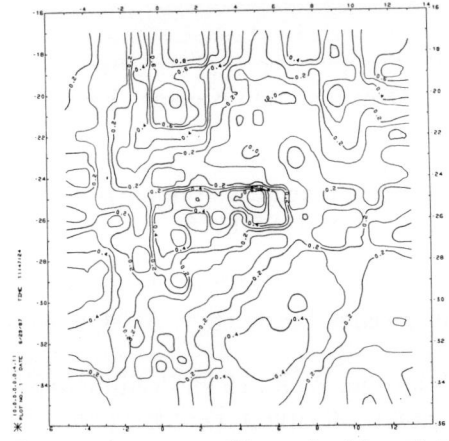

Figure 7. Comparison map (F-type) with code (0,0,0,0,0,0,4,1).

CONCLUSION

The algorithm allows a comparison of any given number of maps. The importance of the input maps may be related to weights because of geological understanding. In addition, the significance of subsets of the input maps can be tested by the same procedure. The resultant comparison map allows contouring, and thus eases geological interpretation.

REFERENCES

Cole, V. B., 1962, Configuration of top Precambrian basement rocks in Kansas: Kansas Geol. Survey Oil and Gas Invest. No. 26, map.

Eschner, T. R., Robinson, J. E., and Merriam, D. F., 1979, Comparison of spatially filtered geologic maps: summary: Geol. Soc. America Bull., pt. 1, v. 90, p.6-7.

Herzfeld, U. C., and Sondergard, M.A., 1988, MAPCOMP - a FORTRAN program for weighted thematic map comparison: Computers & Geosciences, v. 14, no. 5, p. 699-713.

Merriam, D. F., and Robinson, J. E., 1981, Comparison functions and geological structure maps, *in* Future trends in geomathematics: Pion. Ltd., London, p. 254-264.

Merriam, D. F., and Smith, P., 1961, Preliminary regional structure contour map on the top of Arbuckle rocks (Cambrian-Ordovician) in Kansas: Kansas Geol. Survey Invest. No. 25, map.

Merriam, D. F., and Sneath, P. H. A., 1966, Quantitative comparison of contour maps: Jour. Geophysical Res., v. 71, no. 4, p. 1105-1115.

Merriam, D. F., and Sondergard, M. A., 1988, A reliability index for the pairwise comparison of thematic maps: Geologisches Jahrbuch, v. A104, p. 433-446.

Merriam, D. F., Winchell, R. L., and Atkinson, W. R., 1958, Preliminary regional structure contour map on top of the Lansing Group (Pennsylvanian) in Kansas: Kansas Geol. Survey Invest. No. 19, map.

U. S. Geological Survey, 1963, State of Kansas topographic map: U.S. Geol. Survey Map, 1:500,000.

Watney, W. L., 1978, Structural contour map: base of Kansas City Group (Upper Pennsylvanian) - eastern Kansas: Kansas Geol. Survey Map M-10.

Yargar, H. L., and others, 1981, Aeromagnetic map of Kansas: Kansas Geol. Survey Map M-16.

Yargar, H. L., and others, 1985, Bouguer gravity map of Kansas: Kansas Geol. Survey Open File Map.

Intrinsic Sample Methodology

Deverle P. Harris and
Guosheng Pan
*University of Arizona,
Tucson*

ABSTRACT

A new and vital notion of sampling, referred to as intrinsic sample (IS), is proposed as a way to increasing the geoscience information employed in the objective, quantitative estimation of one or more mineral endowment descriptors, for example, number of deposits. An intrinsic sample consists of genetically related geologic objects or their geofields. The set of ISs on a control area constitutes a sample to support the estimation of the relations of one or more mineral endowment descriptors to geology.

Important elements of the IS methodology include (1) the use of genetic relations to integrate diverse geodata, (2) the synthesis of relevant information from multiple geodata sets, and (3) the optimum discretization (cutting) of synthesized fields to delineate geofields and intrinsic samples.

The Walker Lake quadrangle of Nevada and California served as a case study for demonstration of the IS methodology. This demonstration produced a map of ISs for epithermal gold and silver.

INTRODUCTION

This is a progress report about ongoing research (Harris and Pan, 1987), the goal of which is improvement in methodology for the objective estimation of one or more mineral endowment descriptors. The scope of this paper is limited to the early activities of an estimation methodology which, when fully employed, could produce probabilistic estimates of one or more endowment descriptors.

The basic notion of intrinsic sample (IS) is developed here. IS is proposed as the unit for which subsequent analysis would yield a probabilistic estimation of one or more mineral endowment descriptors. Delineation of intrinsic samples is itself a major task and achievement, one which requires geoscience and considerable analysis of diverse geodata.

The Walker Lake 1° by 2° quadrangle of Nevada and California is used as a case study to ensure that the ideas advanced and methods developed are not too academic or impractical, and to serve as a real-world demonstration, based upon actual geodata, for example geochemistry, aeromagnetics, gravity, hydrothermal alteration, structural fractures and faults, and geologic maps. This case study is restricted to the epithermal gold-silver geologic environment of the Walker Lake quadrangle.

Demonstration of the IS methodology includes an IS map for the Walker Lake quadrangle, showing the epithermal gold-silver ISs and the geofields used in their delineation.

INTRINSIC SAMPLE (IS) THEORY

Concept: genetically related geologic objects

The concept of intrinsic sample (IS) is clearly elucidated by its antithesis, an artificial sample. As an example of the latter, consider the placement of a two-dimensional grid over a geologic map and the quantification of geology by selected measurements, such as percentage of cell (intergrid) area occupied by a specific rock type, number of faults, number of mines, etc. Here, the grid of cells constitutes a sample, and there is implied *a population of cells* defined by multiple geologic measurements, that is a *multivariate population of cell attributes.* Thus, by seemingly innocuous and practical methods of quantification, geology and mineral occurrence have become attributes of cells.

In contrast to the population of cells having multiple attributes, consider a population in which *each member* is a *set* of *genetically related geologic objects*, for example, igneous intrusive and associated altered host rock, and each member is described by features of the related geologic objects. In this situation, geology and mineral occurrence are attributes of a set of related geologic objects. This latter scheme employs as the reference for observation and quantification, features which are *intrinsic* to the Earth's crust, not artificial.

Working definition - associated geofields

Conceptually, genetically associated geologic objects define intrinsic samples. However, because geologic objects typically are distributed in three-dimensional space and *direct* information on the third-dimension is especially meager, a more useful working definition of intrinsic sample is that it is defined by genetically related geologic objects or their geofields.

Suppose, for example, that the genetic model includes igneous intrusion at shallow to moderate depth. The most prospective regions for mineral deposits include regions in which igneous intrusives exist at shallow depths but are not exposed. To be useful in the appraisal of undiscovered mineral resources, the definition of intrinsic sample must be such that unexposed, but present, igneous intrusives can be identified as components of intrinsic samples. To meet this requirement, delineation of intrinsic samples is based upon fields of geoscience information, for example, magnetic flux density, geochemistry, and gravity, that give evidence of the existence of a geologic object, for example, an igneous intrusive, at depth. Figure 1 shows an idealized model of an intrinsic sample, its geofields, and intrinsic sample boundaries delineated by two different methods: synthesis and union.

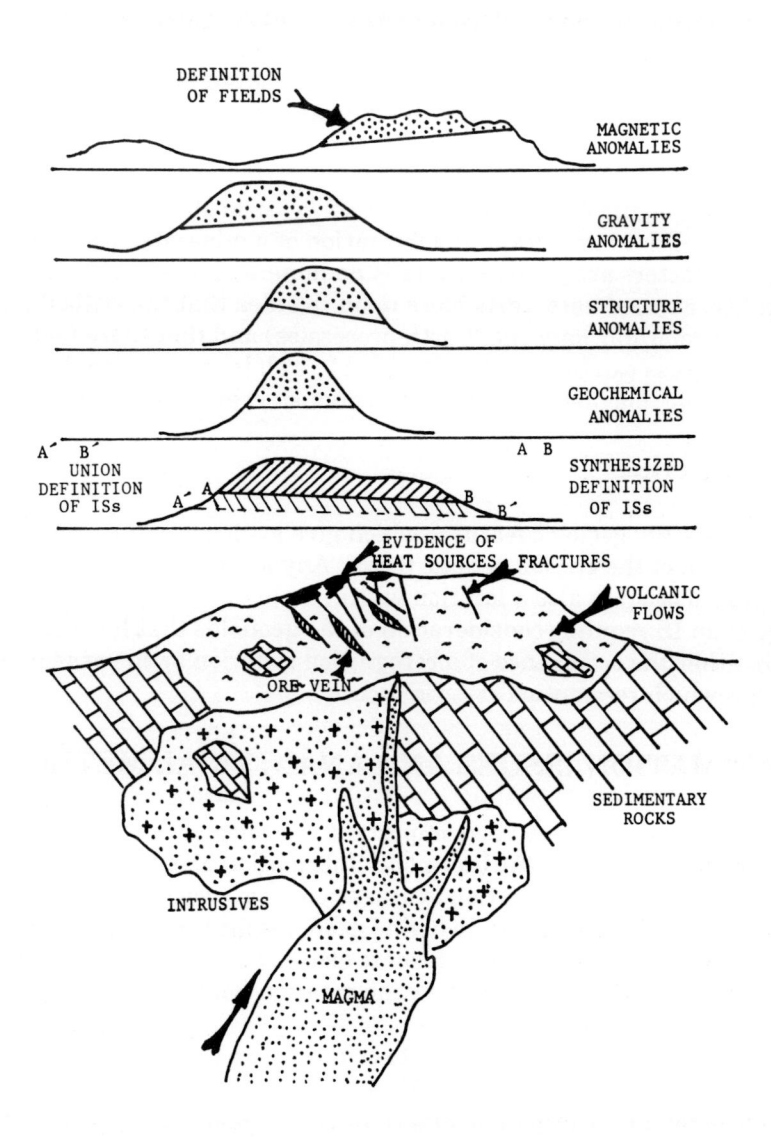

Figure 1. Idealized model of intrinsic sample.

Critical genetic factor

The working definition of IS as the union or synthesis of genetically related geologic objects or their geofields places great importance on genetic models in the new, improved quantitative and objective methodology. As the scope of this paper covers only the delineation of ISs, the contribution and use of genetic models goes far beyond its use as here examined. In fact, the delineation of ISs is predicated upon only one genetic factor, which is referred to as the critical genetic factor. Essentially, the geologic objects or geofields which comprise the IS are identified by this critical factor. The other factors of the genetic model are employed later in the estimation of endowment descriptors for ISs, but only the critical genetic factor is used to delineate ISs.

The idea of critical genetic factor does not rest solely upon one factor being more important or critical than another in the formation of a mineral deposit, because unless all genetic factors are present, there is no mineral deposit or mineral endowment. Criticality, as used here, rests more upon the idea that the critical factor arises from few, preferably only one, Earth process(es) and that those features formed by that process can be detected reasonably well by conventional sensing technologies, for example, magnetics, gravity, geochemistry, and geologic mapping.

Recognition criteria

Recognition criteria are geologic features which give evidence of the previous existence and operation of the critical genetic factor. Any location at which one or more recognition criteria occur is also a location within an intrinsic sample. Delineating the boundary of an IS requires consideration of the geofields that have coordinates in common with either the occurrence of a recognition criterion or one or more other geofields within which recognition criteria occur.

INFORMATION ISSUES AND ANALYTICAL METHODS

Spatial variation of geodata may be as important in geological analyses and decision as the data values themselves. Furthermore, information about a decision variable that is most useful may be information synthesized from two or more data sets. This is especially so in the use of geodata to define geofields for the delineation of ISs. Consequently, data are analyzed in their most elementary form for important spatial variation which can be synthesized to new information fields.

WTMC, weighted and targeted multivariate criterion, was determined to be useful for the synthesizing of information. WTMC combines two or more criteria in a weighted linear equation to produce a single score. Estimation of the coefficients of the variables in the WTMC equation considers a priori weights of the variables and samples. This score differs from the usual regression estimate because a priori variable weights and sample (location) weights are considered in the estimation of parameters of the equation. A WTMC equation was used to combine 14 geochemical elements to a single geochemical score. A priori variable weights were determined from a factor-like analysis (π_{ijk}) based upon 3-variable correlations. Data weights consisted of the sums of gold and silver concentrations.

Replacing a grid of cells by intrinsic samples presents a data problem similiar to the background-anomaly separation problem routinely dealt with by geochemists and geophysicists. The sizes of geofields are determined by the cutoff values used to define the fields. To avoid the arbitrary selection of cutoff values, two techniques were employed: optimum discretization (OD) and coherency analysis (CA). OD was applied to the WTMC geochemical score, seeking those cutoff values of the score and of gold-silver mineralization which maximized the probability for their joint occurrence on known epithermal gold-silver districts. Once determined, the WTMC cutoff value was used to cut WTMC scores throughout the quadrangle, thereby delineating geochemical fields.

Coherency analysis (CA) was employed to identify the spatial sizes of gravity and magnetic anomalies which are most coincident in space with the WTMC geochemical anomalies. Following a Fast Fourier Transformation (FFT) to the two-dimensional frequency domain and appropriate filtering, cross correlations of various frequencies of the polar spectra were computed, seeking that range of frequencies having statistically significant correlations. A band-pass filter was used to isolate the selected frequencies, following which the filtered frequencies were inverse transformed (FFT^{-1}) to their spatial equivalents.

INTRINSIC SAMPLE CASE STUDY - WALKER LAKE QUADRANGLE

Features of epithermal gold and silver intrinsic-sample analysis

Genesis in Brief. Important features of the genetic and occurrence models for epithermal gold-silver deposits are the "intrusion of a silicic magma (heat source) into a pile of volcanic and (or) sedimentary rocks and the presence of porous rocks which allow the development of convecting geothermal waters" (Busby and Menzie, 1986, p. 77). Geothermal processes that influence ore deposition include the following (Busby and Menzie, 1986, p. 77): "(1) circulating, hot alkali chloride waters, (2) boiling of these waters in the near surface environments, and (3) mixing of ascending alkali chloride waters with descending, oxidizing either acid or neutral meteoric waters." Thus, in the current view of genesis, the source of the gold and silver may be either the intruding magma or overlying or adjacent volcanic rocks; gold and silver may be leached from volcanic rocks by formation and meteoric waters that are boiling as a result of the heat generated by the shallow magmatic intrusive. The intrusion of a magma to a shallow depth is vital either for hydrothermal emanations of gold-silver and fluids or for the heat and fluids required to drive a shallow, boiling-water geothermal system.

Critical Genetic Factor. Heat source is selected as the critical genetic factor because of the foregoing reasons plus the fact that geologic objects formed by shallow intrusives provide reasonably good signatures that permit their detection. Preserved subareal volcanic terranes exhibit surficial land forms that facilitate identification of the previous existence of the critical factor (heat source); these include stratovolcanoes, calderas, moors, lava flow dome complexes, radial fracturing, and porphyritic to glassy igneous intrusives. Furthermore, under certain conditions magnetic and gravity fields may give evidence of the geologic objects that comprise the subareal volcanic terrane.

Recognition Criteria for Heat Source. Delineation of ISs requires first the use of recognition criteria. These are "deterministic criteria," inferring that their presence establishes with complete certainty that the location of the criterion is contained within an intrinsic sample. In general, recognition criteria (RC) are dictated by the critical genetic factor (CGF).

In the situation of epithermal gold-silver deposits, recognition criteria selected were for heat source:

>Epithermal ore or prospect (epithermal mineralization)
>Outcrop of a Tertiary intrusive
>Presence of hydrothermal alteration
>Geochemical anomaly consistent with epithermal environment
>   and processes
>Caldera, stratovolcano, moors, and subvolcanic bodies

Rationale for that presented, we believe, is self-evident. If any one of these criteria is present, its location is considered to be within an intrinsic sample. Of course, more than one criterion may coincide or overlap; such is the situation for known epithermal gold-silver deposits.

Geology fields

Geology fields considered here consist of mineralization and of those lithologic bodies that give evidence of a previously existing heat source, except for hydrothermally altered bodies which constitute separate fields. The role of geology fields in this study is limited to the identification of intrinsic samples. Some geologic information, for example, the outcrop of a Tertiary intrusive is direct prior knowledge about the presence of ISs; consequently, by definition, the region of outcrop is within an IS — such information constitutes a recognition criterion for IS. Other recognition criteria for the critical genetic factor (heat source) include volcanic systems, hydrothermal alteration, and mineralization. Their recognition had been made from U.S. Geological Survey geologic maps, exploration drilling, and LANDSAT photography. Consequently, the delineation of geology fields was achieved simply by plotting observed outcrops of relevant lithologies on a map. Each isolated outcrop constituted a field.

Hydrothermal alteration field

Relationship. Because hydrothermal alteration originates from chemical reactions between hot fluids and host or wall rocks, its occurrence indicates the past presence of heat source, which is the critical genetic factor for the definition of an intrinsic sample (IS). Alteration halos in either host rock or wall rock are formed by interaction between hot rock and hot solutions that may or may not carry useful gold and silver. Mineralization may not be present, or it may be at depth underneath altered halos. Regardless, alterationrequired a heat source, and its presence is sufficient for the identification of an IS.

Walker Lake Information. A map of hydrothermal alteration was produced for the Walker Lake quadrangle through field evaluation of a limonite map that was compiled from digitally processed images from the LANDSAT satellite Multispectral Scanner (MSS) (Rowan and Purdy, 1984). According to Rowan and Purdy, limonitic bedrock areas were extensively evaluated in the field, each area examined being categorized as either altered or unaltered. In a few areas of limonitic sedimentary and metasedimentary rocks, the origin of the limonite was not clear, and a decision as to whether it was a product of hydrothermal alteration was based upon comparison of the normal mineral content and texture of the lithologic unit with that of the limonitic exposure. Hydrothermally altered rocks that are not limonitic, such as bleached silicified and argillized rocks, were not distinguished consistently in the CRC images (Rowan, Goetz, and Ashley, 1977).

Geochemical fields

Data Analyses. The geochemical data available in the region included analyses of 815 rock samples, 1,116 stream-sediment samples, and 1,005 bulk-sediment nonmagnetic heavy-mineral-concentrate samples.* All three types of samples had been analyzed for Fe, Mg, Ca, Ti, Mn, Ag, As, Au, B, Ba, Be, Bi, Cd, Co, Cr, Cu, La, Mo, Nb, Ni, Pb, Sc, Sn, Sr, V, N, Y, Zn, Zr, and Th using a six-step semiquantitative emission spectrographic method. Some elements with lowdetection levels, such as Zn, Sb, Cd, Bi, Au, etc., also were analyzed by using atomic-absorption spectrometric analysis.

Rock samples were utilized in this study chiefly for recognition of anomaly sources of different elements; they were not employed as an input to quantitative analysis, because the number of the samples is insufficient for a spatial analysis in such a large region. Measurements of heavy-mineral concentrate were not considered fully here either, because their analytical precision is lower than it is for the other two types, and the number of samples also is less than that for stream sediments. The statistical analyses described in this paper were applied only to stream-sediment data. Concentrations of elements in a stream sample constitute a composite sample of the terrain that feeds the sampled drainage tributary. Consequently, the statistical analysis of element associations and spatial patterns was preceded by drainage basin and transportation analyses in an attempt to associate concentrations with sources.

Control Area for Geochemistry. Description of geochemical fields requires the quantification and estimation of relations between gold and silver deposit occurrence and geochemical values of gold and silver and other associated elements. In order to have

---

* A computer tape containing analyses of these samples is available from NTIS.

as specific a relation as possible, it was necessary to specify a control area that facilitates the estimation of the desired relations. The control area selected for epithermal gold-silver deposits is located chiefly at the Bodie 15' quadrangle in the southern part of Walker Lake. The number of sample points in this area is 92, which is sufficient for any ordinary statistical analysis. This control area is characterized by the following geologic features:

> Anomalies of gold and silver (greater than their detection levels: gold is zero and silver is 5 ppm) are almost completely consistent in their spatial distribution.

> Host rocks of this deposit type are chiefly Tertiary dacite, rhyolite, and some other acidic intrusives.

> Most gold-silver deposits occur in extensively faulted areas. Localized (contrasted with regional) faults are classified into two groups, NNW and NE, and in some localities seem to be associated with the occurrence of Au-Ag deposits in this area.

> The area has several strong lead anomalies which are spatially associated with gold and silver anomalies. This provides some evidence for common genesis by volcanism.

> The area has no copper and molybdenum anomalies, and only weak anomalies of zinc, which indicates good separation of these two groups.

> Strong argillaceous and siliceous alterations are associated with the occurrence of gold-silver deposits.

Statistical Analysis and Geochemical Fields. π_{ijk} Model and Analysis of Elements Associations. Four stages of analysis are summarized in this section on geochemical fields. The first stage investigates the elements for inherent relationships and associations by employing a form of factor-like analysis referred to as the π_{ijk} model, which considers the relations between three elements and some nonlinearities (Harris and Pan, 1987). One of the important objectives of the π_{ijk} model is to determine those associations of elements that are the best indicators of epithermal gold and silver deposits. Because the elements of interest (Au,Ag) usually have low detection levels, use of these concentrations alone as mineralization indicators may lead to an underestimate of potential targets, especially when potential resources are hosted in rocks at depth and the main elements diffuse slowly. Locally, no anomalies of these elements can be detected. In certain geochemical environments, genetically based associations of Au and Ag with some other elements may be better indicators of mineralization or of potential targets.

Results of the π_{ijk} type analysis are summarized, as follows:
> A typical feature shown in the π_{ijk} analysis is the negative association of two groups (Au,Ag,Bi) and (Cu,Mo).

Element B also is closely associated with Au, even though it is even more closely associated with Mo. B is an element associated with an epithermal environment favorable to the formation of gold and silver deposits.

Zn and Zr also are associated with gold-silver mineralization. They possibly represent an environment of intermediate (chemically) volcanism.

Pb and Be also have a strong association with gold and silver. This result is consistent with qualitative observation, and it also is acceptable from the genetic point of view, because Pb is typical of some intermediate-acidic volcanic rocks.

On the basis of the π_{ijk} analysis, initial (a priori) weights were assigned to the 14 elements; these weights, which are shown in Table 1, were inputs to Stage 2 of the analysis.

Table 1. Prior and Posterior Weights

No.	Variable	Prior[1] Weights	Coefficients[2]
1	Au	0.3333	0.4955
2	Ag	0.3333	0.7408
3	Cu	0.20	0.0646
4	Pb	0.2667	0.3820
5	Zn	0.1333	0.0742
6	Fe	0.0667	-0.0068(*)
7	Ca	0.1333	0.0398
8	Sb	0.2667	0.1432
9	Zr	0.1333	-0.0320(*)
10	V	0.0667	-0.0163(*)
11	Bi	0.2667	0.1391
12	Mo	0.0667	-0.0043(*)
13	Be	0.0667	0.0041(*)
14	B	0.2667	0.0850

[1] Based upon π_{ijk} analysis.

[2] Based upon WTMC analysis.

WTMC Model and Geochemical Scores. The second stage analysis uses these a priori data weights as inputs to a newly devised multivariate model, referred to as Weighted and Targeted Multivariate Criteria (WTMC). In addition to the a priori elements weights from the π_{ijk} model, WTMC employed observation or sample weights: The weight given the ith sample is the sum of its gold and silver concentrations, $Au_i + Ag_i$. Both sample and element weights were considered by the WTMC technique in the estimation of the parameters of a linear equation that describes a geochemical score as a function of the 14 elements. The coefficients, which also are referred to as posterior element weights, are shown in Table 1. In this table the elements listed with (*) have been eliminated in the subsequent calculation of weighted scores, since their coefficients are small. Thus, for epithermal gold-silver deposits, the equation giving the magnitude of the geochemical field, $WTMC_i$, at the ith sample location is the following:

$$WTMC_i = 0.4955 w_{Au\_} Au_i + 0.7408 w_{Ag\_} Ag_i + 0.0646\_ w_{Cu\_} Cu_i + 0.3820 w_{Pb\_} Pb_i +$$

$$+ 0.0742 w_{Zn\_} Zn_i + 0.0398 w_{Ca\_} Ca_i + 0.1432\_ w_{Sb\_} Sb_i +$$

$$+ 0.1391 w_{Bi\_} Bi_i + 0.0851 w_{B\_} B_i ,$$

where

w_{Au}, w_{Ag}, w_{Cu}, w_{Pb}, w_{Zn}, w_{Ca}, w_{Sb}, w_{Bi}, and w_B are the prior weights.

Evaluation of this equation on each sample produced a set of geochemical field scores. The polar power spectrum of these scores reveals the need to filter the scores to remove noise (random error). A filter was designed accordingly, and the geochemical field scores were taken as the inverse FFT of the low-pass part of the frequency spectrum.

Optimum Discretization to Delineate Field Boundaries. The third stage determines optimum discretization or "cutting" of the scores so as to optimize the match on the control area of the resulting geochemical fields with known epithermal gold and silver mineralization. Figure 2, which shows entropy correlation plotted against the geochemical field scores, WTMC, indicates an optimum cut of the geochemical scores at 520.0

Fields Outside of the Control Area. The fourth and final stage in the geochemical analysis is to use the WTMC equation to generate WTMC scores for each sample location outside of the control areas. These scores then were cut at the optimum level, as predetermined on the control area, to give epithermal gold-silver geochemical fields throughout the Walker Lake quadrangle.

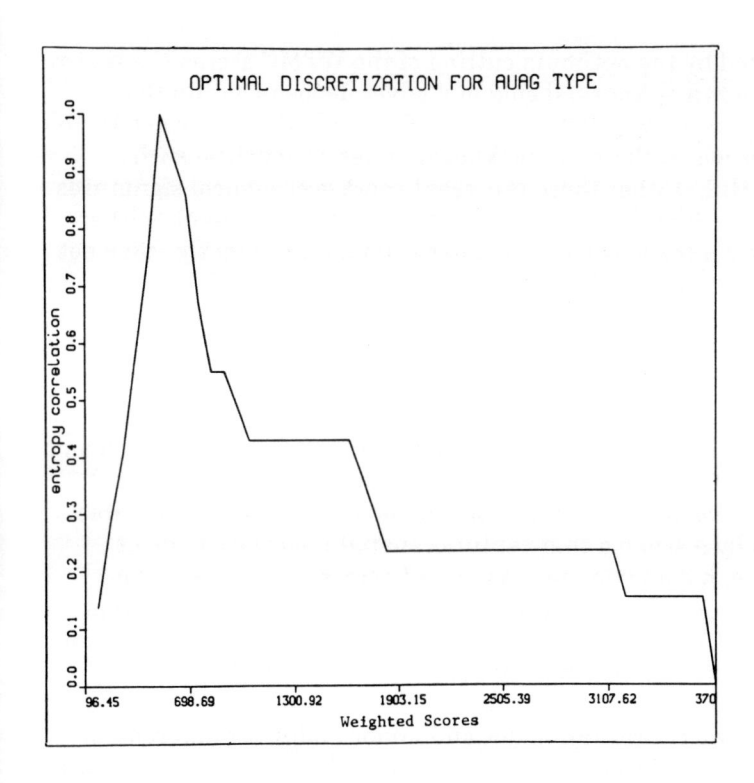

Figure 2. Optimal discretization for AuAg deposit type.

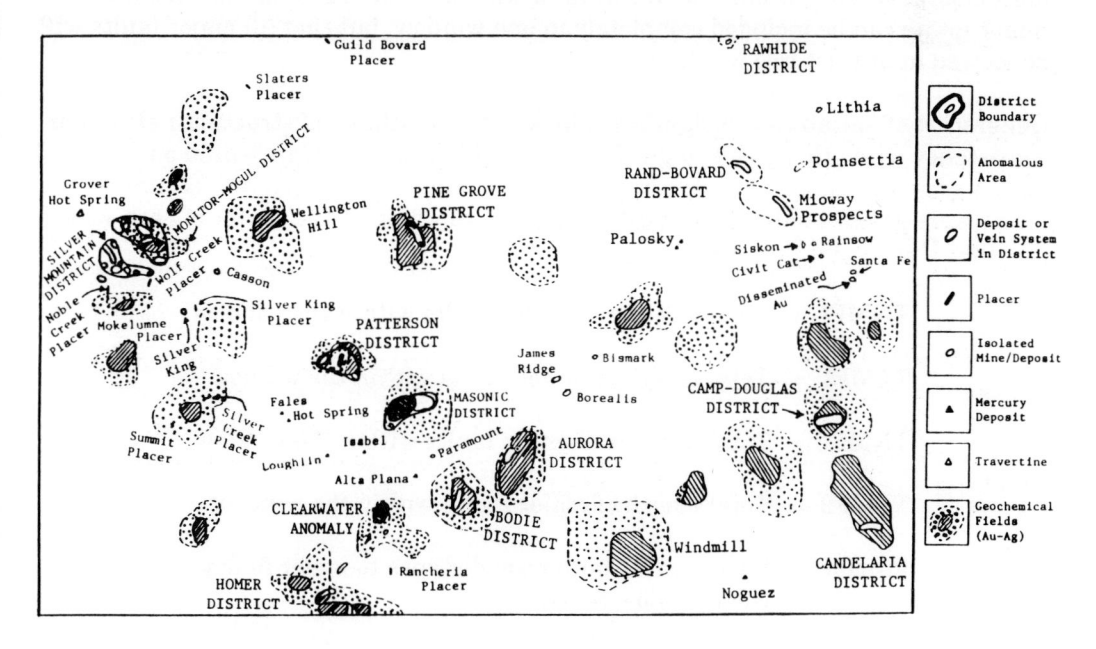

Figure 3. Delineation of geochemical fields for gold-silver type deposits by optimum discretization (cutting of scores).

Geochemical fields predicted by the optimum cutting of the WTMC scores are shown in Figure 3. Most of the known epithermal gold and silver deposits within the Walker Lake quadrangle coincide well with these geochemical fields. Of considerable interest are those fields for which there are no known mines or districts, such as Windmill and Wellington Hill. Either these represent relict geochemical signatures of deposits long since destroyed and removed by erosion, or they represent epithermal Au and Ag deposits at depths which do not outcrop and, consequently, have not yet been delineated by exploration drilling.

Structure fields

Selection of Data Window and Generation of Variables. Structural fields, similiar to geophysical fields, should be analyzed for relevant information by data enhancement and information synthesis, stressing the relations between faults and epithermal mineralization. To generate variables useful in describing these relations requires sampling of the fault data by a scheme that captures spatial information about structure. For this purpose, a data window was moved across the map and the numerical value for each of ten selected features was recorded at the window center. In this way the influence of a region's (window area) structure is ascribed to a point, creating a "field potential"-like expression of structure for each variable.

Faults were classified into two categories, major and minor faults, for different magnitudes of faults may reflect different types of genetic relations. In this situation 7.5 miles was selected as the cutoff for major and minor faults, chiefly because the maximum axial length of known mineral districts is about 6.0 miles. On the basis of fault classification, a 5-mile square window was selected. This size assures that minor faults can be included completely in one window, but that all major faults will be viewed in at least two windows.

Generation of variables is a significant task in the synthesis of structure fields. For consistency with other fields, digitally described faults were represented on a 55 x 55 matrix of grid intersections. With the aid of a computer program, ten features of the faults were generated from the digital data and recorded at the center of each window (see Harris and Pan, 1987).

TNMF1 = Total number of minor faults within the window

TLMF1 = Total length of minor faults within the window

TNMF2 = Total number of major faults within the window

TLMF2 = Total length of major faults within the window

RMDM = Reciprocal of minimum distance to major faults from window center

NFNE = Total number of faults trending northeast (15 - 75°)

NFNW = Total number of faults trending northwest (285 - 345°)
NFEW = Total number of faults trending east-west (75 -105°)

NFSN = Total number of faults trending south-north (345 - 15°)

TNFI = Total number of fault intersections within the window

Synthesis of Structure Fields. The same area used in the synthesis of geochemical fields was selected as a control area for structure fields. The number of data points in the control area is 324. The data of the ten measurement variables were standardized to have equal means and deviations so as to remove the impact of different magnitudes on measures of correlation.

The geochemical field for epithermal gold and silver was selected as the background variable for the synthesis of structure fields because geochemical anomalies have been observed to be highly coincident with epithermal gold-silver deposits. Unit weights for the ten variables were employed as a priori weights in the WTMC analysis, chiefly because limited prior knowledge did not indicate any one feature as being of more importance than others. Coefficients of the fault variables in the WTMC equation are listed in Table 2.

Table 2. WTMC Results for Fault Variables

Measurement	Coefficient
TNMF1	0.1404
TLMF1	0.1000
TNMF2	0.1037
TLMF2	0.1049
RMDM	−0.0431
NFNE	−0.1577
NFNW	−0.1415
NFSN	−0.7736
NFEW	−0.7158
TNFI	−0.0660

The eigenvalue for the vector was calculated to be 0.1452, which equals the maximum correlation coefficient between the measurements and the background anomalies in the control area.

Structure fields for grid intersections outside of the control area but within the Walker Lake quadrangle were generated by computing scores for each grid location, using the given coefficients.

Frequency Analysis of Structure Fields. Similiar to other geofields, structure fields also may be enhanced by frequency analysis. To facilitate data enhancement, a low-pass filter was designed to remove noise, that is, energy at frequencies higher than 0.41 CPM. Regional features were removed so as to enhance local structure features by passing the frequencies higher than 0.03 CPM in the filtered field. The second vertical derivative of this high-pass structure field was computed in the frequency domain and then transformed to the space domain.

Gravity fields

The Third Dimension. A serious limitation in traditional methodologies for resources appraisal has been the lack of information on the third dimension, that is, subsurface geologic objects. Heat source, the critical genetic factor for epithermal gold and silver deposits, may be concealed, even though one or more recognition criteria give evidence of its existence. In such situations, features of the igneous body that was the heat source may be revealed by the use of geophysical fields, gravity being one of the most important for this purpose.

Isolating the gravity field caused by a specific subsurface source is difficult unless simplified by additional information, such as density and depth to source, because gravity measured at the surface above a specific source consists of the influence of objects above, below, and adjacent the specific source.

There is a natural inclination to try to separate a gravity field into deep and shallow components, but such separation is problematic if it is based only upon properties of the field itself. Consequently, separation by depth was rejected as a way of identifying gravity fields of heat sources. Instead, gravity fields were delineated by isolating the fields having a size range that is most coincident with mineralization.

Identifications of the Sizes of Geologic Objects Most Related to Mineralization. Identification of those objects that were actually involved in mineralization is impossible in an absolute sense. However, a useful approximation can be made by identifying those fields of the size range that is most coincident with mineralization fields. This was achieved by associating gravity fields with prior knowledge about mineralization of known mineral deposits or districts and their gravity anomalies; this knowledge is critical to interpretation of gravity, because these districts are themselves *known intrinsic samples*. Because of data limitations, this study used geochemical field as a proxy for mineralization.

Delineation of the gravity fields which may be the result of subsurface heat sources was made by identifying the frequency bands of gravity and geochemical fields having statistically significant correlation (coherency). A band-pass filter was used to suppress the gravity components for frequencies not within the coherent band.

Coherency of Gravity and Geochemical Fields. Prior to the coherency analysis, the regional components of gravity were suppressed by applying a high-pass filter in the frequency domain: responses for frequencies below 0.04 CPM were suppressed.

The auto-power spectra of the geochemical and gravity fields and their cross-power spectrum were smoothed using the Tukey-Hamming window (Priestley, 1981). After these smoothed spectra were transformed to polar (one-dimensional) form, the coherency between the two fields was computed and the statistical significance of each computed coherency was calculated with the aid of a small FORTRAN program. The result showed that a coherency of 0.22 is statistically significant at the 10% significance level (see Fig. 4). On the basis of this result, a band was selected from 0.0 to 0.525 RPM, which consists chiefly of the low-frequency part of the high-pass field. The gravity field for this band was produced by repeated filtering. Several features can be observed from the map of the filtered response:1

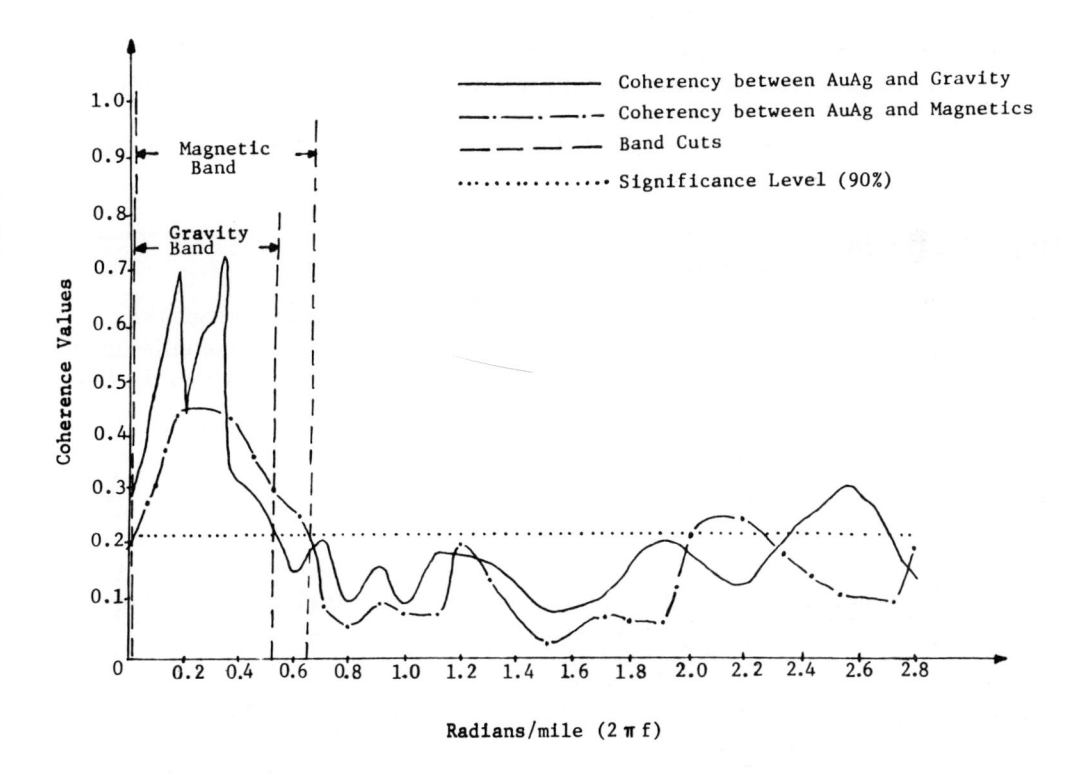

Figure 4. Coherency analysis between geochemical and geophysical fields.

(1) The gravity anomalies are enhanced. This enhancement is useful because it leads to a more consistent delineation of IS boundaries.

(2) Gravity highs are closely associated with geochemical highs and, thus, the mining districts, such as Aurora, Camp Douglas, Candelaria, Monitor-Mogul, etc.

(3) Some of the gravity anomalies are pronounced in contrast to others or to background, whereas some of the linearities which showed clearly in the high-pass field are lost in this band-pass field. Even so, we have gained more than we have lost because of the enhancement of those gravity fields of geologic objects that are most coincident with mineralization. This is a valuable contribution to the delineation of intrinsic samples.

Delineation of Gravity Fields. Delineation of boundaries of gravity fields of the band-pass frequencies is needed for the delineation of ISs. This was done by computing the second vertical derivative of the filtered gravity field. Let d and v be the horizontal distance between the zero line and the center of an anomaly, and the center value of the anomaly. Then, the gradient might be described by v/d. Generally, the larger the ratio v/d, the more the zero line should be corrected, and vice versa. This idea may share some properties in common with the standard deviation rule usually used in geochemical studies in which a critical value is zero ± one or two deviations. This idea, in fact, was adopted in this study and yields both gravity highs and lows (see Fig. 5).

Magnetic fields

Role of Magnetic Fields. The direct information in magnetic fields is about the magnetic susceptibility of rocks, and all interpretations of magnetic anomalies must be based upon this feature. This feature reflects the lithologies from which the fields originate and thus is useful for distinguishing boundaries of geologic units.

Coherency of Geochemical Field with High-Pass Magnetic Field. As noted previously, a major task in the delineation of intrinsic samples is the "optimum" synthesis of available and relevant geoinformation. The goal of this synthesis is to locate and enrich "jointness" features of different geofields. Such information is useful in the identification of boundaries of intrinsic samples. Coherency analysis is useful for this purpose. Accordingly, a high-pass magnetic field was correlated in the frequency domain with a priori information on mineralization (background field), that is, gold-silver geochemical anomalies.

Delineation of Magnetic Fields. As with gravity, the second vertical derivative was used to delineate the boundaries of the magnetic fields of the geologic objects (see Fig. 5). If a constant critical contour is used to define the fields, the scales of the defined units differ considerably — some of them are large and others are small, reflecting the heterogeneity in lithologic formations.

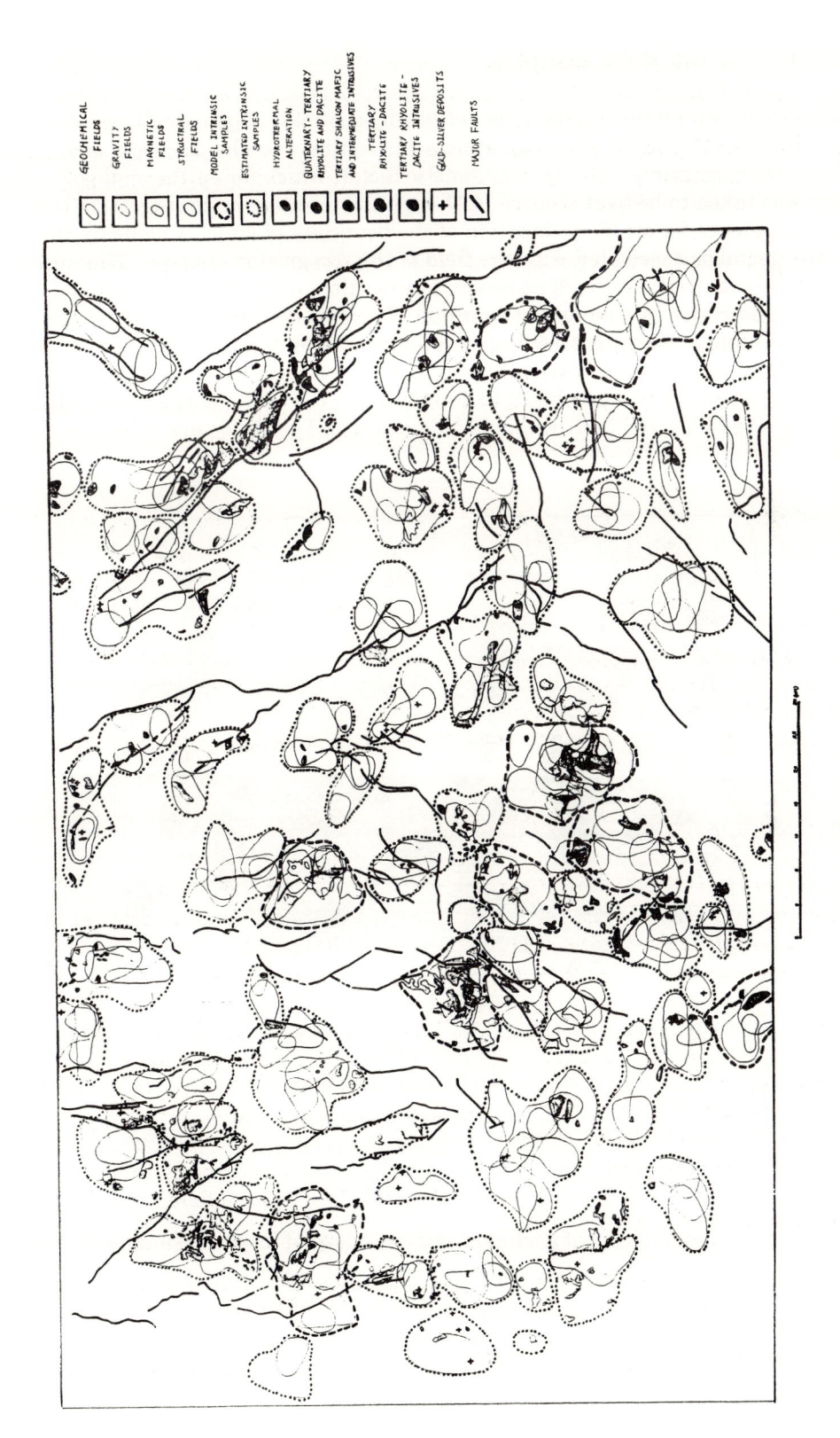

Figure 5. Delineated ISs by union of different existing geofields in Walker Lake quadrangle.

Delineation of intrinsic samples

Recapitulation. An intrinsic sample is defined by geofields of genetically related geologic objects. The delineation of a known IS requires the identification of at least one recognition criterion of the critical genetic factor, which for epithermal gold-silver deposits was taken to be heat scource. Although the presence of a recognition criterion gives evidence for an IS, delineation of the boundary of an IS is made by resolution of the geofields associated with the field of the recognition criteria. This resolution can be made variously, such as union, intersection, or synthesis. Figure 6 is a schematic illustration of delineation by union. Because the intrusive is a recognition criterion, the field of the intrusive lies somewhere within the IS. The union of the geofields associated with the intrusive field in Figure 6 delineates the boundary of the IS. This procedure was employed to delineate the ISs of the Walker Lake quadrangle; these samples and their recognition criteria and geofields are shown in Figure 5.

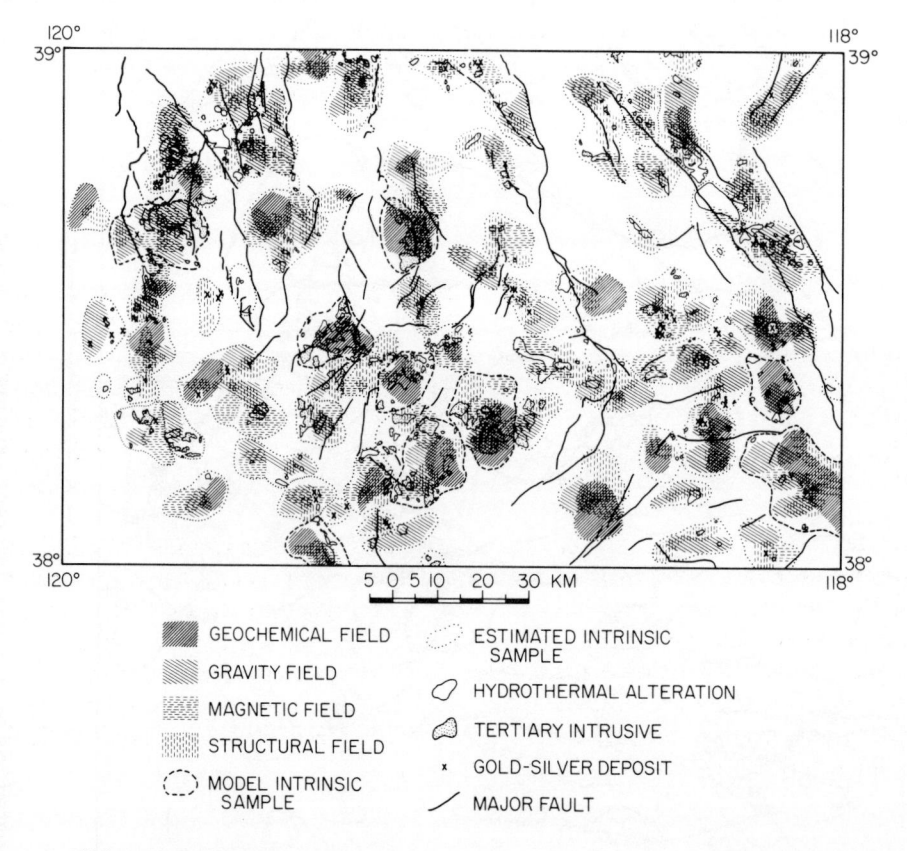

Figure 6. Schematic diagram of geofields and boundary of intrinsic sample (IS) as union of geofields.

Intrinsic Samples of the Walker Lake Quadrangle - Union of Geofields. Intrinsic samples of the Walker Lake quadrangle that were identified by the union approach (Fig. 5) have several features. First, all of the known mineral districts are captured well by the union of all four different fields, except for the Candelaria district, which as delineated is larger than the explored area. The sizes of the ISs range in size from 8 to 25 square miles. The shapes are complex, ranging from almost a circle to a narrow ellipse.

Second, most of the estimated ISs are hosted in Tertiary andesite-latite flows and tuffs, rhyolite-dacite flow, as well as other Tertiary intermediate and mafic intrusives and subvolcanic rocks. The Tertiary intrusives have been identified as evidences of heat sources, which either introduced gold and silver through fluids emanating from the intrusive or activated gold, silver, and possibly lead, present in the Tertiary volcanic flows and redeposited them in veins along fractures or shear zones.

Third, the estimated ISs seem to form several alignments that are clearly associated with structural zones. Some are located in extensively faulted areas, whereas others are distributed along major fault zones and zones of structural weakness at the intersections of major and minor faults.

SUMMARY STATEMENT

This paper presented a "first cut" of a new methodology for the objective, quantitative estimation of one or more mineral-endowment descriptors. Specifically, this paper dealt with only one element of this methodology, the basic observational unit. Intrinsic sample (IS), which is advanced as that basic observation unit, is a fundamentally new and vital notion of sampling. That samples are termed intrinsic implies that an entity described by geologic and resource information is natural. More specifically, an intrinsic sample consists of genetically related geologic objects or their geofields.

The new methodology requires first that such intrinsic samples be identified; then, these samples are described by geoscience and resource information; this information on the set of intrinsic samples within the control area is analyzed by multivariate mathematical techniques for objective, mathematical relations of geoscience information to mineral-endowment descriptors. Finally, the multivariate relations are evaluated on the geoscience information of ISs outside of the control area to give estimates of one or more mineral-endowment descriptors. This study dealt with only the *first step*, the delineation of intrinsic samples.

The basic steps leading to the delineation of intrinsic samples (ISs) is summarized as follows:

- Identification of a suitable genetic model

- Identification of the critical genetic factor

- Identification of recognition criteria for the critical genetic factor

- Processing of geochemical, gravity, magnetic, structure, and selected geologic data and the synthesis of relevant geoscience information

- Delineation of geofields of each data set, including information variables

- Delineation of intrinsic samples through the synthesis of all geofields or through the union of geofields

Genetic considerations are fundamental in the synthesis of geofields. For epithermal gold and silver deposits in the Walker Lake region, "heat source" was considered to be the critical genetic factor. Other types of deposits may have different critical genetic factors for the identification of ISs.

Especially important features of the IS methodology are (1) the use of genetic relations to integrate various data sets, (2) the synthesis of relevant information from multiple data sets, and (3) the optimum discretization of synthesized fields to give geofields and ISs. This procedure is implemented through the following generalized steps:

- Digitization of all data sets

- Treatment of each data set individually

- Use of genetic relations to extract important "jointness" features of data sets, creating information variables

- Synthesis of treated data sets and information variables to an a priori weighted and targeted field response or score

- Optimum discretization of synthesized score(s) for the delineation of geofields or ISs.

The last three of these steps may be applied more than once for the delineation of information variables and geofields, and more than once in the delineation of ISs.

This methodology stresses the importance of steps 3, 4, and 5 in the delineation of intrinsic samples. Two new statistical methods, WTMC and OD, are useful especially in completion of these steps. Another important and powerful, but not new, tool for these steps is analysis in the frequency domain of the coherency of fields of two data sets. Coherency analysis is particularly useful in selecting a high-information component of a complex data set. For example, mineralization fields can be used to extract that set of frequencies (or equivalently, wavelengths) of gravity that are associated most closely and useful in describing and predicting ISs. Because coherency analysis is rich in information about "similarities" between fields, the derived fields share common genetic relations and geological similarities.

As stated in the introduction, this paper reports on ongoing research. Accordingly, were this study to be redone, some things would be done differently. One of these would be to use a measure of mineralization density or a "field measure" of mineral occurrence as the background variable for information synthesis and for optimum discretization. While conducting the research, the decision to use a geochemical measure as a proxy for mineralization was based upon (1) the nonavailability of a measure of mineralization density and (2) the discrete nature of mineral occurrence. Although mineral-occurrence data were available, their discrete nature presents difficulties in application of optimum discretization and coherency analyses for the generation of information variables and the delineation of geofields and ISs. Because of these difficulties a geochemical measure was used as a proxy. Although this measure is a good proxy and probably yields similar results, it is not conceptually "clean", because geochemical concentrations are themselves important fields to be synthesized. Conceptually, a better solution may be to generate a "field measure" of mineral occurrence by moving a data window across the quadrangle, recording at the center of each position of the window, the sum of all gold or silver mines and prospects. Then, the OD technique could be applied to determine the cut of this "field measure" and the cuts of the geoscience information field for which joint occurrence is maximized.

A second improvement on the study would be the delineation of ISs by synthesis of all geo- and -information fields and optimum discretization using all recognition criteria. At the time this study was performed, such analysis could not be completed because some of the recognition criteria had not been digitized.

A third improvement would be to investigate the effect of anisotropy in the mineralization, gravity, and magnetic fields on the coherent range of frequencies used to design the band-pass filters. The presence of anisotropy was ignored in this first cut. Allowing for anisotropy probably would broaden slightly the band-pass filters, but because ISs were delineated as the union of geofields, it is doubtful that anisotropy would exert significant change in the size or shape of the ISs reported in this study.

Finally, the authors believe that, based upon the methodology described in this paper, mathematical models to estimate the magnitude(s) of one (or more) mineral endowment descriptor(s) for unexplored ISs should be more powerful estimators and should be more acceptable to the geoscientist because of the greater amount of geoscience information that they employ. Much of the increase in geoscience information takes place in the process of delineating ISs.

ACKNOWLEDGMENTS

The research reported in this paper was made possible by a grant (#14-08-0001-G1399) from the United States Geological Survey. Grateful acknowledgement is made of this support and for the data provided by various U.S. Geological Survey offices and personnel. Special acknowledgement is made of the guidance provided by David Menzie, Lawrence Drew, and Glenn Allcott.

Although this study could not have been made without the support and assistance acknowledged, responsibility for the ideas and procedures developed rests solely with the authors. These ideas and procedures are at the formative level and require additional study and development. Accordingly, they do not imply official positions or methods of the U.S. Geological Survey.

We wish to acknowledge contributions to this research project made by John Earnhardt, Tetevi Wilson, Robert Vagnetti, Yinghong Miao, and Keith Long, all of whom are graduate students in mineraleconomics and assisted in data acquisition and processing. Finally, we are appreciative of the excellent assistance of Alice Yelverton in preparation of the manuscript.

REFERENCES

Busby, W., and Menzie, D., 1986, Delineation of mineral deposits in subaerial volcanic rocks: unpubl. manuscript, U. S. Geological Survey, p. 72-114.

Harris, D. P., and Pan, G., 1987, An investigation of quantification methods and multivariate relations designed explicitly to support the estimation of mineral resources-intrinsic samples: Report on research sponsored by U.S. Geological Survey Grant #14-08-0001-G1399 (Sept. 1987), 200 p.

Priestley, M. B., 1981, Spectral analysis and time series, v. 1 and 2: Academic Press, New York, 890 p.

Rowan, L. C., Goetz, A. F. H., and Ashley, R. P., 1977, Discrimination of hydrothermally altered and unaltered rocks in visible and near-infrared multispectral images: Geophysics, v. 42, no. 3, p. 522-535.

Rowan. L. C., and Purdy, T. L., 1984, Map of the Walker Lake 1° by 2° quadrangle, California and Nevada, showing the regional distribution of hydrothermally altered rocks: Dept. of the Interior, U. S. Geological Survey.

Evaluation of the Gold Potential of the Bohemian Massif

J. Janatka and P. Morávek
Geoindustria Praha

ABSTRACT

As a result of geological research and mineral-resource appraisal of the Bohemian Massif, comprising both archival and field studies, nearly 1,500 gold occurrences and indications of various types of mineralization were collected, Data from nearly 80 gold deposits mined in the past were used in order to classify gold mineralization using variables characterizing lithology, mineral assemblage, tectonic position, etc. The classification, using cluster analysis, resulted in distinguishing two main types of gold mineralization: (1) mineralization in volcano-sedimentary terrains intruded by granitic rocks and (2) mineralization in metamorphic crystalline terrains.

Evaluation of gold potential was carried out in two ways—empirically, considering all local conditions, and by multivariate analysis.

In the multivariate approach, the entire studied area was divided by a network of elementary cells, and a control set of cells was formed. The control cells covered four gold districts in different geological environments and several nongold-bearing areas. In the control set, relationship of gold mineralization and complexity of geochemical, and geophysical indications was defined by multiple linear regression. Extrapolation of the defined relationships resulted in the appraisal of the gold potential of the Bohemian Massif.

INTRODUCTION

The Bohemian Massif is one of the most important regions of gold mineralization in Europe. It can be divided into several units according to its geological structure; among them, the Jílové Belt - the Upper Proterozoic volcano-sedimentary complex - is the most important region for gold mineralization.

Gold mineralization in the Bohemian Massif was considered for tens of years to be connected genetically with granitic rocks. Extensive research of their chemical composition resulted in their being divided into several metallogenic groups accomplished by cluster analysis (Sattran and Klomínsky, 1970); one group was classified as gold-bearing granites.

The rising price of gold in early 70s has resulted in renewed interest in the gold mineralization of Bohemia. An extensive exploration program named "Geological research and mineral resource appraisal of prognostic regions of the Bohemian Massif" was undertaken; evaluation of gold potential was one of its main objectives.

DATABASE OF GOLD OCCURRENCES

During the first period of the study, data from nearly 1,500 different gold occurrences were collected. The data, comprising results of both archival and field studies, characterized geological and geographical position, mineral assemblage, alteration characteristics, etc., and quantitative parameters such as tonnage, extent, grade, depth, and thickness.

Most gold occurrences in the Bohemian Massif are placer deposits. The second most abundant type is represented by primary gold deposits and mineral indications. The two groups are followed in number by geochemical anomalies, including indications from the regional panning (heavy minerals) prospecting which covered the whole area of the Bohemian Massif with average density of 1 sample per square kilometer (Tencik, 1982).

CLASSIFICATION OF GOLD DEPOSITS

In conjunction with the field and archival studies, intensive research of gold metallogeny was carried out. The research was influenced by advances in understanding of mechanisms of gold accumulation and transport and by substantial change of presumed gold source rocks towards mafic rocks of the greenstone-belt type.

The new genetic theory of gold mineralization (Morávek and Pouba, 1984) indicated a necessity to accomplish a new classification of gold mineralization in the Bohemian Massif with data from nearly 80 gold deposits being used. The database information was transformed into 31 binary variables indicating the presence of various types of rocks, characteristic ore minerals, morphological type of mineralization, tectonic position, etc. The classification was made using cluster analysis (BMDP software package). As a result, two principal groups of gold mineralization were distinguished:

- mineralization in low-grade terrains (areas of occurrence of volcano-sedimentary rocks usually intruded by granitoids);
- mineralization in high-grade crystalline terrains.

In addition, several other specific groups of mineralization as well as subtypes of both previously mentioned principal groups have been distinguished (Morávek, 1985; Morávek and Pouba, 1987).

EVALUATION OF GOLD POTENTIAL

Evaluation of the gold potential of the Bohemian Massif was carried out in two ways:

- empirically, on the basis of similarity with well-developed deposits, considering all local conditions;
- by multivariate analysis.

The multivariate approach was influenced widely by the work of Agterberg (Agterberg and others, 1972; Agterberg, 1981,etc.). Initially, the entire area studied was divided by a network of elementary calls 5 x 5 km (2,909 total), and a set of control cells was formed. The results of the classification of deposits indicated the necessity to use at least two metallogenic models of gold mineralization. Therefore, the control cells covered two gold districts in low-grade terrain (Jílové, Zlaté Hory), two districts in high-grade terrain (Roudny, Kasperské Hory), and a subset of sterile-ballast cells, without even the slightest indication of gold mineralization. In all the control cells, about 70 semiquantitative and quantitative variables were determined. The variables characterized lithology, type and age of host rocks (21 variables), petrometallogenic types of rocks (14 variables), tectonic setting (7 variables), metamorphism (6 variables), geophysical data- gamma-activity, magnetic intensity, Bouguer anomalies, gravity gradient in eight different directions (11 variables total), anomalous concentrations of characteristic associated minerals detected by panning prospecting (12 variables). Correlation and factor analysis were applied on four data sets, represented by control cells form individual gold districts (12 cells each), each time combined with ballast cells (50 cells).

The results supported the forming of two large control cells according to the individual metallogenic models and helped to diminish the total number of variables: only variables with an absolute value of correlation coefficient with the dependent variable >0.25 were taken into account.

General lack of the data on gold production from individual ore structures of the districts did not allow use the tonnage of gold as a dependent variable. Therefore, the spatial extent of mineralized structures was quantified; for example for linear objects, 100 meters of length was taken as a unit.

In the next step, 22 selected variables were quantified in the Bohemian Massif area. All the variables were transformed into 0 - 4 range (for lithological variables, maximum value, 4, was given to rocks sharing the entire area of a unit cell). Furthermore, new, "artificial" variables were constructed, usually by the addition of original variables (e.g. rock type + anomaly area from panning prospecting). In the instant of coexisting rock contacts, value of the new variable was set equal to that quantifying the presence of the minor rock type. This described operation was induced by the presumed higher significance of two coexisting geological phenomena (e.g. bimodal volcanics + geochemical anomaly) compared to isolated variables (e.g. tectonic structures). This operation resulted in the creation of 2,909 x 255 matrix.

As a method of estimation of the gold potential, the multiple linear regression (BMDP software package) was used. The regression was processed in two ways for each metallogenic model - both with and without taking into account the results of the panning prospecting. The two methods of solution were implied by the fairly high number of occurrences of gold mineralization undetected by regional panning prospecting (approximately 30 percent of all the occurrences, regardless to their size - Morávek, 1985).

The sum of the function values in the individual control cells covering the model districts was compared with total amount of gold (sum of both previously mined and in current ore reserves). In the situation of the ballast control cells, the calculated function value (close to 0) was set to zero tonnage of expected ore reserves; their value in all other cells was calculated by linear interpolation.

This procedure was used for individual models as well as for summary solutions (Figs. 1, 2); in the latter situation, the addition of the function values calculated for each metallogenic model yielded summary solutions for both ways of processing (Fig. 3).

RESULTS

Both solutions of the individual metallogenic models are similar in shape; nevertheless, the panning prospecting helped to more detailed contouring of prognostic areas (Figs. 1, 2). Placer deposits, although being valuable in the control set, had to be excluded from the processing because of the distance from presumed source regions in southern Bohemia.

The dependence of distribution of gold mineralization on raw geophysical data, such as gravity gradient, gamma-spectrometry, etc. is generally insignificant. Far better results were achieved in the synthetic interpretation of geophysical data, which helped to delineate a significant zone of inhomogeneity referred to as the "global tectonic zone" (Krs, 1983).

In the low-grade model, 16 variables were used in the linear regression equation. The most important ones are as follows (correlation coefficient listed in the brackets): Au panning anomaly + global tectonic zone (0.911), bimodal volcanics + NE tectonics (0.890), granitoids of Au-metallogenic group + global tectonic zone (0.892).

In the high-grade model, 13 variables were used in the linear regression equation. The most important ones are as follows: Au panning anomaly (0.859), E-W tectonics (0.724), varied sequence of crystalline rocks (0.614).

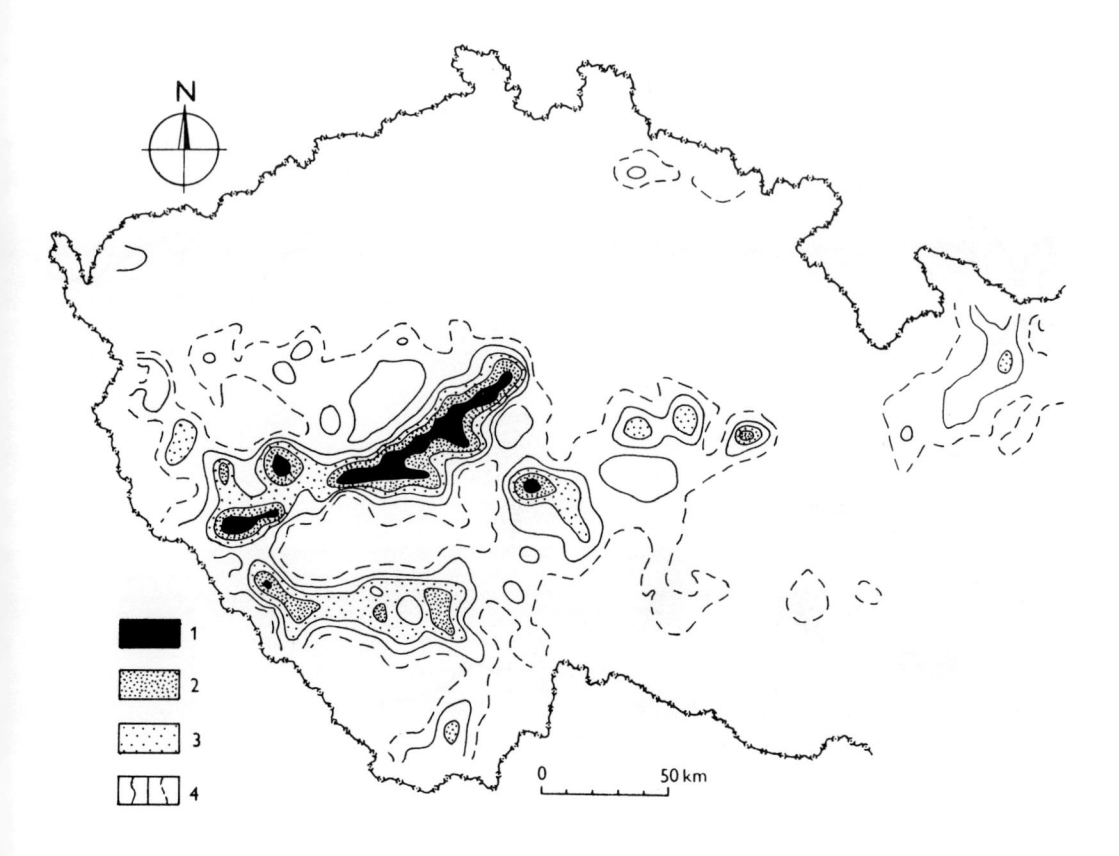

Figure 1. Low-grade model with respects to results of panning prospecting with probability contours of presence of deposit with reserves over 10 t Au. Probability values: 1- over 0.88, 2- over 0.7, 3- over 0.5, 4- equal to 0.3, and 0.1 respectively.

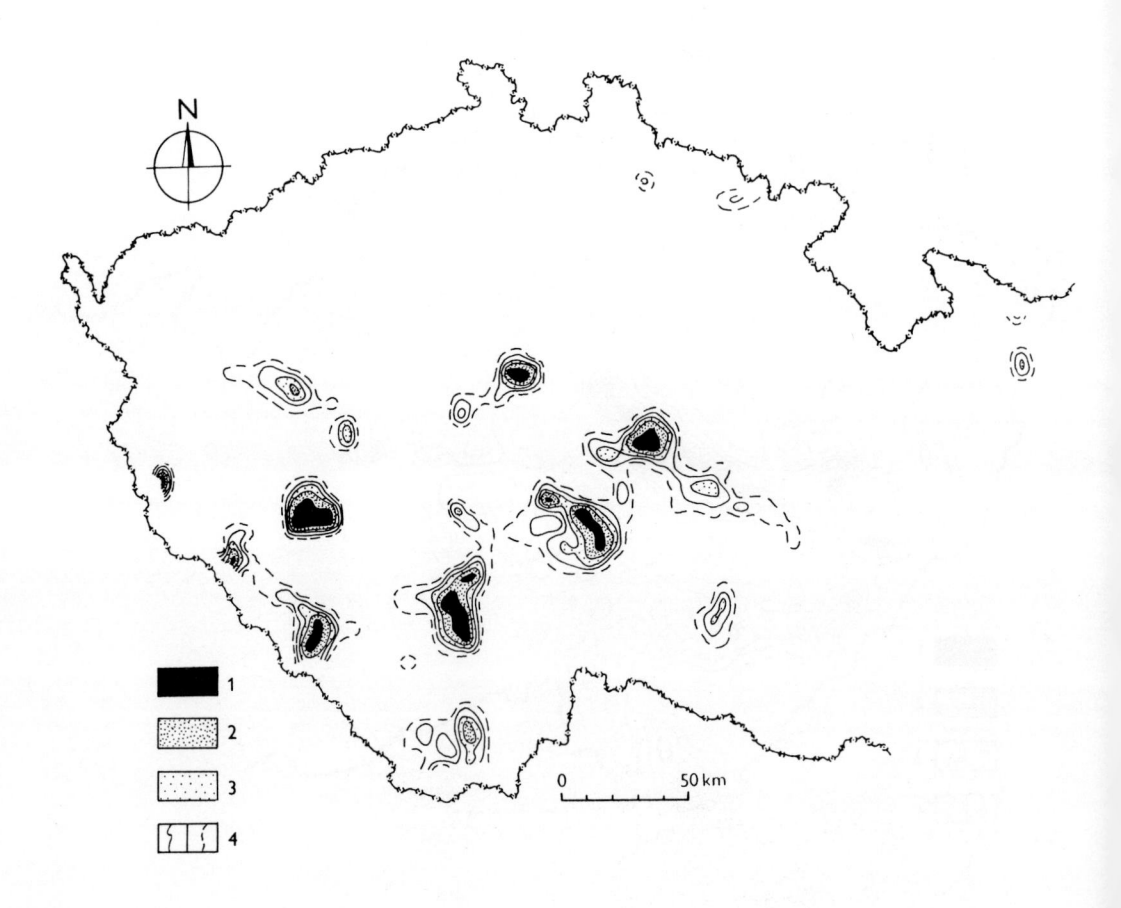

Figure 2. High-grade model with respect to results of panning prospecting with probability contours of presence of a deposit with reserves over 10 t Au. Probability values are same as in Figure 1.

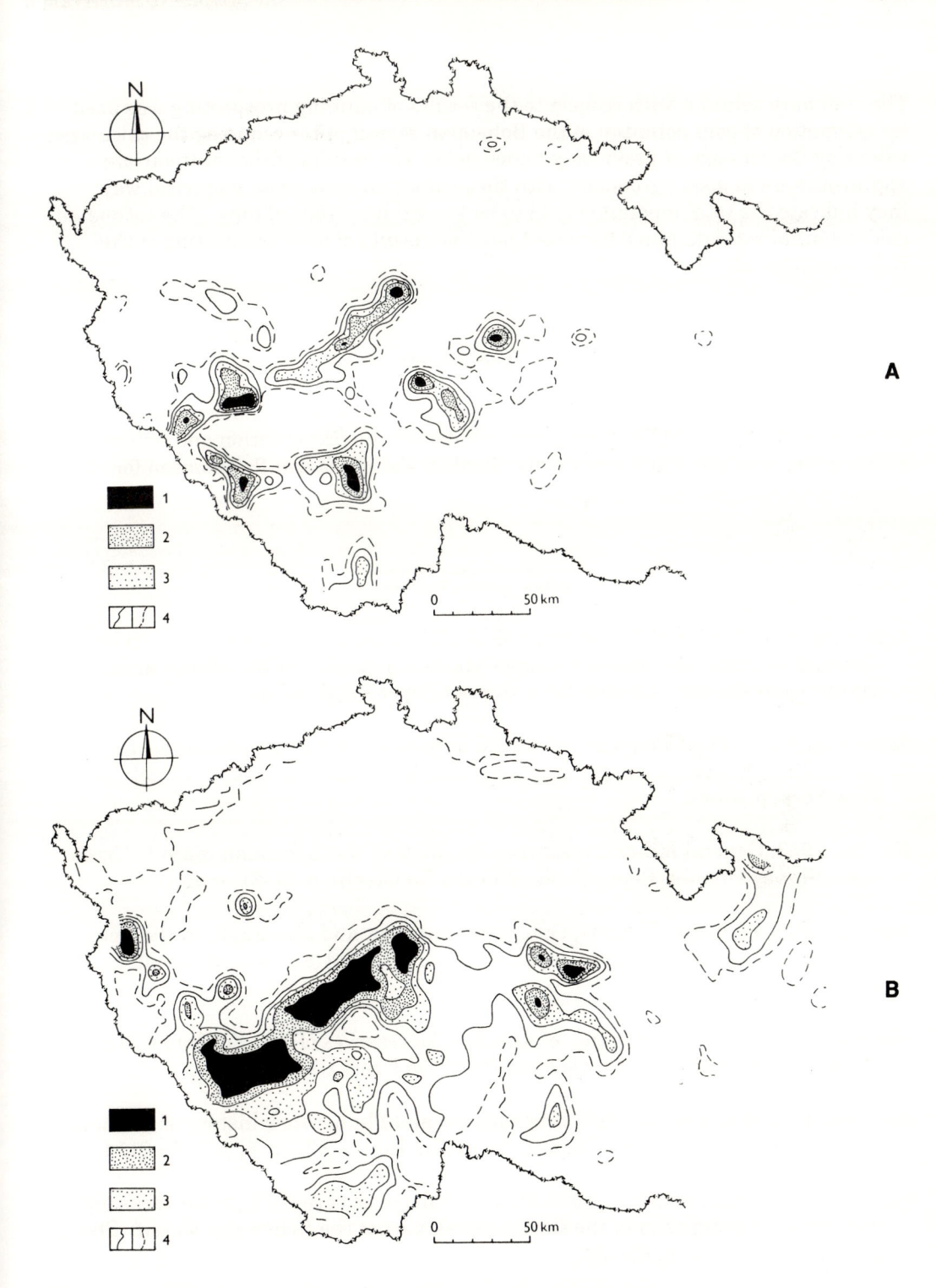

Figure 3. Summary solution: A) with respect to panning prospecting, B) without respect to panning prospecting, with probability contours of presence of deposit with reserves over 10 t Au. Probability values are same as in Figure 1.

The summary solution with respect to the results of panning prospecting was used for estimation of gold potential of the Bohemian Massif, after reducing the calculated values by the tonnage of previously mined gold. The results of the multivariate appraisal were in good agreement with the evaluation done in an empirical way: they indicate the total amount of gold to be several hundreds of tons. The estimated gold potential was sustained by recent positive results of gold prospecting of the Bohemian Massif.

ACKNOWLEDGMENTS

We are extremely grateful to G. Gaál, D. Merriam, and other members of organizing committee of COGEODATA symposium, who made the presentation and publication of this paper possible. Furthermore, we thank J. Kalnejais and R. Thomson for critical reviewing the manuscript.

REFERENCES

Agterberg, F.P., Chung, C.F., Fabbri, A.G., Kelly, A.M., and Springer, J.S., 1972, Geomathematical evaluation of copper and zinc potential of the Abitibi area, Ontario and Quebec: Canada Geol. Survey Paper 71-41, 55 p.

Agterberg, F.P., 1981, Computers as an aid in mineral resources evaluation, *in* Merriam, D.F. ed., Computer applications in the earth sciences: Plenum Publ. Co., New York, p. 43-62.

Krs, M., 1983, General mineral resources appraisal of the Bohemian Massif: Czech. Geol. Survey Unpubl. Rept., Geofond Praha (in Czech), p. 20-21, map.

Morávek, P., and Pouba, Z., 1984, Gold mineralization and granitoids on the Bohemian Massif (Czechoslovakia), *in* Foster, R.P. ed., Gold '82: A.A. Balkema Publ., Rotterdam, p. 713-729.

Morávek, P., ed., 1985, Evaluation of possible Au reserves of the Bohemian Massif: Geoindustria Unpubl. Rept., Geofond Praha (in Czech), 172 p.

Morávek, P., and Pouba, Z., 1987, Precambrian and Phanerozoic history of gold mineralization in the Bohemian Massif: Econ. Geology, v. 82, no. 6, p. 2098-2114.

Sattran, V., and Klomínsky, J., 1970, Petrometallogenic series of igneous rocks and endogenous ore deposits in the Czechoslovak part of the Bohemian Massif: Sbor, geol. Ved, LG, v. 12, p. 65-154.

Tencik, I., ed., 1982, Panning prospecting of the SW part of the Bohemian Massif: Geoindustria Unpubl. Rept., Geofond Praha (in Czech), 128 p.

Comparison of Subjective and Objective Methods in Quantitative Exploration: Case Studies

Claudia Kliem, and
Theodoros Petropulos
Freie Universitaet Berlin, Berlin

ABSTRACT

A comparison between various approaches in quantitative mineral-resource appraisal will be tried on the basis of two case studies in NW Turkey (Biga Peninsula) and NE Greece (E. Rhodope). Both test areas have a history of exploration and small mining activities for base metals, carried out by the domestic geological surveys as well as by foreign companies.

In planning an assessment of mineral wealth in a region, the determination of the actual information level is the starting point. The intensity of exploration work carried out, the degree of uncertainty in interpreting the mineralized phenomena and the presence/absence of known deposits/mineralizations in the study area are some of the most important factors influencing the whole analysis.

An important step is the quantification of the geological knowledge (maps, reports, assays). Several critical issues are incorporated into this phase of analysis, briefly stated the objective-subjective integration of geodata. The result of this phase is a set of variables which are recorded manually or automatically in any form such as binary, ternary, or according to an evaluation scheme or from probability distributions.

A variety of estimation techniques is well described in numerous excellent publications covering the whole spectrum from simple deterministic computations to the recently developed expert systems. Aim of this paper is to discuss the interplay between the exploration status and the next analytical steps. A focus point is the quantification of geodata and particularly the description of the characteristics of the objective and subjective approaches.

INTRODUCTION

The estimation of the ore favorability index over an area considered that is done before the selection of target subareas for further exploration, has been a general practice, in circumstances where the decision to continue an exploration project is given, and thus its planning depends only upon geological factors. The ore favorability mapping enables also a ranking between the candidate subareas but it does not give any type of information about the number of deposits expected, the tonnage, grade, or price in dollars. It corresponds to the geological resource models (Harris, 1984), namely the models where attention is given only to the geology of the region under study as a basis for appraising the expected resource/endowment, whereas the economical potential may be naive, confused, or nonexistent.

On the other hand, the purpose of mineral-resource assessment is to obtain estimated number of mineral deposits of different types (usually in a relatively large area), grade, size, and other characteristics of them (Agterberg, 1986). This is the situation where it is attempted to estimate the economical potential of the mineral endowment of an area, basing on a data set which consists predominantly of geological variables but their evaluation is made with respect to economical aspects.

Although there are differences between mineral-resource assessment and ore-favorability estimation there a common starting point exists concerning the acquisition of geological knowledge. This is followed by the subsequent of knowledge valuation in order to forecast the undiscovered, regardless whether it is the favorability for ore occurrence or number of deposits. For this reason it is believed that although the following case studies have the focus point on the favorability mapping, a discussion about the various techniques applied remains actual also in the situation of the predominant term and practice of mineral-resource assessment.

COMPARISON AND DISCUSSION

Two different approaches for the estimation of the ore favorability index are carried out in two test areas (Fig. 1):

(1) Eastern Rhodope / NE-Greece (polymetallic deposits), and
(2) the old Pb-Zn-mining district of Biga Peninsula / NW-Turkey

The first approach belongs to the subjective methods and has its focus point in the quantification of geological information/knowledge according to the current ore-discovery model. This was done for both regions. On the other hand, the construction of a decision structure in the objective approach is based on multivariate analysis of automatically formalized map information (Biga Peninsula). The selection of the proper type of approach depends on

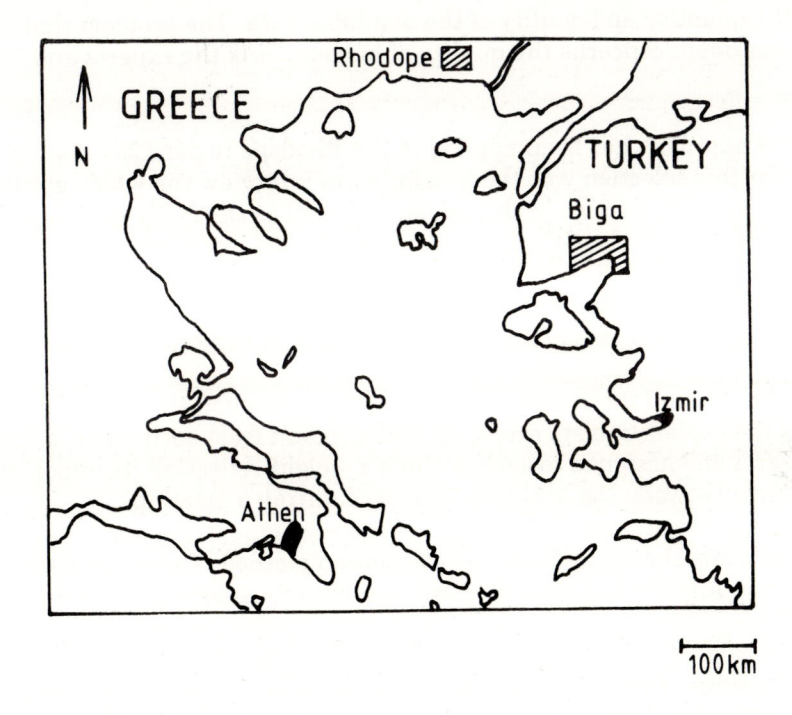

Figure 1. Locality map of test areas in southern Biga Penninsula/NW Turkey and eastern Rhodope/NE Greece

(1) the state of available data (quality and quantity),
(2) the possibility to formalize the current ore-deposit model for the region, and
(3) the presence of known occurrences of mineralizations as training areas

One may select the objective (multivariate) approach for the prediction of favorable areas if

(1) the knowledge about the test area is restricted: relationships and indicators reflecting geological processes and ore genesis are to be pointed out by analyzing the training areas, and if
(2) in later phases of exploration a great amount of measurements is available such as geophysical, geochemical, remote-sensing data, etc.

The subjective approach

In early phases of exploration the ore-favorability estimations can be done by subjective approach which consists mainly in a subjective valuation of the geological situation by geologists and other exploration experts. The formulation of the model is of greatest importance. A main characteristic feature of this method is its ability to incorporate the opinion of the experts, the particularities of the specific exploration project and the quantity and quality of the available data. The problem that arises in subjective assessment concerns the quality of judgement of the experts and its variability.

This subjective approach has been applied in the Rhodope region (25x30km^2). The main reason for this selection was the possibility to interview the whole group of explorationists who were active in this project, as well as the fairly good understanding of the geological controls of the base-metal mineralization in the area.

The aim of the analysis is the delineation of favorable areas for further exploration work on the basis of the information level of the regional exploration phase.

In order to define/formulate the ore-discovery model a method similar to Delphi was used. Each of the explorationists prepared, independent from each other, a list of favorable criteria and assigned to each of them a weight reflecting its importance. Minor differences between the lists were removed during a meeting.

The Monte Carlo simulation was used as estimation method. The state of the criteria was valuated for each cell by frequency distributions of a simple form (uniform or triangular).

Because the objective of the study was the integration of all types of information such as geological, tectonical, geochemical, and geophysical data the Delphi procedures were more valuable in defining the ore-favorability model than a big number of variables (subjective estimation of the ore-favorability index or number and quality of expected deposits). The Figures 2-4 demonstrate the effect of the use of the Delphi method. Figure 2 represents the results of the final ore-favorability estimation, the Figures 3 and 4 show the favorability indices derived from the subjectively defined model by only one expert. The comparison denotes the lack of the "super expert".

This way of data treatment approximates the way of thinking of the geoscientists because uncertainties can be considered in form of distributions. It becomes more important when the criteria are relevant to many data sources and it also is a significant topic in the application of the developments of artificial intelligence such as Prospector (Duda, 1980).

Subjective methods regard all evaluation units to be independent from the others, and thus it is possible to compare different parts of the same metallogenetical province. The variable of interest (the favorability index) is not derived from the variability within the given data set as it is done by multivariate statistics. The validity of the results from multivariate analysis which provide relationships and ranking between the cells is limited to the extension of the test area only.

With subjective definitions it is not possible to detect other structures than these described by the model. In objective circumstances one also can establish new or unknown relationships and construct several prediction structures which are present in an automatically recorded data set.

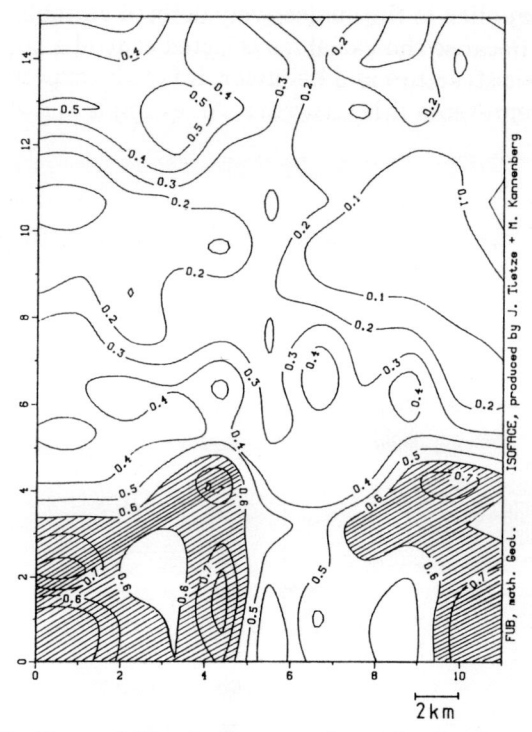

Figure 2. Favorability indices produced by "Delphi" method.

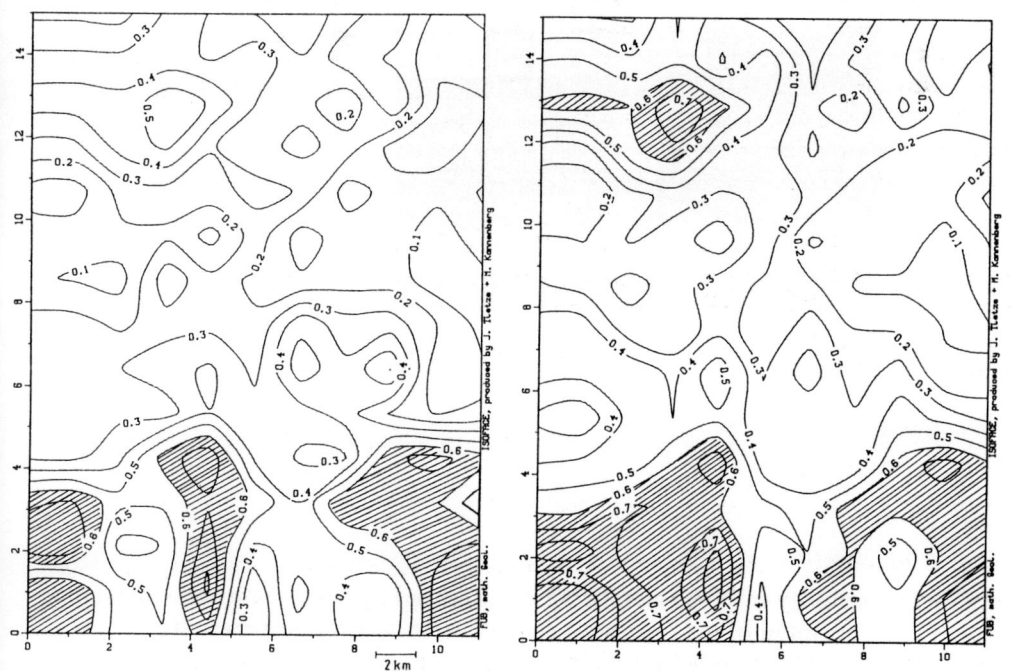

Figure 3. Favorability indices produced by one expert.

Figure 4. Favorability indices produced by another expert.

Multivariate analysis also can be applied to the subjectively defined variable set to delineate the favorable subareas. Because the variables selected control a specific mineralization with a characteristic structure and behavior, it is to be expected that at least one cluster or one factor represents this structure. An example is given by applying cluster analysis.

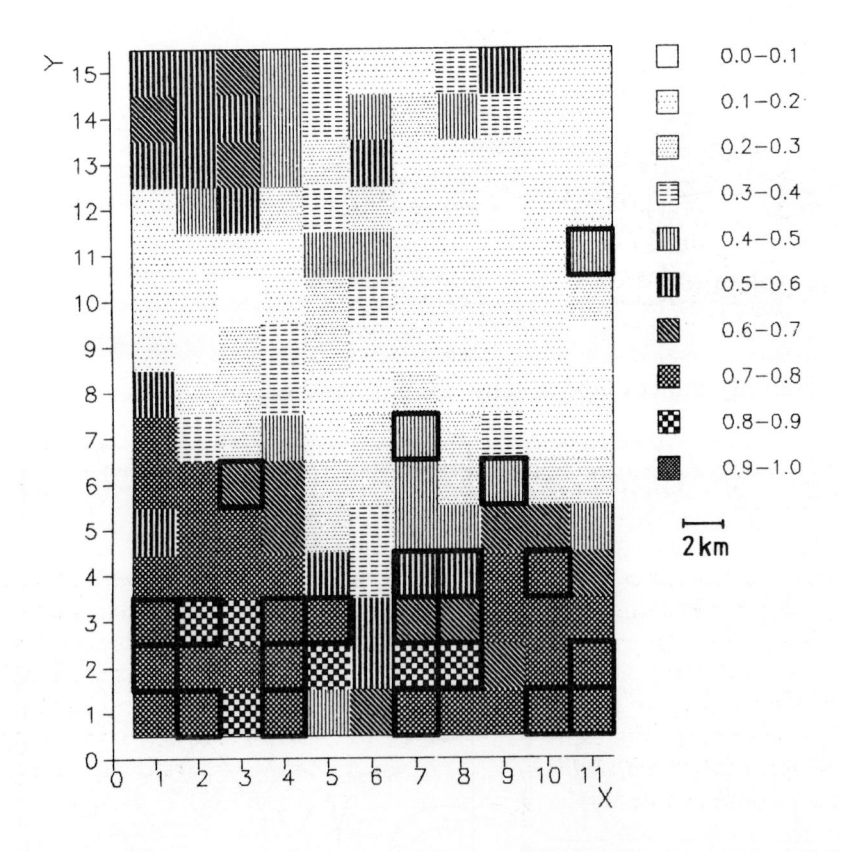

Figure 5. Favorability indices (membership) produced by fuzzy-c-means-clustering. Bold outlined boxes represent cells with known mineralizations.

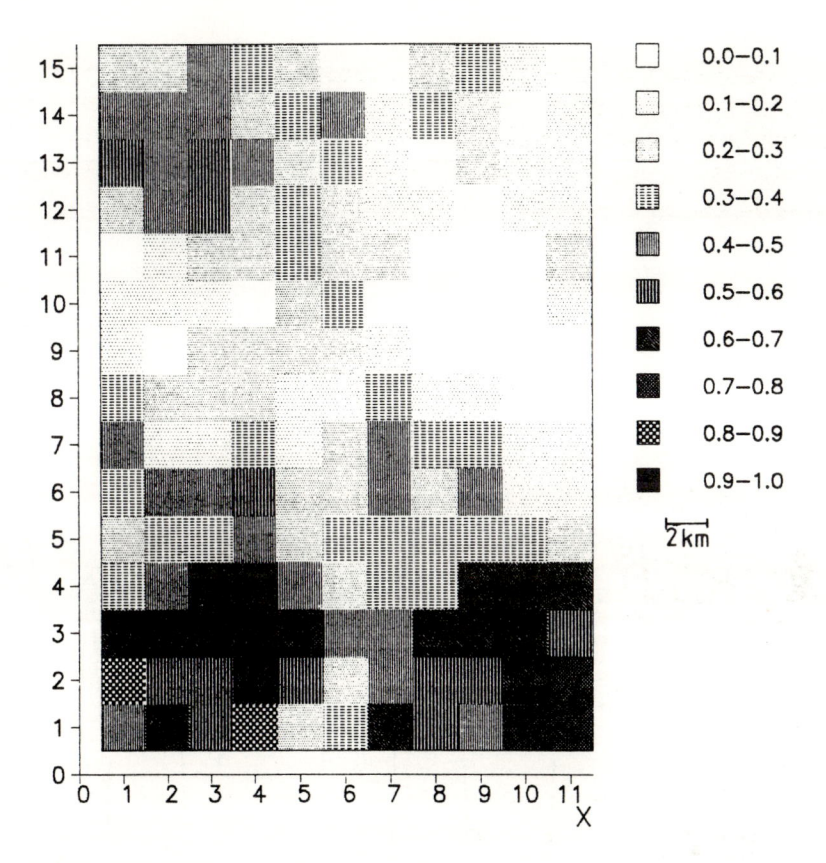

Figure 6. Favorability indices (Bayes' estimates) produced by Monte Carlo simulation.

The Fuzzy-c-means procedure (Bezdek, Ehrlich, and Full, 1984) provides for Rhodope a two cluster solution (Petropulos and others, 1987). The characterization of one of them as the 'favorable cluster' was possible by its centroid vector indicating the well-defined structure of the ore-deposit model applied. Figure 5 shows the spatial distribution of the memberships of the cells to the 'favorable cluster' which can be compared directly with the results of the Monte Carlo simulation (Fig. 6). The results of the subjective approach are sensible only to the model definition (to the differing weights) not to estimation methods.

For Biga, 6 different methods were used to delineate favorable areas with a subjectively defined criteria set. The results of all methods remain the same (Fig. 7). The curves show the favorabilities for the 54 10x10km^2 cells:

(1) the absolute values differ slightly for regression analysis, factor analysis, Fuzzy-clustering, and Monte Carlo simulation, whereas the discriminant scores show highest variability, but
(2) the ranking between cells is the same, and
(3) relocation procedure from CLUSTAN program package (Wishart, 1978) provides a distance measure that also can be interpreted as favorability index.

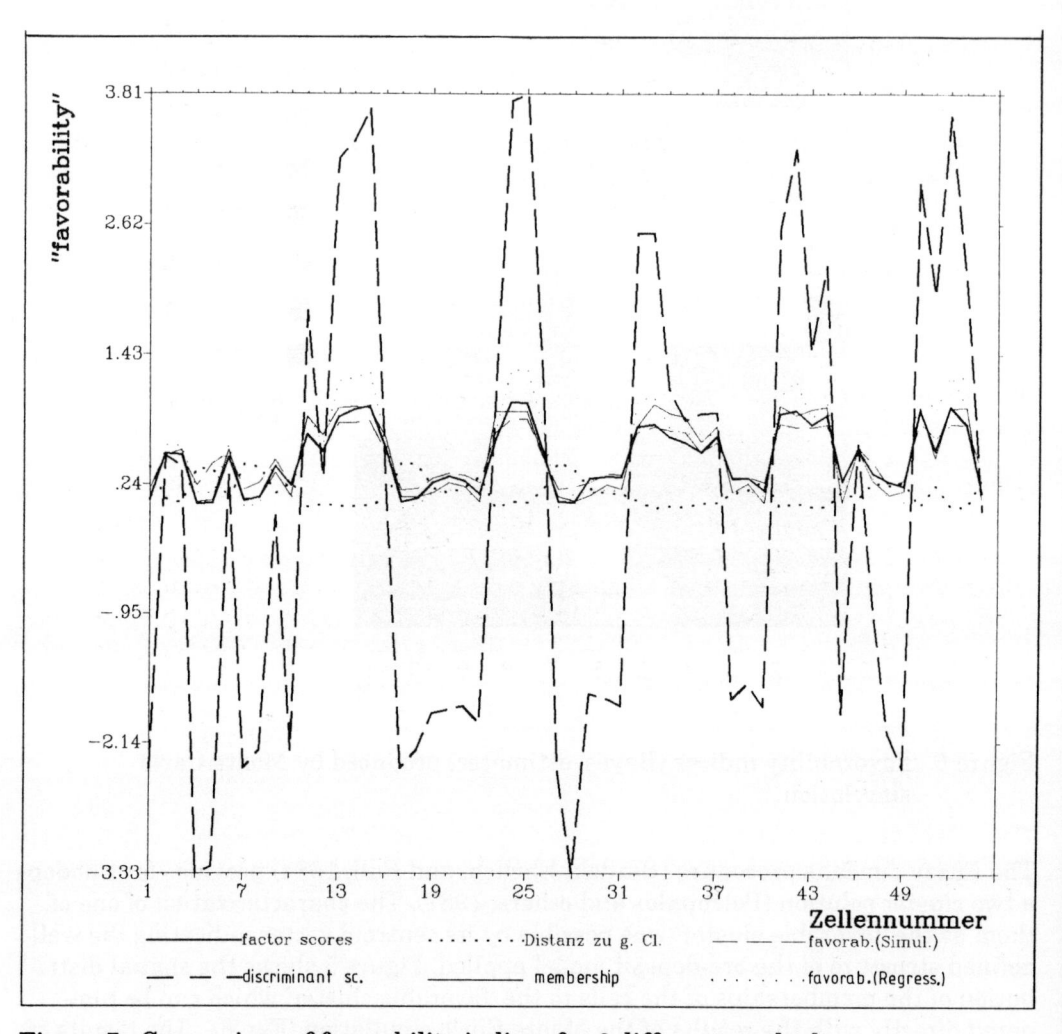

Figure 7. Comparison of favorability indices for 6 different methods.

The objective approach

Harris (1984) points out the main disadvantage of the objective approach. It lies in the strongly simplified data which originates from the automatic recording of maps and transformation into binary or ternary data sets. The values describe presence and absence or favorable and unfavorable attributes whereby a lot of information get lost. In contrary, the subjectively defined variables can be more sophisticated.

For Biga Peninsula all known and also some new acceptable relationships and indicators could be extracted from binary variable set (extracted from geological, tectonical, and geochemical maps) for 1x1km^2 cells. The size of the test area is about

$60 \times 80 \text{ km}^2$. The main task was the evaluation of the mineral endowment and to prove the applicability of multivariate techniques in:

(1) distinguishing the different types of mineralization represented in the data set. Digitizing of map information leads to recording every existing type of mineralization,
(2) the derivation of relationships between training areas and the controlling variable set,
(3) the reduction of variables by the delineation of relevant indicators for each type, and
(4) to insert these indicators into model functions for the prediction of undiscovered occurrences.

Probabilistic and multivariate statistical methods such as association and characteristic analysis were applied (see also Kliem and Burger, 1987). Some results are to be presented here.

(1) Association analysis lead to a misclassification of the types from different training areas. This is shown in Figure 8. The separation of groups depends only on the presence of that variable with the highest overall associativity without regarding the similarity or dissimilarity between the remaining variables. Two different mineralizations (1=contact metasomatic Pb-Zn; 2=porphyry copper ores) are classified into the same class which is separated from the whole population by the presence of Pb-anomalies.

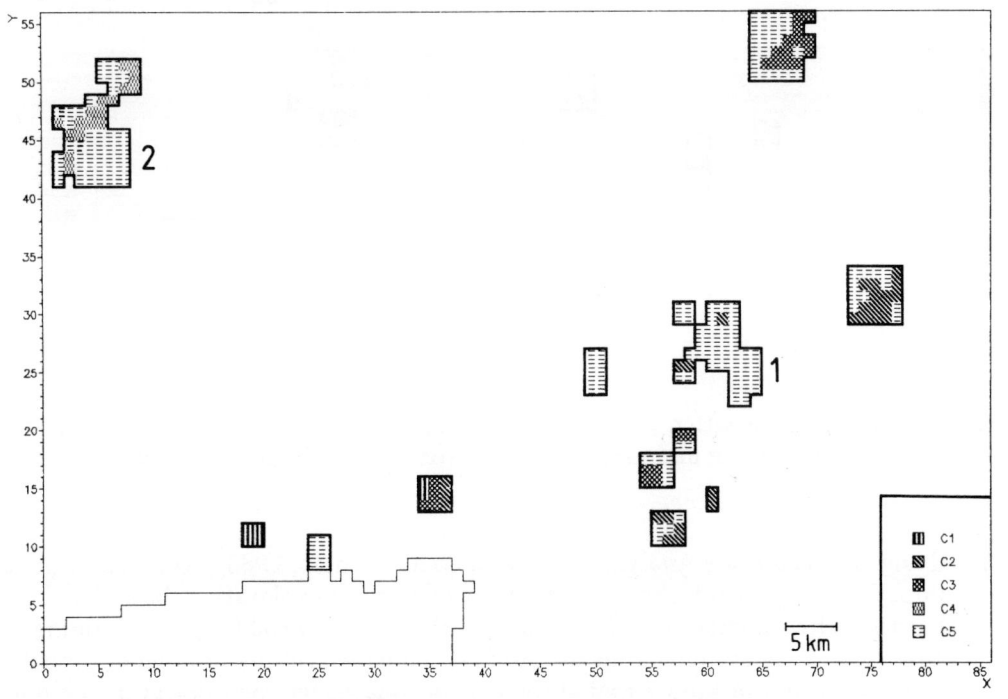

Figure 8. Results from association analysis. Distribution of all training cells (size 1×1 km^2) to 5 clusters. Areas 1 and 2 contain mineralizations of different types.

(2) The so-called iterative relocation procedure from CLUSTAN which optimizes the memberships of the cells to the clusters provided the four main types of mineralization on Biga (Fig. 9). The region in the northwestern part of the test area was recognized to be different from the others. Each cluster was characterized by the conditional probabilities for the binary variables and their ratios R of the conditional to the a priori probability. Probabilities higher than 0.1 and R greater 1 were used for extrapolating the contact metasomatic type from cluster 3 to the entire test area (Fig. 10).

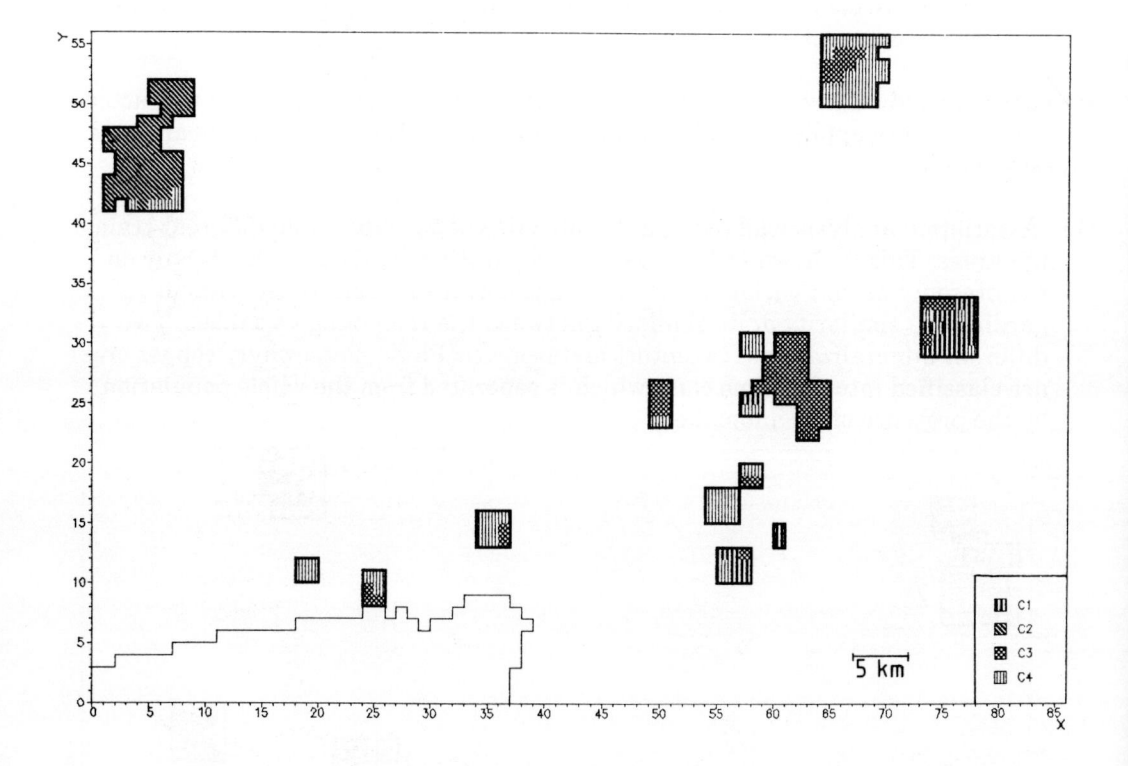

Figure 9. Results relocation procedure. Distribution of all training cells to 5 clusters characterizing main types of mineralization on Biga Peninsula.

(3) Using Characteristic Analysis (McCammon and others, 1983), which presents a simplified regression technique for binary or ternary exploration data, the characteristic weights are calculated from the binary variable sets of training areas. They consist in loadings for the first component of the product or probability matrix or in the square root of the sum of squares for each row of the product matrix. The localization of possible Pb-Zn mineralizations of contact metasomatic type on Biga by a multivariate favorability formulation is shown in Fig. 11.

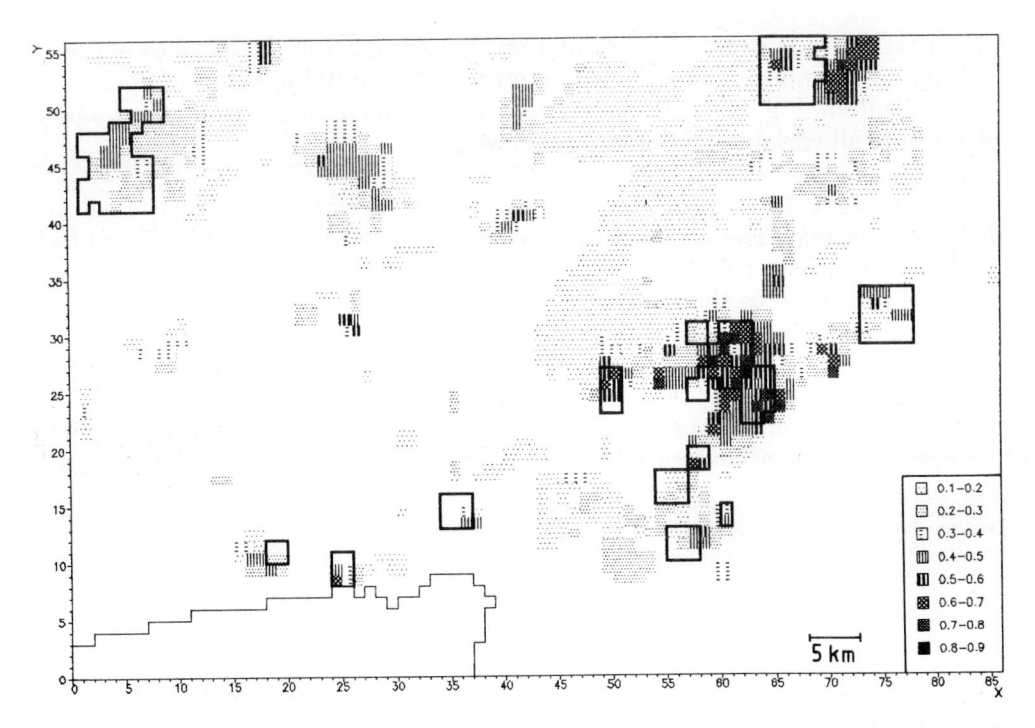

Figure 10. Favorability indices for cluster 3 (contact metasomatic type) extrapolated to entire test area.

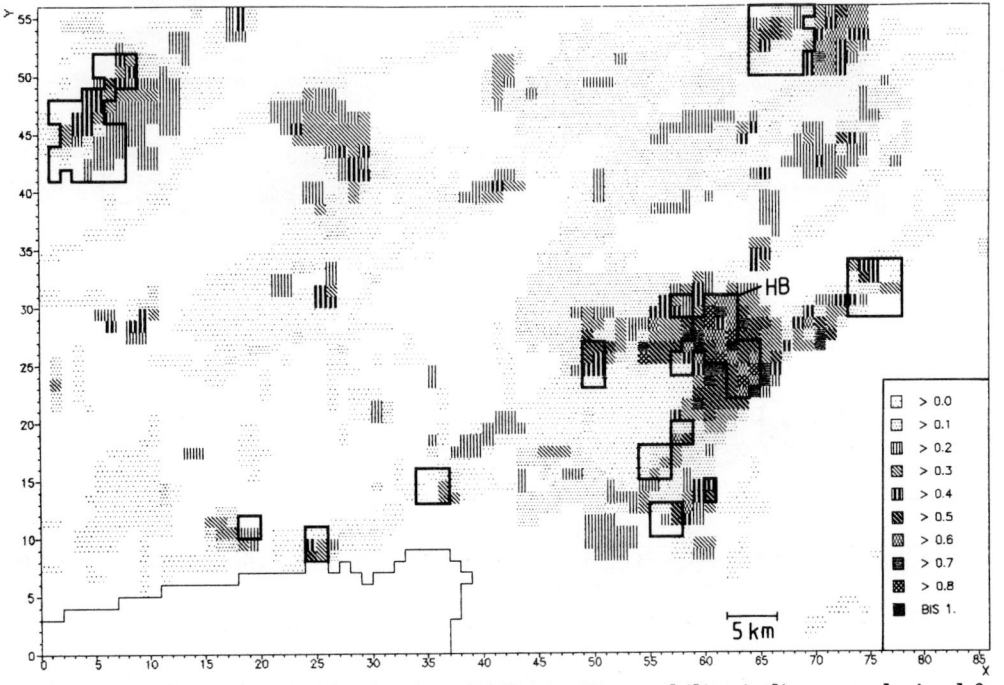

Figure 11. Results from characteristic analysis. Favorability indices are derived from training area HB (HanderesidBagirka) which contains contact metasomatic type of Pb-Zn-mineralization.

(4) The next example shows the distribution of the number of 1-1 matches in the variables by comparing all cells with a model cell containing a known Pb-Zn mineralization (Fig. 12) This model is constructed as unweighted linear combination where all variables obtain the same weight.

(5) A logical model also can be defined for the binary variable set in order to indicate favorable settings. The presence of a favorable geological unit within the contact zone of an intrusive body form the condition. This subjectively defined decision model is presented by the number of variables which agree with the given model (Fig. 13). The highest value if the condition is true is (5).

Based on a given variable set, the results, however are insensible to the estimation methods. The multivariate statistical modeling provides also the searched structure but for practical application a careful interpretation of the more or less "black box results" is necessary.

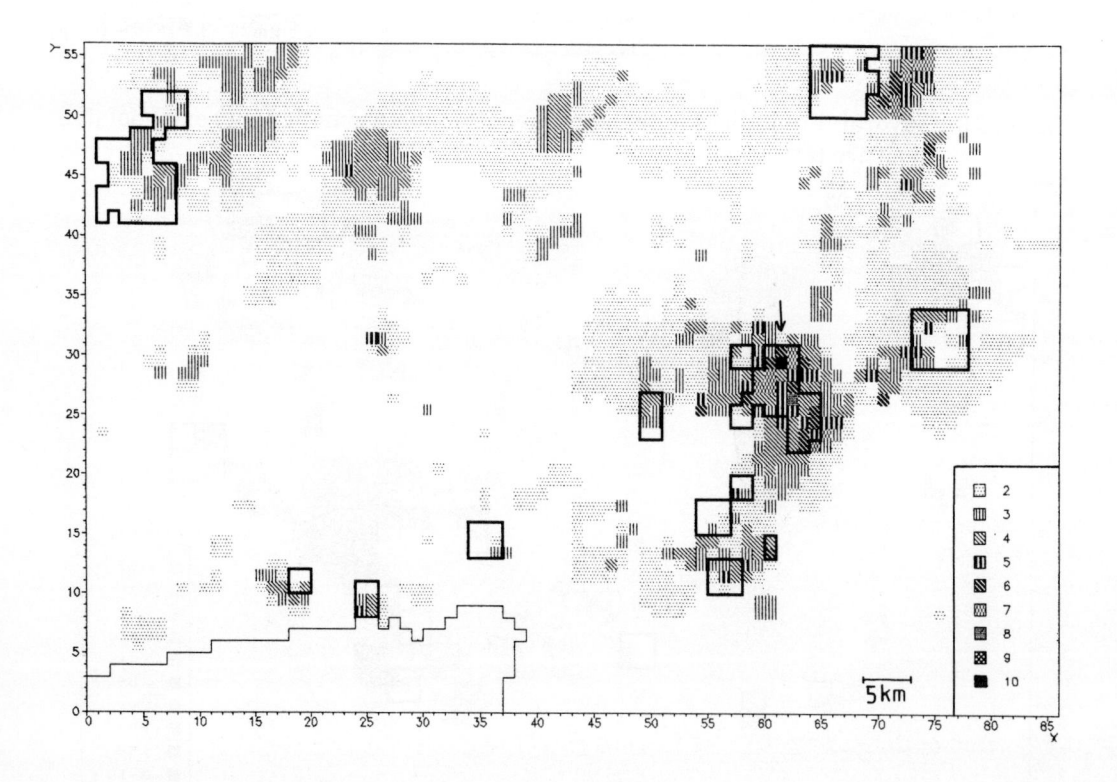

Figure 12. Results from unweighted linear combination. Favorability is calculated by counting 1-1 matches in variables between all cells and black model cell.

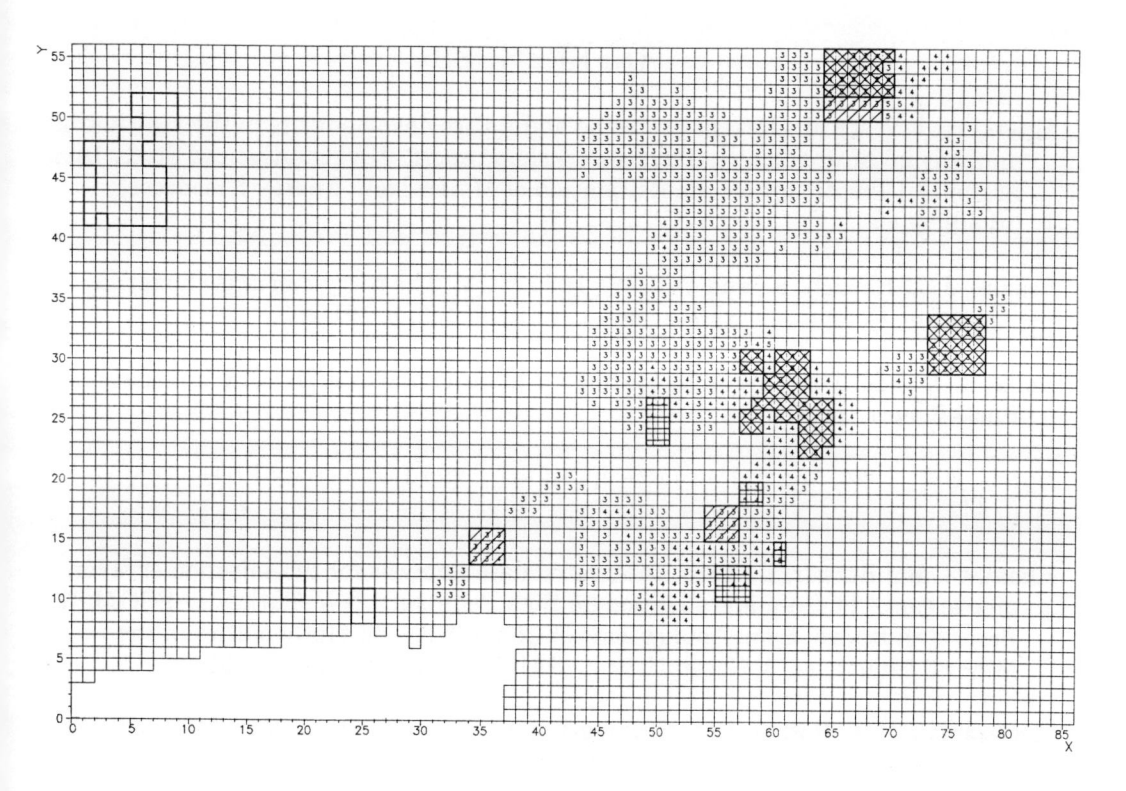

Figure 13. Logical favorability model. Numbers show degree of association between all cells and logical model (▨ = contactpneumatolytic, ▩ = contact-metasomatic, and ⊞ = hydrothermal type of mineralization).

CONCLUSION

Different approaches were applied up to now in order to estimate undiscovered mineralizations. They have different main ideas: the subjective methods reflect the judgements of explorationist more than objectively derivable facts such as variance analytical methods do. The analytical techniques ignore the subjective judgements and also are objective. They model the information recorded, usually automatically, and also are able to extract the parameters relevant to the problem of mineral resource/endowment evaluation.

The selection of the proper approach remains in the hand of the analyst. The degree of acceptance of the results by the geologists and a type of cross validation of the different methods is the only guarantee for the goodness of the analysis.

REFERENCES

Agterberg, F.P., 1986, Current problems and future developments in the multivariate analysis *in* Multivariate analysis workshop report: NATO Advanced Study Inst. Statistical Treatment for the Estimation of Mineral and Energy Resources, unpublished.

Bezdek, J.C., Ehrlich, R., and Full, W., 1984, FCM: The fuzzy-c-means clustering algorithm: Computers & Geosciences, v.10, no. 2-3, p. 191-203.

Duda, R.O, 1980, The Prospector system for mineral exploration: SRI International, Final Report, p. 102.

Harris, D.P., 1984, Mineral resource appraisal: Oxford University Press, New York, p. 442.

Kliem, C., and Burger, H., 1987, Integrated data analysis for the selection of favorable areas of Pb-Zn mineralization on Biga Peninsula, NW Turkey: APCOM 87. Proc. 20th Intern. Sym. Applications of Computers and Mathematics in the Mineral Industries, 3, Geostatistics: SAIMM, Johannesburg, p.101-111.

McCammon, R.B., Botbol, J.M., Sinding-Larsen, R., and Bowen, R.W, 1983, Characteristic analysis 1981: Final program and a possible discovery: Jour. Math. Geology, v. 15, no. 1, p.59-83.

Petropulos, T., Burger, H., Constantinides, D.C., Demetriades, A., Katirtzoglu, C., and Michael, C., 1987, Estimation of the ore favorability index for polymetallic deposits in the Essimi-Kir-ki Region, NE Greece: Proc. Sym. on Geology and physical geography of the rhodope massif: Smolen/Bulgaria.

Wishart, D., 1978, CLUSTAN-User-Manual: Inter-University/Research Councils Series Rept. 47, Edinburgh Univ., 164 p.

Analysis and Integration of Reconnaissance Data in a Mineral-Resource Assessment of Austria

H. Kürzl
Joanneum Research Association, Leoben

ABSTRACT

In exploration for mineral resources, reconnaissance data offer a specific type of information to be used mainly in regional investigations. According to their quality and regional density they require certain types of treatment to be suitable for integration studies. For one mineral-resource assessment program of Austria, covering approximately 1.500 km^2 of Alpine area, several types of geological information have been gathered and stored into a digital Geographic Information System (GIS). These data consist of subsurface and surface geological information, aeromagnetic data, and geochemical data and information about all known mineral deposits and occurrences in the specific area. In addition an update of mining claims and leases are maintained in the GIS.

For the aeromagnetic as well as the geochemical data, a variety of data-analysis tools have been applied to generate meaningful derived variables suitable in integration studies primarily to outline favorable target areas for mineral exploration. For the aeromagnetics gradients, regional and local variability and some ratios have been calculated. Based on regional geochemical groups, factors have been calculated to outline indicative multielement patterns in the areas. The geological information has been coded according to a detailed key allowing for a selective retrieval of different attributes such as rock types, age, and mineral assemblages. The integration study was carried out mainly in an exploratory way by using a variety of the data treatment and overlay facilities offered by the GIS. By using these techniques a variety of maps was prepared which reveal regional features such as differentiation in homogeneous rock sequences or local signatures indicating a relation to unknown mineralization. The assessment itself was carried out by interpreting the composite maps and by numerically calculating favorability based on the local features observable around known mineral occurrences.

INTRODUCTION

Based on a governmental program established in 1980 for mineral resources research in Austria, a mineral-resource assessment program (MRA) has been developed during the last few years by Joanneum Research Association, Mineral Resources Research Division (MRRD), Leoben. The first local study, covering approximately 1.500 km², has been completed recently with ongoing activities covering an area of about 6.000 km².

The primary aim of the government funded research program was to stimulate mineral exploration activities by the industry in providing and presenting recent regional reconnaissance data and documenting Austria's minerals inventory based on archival information. One further result of the research is the evaluation of the mineral-resource potential of certain areas, especially where conflicting interests concerning land use exists.

However, during the last year the priorities in economic politics have changed. Especially under the influence of the minerals commodity market and prices, the domestic supply of mineral raw materials became a secondary matter of concern for the central agencies. Therefore it was necessary to demonstrate that a mineral resource-assessment program is an important part of a larger system and can be intergrated easily with other spatial information and can provide an important background information, usually essential in regional studies.

A data-processing system was developed to meet these important basic objectives of the MRA. Additionally, it became apparent that the MRA should be multipurpose, handling multiple commodities, and tailored to the client's specific needs. The analysis performed by the MRA represents an integral part of the regional natural resource potential evaluation. The data-processing system used also should permit the analysis of the data for the MRA in a formalized and systematic process and should become part of an ongoing program of periodic updates and reassessments. In view of these requirements, no single computer package has been able to meet all these needs at once and it was necessary to determine a combination of software, derived from a number of different systems, structures, and sources. The ideal system should have available the following features and capabilities: Regional database and management system:

- Data analysis tools according to the specific data types;
- Documentation and thematic mapping;
- Integration studies (eg. mineral exploration, land use planning) and
- Qualitative (quantitative) assessments of the mineral-resource potential of certain areas.

Locating and developing the required system-modules was influenced not only by the necessary functions and reports, but also heavily by the data types; their primary availability and optimal way of processing. Therefore the next section will give an overview of the basic data sets involved in the system.

THE STUDY AREA AND BASIC DATA SETS

The initial study area, which represents the necessary test data to develop and implement the system, lies in the Niederen Tauern in the province of Styria (Fig.1). To introduce a systematic approach at the beginning of the program, data capture was done according to 1:50,000 scale map sheets. The study area covers three map sheets with an area of about 11,500 km^2. The area covers the eastern part of the Woelzer Tauern and nearly the whole of the Seckauer Tauern.

The area is dominated geologically by the Woelzer mica schist unit, which consists mainly of Paleozoic metasediments intercalated with amphibolites, marbles, and swarms of pegmatites. To the north, a Paleozoic low-grade metamorphic nappe of metapellites and greenschists overthrusts the Wölzer mica schist. To the east the Seckauer metagranites occur, partly covered by a series of Permian quartzites, dipping under the Woelzer mica schist. The types of known mineral occurrences and deposits are dominated by veins associated with As-Au and Pb-Ag. In the vicinity of the overthrust plane of the two Paleozoic nappes base-metal deposits dominated by Cu have been mined. At the present time, no operating mine in the area exists.

Because of space limitations, only two of the three map sheets are presented in the figures. The examples of data analysis and integration studies are restricted to the Woelzer mica schist formation to preserve clarity and simplicity in presentation.

The mineral inventory file

For a country with an old mining tradition like Austria, a variety of historical documents and records concerning mineral deposits and occurrences are available. However, they are in a heterogeneous form, mainly descriptive and without reliable quantitative information. A documentation has been compiled concerning all known occurrences in systematic form, but no recent investigations have been carried out. Therefore all the mentioned localities in these files have been ground checked and a modern geological description added. In addition, a comprehensive bibliography has been created by adding recent publications, research reports as well as the unpublished literature from universities and government agencies.

Geology

Another important information set is represented by geological maps. The documentation scale selected was 1:50,000. However only a few areas in Austria are covered with modern 1:50,000 geological maps; but a variety of manuscript maps in different

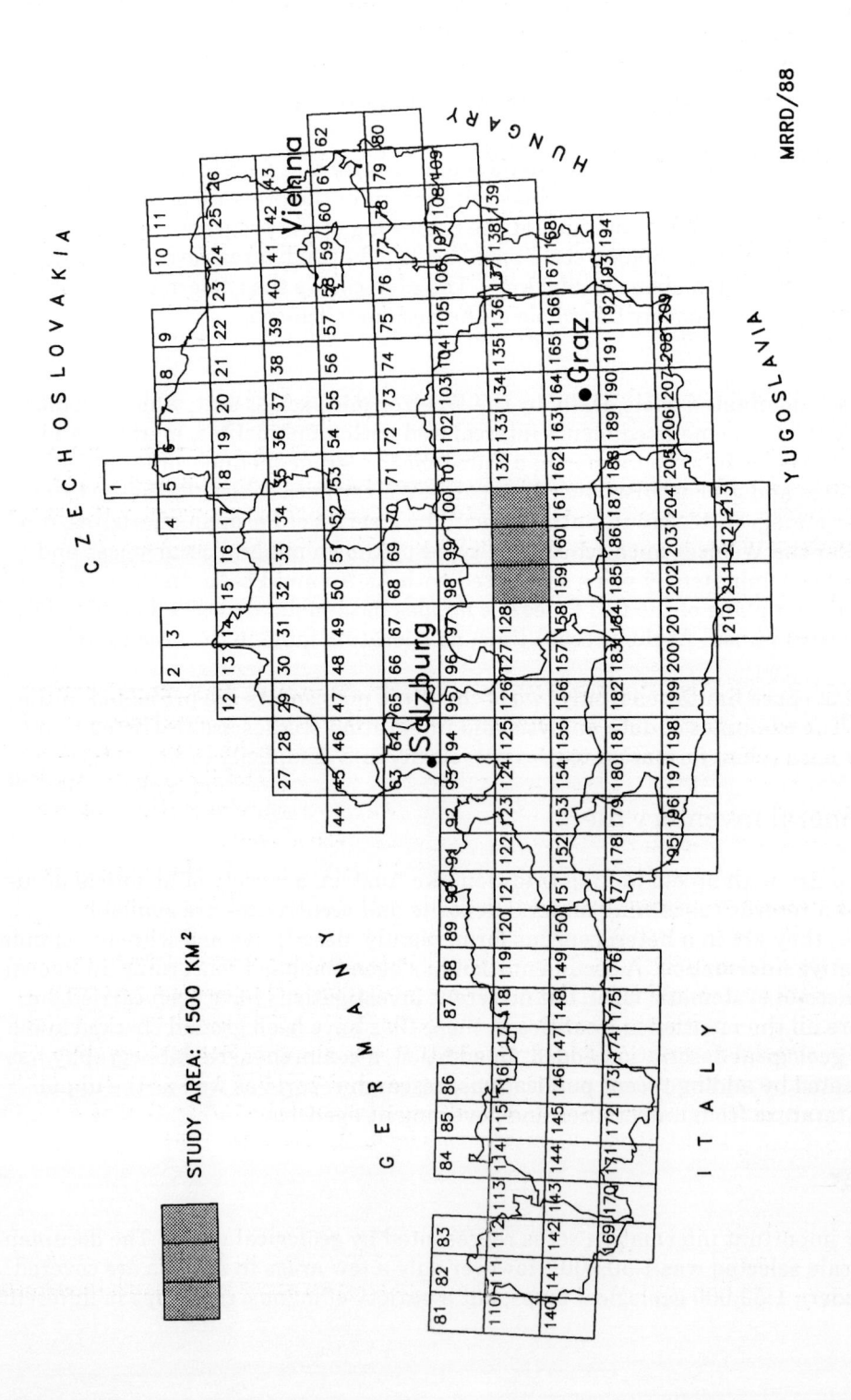

Figure 1. Location of study area in Austria for inital mineral-resource assessment program.

scales are available. Because of the historical development of geological mapping in Austria, no systematic approach to regional geological information exists. In the data capture and preparation phase, the information had to be generalized and brought into a systematic scheme (general legend) before digital map files for the GIS were created. The basic geological information digitized consists of bedrock geology, Quaternary geology, and regional structural features encoded according to their respective characteristics.

Geophysical and Geochemical Data

Recent quantitative data, such as aeromagnetic and regional geochemical stream-sediment (-80 mesh) data, represent further information layers in the GIS. Their availability in raw data form represents only a small part of their inherent information so that preprocessing and data analysis becomes essential, which will be described further in a separate chapter.

Mining rights

A specific layer of data in the GIS with important legal aspects are mining claims and leases. A file containing information regarding all existing mining rights in Austria has been developed by the MRRD. This database was established in the last 5 years and serves as an important source of information to the Austrian mining authorities.

Geographical data

In addition, a digital topographic surface model is included in the GIS and is presented in combination with hydrological information and various geographical background attributes including names of townships, mountains, and rivers. It was not the intention, however, to establish and maintain complete digital geographical maps. As part of a comprehensive information system they are essential, but are supplied by the responsible governmental agencies respectively.

Miscellaneous information

Data not yet included in the GIS are regional structural features (lineaments) derived from photogeological or remote-sensing interpretation. Regional gravimetric data would be desirable but are not yet available for the study area. Both data types could represent important information especially for the evaluation of the mineral-resource potential in integration studies. Their actual use in the search for metallic ores in this specific instance cannot be evaluated because no similar studies for comparison exist in geological environments such as in the Eastern Alps.

THE GEOGRAPHICAL INFORMATION SYSTEM

One of the main features of a GIS is its capability to capture, store, handle, and integrate spatially related data and facts. Experience was gained by inhouse development, especially for geological data of a GIS. Two years ago, a system evaluation between our own package and systems becoming commercially available for minicomputers showed, that an effort of at least two man-years would be necessary to catch up with the capabilities offered by the marketed systems. Therefore such a system was purchased one and one half years ago to serve as the central core of a database for the MRA-program.

Naturally, commercially available GIS systems must be generalized to be applied to a variety of problems where spatially related data have to be treated and therefore cannot be as easy to use and comfortable as a system developed for geological data alone, which does not yet exist on the market. Experience has shown that geological information does fit well into the formal structures of a GIS. Once the necessary structures have been established, geological maps can be digitized and coded easily and in a relatively short time by trained auxiliary personnel.

In context to this project, the GIS has been used mainly for data capture and editing as well as for the establishment of the relational database and its data-management system. Further important modules are the interactive cartographic subsystem, the thematic integration facility and a set of programs to manage, store, and analyze spatial information available on grid files. The geological and geographical data has been treated exclusively with the GIS. For all other data types, separate databases and analysis modules exist. The newly derived variables and conditional information from these modules have been integrated in the GIS system.

Concerning the geological information on maps, the geographical appearance of a certain lithology is defined as a polygon, which is related to the following attributes:

- color (pattern) code for graphical output;
- regional tectonic domain;
- stratigraphy;
- general rock type;
- mineral association with a content more than 10%;
- accessory minerals with a content less than 10%; and
  rock name and texture.

In addition to bedrock geology, the surficial geology has been captured as a separate information layer defining Quaternary lithology such as recent unconsolidated material in valleys, gravel terraces, and glacial material. Figure 2 is an example of a map display of digital geological information. The display corresponds to the original 1:50,000 map and shows the surficial formation of the quaternary. Another layer defines the regional tectonic boundaries and local faults in form of different coded line types. As further detailed structural information is scarce, no further types are defined yet (e.g. dip and strike, fold axis, etc.). Figure 3 displays the digitized struc-

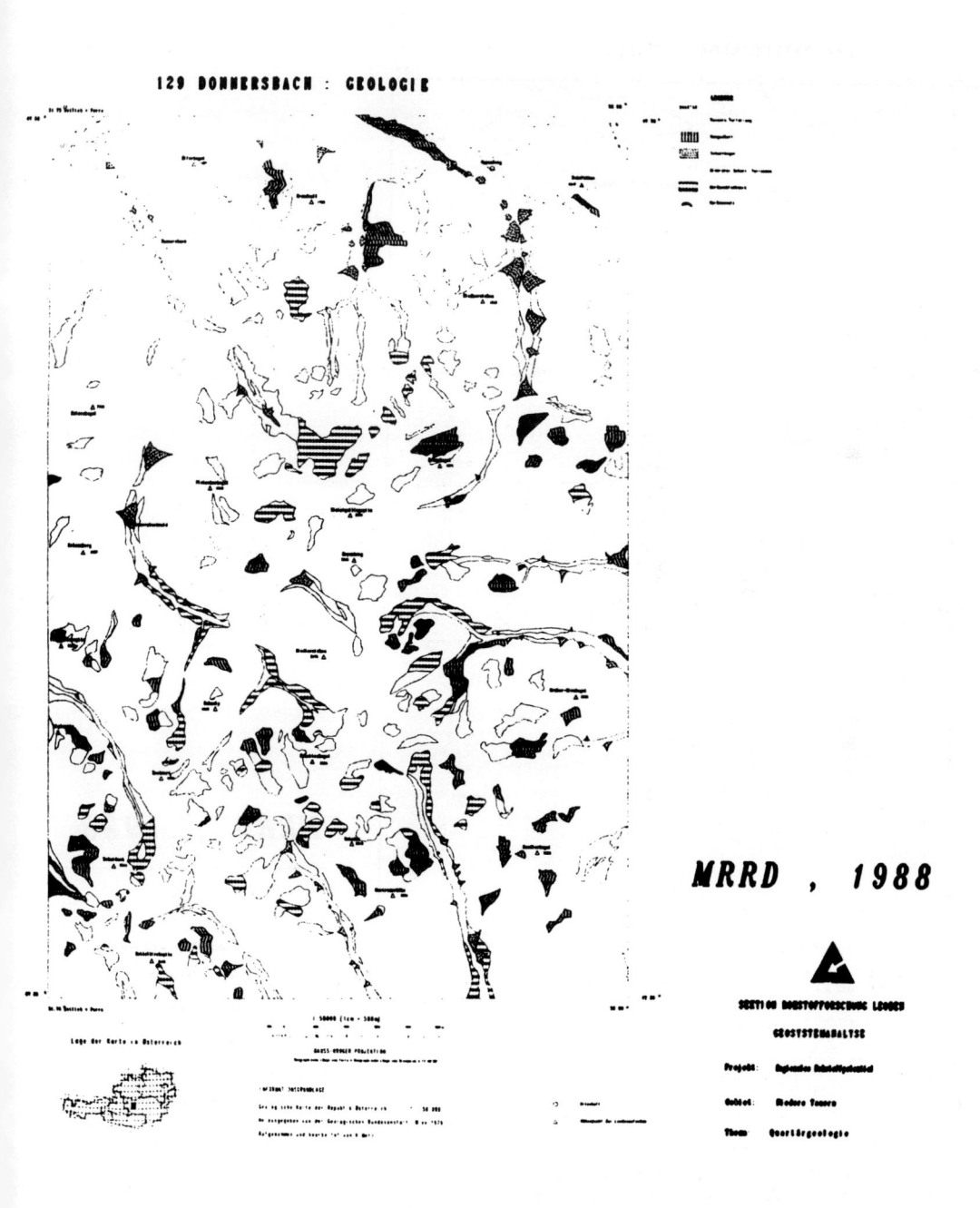

ANALYSIS OF DATA IN ASSESSMENT PROGRAM OF AUSTRIA

Figure 2. Map display of digitized Quaternary lithology (original scale 1:50,000).

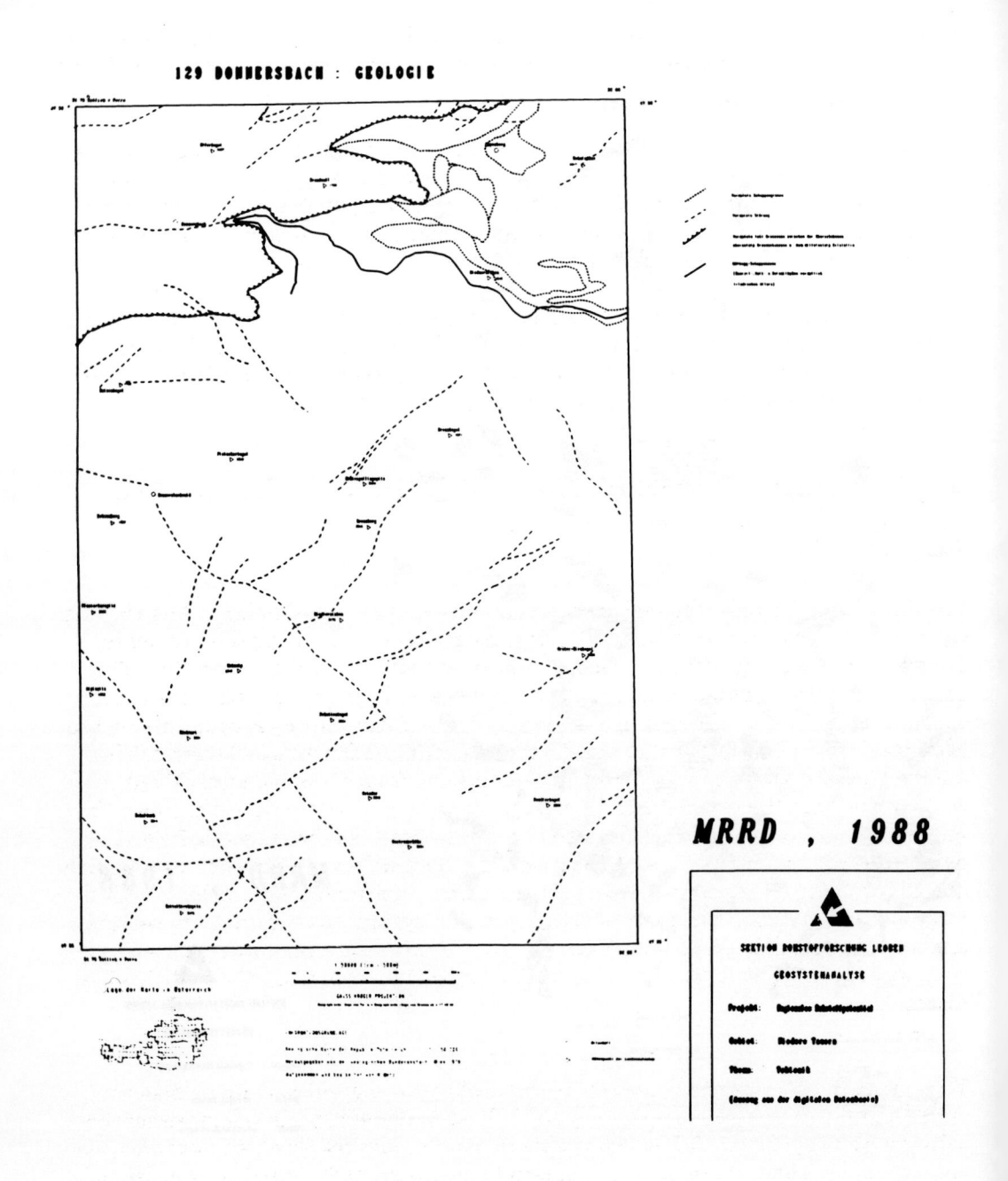

Figure 3. Map display of digitzed major structural features (original scale 1:50,000).

tural information on a computer generated map with a original scale of 1:50,000. All this information is stored in vector form and related to the attribute files, established interactively.

The GIS also facilitates an easy way of handling hydrological data such as drainage systems. This information, in combination with the digital terrain model which was made available via the GIS in a 50 x 50 meters grid form, was used to establish geochemical surfaces based on the stream-sediment data and their related catchment basins, which will be described in detail in the following chapter.

Communication with other databases and management systems has been facilitated by the development of specific interfaces allowing direct access to files. Files are reformatted in advance into a systems readable form, based on a selective retrieval, so that only the required information is incorporated into the data set.

ANALYSIS OF GEOPHYSICAL AND GEOCHEMICAL DATA

Aeromagnetics

The data acquisition for the regional aeromagnetic survey (total field) in Austria was carried out in cooperation by several governmental agencies during the last years. The availability of the measurements has been in form of raw flight-line data. The profiles were flown 2 km apart with an average sampling density of 65 meters. Because of the extreme changes in topography over Austria, the measurements had to be carried out in 6 different flight altitudes ranging from 800 to 4,000 meters. The data in the study area of the Niederen Tauern come from the 3,000 meter flight horizon. By checking the raw data, differences between the absolute values between the individual flight lines had to be leveled in order to evaluate the magnetic field. The data exhibited considerable noise. Based on the general layout of the survey a heavily biased sampling scheme was created raising certain questions and problems with the interpolation method. The most effective gridding system has been determined to be a method by Hood, Holroyd, and McGrath, (1977) which involves fitting smooth, continuous interpolation functions along parallel lines perpendiculary to the flight-line direction.

These specific circumstances and the additional governmental contract to map the entire survey at the 1:50.000 scale by MRRD led to the inhouse development of the software package MAGPAC, which is able to handle, connect, level, analyze and present regional (airborne) geophysical, especially magnetic, data. MAGPAC consists of the following components:

- A data-management system;
- Connection of individual profile parts into consistent profile series;
- Interactive leveling of the profile set;
- Linking of profile sets to the control profiles;

- Removal of noise from the measurements (approximation by splines);
- Gridding of data by interpolation with bicubic-splines;
- Data filtering of various types; and
- Contouring and plotting of established magnetic fields or derived variables.

For a large program it is intended to apply linear filter operations, to perform, for example, downward/upward continuation and magnetic pole reduction. For the initial study only gradients, regional and local variability and a ratio of the latter two have been calculated to gain a better resolution of the total field, and more meaningful variables for resource-assessment studies.

Experience in the study area shows that the data must be edited carefully in advance to gain a reliable presentation of the total field. The regional variability in Figure 4 as an example is based on the coefficient of variation by using resistant parameters for central tendency and variance, similiar to median and hinge spread as described by the 5-number summary of Tukey (1977). The map exhibits regional trends and structures not being obvious in the original residual field. A significant boundary of variability can be observed dividing the map into two different areas with high and low regional variations and different trend directions. These features follow only partly known lithological-tectonical characteristics of the area. Especially the mica schist formation reveals more internal structure than assumed until now and need definitely a reinterpretation. The white spot in the northwestern part of the map represents the signature of a serpentinite body, an exotic lithology and a magnetic outlier. The significant extension of the regional anomaly assumes a larger body than observable on outcrop. The area with this high magnetic anomaly has been clipped for technical reasons, thus drawing attention to map features with more importance in the resource appraisal. The residual field and four derived variables have been integrated into the GIS for further integration studies. These are described next in detail with examples based on the regional and local fields of variability.

Geochemistry

The regional stream sediment survey (-80 mesh fraction) of Austria covers an area of about 40,000 km^2 of the Eastern Alps and the Bohemian massif, which is about one-half about Austria. The Calcareous Alps and the sedimentary basins have not yet been sampled. The fine fraction was analyzed for 36 elements, including a variety of major and accompanying trace elements, as well as 11 ore indicative elements (metals). MRRD has developed its own data analysis and mapping package GCP to be able to treat and interpret this large amount of data (Kürzl, 1988). The GCP is a command language driven package with a database management system, various statistical modules and an interactive mapping facility. The most important features are the implemented exploratory graphics and the robust and resistant techniques in univariate and multivariate statistics.

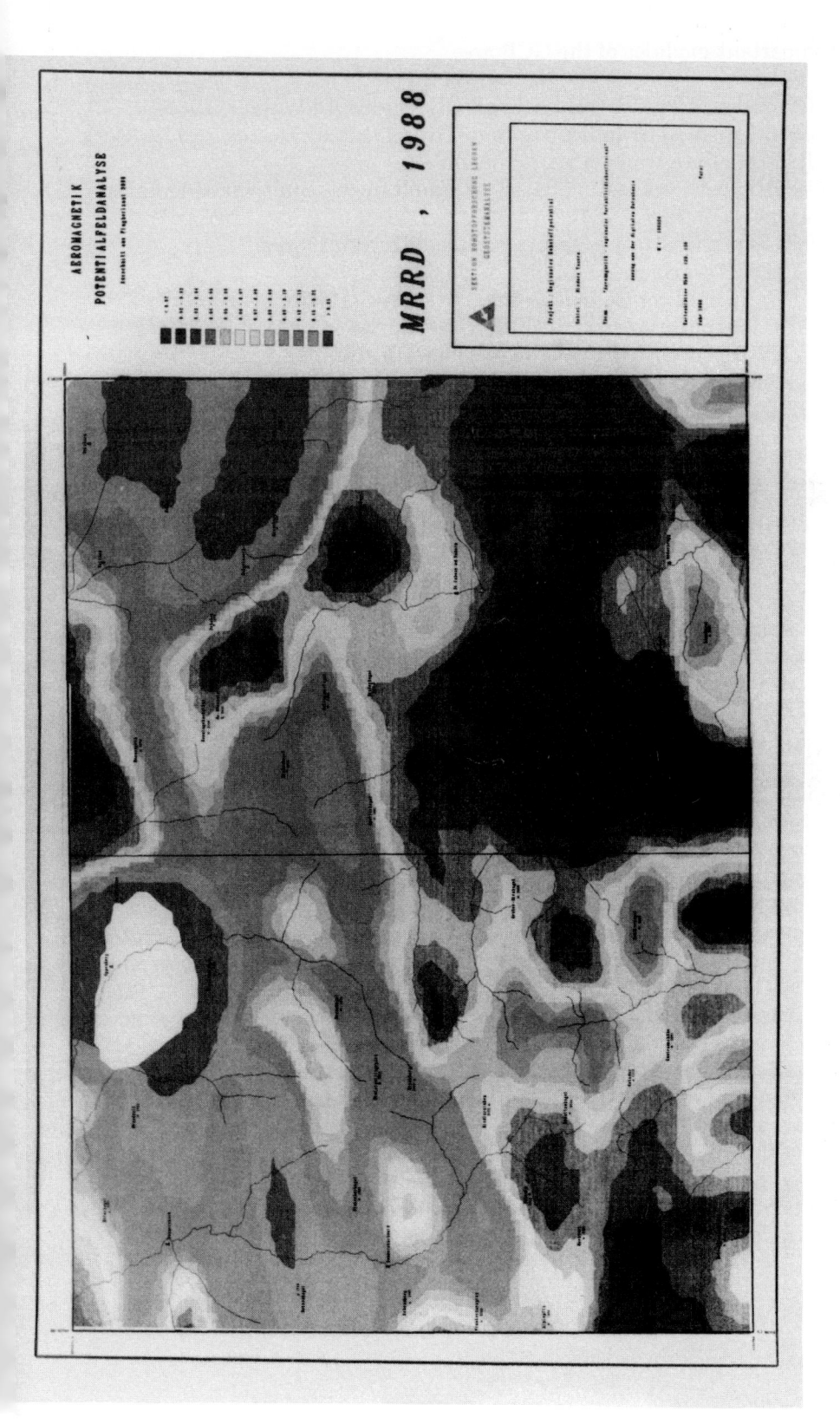

Figure 4. Analysis of aeromagnetic data: map showing regional variability of residual field (orginial scale 1:100,000).

The most important modules of the GCP are:

- LCP (Laboratory Control Package) (Reimann and Wurzer, 1986);
- SGP (Statistical Graphics Package) (Kürzl, 1988b; Dutter, 1983,1987a);
- TRAFO (a data transformation module);
- TEST (hypothesis testing based on simultaneous multivariate analysis of variance) (Wurzer, 1984);
- MVG (a multivariate graphics package) (Kürzl, 1988c);
- CLUSTER (various techniques of cluster analysis);
- FAC (principal component and factor analysis) (Wurzer, 1988);
- CANC (canonical correlation analysis) (Karnel, 1986);
- BLINWDR (robust multiple regression) (Dutter, 1987b);
- CORRES (correspondence analysis) (Karnel, 1988);
- GEODISP (mapping of raw data and results in various ways, point sources or generation of surfaces).

A combination of these modules, not described in detail in this paper, has been used to prepare the raw data and to create statistically derived variables especially suitable for this investigation.

For regional integration studies it is desirable to be able to reduce this large amount of variables to a few significant ones and to represent them in form of geochemical surfaces. For the effective reduction of dimensionality, a robust principal components analysis (Wurzer, 1988) has been used.

The data set of the study area has been divided into meaningful homogeneous geochemical/geographical regions generally corresponding more or less to the regional tectonic domains of the area. All these groups exhibit a distinct geochemical signature obviously deriving from different geological/geochemical processes. Therefore, a separate multivariate analysis of the survey data was performed for each regional group in this study.

Two variable sets exhibiting meaningful variability were extracted and treated separately by the PCA. One set showed a distinct relationship to lithology; the other one showed a relationship to the mineralization processes. To be useful in integration studies, these sets of derived variables have to be presented in form of geochemical surfaces and transfered to the GIS. For the generation of these surfaces GIS features have been used extensively. The catchment areas of the individual samples have been defined by using the digital terrain model and the information concerning the drainage systems. After each sample has been assigned to a catchment basin, the coordinates of the samples have been shifted to the center of gravity of the assigned areas. This led to a homogeneous geographical distribution of the samples. The sample points have been triangulated and the values interpolated to create the geochemical fields, which have been smoothed and presented in vector form.

Figure 5 shows the results of this procedure for one regional geochemical group (Woelzer mica schist unit) and one significant indicative factor with As as the dominating element. In addition to the geochemical field, the most important drainage systems, geographical information (names of settlements and mountains) and the known mineral occurrences (stars) are displayed on the map as background information. The class selection for the geochemical fields follow boundaries as established by Tukey's (1977) five number summary and the resistent outlier definitions. This always gives 50% of area for the inner class and approximately 25% for the two adjacent classes, depending on the amount of outliers present. Experience with these types of intervals shows that the technique is able to reveal regional trends and variability and local behavior (anomalies) in a simple but efficient and satisfactory way, and it therefore has been used exclusively throughout this study in such situations.

Integration Studies

As discussed, a variety of different information layers have been established to be readily available for integration studies. Different strategies have been suggested and techniques applied to perform this task by the computer. One of the first developments for data integration resulted in the NCHARAN-program (McCammon and others, 1984) which is already in use. In this program, the individual variables have to be encoded into a binary (ternary) distribution of favorability for the cells of the rastered area with respect to a particular deposit model. Based on the multivariate behavior of the model cells, weights are assigned to the respective variables and favorabilities for certain deposit models are calculated for all the cells based on a measurement of similarity to the model cells. In general, the numerical approach is based on a variety of assumptions and simplifications so that many uncertainties enter into the model. Especially the necessity of the binary coding and assignment of the values to raster cells create a coarse-grained picture of the underlying process. This coding is derived from information which exhibits generally a higher resolution and, in addition, a significant amount of meaningful variability through encoding is lost.

Another approach, used in this study, follows the ideas of Exploratory Data Analysis (EDA). This is especially supported by the recent technical developments of data-processing systems, which facilitates high-quality graphical display on large terminal screens as well as on color plotters or hardcopy units at variable scale and resolution. Large memories and high-level programming further supports interactive work on graphical terminals with short response time required.

This has provoked the quick introduction of EDA to these type of problems significantly (Kleiner and Graedel, 1980). It does not necessarily replace the numerical statistical methods but represents an additional tool and offers ways for interpretation to which, geologists are especially used. In the following few examples it is demonstrated how this can work and how a GIS can support this way of data integration.

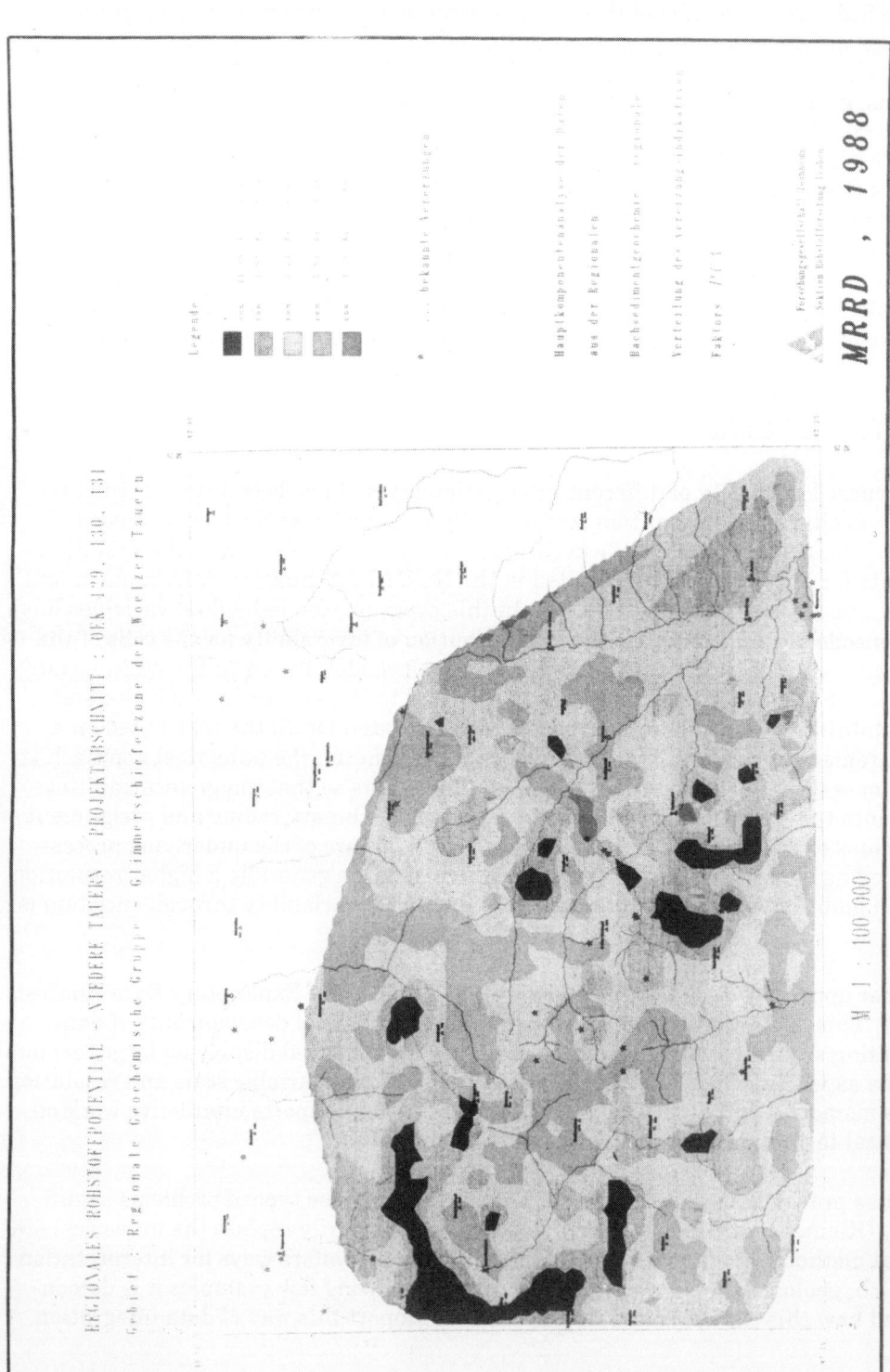

Figure 5. Analysis of geochemical data: indicative as factor presented as geochemical surface related to mica schist formation (orginal scale 1:100,00).

In these exploratory studies the GIS allows for the generation of thematic composite maps out of different information layers by predefined criteria. For example this facilitates, line or polygon overlays or a selection of certain variable combinations based on Boolean algebra. Further techniques are selected retrieval and aggregation of polygons with similar overlays or overplotting of multiple layers. Based on these user defined criteria, geographical displays can be generated which directly allow, for the study of spatial relations of different features to known mineral occurrences, structures, lithology, or even morphology. Figure 6 shows such an integration map. Here the outliers of the first four ore indicative factors have been displayed and overplotted in different colors and shades. The background consists of the structural features (different line types), the known mineralization (stars) as well as of the dominating drainage system and the locality names.

In such a geographical display the spatial relation between the different factor anomalies can be studied. In addition, their regional pattern can be compared with the main direction of structures displayed. We also can see directly which known mineralization has generated a significant anomaly, and which form and type they reveal. Areas with a similar anomaly type can be recognized visually and may be a hint to undiscovered or concealed deposits.

In Figure 7, in addition to the mentioned map features, the local variability of the aeromagnetic field has been displayed.

To preserve clarity, the factor anomalies have been aggregated to one single pattern. This allows to relate the local magnetic signatures to local geochemical multivariate behavior and to all the other geological information displayed. Many of the indicated directions coincide in the two different geofield and follow the trends of the regional structure. However, detailed structural (lineament) information, which were not available at the time of the study, could give additional clues to the mapped situation and indicate further areas of interest.

A similar thematic map is shown in Figure 8. Here the regional variability of the aeromagnetic field is overplotted by the regional deviations (approximately upper quartile of the scores distribution) of the first four significant lithology related factors. They can be distinguished by the different hatching, and also overlaps are clearly visible. A compact pattern arises in the northern part of the mica schist formation which coincides well with the structures indicated by the underlying variability field of the aeromagnetics. In this situation, we can infer that the uniformly mapped mica schist formation exhibits internal structures and facies changes not known before.

Further examples of these exploratory integration studies could be the overlay of lithology related and ore indicative factors, selected retrieval of certain rock types, similiar amphibolites and overlays with local aeromagnetic anomalies or a combined retrieval where certain conditions of the respective information layer have to be met. If a well-defined deposit model has been selected, conditions can be identified which must be met within our set of variables and these can be refined subjectively. Direct

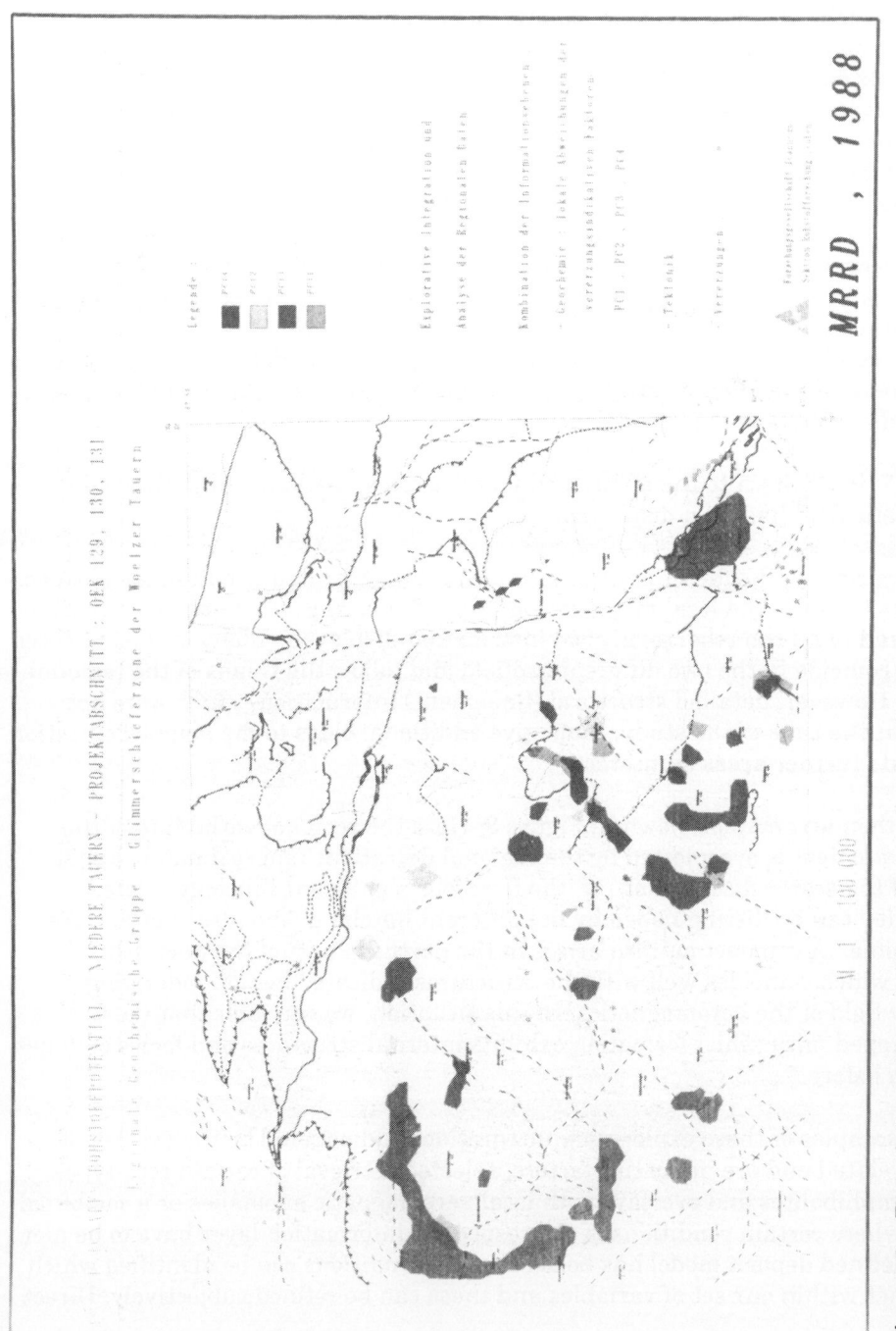

Figure 6. Data integration studies: thematic map display of 4-overplotted ore indicative geochemical factor anomalies integrated with regional structural features (different line types), known mineral occurrences (stars). Background information consists of major drainage system and locality names (original map scale 1:100,000).

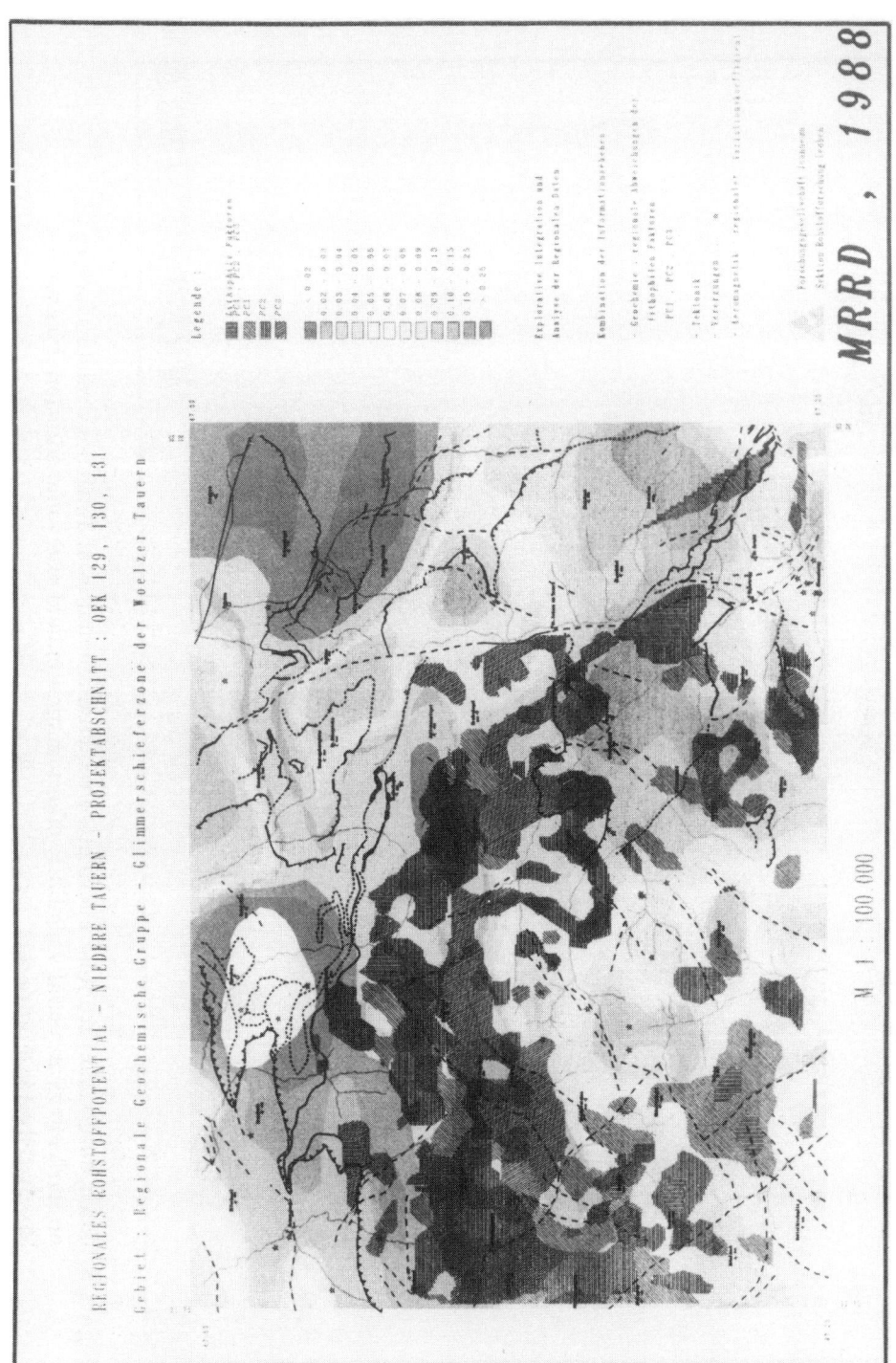

Figure 7. Data integration studies: aggregated ore indicative factor anomalies underlying by aeromagnetic field of local variability and some additional features as described in Figure 6.

Figure 8. Data integration studies: overplotted regional anomalies of 3-lithology related factors underlying by regional variability of aeromagnetic field and some additional features as described in Figure 6.

search for such deposit types within our multivariate data can be accomplished interactively. Currently this is the limit of the present system. The detailed interpretation of the maps produced, already give a variety of important results and allows for the subjective evaluation of the area. It also lays a basis for the application of further numerical-statistical methods to infer indications of mineral deposits and target areas for further follow up along with certain probability measures.

As already indicated, there are several possibilities to perform this task and although a technique which guarantees reliable results has not been established. In this study, only one method has been tested until now using the program FINDER (Singer, 1985) which uses a combination of the area of influence procedure (Singer and Drew, 1976) and Bayesian statistics (Raiffa, 1968). Figure 9 shows a clipped digital geological map section. The section is dominated by the mica schist formation (light gray). Overplotted hatched cells indicate favorable areas for mineral deposits. The deposit type investigated has been a structurally controlled Au-Arsenopyrite type. The control area occurs in the middle of the lower part of the map. It is indicated with two stars and several hatched cells in the surrounding.

The method allows for integration of up to four geological, geochemical, and geophysical variables. For this study two geochemical variables and one geophysical variable have been selected. In the control area, which is considered to be mineralized, a frequency distribution for each variable, has to be estimated. For the other area which is considered to be barren, frequency distributions also have been estimated. The a priori probability was considered to be 0.005 and the area of influence 500m for circular targets. The results show that with the selected parameters a discrete discrimination between barren and favorable areas can be gained. Recent followup work in one of the two indicated areas has led to the discovery of a mineral showing, which is now under detailed investigation.

Further developments and concluding remarks

Based on regional survey data and geological investigations, modern computer techniques offer several new ways to perform regional resource-assessment studies. However, it never can be a standard procedure. Much depends on the type of information available, the principal scope of the study, the size of the area investigated, the geological factors and the types of mineral deposits expected as well as the computer and software facilities available. A GIS will become a basic part of such systems. A GIS allows for a more versatile use in the MRA of the original data and derived results and supports the transformation of MRA into an ongoing long-term activity. By working on such programs such as the study presented here, several problems have been encountered which might be valid for other similar projects. One of them is that reliability and suitability of the available basic data sets is not always guaranteed. In addition the data need significant preprocessing to be integrated in such studies. Geological map information may lack or ignore important aspects of economic geology. This leads to the necessity of complementary metallogenetic studies and field checks to fill this important gap.

Figure 9. Data integration studies: map section of digitized composite geology overplotted by cells which indicate favorable target areas for structurally controlled Au-As deposit type (original map scale 1:100,000).

Further developments following the given techniques and studies will be the establishment of a better link between qualitative/quantitative information. This will lead to the creation of knowledge-based systems, and should allow the generation of user-friendly assessment models and user-oriented results. Besides that, numerical-statistical techniques should be developed which will meet the specific local requirements and allow the expression of the results also in probalistic terms.

The exploratory interactive work with the multivariate data fields seems to represent a significant part in a computerized MRA and supports a new approach to the treatment of regional data and interpretation. It indicates a development which is in line with the latest computer techniques and will lead to comfortable, user-friendly information systems with short response times and a broad variety of geographical and mapping facilities.

ACKNOWLEDGMENTS

The study was jointly funded by the Austrian Ministry of Commerce and the provincial government of Styria. Special thanks are to Prof. J.Wolfbauer, head of Mineral Resources Research Division, Joanneum Research Association, for his steady support of the work. The author also wants to thank the entire working group for its special assistance and effort to develop and perform the MRA-program. Statistical advice was given by Prof. R. Dutter of the Technical University Vienna and support in data processing by F. Wurzer and L. Höbenreich and W. Wassermann. For stimulating discussions and suggestions the author wants to thank especially H. Peer, C. Reimann, and D. McCarn.

REFERENCES

Dutter, R., 1983, COVINTER: A computer program for computing robust covariances and for plotting tolerance ellipses: Graz Technical University, Institute for Statistics, Res. Rept. No. 10, 66 p.

Dutter, R., 1987a, Robust statistical methods applied in the analysis of geochemical variables, *in* Contributions to stochastics: Physica-Verlag, Heidelberg, p. 89-100.

Dutter, R., 1987b, BLINWDR: A FORTRAN programme for robust and bounded influence regression, *in* Statistical data analysis based on the L1-Norm and related methods: North-Holland, Amsterdam, p. 139-144.

Hood, P.J., Holroyd, M.T., and McGrath, P.H., 1977, Magnetic methods applied to base metal exploration, *in* Geophysics and geochemistry in the search for metallic ores: Geol. Survey Canada, Econ. Geology Rept. 31, p. 77-104.

Karnel, G., 1986, Robust canonical correlation (implementation and application); short commun: COMPSTAT 86, Rom; Physica Verlag, Heidelberg, p. 123-124.

Karnel, G., 1988, Robust canonical correlation and correspondence analysis: Proc. First Intern. Cont. on Statistical Computing, (Cesme, Turkey), in press.

Kleiner, B., and Graedel, T.E. ,1980, Exploratory data analysis in the geophysical sciences: Rev.Geophysics and Space Physics, v. 18, no. 3, p. 699-717.

Kürzl, H., 1989, Data analysis and geochemical mapping for the regional stream sediment survey of Austria: Proc. 12th Intern. Geochemical Symposium (Orleans): Jour. Geochemical Exploration, no. 32, p. 349-351.

Kürzl, H. (1988): Exploratory data analysis - Recent advances for the interpretation of geochemical data: Jour. Geochemical Exploration, no. 30, p. 309-322.

Kürzl, H., 1988c, Graphical displays of multivariate geochemical data on scatterplots and maps as an aid to detailed interpretation: Proc. IAEA Technical Committee Meeting, Geological Data Integration Techniques, IAEA - TECDOC - 472, Vienna, p. 245-272.

McCammon, R.B., Botbol, J.M., Sinding, R.L., and Bowen, R.W., 1984, The new characteristic analysis (NCHARAN) program: U.S. Geol. Survey Bull. 1621, 27 p.

Raiffa, H., 1968, Decision analysis.- introductory lectures on choices under uncertainty: Addison-Wesley, Reading Massachusetts, 309 p.

Reimann, C., and Wurzer, F., 1986, Monitoring accuracy and precision — improvements by introducing robust and resistant statistics: Mikrochimica Acta 1986 II, No.1-6, p. 31-42.

Singer, D.A., 1985, Preliminary version of FINDER, a Pascal program for locating mineral deposits with spatial information: U.S. Geol. Survey, Open File Report 85-590, 12 p.

Singer, D.A., and Drew, L.J. 1976, The area of influence of an exploratory drill hole: Econ. Geology, v. 71, no. 3, p. 642- 647.

Tukey, J.W. ,1977, Exploratory data analysis: Addison Wesley Publ. Co., Reading, Massachusets, 688 p.

Wurzer, F., 1984, Simultane multivariate varianzanalyse - Untersuchungen mit Bootstrap, unpubl. masters thesis, Technical University, Graz, 96 p.

Wurzer, F. ,1988, Application of robust statistics in the analysis of geochemical data, in Chung, C.F., Fabbri, A.G., and Sinding-Larsen R., eds., Quantitative analysis of mineral and energy resources: NATO ASI Series C, Reidel Publ. Co., Dordrecht, 738 p.

Region - SCANDING - Mineral Forecasting Computer System

E. A. Nemirovsky
International Research Institute for Management Sciences, Moscow

ABSTRACT

A man-computer Region system - SCANDING - is a further development of the Region - system, well known in USSR and other socialist countries. This system supports a special technology of analysis and interpreting geoinformation for solving mineral forecasting problems.

A new member of the Region family - SCANDING - is a software environment for the IBM PC/XT, AT, or compatibles. It gives geologists user-friendly interactive tools for geological data processing based on pattern recognition, cluster analysis, and heuristic modeling methods.

With SCANDING geologists can more effectively realize traditional procedures of mineral forecasting based on extract additional useful information from geological data.

Nowadays success in human practical activities becomes more and more dependable on information provision from one's ability to make fast and proper decisions, that is from enhancing the researcher's creative work efficiency. It is extremely important not only acquire all necessary information but to use it in the most efficient way. In this respect the tasks of natural-resources study, mineral-deposits predicting, and prospecting are not excluded. Great volumes of new data are produced remote-sensing methods; simultaneously volumes of drilling and field data become available. More and more diverse data is produced from geophysical surveys and other disciplines such as geomorphology, metallogeny, geodynamics, etc. Using different types of information separately in principle may not yield significant positive results. Without computer processing one may fail to utilize all data available, may interpret some information improperly, and suffer loses in working time.

Up to now mathematical methods introduced in applied geology have been carried out presumably along the way of classical methods and algorithms application. Such an approach was used for solving a broad variety of different tasks, although initial data did not always correspond to conditions of successful utilization of these algorithms. So, there is a need to create an absolutely new geoinformation technology based upon a systems approach and automation of procedures related to collecting, transforming, storage, and analysis of data about natural systems and diverse knowledge in the geological sciences.

In applied geology there is an urgent necessity to elaborate and introduce new computer methods of geological predicting, prospecting, and evaluating mineral resources based on the use of informatics and cybernetics. This new and quickly developing trend of theoretical and applied investigations plays an increasing role upon the character and methodology of scientific and practical activity.

Information computer technologies now are introduced in various fields of human activity. Their success in "precise disciplines" where processes and values can be measured quantitatively and therefore calculated using different mathematical algorithms is known.

Nowadays computers and computer methods also are introduced in the geological sciences. To make this process more efficient it is necessary to keep in mind one of the main earth sciences' peculiarities — ill-structure of almost all geological problems, which for a long time prevented the natural sciences transition from a descriptive to quantitative science. Such progress becomes possible now because of the following principal prerequisites:

- increasing "computer literacy" of geologists;
- creation of man-machine computer systems with advanced interface (both verbal and graphical);
- development of territorial geoinformation databank creating methods; and
- development of intellectual database building tools.

This enables one to accumulate, store, and utilize expert knowledge, for example about the nature of ore mineralization, mineral deposits, formation processes, etc.

Geological predicting is a complicated task, primary because of a shortage of necessary field data, and requires the use of an expert's knowledge and practical skill in combination with computer application while making decisions. New computer systems open broad possibilities not only for combining abilities in data processing with geologist's practical skill, knowledge, and intuition but allows one to perform this process at a qualitative level. This includes partial formalizing of notions and qualitative evaluation of individual features and criteria. It is worth emphasizing that heuristic methods in applied geology can be considered not only possible but essential. Intuition or heuristics is considered as a subconscious skill accumulated by an expert during his practical activity, an ability to create the whole finite image using separate details. This feature is particularly important in geological predicting for the latter usually is carried out under conditions of information deficit (this circumstance by itself defines the notion "prediction"). The information deficit can be overcome only by accepting one of the possible working hypotheses generated by the geologist.

Thus, technology of "computer prediction of mineral deposits" should include at least two main components: sophisticated hardware and software and advanced man-machine methods. Modern computer systems especially designed for geological prediction tasks can be regarded as the first component. The second constituent is regarded as the combination of different man-machine approaches and methods for complex interpretation of complex geological, geophysical, geochemical, remote sensing, and other information. Such methods are not restricted by the frames of formal logical and pure arithmetic operations performed by computer. They allow the use of human creative abilities, that is simulate all principal merits of traditional geological prediction methods (Chumachenko, Vlasov, and Marchenko, 1980).

Research teams of specialists from organizations of the Ministry of Geology of the USSR and the International Research Institute for Management Sciences (IRIMS) have developed a new man-machine technology for mineral-deposits predicting. This technology fully meets the described requirements and illustrates positive results while being tested. The technology combines traditional methods of geological predicting and diagnostic tasks solving with broad use of modern achievements in informatics. Man-machine system "Region - SCANDING" comprises a kernel of technology (Marchenko, Nemirovsky, and Seiful-Mulukov, 1986; Chuumachenko and others, 1988). SCANDING is a Russian short title for System of Complex Analysis of Ground and Remote Information for the Geological Purposes.

Modern trends in applied informatics were used while creating the system:

- the system should be user-friendly, because geologists as a rule do not have special skills in using mathematical methods and operating computers;
- the system should provide facilities for interactive data accumulation, management, and processing with results;
- the system should be equipped with special facilities for man-machine interaction both in verbal and graphic forms. Graphic form of geodata presentation is more usual for the geologist;
- the necessity to accumulate and utilize expert geological models for generating prediction decisions.

While building the system two main requirements were taken into account:

- the system should be accessible to a wide circle of geologists. For this purpose it should be designed to run on mini- and-microcomputers;
- the necessity to provide the technology succession during transfer to new minicomputers and PCs. For this purpose the software should be written using one of the most widespread high-level programming languages.

"Region -SCANDING" is implemented on a minicomputer and also a PC. Programming language is FORTRAN-77 (MS-FORTRAN for PC). The system performs following functions:

- synthesizes new knowledge about regularities of mineral-deposit distribution and make decisions based on this knowledge;
- interactively analyzes different versions of prediction decisions (both in verbal and graphic forms);
- outputs results in traditional graphic forms.

Technology main functions are supplied in the system owing to the existence on a level with conventional units (territorial databank, user communication unit, etc.) newly developed units and improving of previously existed ones.

"Region - SCANDING" dialogs are developed on the base of the critical analysis of established geological predicting man-machine technology. The dialogs simulate almost all information-processing functions. Such approaches relieve the geologist from the necessity to keep in mind the details of different data conversion and processing procedures and other service information which is not concerned immediately with the process of geological data analysis and data interpretation. All this facilitates the system being mastered by the beginner and creates a rather comfortable environment for the work. Dialog schemes independent from applied software simplifies modifying and development of dialog unit and allows alternative dialog schemes to be used for performing the same functions. Besides, the system can communicate in English or in Russian depending on user's options.

The system has five main operation modes (implemented as functional units supported by databank and common software): "Acquaintance with the system," "User training," "System demonstration," "System running," and "System development". In "Acquaintance with the system mode" the user is presented with a short explanation, which introduces the main concepts, evidence, and definitions. While working in the "User training" mode, the user learns to run the system. Subsequently this unit is expected to develop in special interactive training system for student geologists. The demonstration mode allows the main system's abilities to be shown through examples. Terminals, digitizers, printers, and supplementary videorecorders and videoprojectors are used. On the bases of the demonstration unit it is expected to develop special videocourse concerned with the development and utilizing of mineral-resource predicting man-machine technology.

"System running" is the principal mode. Interacting with computer, a skilled user can crate and modify territorial and object databanks, including geological, geophysical; and other map data; build digital models of investigated territories; create and modify dialog subsystems; obtain information concerning the system's state; process data by applying statistics, pattern recognition, cluster analysis, and heuristic modeling; and output results in graphic form, generate graphic hardcopies in the form of maps, diagrams, etc.

"System development" is designed for system and applied programmers and provides facilitates for creating, debugging, testing, and connecting new software routines; system supply of data and knowledge bases; distribution of computational resources and quotas; yielding and analyzing statistics about running the system.

In general, the proposed architecture and software permit a higher level all geological predicting processes: from data acquisition to decisions making. These processes envisages the geologists participation at all stages of data processing, analyzing, and interpreting, which undoubtfully leads to a greater utilizing of the human's creative abilities.

REFERENCES

Chumachenko, B. A., Marchenko, V. V., and Nemirovsky, E. A., 1988, Region-SCANDING - mineral deposits predicting computer system: IRIMS Publ., 86p. (in Russian).

Chumachenko, B. A., Vlasov, E. P., and Marchenko, V. V., 1980, System analysis in geological evaluation of territories' ore-prospects: M., Nedra, 248 p. (in Russian).

Marchenko, V. V., Nemirovsky, E.A., and Seiful-Mulukov, R. R., 1986, Applied geocybernetics: VINITI Publ., Physics of the Earth, v. 9, 164 p. (in Russian).

Man-Machine Analysis of Geological Maps

V. V. Marchenko and
E. A. Nemirovsky
International Research Institute for Management Sciences, Moscow
U.S.S.R.

ABSTRACT

A geological map can be transformed into discrete form by the computer. Each pixel "s" of the map "S" (s∈ S) is described by its position with regard to the nearest contacts of all geological map features {x}.

$$f\{r(S)\};\ r(p) = \inf_{q \in w} p\ (p,q) - \inf_{q \in w} p\ (p,q)$$

where $p\ (p,q)$ — distances between pixels center (p and q)
w, w — presence or absence of geological map features {x}.

Thus we can evaluate statistically the spatial distribution of geological formations, structures, magmatic complexes, mineralized sites, and so on. This approach enables us to realize heuristic simulation and computer classification. The proposed method of geological model development uses a specialist's theoretical knowledge and experience. Consequently this model is used for computer classification utilizing information from geological, geochemical, and remote-sensing regional bank.

Cartographic data because of its tremendous informative capacity plays a significant role in natural-resource investigations. For example, one standard geologic map sheet stores up to 10 million bits of information concerning time (i.e. different formation ages), space (i.e. spatial interrelations of geological units), and substance (i.e. chemical composition and physical properties of rocks and minerals). Cartographic data contributes from 40% to 100% in prediction decisions elaborated on at different stages of a geological investigation. Conventional maps continue to remain the best geoinformation transfer from person to person. At the same time cartographic data in it habitual form can not be processed by a computer. So, it is necessary to develop forms of map representation that will enable their man-machine analysis. Nowadays computerized analysis of cartographic information is carried out mainly using the following scheme. Undigitized maps are subdivided into equal-size (usually square-form) cells-pixels and within each pixel values of different features from the map are measured. In that way cartographic information is coded in a binary way. However, such an approach compared with a visual analysis of a map by person does not provide full and detailed geoinformation. While examining a map or other pieces of cartographic information, a person sees that a feature is available or absent in particular cell and the interrelations with surrounding cells. Such distinction in geoinformation apprehended by human beings and the computer leads to a significant loss of useful information during its analysis. It is known from practice of geological predicting that a separate feature influence upon the mineralization processes is not restricted only by the area of it location but is in complex spatial relationship with the influence of other features. The same is true for the processes of sedimentation, soil formation and erosion, vegetation cover, etc. How can this discrepancy between geoinformation analysis based on it human visual apprehension and computer analysis be overcome?

On the one hand a person is able to conduct a more detail cartographic geoinformation analysis (in terms of the examined object environment evaluation). But, as a rule, the person carries out this analysis at a qualitative level or takes into account only gradations reflected in the map. On the other hand, computerized processing provides more discrete quantitative analysis, image transformation, and synthesis, that is the job is carried out beyond natural human abilities.

The approach described here eliminates many of the mentioned defects and can be used successfully for cartographic geoinformation man-machine analysis (Marchenko, 1982).

Let us now introduce two notions. *Cartographic feature* is a notion where we comprehend some particular geologic formation reflected on a geologic map. The cartographic feature can determine the geologic structure of the examined territory and describe its evolution. It is included in the map explanation as an independent unit and can be of square, linear, or pointwise distribution. Other, nongeologic maps naturally will have their own cartographic features reflecting their specific characters (tectonic, geographic, soil, forest, topographic, etc.).

Cartographic value of particular feature in a cell which has numerical characteristics, describes the precise spatial position in regard to the area of the give feature.

The value of a function: $r(p) = \inf_{q \in w} p(p,q) - \inf_{q \in w} p(p,q)$

is taken as characteristic of an elementary cell P(x,y) with respect to particular cartographic feature. Here

w	-	is an area of the given cartographic feature distribution;
w	-	is an area, where the given feature is absent; and
P(p,q)	-	is the distance between p and q cells centers.

It is assumed that the boundary of an area of geologic cartographic feature distribution (its contact) belongs to the same area.

The function r(p) is described in terms of theory-sets operations. Here inf is the value that determines the minimum distance from particular point up to the contact of examined geological cartographic feature.

This function determines the shortest distance between any specified center of an elementary cell p and an area of the considered cartographic feature distribution. If point p belongs to the area of considered cartographic feature, then the distance is taken with the sign"+". If point p is out of the feature's boundaries then corresponding value is taken with a negative sign.

Cartographic values are computed using specially developed software. Theses values can be normalized, which opens the scope for complex analysis of heterogenous cartographic geoinformation (geological, geographical, soils, and other numerical information, that is potential geophysical fields).

Cartographic geoinformation is stored in computerized territorial databanks in the form of two arrays. The "source" array includes cartographic data in original, binary form. The second array contains transformed data which quantitatively characterizes the spatial position of each elementary cell on the map with respect to the nearest profile of all cartographic features. In describing elementary cells in the form of two arrays, the following main goals can be achieved. On the one hand, source cartographic data can be sorted in the original form, on the other, broad opportunities are open for cartographic geoinformation quantitative formal-logical, man-machine analysis. Computerized analysis of cartographic data transformed in that way allows one to utilized available geoinformation at a broader scale. This approach gives an opportunity for repeated and multivariate use of cartographic geoinformation.

While utilizing this approach it is possible to obtain different statistical characteristics of separate cartographic features in the boundaries of the examined territory. This circumstance opens broad prospects for realizing cartographic data analysis at a qualitatively new level of detail, employing mathematical methods and modern data-processing hardware.

For example there is the possibility for a quick compilation of samples and geological objects in the examined territory; also for the statistical analysis of parameters of separate cartographic features and determination of their informative values in solving particular tasks of pattern recognition, mineralization, locating regularities, etc. It is possible also to reveal statistically significant complex geological anomalies, carry out standardless taxonomical regioning, analog search, pattern-recognition, and heuristic simulation (Marchenko, Nemirovski, and Seiful-Mulukov, 1986).

So the new approach for cartographic data digitizing offers opportunities for advanced computerized analysis of different situations taking into account geological objects and spatial positions with respect to all cartographic features. In other words, computer-analysis "logics" approaches human logics of complex geological situations visual recognition. This approach has been tested in different geological regions for several years (Marchenko, 1988a).

Geological maps are digitized and widely used in the USSR geological community and in other countries. Using these digitized maps, many ore-predicting and evaluating task were solved successfully in many regions (Kazakhstan, Ukraine, Caucasis, Yakutia, Baikal region, Carpathian-Balcan, Rodopa etc.). The work was carried out at a broad scale range from 1:25000 to 1:1000000) and concerned with different types of ores and nonmetallic products (copper, tin, complex ores, precious metals, phosphorites, and others). This activity also included geological mapping tasks, It was determined that even analyzing obsolete information (30-40 years old), new promising previously unknown areas can be revealed. Heuristic simulation of different situation also can be carried out based on geologists theoretical knowledge, practical skill, and intuition with subsequent computerized search of such situations using information from regional databanks of cartographic geological data. Heuristic rules elaborated on by skilled geologists and tested in practice can be stored in the "intellectual" databank and widely used in practice (Marchenko, Nemirovski, and Seiful-Mulukov, 1986).

Thus the following conclusions can be drawn:

(1) Diverse cartographic information can be digitized by computer for subsequent man-machine analysis. Each elementary cell's spatial position can be characterized with respect to the nearest profile of all cartographic features on a given map. Simultaneously it is possible to obtain various statistical parameters of cartographic features.
(2) Although a man-machine analysis of cartographic information is transformed, it seems to be possible to carry out complex searches, recognition, and examination of quantitative regularities in their spatial distribution on a map. This approach allows a diverse map "synthesis" and a complex interpretation.
(3) Man-machine analysis of diverse digitized cartographic information gives way for a sharp increase of specialists creative work productivity whereas solving particular tasks of natural-resource investigates and environmental protection.
(4) This approach opens new qualitatively possibilities for future geological investigations based on broad use of information technology (Marchenko, Nemirovski, and Seiful-Mulukov, 1986).

REFERENCES

Marchenko, V.V., 1982, Geologic maps man-machine analysis: Sovietskaya geologiya, v.7, p. 13-26 (in Russian).

Marchenko, V.V., 1988a, Transforming cartographic geoinformation in descrete form applicable for man-machine analysis: Izvestiya Acedemii nauk SSSR (ser. Geology) v.5, p. 115-124 (in Russian).

Marchenko, V.V., 1988b, Man-machine methods of geological predicting: M., Nedra, 240 p. (in Russian).

Marchenko, V. V., Nemirovsky, E. A., and Seiful-Mulukov, R. R., 1986, Applied geocybernetics: VINITI Publ., Physics of the Earth, v. 9, 164 p. (in Russian).

GEONIX — an UNIX-based Automatic Data-Processing System Applied to Geoscience Information

S. Sauzay, H. Teil, M. Vannier, and L. Zanone
Ecole des Mines de Paris, Fontainebleau

ABSTRACT

GEONIX, developed by the IGM, is an integrated, comprehensive data-processing system for the simultaneous treatment of documentary, technical, and scientific data. It provides a basic set of easily learned, convivial instructions for storing, retrieving, and processing the different information encountered in a geoscience inventory, such as bibliographical, geochemical, or mineral-deposit data. Textural, numerical, and structured data, and also predefined name lists can be introduced into any GEONIX file, which can be processed as an entity, as preselected subfiles or in association with other files. In addition to all the usual functions of a documentary system, GEONIX offers an invaluable tool because of its mathematical and graphical possibilities. Mathematical expressions are formulated freely, and results subsequently used to produce histograms, 2-D or 3-D graphs, together with numerical data coming from different integrated files. Written in the C language for a UNIX (or UNIX-like) operating system, it is portable through mainframes to microcomputers in a multi-tasking, multiuser environment. Because of its modular conception, other features have been introduced easily into the system, such as interface software for a digitizer or a plotter. Actually, GEONIX is in use at various geological or mining surveys, and at the Centre de Géologie Générale et Minière for developing an African mineral-deposit databank as well as thermodynamical and geochemical files.

INTRODUCTION

For a period of twenty years, the IGM has been collecting and computer-processing worldwide, mineral-deposit data for specific geological and statistical studies. Until a few years ago, a mainframe was used for the data processing, but the arrival of low-cost microcomputer technology opened up new horizons for the handling of geological information. After studying the software available on the market, the IGM decided in 1985 to develop its own software for handling both documentary and scientific data, and thus created GEONIX, an automatic data-processing system (ADP).

For GEONIX, the UNIX operating system was selected because of its portability through mainframes to microcomputers and portables, and also because of its increasing acceptability in the computer market. Another important factor governing this choice was that UNIX is able to work in a multiuser and multitasking environment, even on a microcomputer. An organization can adapt therefore its hardware according to its needs beginning with a monouser microcomputer installation and subsequently adding terminals to the basic setup as required. The C language was used for writing GEONIX because of its close association with the UNIX system and the highly sophisticated set of development tools. A modular, integrated system was created, allowing additional special applications to be easily incorporated.

At the IGM, GEONIX was developed on a mainframe as well as on mini- and microcomputers (compatible AT). Graphic features were added to the system on an OLIVETTI M28 microcomputer whose particularity is its incorporated standard graphic screen in high resolution (640x400) as well as in normal "compatible" definition (320x200).

With the necessary UNIX-like (XENIX) operating system, the minimal recommended configuration for an efficient use of GEONIX is a RAM of 750 ko to 1 Mo (eg. 128 ko are reserved for GEONIX with a 80286 INTEL processor), and a mathematical coprocessor (as INTEL 80287).

This system now is used in multipost and multiuser mode, mainly on microcomputers, by several research groups in the Centre for storing, retrieving and processing their data, and by some national geological and mining institutions, particularly in West Africa.

SYSTEM OBJECTIVES AND APPROACHES

The main objective of an ADP system for a geoscientific organization is that all types of data encountered in a geoscience inventory can be handled and processed in one comprehensive system. Versatility and flexibility therefore are important so that the same basic set of instructions can be used not only to handle documentary or technical files but also the more complex files containing scientific data. These instructions need to be convivial so that a noncomputer specialist can use them easily and progressively, applying the more sophisticated features as knowledge of the system improves. Moreover a system, if it is to be of interest to researchers, requires to be open-ended with the possibility of handling complex studies and integrating individual applications.

Geological files are particularly difficult to process because of the extreme complexity of the concepts handles. A simple concept such as that of a "vein" may be described by widely different parameters such as thickness, inclination, direction, paragenesis, age, or host-rock. All these variables may need to be associated with the notion "vein" in certain applications such as mineral-deposit files. This structured aspect of data is completely opposed to the "independent" data variables represented in files such as accounting (customer, invoice, and address files). In these file types, the relationship between variable and value is expressed most simply (eg. in the form of "name=Smith, price=500, quantity=10), whereas in geological files, it can range form an elementary description to a two-way data table or an even more complex structure. For example, it is usual for a mineral deposit to contain several veins, each with a different thickness or paragensis, or for basaltic flows to have various compositions.

Just as the objects described in geology are complex, so are the relationships that can be built between these objects. Again, a comparison with an accounting type file helps to illustrate this feature. With these files, certain relationships are so general that it becomes interesting to integrate them into the system, for example, the customer code in the reference file and invoice file automatically associates name, address, and amount to be invoiced, and prepares the invoices to be sent. In geology, the procedures are not at all so simple, and the relationships that may be required between geological files are completely different. They may involve complex parameters (Cu/Ni ratio in the sulphurs, CIPW norm of the associated basalt, presence or absence of recent granitic intrusions). Parameters of interest, which may require fairly complete calculations, usually are defined or redefined during a research study, so that they cannot be implemented a priori into the system as a general purpose geological tool.

In addition, geological applications are distinguished by the necessity to have sufficiently important graphic procedures, such as use of a digitizer, and screen viewing with subsequent printed or plotted output for one, two- or three-dimensional diagrams (eg. histograms, ternary, or block diagrams). On the contrary, the "pie-chart" is practically of no value.

Finally, it is a great asset in a system to have functions for the interpolation of spatial data, or for statistics (means, standard deviations, multivariable analysis), associated with the previous graphic functions.

FILE DESCRIPTION

Several data types can be handled and processed by GEONIX (textual, numerical, dates, structures), using a complete alphabet (including accents and special characters) so that files can be created in different languages.

The user-defined fields for each variable form the analytical part of a record which has a full-screen data presentation for ease of viewing. There is a limited number of fields per record, and no restrictions on field length. Each field is given a rank number which is used, as well as its label, in mathematical or other functions to call or treat its value. The record format is fixed by the user, taking into consideration the different data acquisition features offered by GEONIX. In addition to the analytical part of the file which can cover one or more screens according to the number of fields, there is an optional screen reserved for a free-text commentary (complementary part).

The various features presented in this article are illustrated by three file types, containing different data formats:
- a bibliographical file for mineral deposits, with analytical fields (as subject matter, county, substance, location) and a complementary part with reference and abstract (BIB, Fig. 1).
- a prospection file with geochemical data (GCHEM, Fig. 2), and
- a petrographic file with mineralogical, physical and chemical fields for experimental liquids (EXPLIQ), Fig.3).

1) analytical part

```
BIB FILE REF N° 3825                              EDITION DATE: 1982

AUTHOR: KOPF P.F.

SUBJECT  Bib# Carto  Crst  Dep# Econ  GGen# GChim  GChro# GMor  GPhys  GStat
         Hist  Inv   Meth  Mine Ocea  Ore#  Pal    Ped    Prev  Pros   Sedim  Synth#
         Teco# Teledet
         OTHERS: Paleomagnetism/
CONTROL  ign  volc sed  res  Ia   In   Va   Vn   UB   VS+  metam# pipe  pegm  vein
         ita# qt   cong ska  grei carb eva  all  pla  lat  fault  cont
         OTHERS: /
ELEMENT  Ag  Al  As  Asb  Au  B   Ba  Be  Ben  Bi  Ca  Car  Cd  Co  Cr  Cs  Cu
         Cya Dia F   Fe#  Fel Ga  Ge  Gem Gra  Gyp Hf  Hg   In  K   Kao Li
         Mg  Mn  Mo  Na   Nb  Ni  P   Pb  PGE  Pyr Rb  RE   Rh  Sb  Sc  Se  Sil
         Sn  Sr  Ta  Tlc  Te  Th  Ti  U   V    Ver W   Y    Zn  Zr
         OTHERS: /

GEOGR. DIV.: West Africa/
COUNTRY: LIB/I.C/GUI/
LOCAL NAME: Mount Nimba/Mount Klahoyo/

LIBRARY  BDM  BGR  BNP# BRF  CEA  CIF  CNR+ EMF  IGM  MUS  SGF
         OTHERS: /
         REF. LIB. #: FK 8578                      LANGUAGE: ENG
```

2) complementary part (second screen)

```
AUTH+TITLE+REF: KOPF P.F., SULLIVAN H.E. + The itabirite of Mount Nimba and the
Liberian orogeny. + Bull. Ass. Intern. Iron Geol., London, 1982, vol. 3, n° 26,
pp. 422-478.


ABSTRACT: The Mount Nimba itabirites and neighbouring occurrences are attributed
by the authors to an epicontinental episode during the Liberian orogeny. This
formation is now interbedded and folded with gneiss and amphibolo-pyroxenites,
and situated on the top of the migmatitic, charnockitic and noritic basement (da-
ted 2.75 Ga).
```

Figure 1. Record extracted from bibliographical file (BIB).

```
GCHEM FILE REF N° 12689        SAMPLE N° XQ 455         SAMPLING DATE 15/11/1982
ANALYST: T.S. FORD             ANALYSIS N° 67872        ANALYSIS DATE 13/01/1983
MISSION NAME: Tutalo           MISSION FILE REF: 321    REPORT FILE REF:    1402
BASIN NAME: Balabong                    1/50000 GEOGR. MAP REF: Balabong South
Basal  X:  12.455     Basal  Y: 26.527     GEOCH. MAP REF: GCh Bala 2
Sample X:  19.432     Sample Y: 30.603

SAMPLE TYPE  soil    stream*  rock(outcrop   borehole  )   water     vegetal
ROCK NAME:/                                BEDROCK: Diorite
        ppm       meth         ppm       meth         ppm       meth        ppm       meth

   Ag    0        AAS    | Cr  55        XRF    | Nb    5        XRF    | Sn    0        AAS
   As    2        AAS    | Cs   0               | Nd   20        XRF    | Sr  800        XRF
   Au    0        AAS    | Cu  35        AAS    | Ni   55        AAS    | Th    7        XRF
   B     0               | Ga  20        XRF    | Pb   15        AAS    | U     2        XRF
   Ba  650        XRF    | Ge   2        XRF    | Pr    0               | V   100        DCES
   Be    0        DCES   | Hg   0        AAS    | Rb   70        XRF    | W     1        XRF
   Br    5        XRF    | La  40        XRF    | Sb    0        AAS    | Y    30        XRF
   Ce   30        XRF    | Li  20        AAS    | Sc   15        XRF    | Zn   75        AAS
   Co   20        AAS    | Mo   1        AAS    | Sm    0               | Zr  260        XRF

OTHERS (el:ppm=,meth=/)  S:ppm=1000,meth=XRF/Cl:ppm=200,meth=XRF/

REGISTER:  Ni:Cu 1.571      r2              r3                r4
```

Figure 2. Record from geochemical file (GCHEM).

```
EXPLIQ FILE REF  168                     AUTHOR: D. Walker et al. (1979)

                                         EXPERIMENT REF: V30-RD8-P12 run 17

                                         INITIAL LIQUID: basalt (MORB)

T(K): 1458              log(fO2): -8.57              P(kbar): .001

        Si    Ti    Al    Fe3   Fe2   Mn    Mg    Ca    Na    K     P     Cr    Ni
        ---------------------------------------------------------------------------
liq     5014  155   1453  0     1096  20    736   1210  257   43    0     9     0

oliv    3879  0     11    0     1671  32    4376  36    0     0     0     5     0

opx     0     0     0     0     0     0     0     0     0     0     0     0     0

cpx     4857  141   543   0     1014  22    1356  1990  46    4     0     41    0

plag    4911  0     3086  0     107   0     46    1526  244   5     0     0     0

OTHER MINERALS: spin/

REGISTER: r90 .127811  r91 0       r92 1         r93 .422935    r94 .391782
```

Figure 3. Record from petrographic (experimental liquid) file (EXPLIQ).

Specific data-acquisition features

GEONIX incorporates special features for data acquisition and data display, well suited for geological purposes.

(a) Items - an alphanumerical field can have many values (denoted items), separated by / . The user can predefine these values and input them in a coded or abbreviated form (to gain space), for subsequent automatic display in the data record. The advantages of such lists are several: easier data acquisition and no typing errors, a checklist on the screen and rapid retrieval which results from the efficient storage form of the items. On data input, the user only has to indicate the presence (*) or absence (-) of an item for binary items, or significant (*), subordinate (+) or absent (-) for a ternary item type (eg. Fig. 2: the field SUBJECT contains binary items and the field CONTROL contains ternary items).

This notion of items also is applied to a numberical field, as in a mining area divided into numbered claims: the presence of claims N°3,5,10 is input in the item form of 3/5/10. Storage of such items is in binary form so data retrieval is rapid.

(b) Structures - as previously mentioned, it important to be able to represent associations between variables in geology. GEONIX provides two possibilities for structure handling: hierarchical for infrequently occurring associations, or tabular (corresponding to a x-y table) for numerical data relating to a set of variables. The user selects the optimum manner for recording these data types;

(1) hierarchical structures: these structures are used mainly to describe hierarchically associated variables, or for objects not represented in a previous tabular structure. Data can be retrieved from these structures for editing and including in calculations, and are written the the form (Fig. 2: the field OTHERS):

object:variable=value, variable=value, .../object:........
S:ppm=1000,meth=XRF/C1:ppm=200,meth=XRF/Se:ppm=0.meth=AAS/...

where object, variable and value represent alphanumerical chains.

(2) tabular structures: x-y tables can be input, stored, and processed by GEONIX, with each column being given an user-specified title and each existent x-y couple a rank number. There are two possibilities for defining the rows: each row refers to a previously defined object whose name is input at the time of file description (all records therefore would have the same variable in the same row), or each row is defined individually for each record. The former input is efficient for geochemical and mineral tables whose row and column names can be predetermined (eg. Fig. 2).

If rows are undefined, as in a petrographic (Fig. 3) or a mineral deposit file, where production data are specified for different substances differing from one record to another, it is possible to write either in a hierarchical form:

Ni 1985:production=20.2,grade=1.5/Cu 1986: production=31.5,grade=2.1/

or in a tabular form (the titles also are included in a file description and visualized with the record):

metal	date	production	grade
Ni	1985	20.2	1.5
Cu	1986	31.5	2.1

File system

Each file is created individually with its own format, although an in-built procedure allows another file to be created with the same format. The limit on a file size will depend only on the hardware in use. Because GEONIX works in a multitask and multiuser environment, a file can be consulted simultaneously by a few users. A record however can be modified only by one user at a time, being locked automatically by the system.

Different relationships may exist between GEONIX files:

- records selected from one file can be related to records from another file by comparing a user-selected field, for example records on certain ore deposits can be selected from a mineral-deposit file, and a bibliographical file consulted directly to retrieve any corresponding references,

- files with the same format can be merged. This allows field geologists to input data on portable computers, and afterwards integrate their files into the parent data-processing system,

- data can be retrieved from one file and integrated into another file, for subsequent processing and graphic output.

GEONIX FUNCTIONS

GEONIX functions are menu-based with one principal and several subsidiary menus for specific functions. From the main menu,, the user selects database-management functions, data analysis or graphical procedures, or user-made applications.

Database-management functions

The usual documentary functions including data capture, correction, display, preselection, edition, sorting, or retrieval are available. Edition in record format or a user-defined tabular format with user-selected fields, as in this list extracted from the BIB file:

Local name	County	Element	Subject	Library	Ref
Obuasi	GHA	Au Ag	Mine Econ	BDM	T 7459
Kalana	MAL	Au Ag Cu	Dep Ore GCehm	CEA	ER 678
Commoner	ZIM	Ag Te	Pros GPhys	CNR	WB 951

For data retrieval, the user is presented with an empty record image on which he specifies the data queries and indicates the search operator (as = < >) for each field. By default, the 'and' operator associates the different fields. For more complex questions, a query language is available incorporating logical expressions ('if, the, else, end, by, to, ...'). Using the regular expressions incorporated into the UNIX system, alphanumerical patterns can be retrieved in both the analytical and commentary parts though searching is slower in the latter.

After retrieval or preselection of records, the corresponding records are stored in a subfile which is considered subsequently as the working file. This important feature of GEONIX enables all subsequent operations (including user made applications) to perform on this working file, which is retained between different sessions of GEONIX unless otherwise modified. The previously nonselected records can form a new file, directly obtained by "inversion" of the actual working file.

Available through a special menu are general file maintenance utilities, such as creating backup copies and restoring them, or formatting diskettes. This menu interface to the basic UNIX commands allows the user to take full advantage of appropriate system functions without having any knowledge of the UNIX command syntax. There also are specific file manipulations such as:
- copying sorted files or sorted subfiles on disk in GEONIX format,
- creating ASCII files or subfiles containing user-selected fields: these files can be stored on diskette in either a UNIX/XENIX or MS/DOS format for introduction into other data-processing systems.

Another menu is reserved for functions concerning the record itself. Automatic data input is possible for a selected set of records, as well as automatic data deletion. Fields can be added, deleted or modified, and their screen representation easily modified. Items can be added or renamed in a binary or ternary list.

Mathematical functions

GEONIX has its own mathematical language enabling the geologist to perform calculations or logical operations on user-selected fields from the actual working file or any other GEONIX file. Computed results can be stored in memory or in register fields for subsequent use by any GEONIX function (further calculations, graphics, editions). Moreover, data can be extracted from structured fields, and records can be selected corresponding to a certain result. The complex logical and mathematical expressions are memorized for future use.

An expression consists of variables associated with operators which any be usual (+ - * /), logical (= < >), or special (log, int, cos, or tg). A variable is either a constant, the value of a field indicated either by its rank number (denoted r1,r2,...) or its name, the content of memory registers (there are 32 available, denoted a0, a1,...), the content of an internal register (denoted f), or the value of the last calculated expression (denoted x). These expressions can be combined with the statements 'if, else, the, end, by, to' for writing small programs within the GEONIX system, without having to use a normal programming language. Simple procedures can be written to determine and print for example minimum and maximum values, or correlation coefficients of variables. Some expressions are:

('Cu'+'Zn')>10 =>   *-if the sum of Cu and Zn is greater than 10, then select the record*

log(r15+r16)/2; x/(x+1); =>f   *-print two expressions and store the latter result (x) in the internal record register (f)*

In the EXPLIQ file, five registers (the field r90 to r94) exist for storing the Fe^{3+}/Fe_{total} proportions calculated from the initial data for each of the analyzed minerals (where total iron is expressed as FeO form). A C program was written to compute and store these values, which were used subsequently in further calculations. To obtain the Y value where $Y=\ln\{(Mg/Fe)_{liq}/(Mg/Fe)_{ol}\}$, the corresponding expression is:

$$y=8{,}3143*r3*\ln(r26/r24*r38/r40/(1.0-r90))*0.1+1600$$

where r3, r26, r38, r40, and r90 are the relative fields in the EXPLIQ file. This value is plotted against another value using graphic functions.

Graphic functions

Numerical data can be extracted from one or more GEONIX files for producing high-quality histograms, x-y, or x-y-z diagrams. Normal resolution diagrams (320*200 pixels) are obtained using the standard screen on a microcomputer, but their features are rather coarse. High-resolution diagrams(640*400) are produced on the standard Olivetti M28 screen, or other specific graphic screen. Available for both resolutions are editing functions for preparing diagrams in graphical report form to be printed in small or large format. User-written titles and scales can be added to the diagrams; diagrams can be overlayed or displaced on the screen; figures can be drawn using the commands 'segment', 'arc', and 'rectangle' with user-specified precisions for their position and size.

The graphic expressions are defined in exactly the same manner as the mathematical expressions, and similarly they can be stored in files for future use. The resulting diagrams also can be saved and recalled when required for editing or image mixing. For frequently occurring standard graphs therefore, the basic edited framework can be prepared and memorized carefully, and then recalled to be overlayed on the diagram produced with a new data set.

Each function has its own special menu, though the editing menu can be displayed by each function. Easy one-character codes activate the different features inherent in each function. Additional features exist for each type of diagram.

(a) Histograms: sixty-four classes are available initially in normal resolution and the double number in high resolution for normal or cumulated histograms. Simple codes enable the user to modify the histogram for better presentation; the number of classes can be reduced by one-half progressively, the ordinate scale expanded or reduced, the original histogram recalled. The total number of samples or the percentage of each class can be indicated automatically.

The user defines the origin and scale of the histogram by mathematical expression, for example the expression for the histogram representing the years referenced in the BIB file from 1900 to the present and with a five-year class interval is ('YEAR'-1900)/5 (Fig. 4).

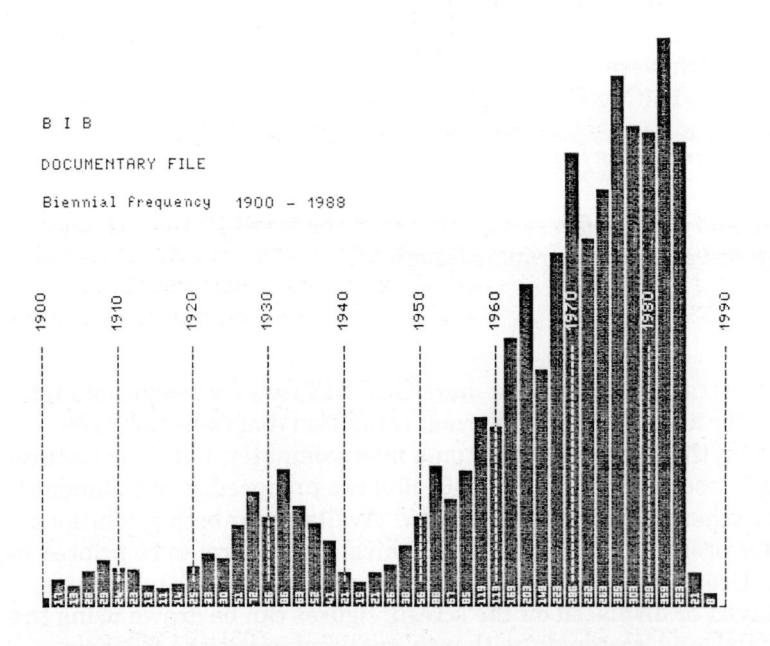

Figure 4. Biennial frequency histogram (form BIB file).

(b) x-y diagrams: user-defined expressions for the x and y variables produce normal or high-resolution scatter diagrams where the distance between two consecutive pixels represents one unit. Using a simple code, isodensity contouring of the points based on a user-defined square grid is possible (Fig. 5). Moreover, points can be joined together following either the abscissa or ordinate axis. This latter feature is used for tracing borehole logs.

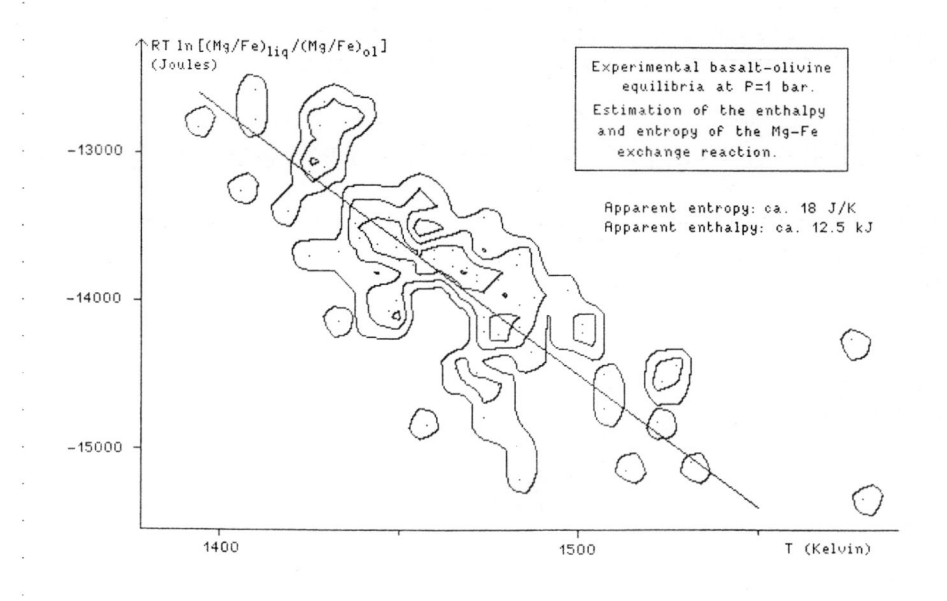

Figure 5. x-y diagram with isodensity contouring (from **EXPLIQ** file).

(c) x-y-z diagrams: in addition to the x and y expressions, the z expression is specified. At the point (x,y), a symbol with size proportional to the z-value can be traced or the rounded z-value can be indicated. All the editing features exist for these diagrams, available in both normal or high resolution. Other specific or more sophisticated features are:

- contouring, which produce z-isovalues. The contour interval can be modified by changing the expression for z (Fig. 6),
- drawing of a block-diagram (Fig.6),
- estimation of a missing z-value in a regular grid, or
- estimation of z at the intermediate points in a grid for more detailed contouring.

Moreover, regularly spaced values can be calculated from a random distribution of points, by taking into consideration the surrounding values and their distance away (a weighted moving average method).

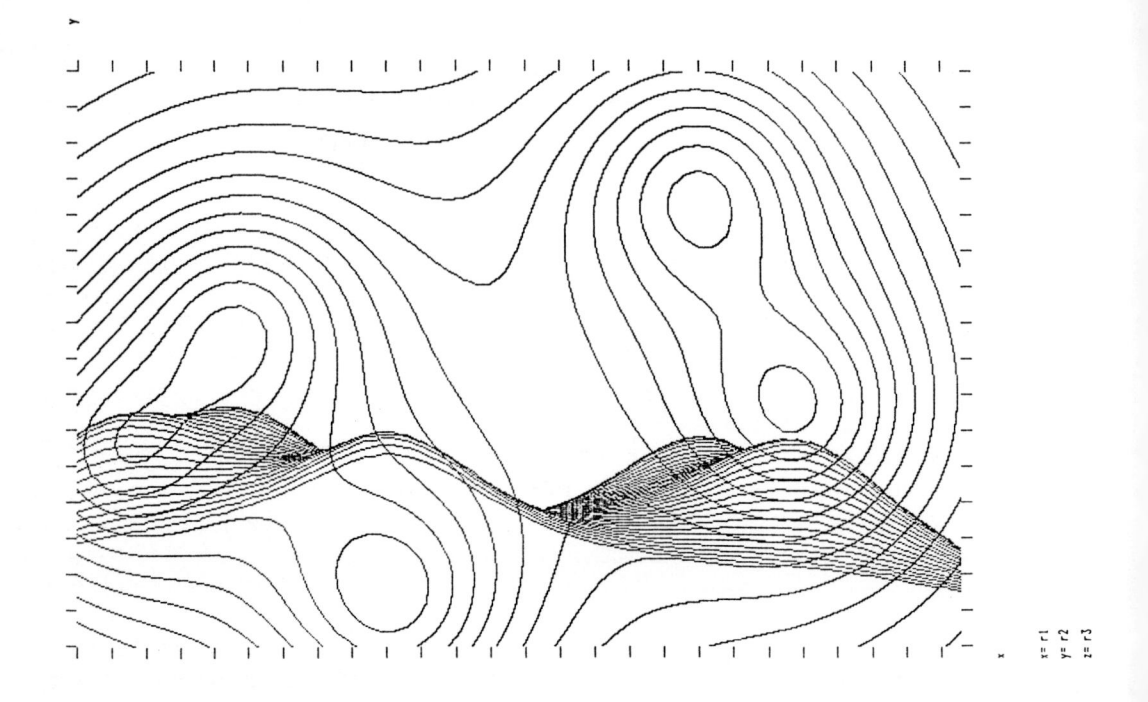

Figure 6. Block diagram overlayed on corresponding z isovalue contouring.

SPECIAL USER APPLICATIONS AND DEVELOPMENT

An interface software for a digitizer has been developed. Its main features are the automatic reproduction of a drawing (eg. a hydrographic pattern) for subsequent high-resolution screen display and printing, and digitized input of coordinates for samples recorded in a GEONIX file. Two searching facilities exist: (a) for records corresponding to samples located within an user-selected zone (its coordinates being digitized); and (b) for samples existing within a give distance from a certain digitized point.

Developments will include interface software for a plotter, and for multivariate statistics such as principal component analysis, correspondence analysis, and cluster analysis.

For special applications, GEONIX provides a set of C language functions, allowing a specialized user or researcher to build his own applications without having to know about GEONIX internal file structures. Such mnemonic codes are not difficult to use within a user-written C program. An user can recuperate data coming from files in other data systems, by writing small interface software using these functions.

AN ILLUSTRATION OF GEONIX

Presented here is a possible sequence of events of how GEONIX is used efficiently for the automatic processing of geochemical data in the GCHEM file (the different functions used are indicated in capital letters).

After data ACQUISITION, any erroneous values are CORRECTED or any duplicate records DELETED. From the main file, records are RETRIEVED corresponding to a certain area. Further RETRIEVAL on this working file is used to select the records input after a certain date.

The hydrographic pattern is digitized for subsequent graphic display and image mixing. The coordinates of the samples referenced in the previously selected records are automatically DIGITIZED and stored in the corresponding fields. The working file is SORTED to produce another file from which certain fields are EDITED in tabular format. An ASCII FILE COPY is made corresponding to this edition and STORED on diskette (in XENIX or MS/DOS format).

The data in the working file are processed using the MATHEMATICAL and GRAPHIC functions. Ratios are calculated for different trace-element relationships and edited. Histograms are made for the different variables and the ratios, and a few are stored for later image mixing (eg. uranium).

Using the high-resolution graphical drawing functions, the previous hydrographic pattern is mixed with the diagram showing the stream sediment sample locations and the uranium values, and also with the uranium frequency histogram (Fig. 7). Regularly spaced uranium values are estimated from the random values based on an user-selected square grid (30*30 pixels), for subsequent contouring (finer grids can be produced for more precise representations). This contour pattern is overlayed on the same hydrographic pattern diagram (Fig. 8).

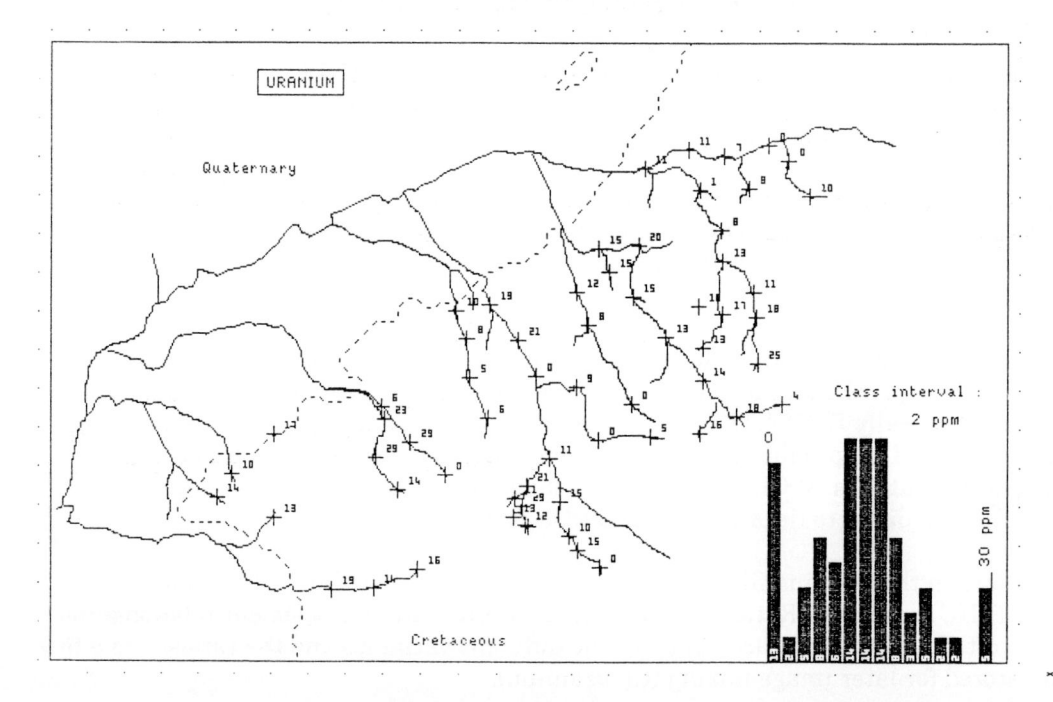

Figure 7. Hydrographic pattern with stream-sediment sample locations and related uranium values (image mixing).

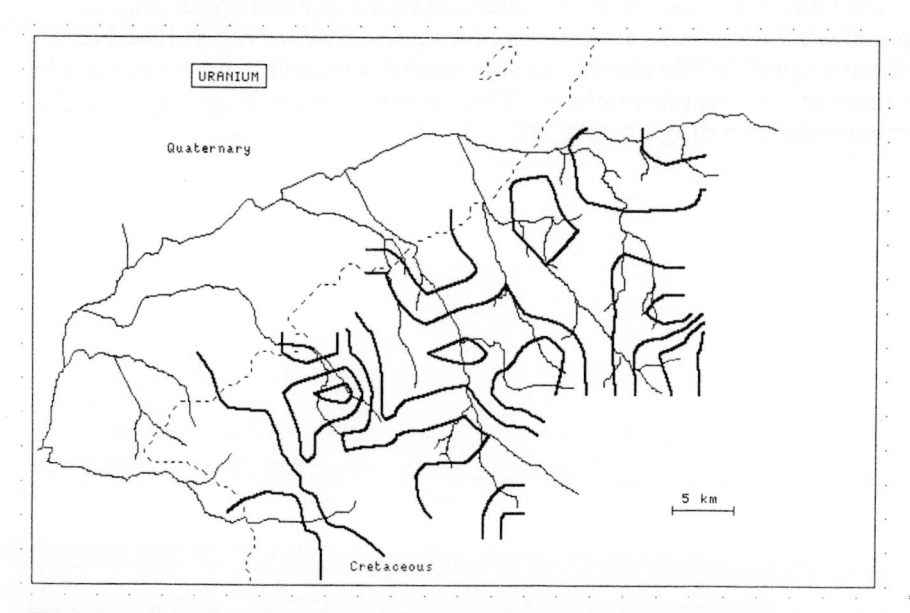

Figure 8. Isovalue contours overlayed on hydrographic pattern.

CONCLUSION

GEONIX provides an extensive and versatile tool for handling the various data types encountered in documentary, technical, and scientific domains. Originally developed for geological and mining research data, it also can be applied suitable in other fields, especially where mathematical and graphical procedures are necessary. Being an integrated and modular UNIX-based system, it can be adapted to the requirements of different users and to the computer facilities available.

ACKNOWLEDGMENTS

We would like to thank researchers and collaborators form the Centre de Geologie Generale et Miniere, especially F. Schneider for the digitizer interface software and the related GCHEM illustration, and Dr. P. Podvin for his criticism and for providing the EXPLIQ figure.

NOTE: UNIX, XENIX, MS/DOS, INTEL, and GEONIX are registered trade marks.

Methods and Techniques of the Prediction of Metallic and Nonmetallic Raw Materials Using Microcomputers in Czechoslovakia

C. Schejbal
University of Mining and Metallurgy, Ostrava

J. Hruska
Intergeo, Praha

ABSTRACT

During the last ten years, various methods and computer techniques have been designed or derived from existing models of mineral-potential assessment in Czechoslovakia. Three centers have reached practical and scientific results of national and international importance regarding recent microcmputers:

(1) Geoindustria in Jihlava - statistical models, particularly those based on pattern recognition, also including comprehensive heuristic principles for regional assessment (system "Prognos"). The construction of a metallogenic-geochemical and geophysical scheme for 1:100,000 to 1:500,000 geological-deposit synthetic maps has been implemented on a H-P 9845 B including graphics for the HP 7580.

(2) Mining geology center for computer applications and geostatistics at the Faculty of Mining and Geology in Ostrava has developed and currently exploits a Geostratistical Software Package "Micro GAD" to solve ore-reserve estimation and sampling problems. A supplemental spatial model on ore-district prediction of volumetric values of metal included applies multivariate statistics. Advanced Geostatistical Methods for Geology and Mining now are being tested for PC -TNS, a compatible Czechoslovak variety of IBM-PC.

(3) Mathematical geology unit at Geological exploration Enterprise in Spisska Nova Ves, Slovakia, using Olivetti PC facilities, has elaborated a comprehensive system of geochemical and structure-metallogenic data evaluation for mineral-potential assessment focused on mineral deposits mapping on a 1:25,000 scale. Statistical methods used are mostly multiple regression with factor or characteristic analysis.

The results of all the mentioned methods and centers applied to various projects in Czechoslovakia and abroad, have been evaluated critically. Limits of application in scale, subject of exploration and computer application have been defined as well as tested in higher probability areas. Positive or promising results have been confirmed, but not in statistically significant figures.

Because of its practical importance, the problem of prognostic assessment of known and newly delineated deposit districts has recently has come into focus of interest in all geologically developed and advanced countries. The selection of promising objects of prospection represents a complicated and, at the same time, vaguely formulated task. The problem can be approached from different methodological viewpoints which, because of the development of theoretical foundations of geological sciences, general prediction techniques, and economic factors, are manifesting incessant dynamism. At present, many prediction models are known (Singer and Mosier, 1981; Schejbal, 1984; Bugayets, Hruska, and Schejbal, 1985), whose potencial applications differ with respect to the dimensions and type of the objects to be assessed, character and extent of input data, etc. Their principal representatives are summarized in Table 1. Let us add that there are different classification criteria of prediction methods (according to their objective, formulation of limit conditions, etc.).

The different types of prediction models also differ in the value of predictions they provide, that is in their revelance (completeness) and probability. Consequently, the suitable model should be selected carefully, so that it is tailored best for the problem to be solved; alternative assessments should be carried out, or a combination of several methods differing in their principles should be employed. The fact is that methodological principles and criteria of assessment hitherto have not been clarified. This complicates practical solutions, hampers mutual comparisons of results and, last but not least, poses a risk of the analysis and its results being influenced subjectively by the experience, erudition, and authority of the person carrying out the analysis. This situation is manifested not only in geology but it is typical for the entire branch of prediction.

There are three groups of prediction assessments distinguished in the CMEA countries, namely:

- estimates of prognostic reserves representing continuations of known deposits (D1),
- estimates of prognostic reserves in mining districts in which known deposits are situated (D2),
- estimates of prognostic reserves in newly delineated districts (D3).

In addition, hypothetical and speculative sources of regional to global metallogenic units (see Table 2) are distinguished.

Different methodological approaches have been used in the assessment of D1 to D3 prognostic reserves in Czechoslovakia. Apart from procedures based exclusively or mostly on manually processed data, mental analyses and expert decisions (metallogenic-prognostic assessments of metallic, some nonmetallic and fuel raw materials of

Table 1. Prediction Models.

Groups	Types
geochemical and geophysical	crustal abundance, quantitative evaluation of local geochemical or geophysical anomalies
structuro-tectonic (geotectonic)	structural-tectonic mineralization pattern
metallogenic	metallogenic-prognostic evaluation (Soviet, French etc. types)
subjective probability (heuristic)	expert estimates, Delphi techniques, simple and comprehensive decision-making rules' models
statistical	frequency distributions, correlation, association and regression analyses, trend analysis, factor analysis, pattern recognition, geostatistical methods, etc.
economic	production analysis, tonnage-grade relationship, econometric models

Table 2. Metallogenic units.

metallogenic provinces

SPECULATIVE SOURCES

metallogenic region

HYPOTHETICAL SOURCES

new district

D3 PROGNOSTIC RESERVES

known district

D2 PROGNOSTIC RESERVES

deposit

D1 PROGNOSTIC RESERVES

the Soviet type, with some features of the French school), attention has been focused increasingly on techniques making use of various mathematical models and computerized procedures. Proposed solutions are based on quantitative interpretations of local geochemical and geophysical anomalies, metallogenic constructions, and the use of a number of multivariate statistical models (distribution analyses, correlation, association and regression analyses, factor analyses), geostatistical methods, theory of information and pattern recognition methods. The prediction assessment procedures are implemented on different types of computers (from large ones to microcomputers).

Recently, the attention in this field has been focused on the development of models and prediction estimates implemented on microcomputers. Three centers have achieved practical and scientific results of national and international importance in this respect.

The team of specialists of Geoindustria and the Central Geological Institute, Prague (Prochazka and Pokorny, 1982; Hruska and Grym, 1983) has designed a system termed "PROGNOS" which is intended for an automated processing of geological and metallogenic data, specifically for a research project named "Prediction Assessment of the Bohemian Massif Ore-Bearing Capacity." The authors' objective was to build up an unified database of geological, geochemical-metallogenic, and geophysical information for the entire territory of Bohemia and Moravia, which would enable to pinpoint prognostic areas and zones as well as to construct a metallogenic scheme for 1:200,000 to 1:500,000 geological deposit synthetic maps. Mostly nonparametric statistical methods, separation characteristic association, and cluster techniques have been applied to individual structural-tectonical units of the Bohemian Massif. The entire region (approximately 70,000 sq km) was divided into a regular 4x4 km grid for the calculation; the 4x4 km cells represented the basis of data matrices for most computer operations.

The PROGNOS system input was an organized data file on 1,397 deposits and ore indications, 3,400 geophysical measurements (both primary and derived), and 3,900 pieces of information drawn from geological maps.

(a) The deposit part of the database contained for each deposit or indications name the following attributes: X and Y coordinates (simple transfer into the 4x4 km grid), appurtenance of the deposit/indications to one (exceptionally more than one) of 16 metallogenic factors, 6 deposit accumulation age categories, 4 types of concentration areas, data on thermodynamic factors, and degree of metamorphism.
(b) The geophysical part included partly interpreted (in the 4x4 km grid) magnetometric values, as well as levels and gradients of gravity measurements.
(c) The geological part of the database was based on interpretations of 1:25,000 to 1:100,000 geological maps and included a classification of 38 rock types (into 5 types of representation in each 4x4 km cell) and ages of mapped geological objects (10 stratigraphic categories).

The processing was performed using HP 9845 desktop computer and BASIC software package. After inputting, filtering, and updating the data, a directory of prediction object and check listings were produced. Metallogenic distribution contingency tables (see Table 3) helped to derive the "activities" of each metallogenic factors which also were presented in cartograms. The first level determination of deposit/indice attributes was performed by association analysis. The accuracy corresponded to the selected 4x4 km grid. However, the first level suffered from a somewhat blurred delineation (insufficient selectivity). The elevation of deposit accumulation of reference objects resulted in more pronounced selective attributes, which then were used to compile a map of "indirect mineralization indices."

The distribution of indirect mineralization indices was tested to obtain areas and factors, such as those of rock types coinciding with deposit accumulations, relations of geophysical variables to deposits, etc.

The results of the first, reconnaissance stage of the ore-bearing capacity prediction in the Bohemian Massif have been extended and made use of on a regional scope (1:50,000 to 1:200,000) in three regions for which more detailed databases are available: in Zelezne Hory District multicomponent statistical methods have been employed, wherreas Maskova and Masin, (1983) use a local anomaly productivity method for geochemical and geophysical data in Western Bohemia: the remaining method uses the concept of the local, probability-based regional value and has been employed in the central and southeastern part of Bohemia.

The Exploratory Geology Deptartment of the Faculty of Mining and Geology, Mining and Metallurgical University, Ostrava, has developed a dialog geostatistical program package termed "MicroGAD", intended for the optimization of sampling systems, prediction assessments of ore-bearing districts and calculations of reserves using statistical and geostatistical methods.

The system allows for:

- building, updating, and transforming purpose-oriented databases, performing an exploratory analysis of data based on robust L and R statistics,
- performing a complete statistical analysis of data, which makes use of the minimization of skewness, simulations of censored data by the Monte Carlo method, verifications of mixed distributions and outliers as well as thorough tests of significance,
- studying the statistical dependence using correlation and regression analyses,
- performing geostatistical structural analysis making use of an interactive graphic interpretation of semivariograms,
- carrying out geostatistical local or global estimates based on point and block kriging methods,
- optimizing observations systems, drawing from geostatistical variance estimates,
- prediction of prognostic reserves in ore districts based on multivariate statistics.

Table 3. Metallogenic distribution contingency table.

SYSTEM PROGNOS   BLOCK PASSPORT: KH-3

BASIC UNIT: Krusne Hory region

BLOCK NUMBER: 3

NO. OF DEP. INDIC.: 16

METALLOGENIC SITUATION IN BLOCK

ASSOCIAT. CLASS	0	1	2	3	4	5	6	ASFM	PV	V1	V2	V3	A	N	+	*	&
Pb,(Ag),Zn,Cu	16	0	0	0	0	0	0	0.00	0	0	0	0	0	0	0	0	0
Cu,Ni,(Co)	16	0	0	0	0	0	0	0.00	0	0	0	0	0	0	0	0	0
W,Sn,(Mo)	16	0	0	0	0	0	0	0.00	0	0	0	0	0	0	0	0	0
Mo,Cu,(W,Bi)	14	2	0	0	0	0	0	1.00	0	0	0	2	0	0	2	0	1
Sn,W,(Li,F)	15	1	0	0	0	0	0	1.00	0	0	0	1	1	0	1	0	1
Sn,Cu,As	12	0	4	0	0	0	0	2.00	0	0	0	4	3	0	3	0	1
Bi,Te,As	16	0	0	0	0	0	0	0.00	0	0	0	0	0	0	0	0	0
Pb,Ag,Zn,Cu	7	4	3	2	0	0	0	1.78	0	0	0	7	6	0	8	0	2
Sb,Ag,As	11	4	1	0	0	0	0	1.20	0	0	0	4	5	0	5	0	1
Bi,Co,Ni,Ag,As	11	5	0	0	0	0	0	1.00	0	0	0	4	5	0	5	0	1
Se,(Cu,Pb,Ag)	16	0	0	0	0	0	0	0.00	0	0	0	0	0	0	0	0	0
U	15	1	0	0	0	0	0	1.00	0	0	0	0	1	1	1	0	0
F	7	5	2	0	2	0	0	1.89	0	0	0	3	7	0	9	0	2
Ba	9	3	2	2	0	0	0	1.86	0	0	0	1	7	0	7	0	1
Au	16	0	0	0	0	0	0	0.00	0	0	0	0	0	0	0	0	0

The program system was implemented on CP/M-80 and MS-DOS operating microcomputers.

The MicroGAD system also is used by geological and mining enterprises concerned with the exploration and exploitation of metals and nonmetals. The experience gained so far shows that it is suitable for detailed predictions, calculations of reserves, and optimization tasks associated with exploration and mining.

The Mathematical Geology Deptartment of Geological Exploration Enterprise, Spisska N.Ves, Slovakia, using Olivetti PC facilities, has elaborated a comprehensive system of geochemical and structural-metallogenic data for assessments of mineral potential, based on 1:25,000 maps of mineral deposits. The statistical methods used include mainly multiple regression techniques with factor or characteristic analyses.

The system is used in prospection works, especially those focused on metallic raw materials.

Apart from the program systems listed here, there are many prediction-associated tasks implemented on microcomputers of different exploratory organizations, which, in addition to statistical procedures, make use of information theory (e.g. in prediction of oil-bearing sections at the southeastern margin of the Bohemian Massif), GUHA automated hypothesis formulation method (prediction of ore accumulations in the Bohemian Massif), pattern recognition procedures based on binary characters (predictions of polymetallic deposits), etc. In these situations, the programs are isolated and do not constitute comprehensive systems.

The results of all the described methods and centers have been employed in various projects in Czechoslovakia and abroad and critically evaluated. Limits of their application in term of the scope, subject of exploration and computer applications have been defined and tested in high-probability areas. Positive or promising results have been confirmed, but not in statistically significant figures.

REFERENCES

Bugayets, A.N., Hruska, J., and Schejbal, C., 1985, Computer methods of regional assessment and prognostic reserves of the ore deposits in the USSR and East european countries: Symposium on computer-aided regional assessment and prediction of ore resources (Alma-Ata), Geoinform CMEA, Geofond, Prague.

Hruska, J., and Grym,V., 1983, Regionalni model stavu a prognozy prirodnich zdroju s pouzitim informaci z pocitacovych databazi (Regional model of the state-of-the-art and prediction of natural resources using computerized databases), *in* Proc. on Symposium Hornicka Pribram ve vede a technice: Sekce Matematicke metody v geologii", (Pribram), p. 14-26.

Maskova, A., and Masin, J., 1980, Matematicke metody prognozovani lozisek s pouzitim geofyzikalnich dat (Mathematical methods of prediction of mineral deposits using geophysical data), *in* Proc. on " 7.celostatni konference geofyziku", Geofyzika, Brno, p. 11-16.

Prochazka, Z., and Pokorny, J., 1982, Progress in geochemical data processing for exploration in the Bohemian Massif (Czechoslovakia), *in* VI. Symposium IAGOD, Tbilisi, p. 267-268.

Singer, D.A., and Mosier, D.L., 1981, A review of regional mineral resources assessment methods: Econ. Geology, v. 76, no. 5, p. 1006-1015.

Schejbal, C. , 1984, Methodology of prognostic reserves assessment in ore districts, *in* Proc. on Computers in earth sciences for natural resources characterisation (Nancy), p. 245-261.

Schejbal, C., Bohac, Z., and Gttner, S., 1987, Mikropocitacovy dialogovy system pro statistickou a geostatistickou analyzu dat (Microcomputer dialogue system for statistical and geostatistical analysis of data), *in* Proc. on Hornicka Pribram ve vede a technice. Sekce Matematicke metody v geologii, p. 670-673.

Use of Characteristic Analysis Coupled with Other Quantitative Techniques in Mineral-Resources Appraisal of Precambrian Areas in Sao Paulo - Brazil

S. B. Suslick and
B. R. Figueredo
Instituto De Geochiencias - Unicamp, Brazil

ABSTRACT

The aim of this work is the application of the method of characteristic analysis in the assessment of the mineral resources in an area of 2400km^2 located in the northeast of the State of Sao Paulo, Brazil. This area comprizes low- and medium-grade terrains of Precambrian age affected by plutonic activity and important faults striking NE. Despite the absence of important mineral deposits, several Fe, Mn, bauxite, Cu, and Au occurrences as well as intrusive bodies and shear zones have stimulated the development of exploration programs in these terrains.

This work was based on chemical analysis for 30 elements of 1900 active stream-sediment samples which correspond to an average density on one sample per square kilometer. The application of the characteristic analysis method was carried out in order to estimate the mineral potential of the area with respect to four types of mineral deposits, namely: deposit associated with basic and ultrabasic rocks, volcano-sedimentary ore deposits, mineral deposits associated with granitoids, and deposits of bauxite.

The geochemical data were treated by a group of techniques, such as: applied statistics for the sampling and analytical errors control, quantification of sampling parameters (e.g. grain size, drainage basin area, etc.) incorporated to the database; geochemical anomaly calculation leading to model cells; definition of the optimum dimension cell on the basis of geostatistic methods; and, finally, the elaboration of mineralization models and model cell selection for the application of the characteristic analysis.

The results obtained in this work were not indicative for the selection of favorable areas concerning mineralizations associated with basic and ultrabasic rocks and bauxite deposits. Nevertheless, they have pointed to a trend, striking NE, which includes the favorable area for volcano-sedimentary deposits and those associated with granite-pegmatite rocks.

INTRODUCTION

The use of characteristic analysis for the integration of the geological data with the aim of assessing the mineral potential of regions and selection of areas for mineral prospection has been proposed by Botbol (1970) and Botbol and others (1978). More recently, McCammon and others (1979, 1983) and Gaal (1984) have suggested innovations in the use of this technique.

In the specific example of regions which have been the subject of regional geochemical surveys, the great advantage of using the characteristic analysis lies on the fact that recognition of an anomaly is more based on its relative concentration than the absolute value of a given element. This is a known fact, but has not been incorporated always in the mineral-exploration data.

Characteristic analysis states that the delimitation of anomalous zone is a function of the correlation of the variable pairs, considering as reference a specific area as a model. An immediate consequence of this type of approach implies in the reexamination of the criteria normally applied in the selection of areas, taking into account that the more interesting targets not coincide necessarily with the high value of one chemical element.

The aim of this work is the application of the characteristic analysis in the assessment of mineral potential of a region, complex from a geological point of view. This region has been submitted to a regional geochemical survey (PRO-MINERIO, 1985), as well as geologic mapping, aimed at the discovery of metallic mineral deposits.

In the stages of selection of variables and measurement of the optimum cell, other quantitative techniques here suggested as an auxiliary in the application of the characteristic analysis, have been used. The different states of work, which are here summarized, have been developed elsewhere (Suslick, 1986). Data on the mineral occurrences, located in the study area of this work, have been stored in the Mineral Occurrence Index of Sao Paulo (IDEM), organized by Figueiredo and Suslick (1988).

METHODOLOGY

Study area

The study area is located northeast of Sao Paulo city, Sao Paulo State, southeastern Brazil (Fig. 1). This region has nearly 2400 km^2 and comprises low-grade schists, medium-grade gneiss, and several types of granitic rocks, all of Archean and Proterozoic age (Fig. 2). Cataclastic zones are expressive, being individualized by

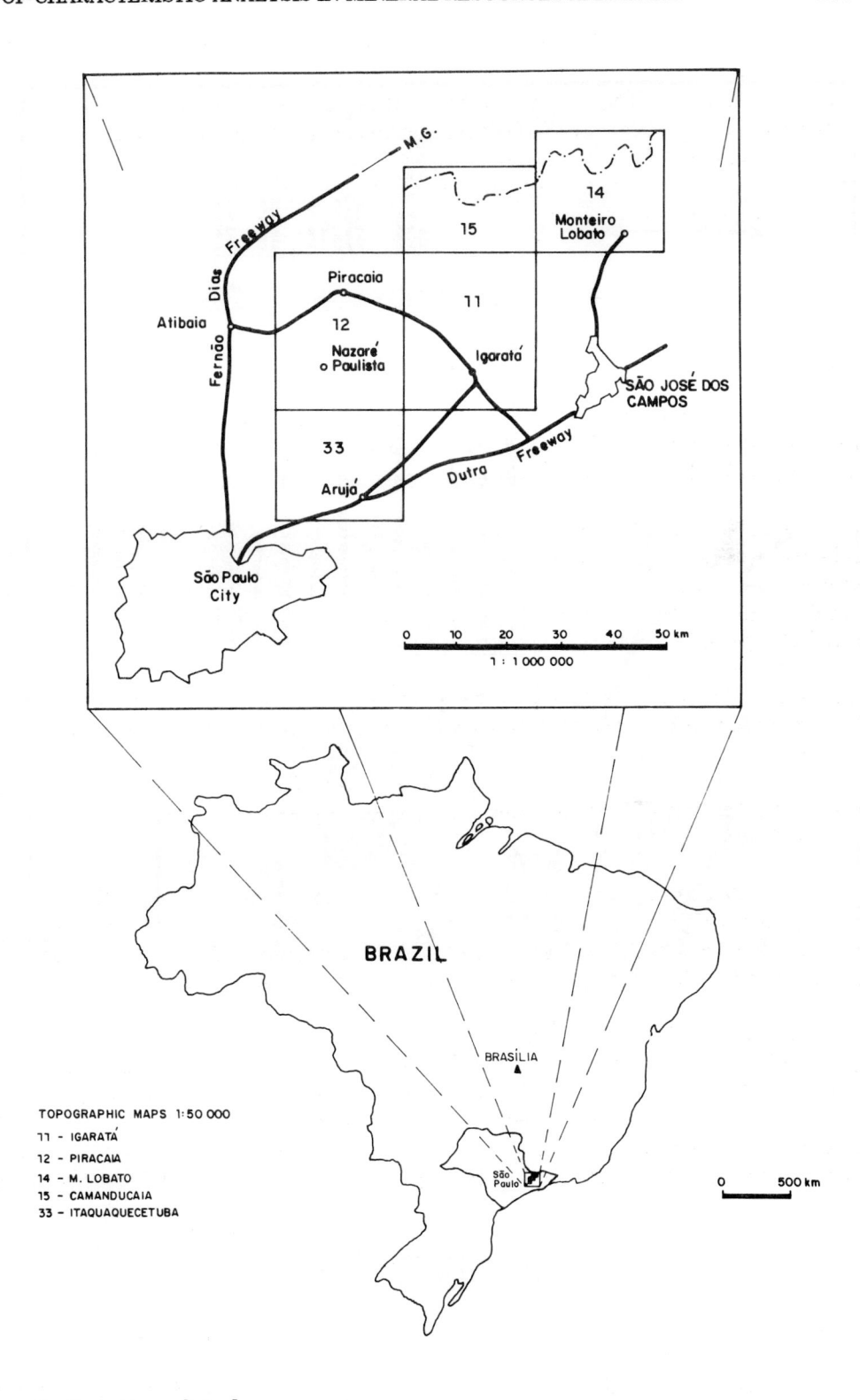

Figure 1. Location of study area.

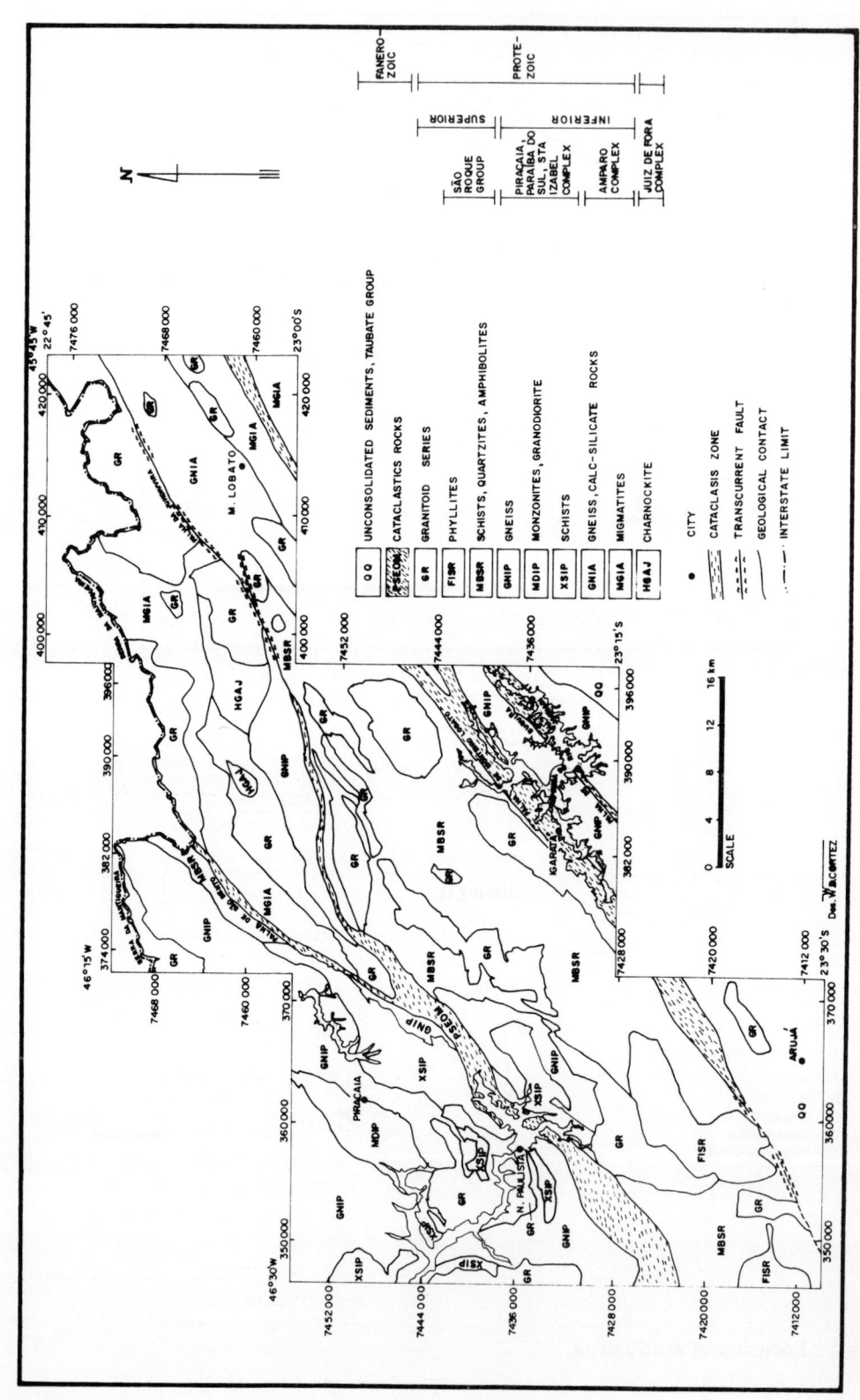

Figure 2. Simplified geological map of Precambrian areas NE Sao Paulo.

great strike-slip faults. In a general way, metallic mineral resources are scarce in this study area, although pyrite and subordinately chalcopyrite occurrences associated with magmatic rocks are known in the northern portion. These occurrences probably are related to a set of fractures, faults, and shear zones striking N-NE. Manganese occurrences, in the form of oxides and hydroxides, have been recorded in the gneiss, notably associated with the quartzose portions. Bauxite occurrences, related to metabasis rocks, occur at the Itaberaba and Pedra Branca Hills, in the southern part of the study area. Noneconomic gold and copper mineralizations probably associated with a volcano-sedimentary sequence also have been identified. The location of the known mineralizations in the study are plotted in the Figure 3.

Database

The data used in this work came form the Regional Geochemical Program carried out by IPT, under the support of PRO/MINEROP/SCT. The study area is covered by five topographic quadrangles (scale 1:50,000) where 1900 active stream-sediments samples have been collected, corresponding to an average density of one sample per square kilometer (being 86 replicates for field control and 164 duplicates for analytical purpose). Afterward, 57,000 geochemical analysis by optical emission spectrography were carried out for 30 chemical elements (Fe, Ca, Mg, Ti, Mn, Ag, B, Ba, Bi, Cd, Co, Cr, Cu, Ga, La, Li, Mo, Nb, Ni, Pb, Sb, Sc, Sn, Sr, V, W, Y, Zn, and Zr). The elements Ag, Be, Bi, Cd, Li, Sb, Sr, W, and Zn were eliminated from the evaluation procedure as they did not register concentrations above the detection limit.

Statistical methods

In the proposed work a group of auxiliary quantitative methods were incorporated in the stages of variable selection and cell measurement used in the application of the characteristic analysis method. The methodology which was followed comprised five integrate stages of work, namely:

Stages	Information
Sampling control and analytical precision	Analysis of replicates and duplicates samples aiming a weight to the trace elements used in the models.
Quantification of the sampling parameters	Selection of sampling parameters, such as: grain size, drainage basin area, pluvial conditions..., which could be included in the models.
Anomaly estimation	Anomaly estimation on the basis of lithotypes aiming to define the model cells.
Regionalization of geochemical data	Definition of a optimum cell size based on geostatistical methods.
Construction of models	Elaboration of mineralization models based on lithological, geological, geochemical...information.

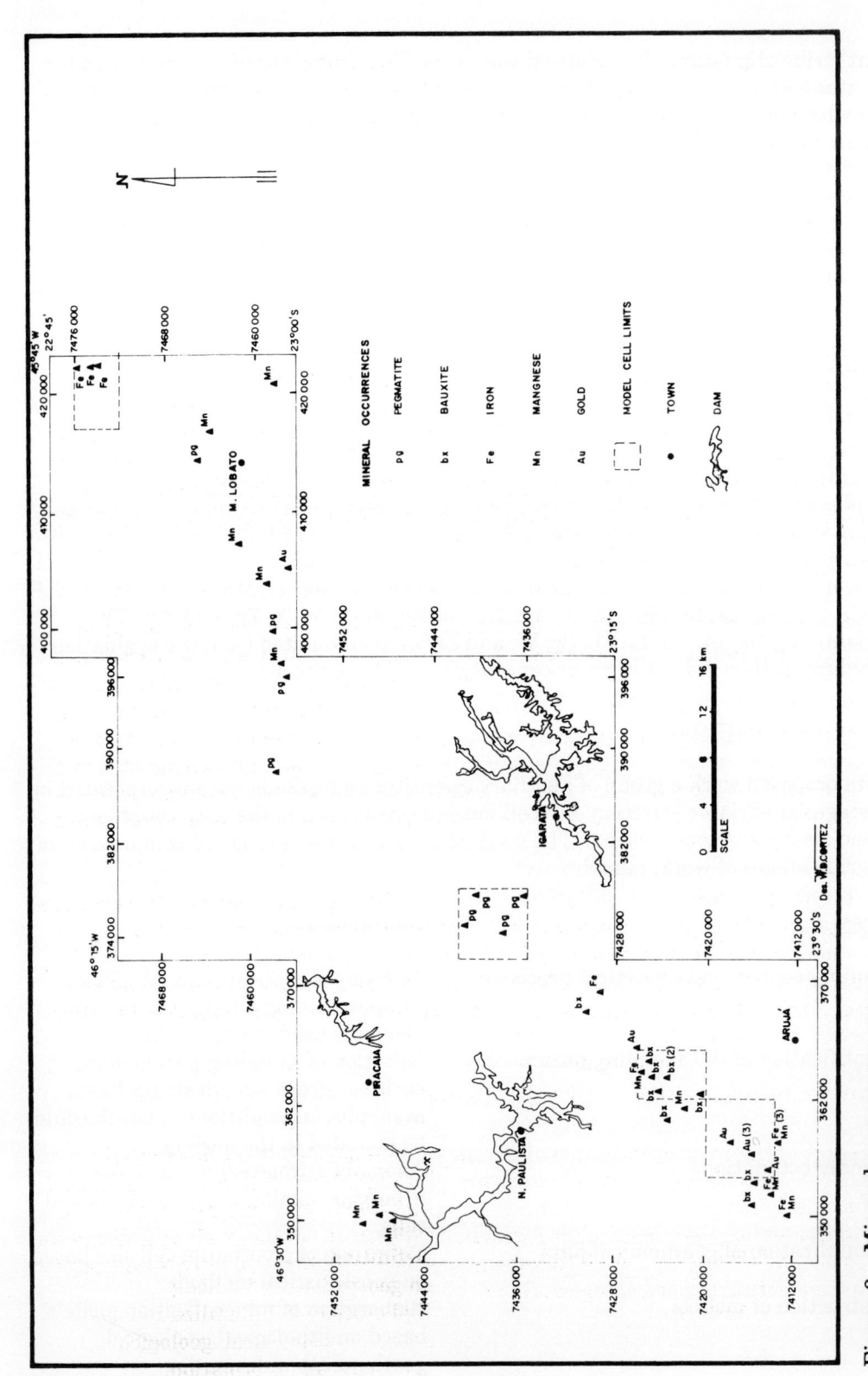

Figure 3. Mineral occurrences map.

In order to accomplish most of these stages, computer programs have been used for the treatment of geological/geochemical data developed by Suslick and Quintanilha (1982) for the IBM 4341 System, with a new versions for the CYBER 470 (Control Data) and another for the IBM PC-XT compatible microcomputer.

USE OF THE AUXILIARY QUANTITATIVE TECHNIQUES

Sampling control and analytical precision

In order to evaluate the total errors or the geochemical heterogeneity of the study area (replicates) as well as the variations derived from the final preparation and analysis of the material (duplicates), it has been used the following methods: graphic analysis (routine versus replicate, routine versus duplicate, ...), analysis of variance — ANOVA (Garrett, 1969) and precision estimation in function of concentration (Thompson and Howarth, 1978; Quintanilha and Suslick, 1982). The results for 21 chemical elements allowed a classification based on five arbitrary category (excellent, good, fair, weak, and problematic) in function of the three previously described methods, as its shown in Table 1.

The assessment of the duplicates samples indicated that the three methods show rather similar results, with some differences in the rank because of the classification subjectivity. The method proposed by Garrett shows some discrepancies concerning the classification in the following elements: Ti for the five quadrangles, La for Itaquaquecetuba, Piracaia and Igarata quadrangles, and Pb in the Piracaia's. This behavior could be attributed to the high dispersion of the data. The same procedures has been used for replicates samples, which have shown as "problematic" elements those that generally present total precision in the spectrographic analysis of geological materials.

The results of the sampling variation (errors) for the chemical elements by replicates and duplicates have been introduced in the characteristic analysis model, using an adjustment factor which was multiplied by eighenvalues of each element, according to the scores defined in Table 2. The conception of adjustment is to assure a higher degree of reliability of the data in function of error, as if it were similar to a quality control of the information.

Quantification of the sampling parameters

This stage of the work consist in evaluation of different component linked to the physical environment (grain size, land uses, contamination sources, drainage types, etc.) in relation to the geochemical final. In summary, it gathers all information to help individualized geochemical background variations from those which indicate the presence of a real anomaly derived form a possible mineralizations. This type of qualitative data is termed categorical.

Table 1. Global evaluation (routine/duplicate) by different methods.

Quadrangle method	Itaqua-Quecetuba A B C	Piracaia A B C	Igarata A B C	Camanducaia A B C	Monteiro Lobato A B C	GLOBAL EVALUATION
ELEMENT						
Fe	5 4 5	5 3 5	5 1 5	5 4 5	5 4 5	4
Ca	5 5 4	5 5 4	5 5 5	5 5 5	4 3 5	5
Mg	5 5 5	4 4 4	2 3 2	5 5 4	5 5 5	4
Ti	4 1 5	4 1 5	4 2 4	4 1 5	5 1 5	3
Mn	4 5 5	4 5 5	4 5 4	4 5 4	5 5 5	5
B	4 5 4	5 5 4	4 5 4	4 5 4	5 5 5	5
Ba	5 5 5	4 5 4	4 5 3	4 5 4	5 5 5	5
Co	3 4 2	3 3 2	4 1 3	3 3 3	2 1 2	3
Cr	4 5 3	4 5 3	5 5 4	4 5 4	3 4 4	4
Cu	3 4 4	3 4 2	4 4 4	3 4 3	3 3 3	4
Ga	4 5 3	4 5 4	4 4 3	5 5 4	4 4 4	4
La	1 3 1	2 5 2	1 4 1	4 5 4	3 4 4	3
Mo	3 4 5	2 5 5	2 4 4	2 2 3	3 3 4	3
Nb	2 4 1	2 4 1	4 4 4	4 5 3	3 4 4	4
Ni	3 4 3	3 4 3	3 3 2	4 4 4	2 3 3	3
Pb	2 4 3	1 4 3	4 5 4	2 4 2	3 4 4	3
Sc	3 4 3	4 5 5	1 2 3	3 4 3	2 3 3	3
Sr	5 5 5	5 5 5	4 5 5	5 5 5	4 5 4	5
V	3 5 2	4 5 4	3 5 2	4 5 4	3 5 3	4
Y	1 3 1	2 4 2	3 5 3	4 5 4	3 4 3	3
Zr	3 5 3	5 5 5	4 5 2	5 5 4	4 5 4	4

CLASSIFICATION SCORE:

1 - problematic
2 - weak
3 - fair
4 - good
5 - excellent

METHOD

A - Average of dispersion measures
B - Log precision
C - Graphical control

Table 2. Weights of elements in characteristic analysis models by duplicate and replicate samples.

ELEMENT	REPLICATE SCORE	DUPLICATE SCORE	REPL.+DUPL. SCORE	QUALITY(X)	MODEL WEIGHT
Fe	5	4	4	5	0.95
Ca	3	5	4	5	0.95
Mg	4	4	4	5	0.95
Ti	3	3	3	10	0.90
Mn	5	5	5	–	1.00
B	3	5	4	5	0.95
Ba	5	5	5	–	1.00
Co	3	3	3	10	0.90
Cr	3	4	3	10	0.90
Cu	2	4	3	10	0.90
Ga	5	4	4	5	0.95
La	1	3	2	15	0.85
Mo	1	3	2	15	0.85
Nb	2	4	3	10	0.90
Ni	2	3	2	15	0.85
Pb	1	3	2	15	0.85
Sc	1	3	2	15	0.85
Sr	4	5	4	5	0.95
V	1	4	2	15	0.85
Y	2	3	2	15	0.85
Zr	3	4	3	10	0.90

CLASSIFICATION SCORE:

1 - Problematic
2 - weak
3 - fair
4 - good
5 - excellent

The studies were carried out for the treatment of categorical data and its link to some chemical element, sampling parameters, and the lithotypes based on the works of Quintanilha, suslick, and Lewis (1984): It and use of 10 trace-elements (Fe, Mg, Ti, Mn, B, Co, Cr, La, and Zr) from 1650 samples of the study area. The geology was simplified into granulites, metabasis rocks, metadiorites, migmatites, and schists regardless the ages and geological units. The description of the data is presented in Table 3, which is divided into two groups: field factors (qualitative character) and the field variable (numerical character). Figure 4 shows the sequence of steps (selection, correlation, and contingency) and the corresponding statistical techniques.

Among the available information, only lithology, grain size, and the land use have real variations and can influence the concentrations. There is no clear interaction among these factors, which might represent a certain advantage because it is unnecessary to consider all these parameters to compensate their interference in the concentration of trance elements. The Cr variation is justified only by the grain size, whereas for clastic linked elements (Ti, Zr, and La) factors such as basin area and grain size are conspicuous in the models, indicating the participation of the environmental control in the behavior of these elements. The Mn variation may be the result of lithological and land-use factors, which can have an important role in the precipitation of Mn oxides. Lithology, grain size, and land use factors are close to Co, Cu, and Fe concentrations, resulting from remobilization and concentration of Mn and Re oxides.

Anomaly estimation based on the lithotypes

In the previous stage it was demonstrated the role of lithology in the regional geochemical data. In this stage the data were grouped according to the main lithotypes with the purpose of defining the significant anomalies (geochemical targets) for characteristic analysis model cells. The main steps are represented in Figure 5. It was selected target areas resultant of the overlapping of geochemical domain of different elements. The location of the targets are plotted on the map (Fig. 6), where the limits were adjusted to squared cells of 2km x 2km. It is important to emphasize that the study area reveals a complex structural and lithological setting not defined clearly yet. This may affect some results in function of the definition of a new geological framework and the discovery of new mineral deposits in the area.
Figure 5. Statistical data analysis used anomaly estimation.

Regionalization of the geochemical data

In this stage geostatistic methods (Matheron, 1970; Journel and Huijbregts, 1978) were used through the variogram calculation in several lithologies in order to estimate a optimum cell size. One of the advantages of the geostatistic compared to several other classical statistical techniques lies on the fact that the former incorporates the spacial concentrations distribution it its conception of geochemical anomaly.

Table 3. Field factors and field variables.

FIELD FACTORS	CATEGORY
grain size (GL)	pebble, sand, silt, clay, organic sed.
field use (FU)	cultivated, uncultivated, forest and other
irrigation (IR)	irrigation area, swamp and other
contamination source (CS)	urban sewerage, industrial, garbase deposit industrial smoke, gas station, civil eng. works and others sources.
water color(WC)	light, darkish
drainage (DR)	very high slope, high slope, moderate and meandering

LITHOLOGY	CODE	NUMBER OF SAMPLES
sedimentes	DD	10
phyllites	FI	93
gneiss	GN	434
granites	GR	417
charnockite	HG	59
metabasic rocks	MB	298
metadiorites	MD	13
migmatites	MG	240
schists	XS	86
total		1650

FIELD VARIABLES	UNITY
pH	--
rainfall (PL)	mm
drainage area (BC)	km2

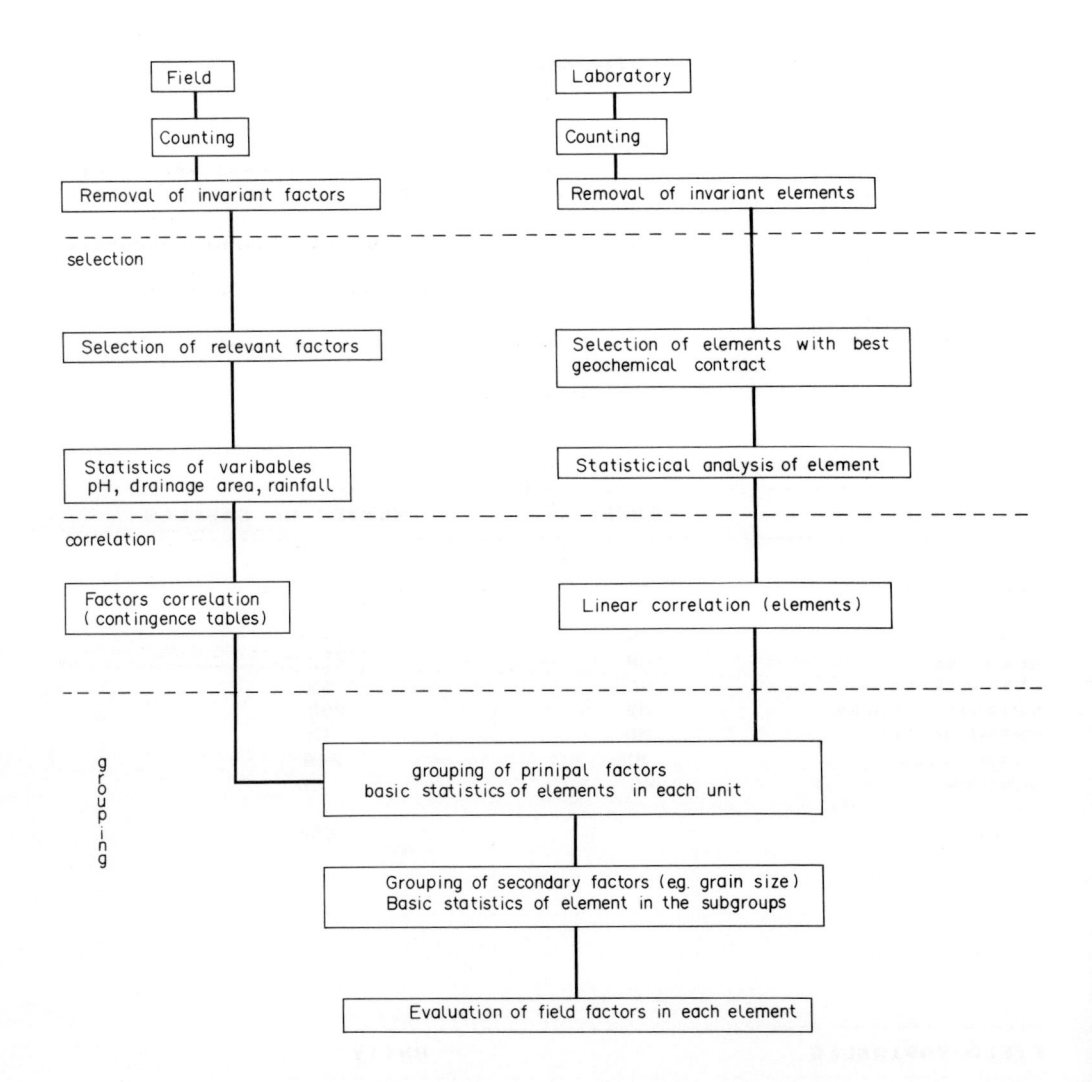

Figure 4. Flowsheet of field parameters evaluation.

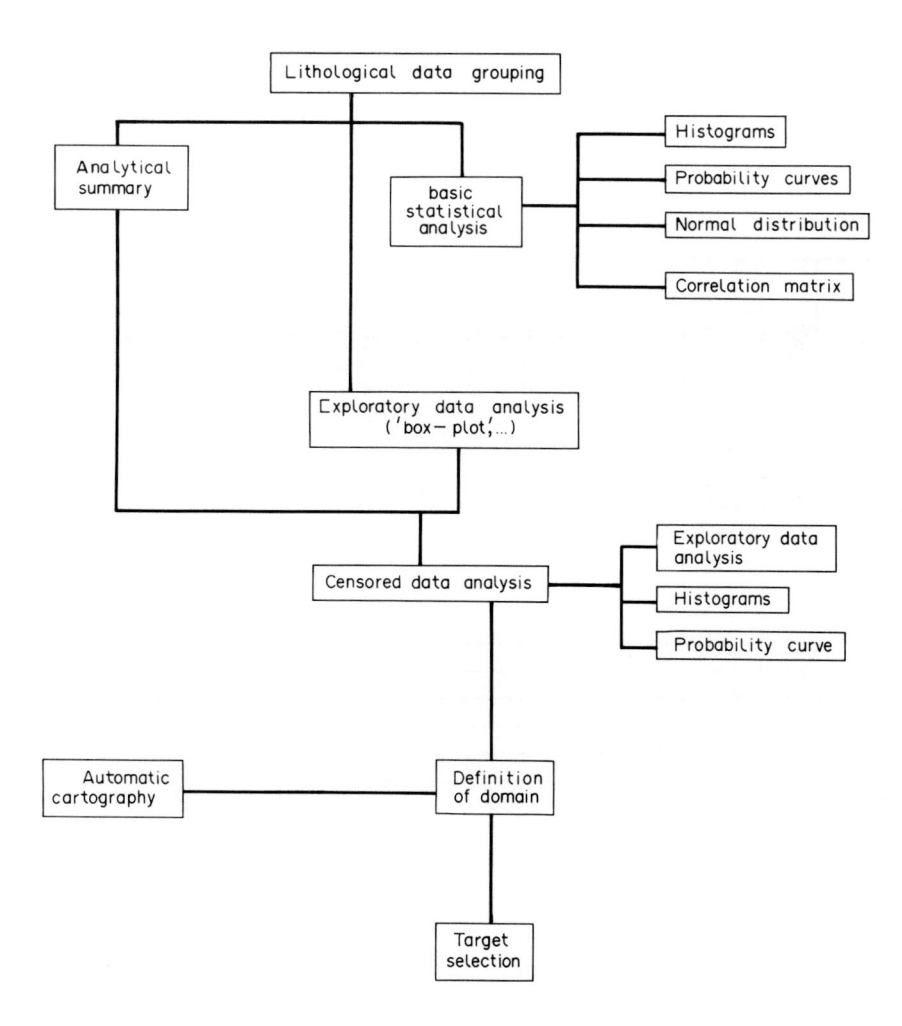

Figure 5. Statistical data analysis used anomaly estimation.

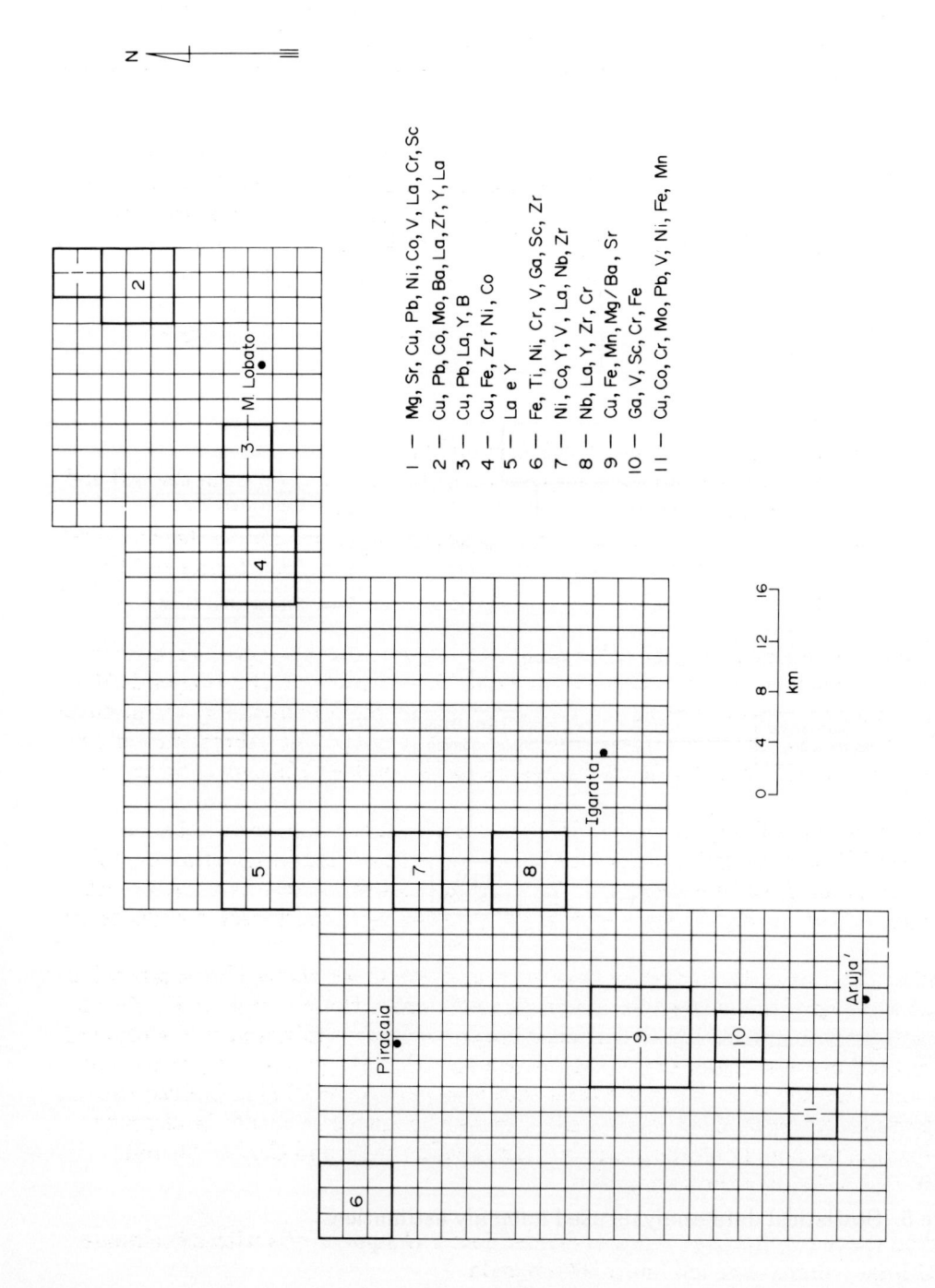

Figure 6. Target area selected by statistical methods.

An optimum cell comprises the selection of a dimension that does not imply in significant loss of geochemical information and allows the detection of geochemical contrasts. Thus it allows the isolation of anomaly from background concentrations. This optimization process should meet to the following criteria:

(a) the variogram, whenever possible, should characterize the background;
(b) the cell dimension should be proportional directly to the size of geochemical anomaly. The conditional probability that a anomaly could not be detected is low.

Regarding these criteria, global variograms were calculated because local lithologies variograms have demonstrated similar results) and four dimension cells analyzed: 500, 1000, 2000, and 3000 meters. It was used for the calculation the concept of cell estimation variance (q^2_E), which is the difference between the actual value (unknown) of the cell and the estimation value through the following expression:

$$\sigma^2_E = E \mid Z_B - Z^*_B \mid^2 = 2\overline{\gamma}(C,P) - \overline{\gamma}(C,C) - \overline{\gamma}(P,P)$$

where:
- C = cell
- P = central point of geochemical sampling;
- $\overline{\gamma}$ (C,P) = mean value of the variogram between a arbitrary pint in the cell and in sampling point;
- $\overline{\gamma}$ (C,C) = mean value of the variogram between two arbitrary points of the cell;
- $\overline{\gamma}$ (P,P) = mean value of the variogram between the sampling points and the point itself, that is, zero.

The estimation variance may be calculated through numerical integration, but for the spherical models, charts generally are available (Journel and Huijbregts, 1978). In order to obtain a mean estimation variance it is sufficient to divide by the number of cell contained informations in each dimension, for example for a dimension of 300 meters, one may obtain 373 cells, 2000 meters corresponds to 769 cells, and so on.

In order to measure the quality of estimates and allow the comparison between variables of distinct magnitudes, it was taken the relative standard deviation multiplied by two and given in percentage. This corresponds to relative interval around the mean (m*) at the significance level of 95%, according to the gaussian hypotheses.

The relation between the precision (2—E/m*) and the dimension cell becomes rather evident when depicted in the diagram of Figure 7, where these variables are shown in monologarithmic scale. From this diagram one may ascertain that the relation of 2000 to 1000 meters network does not cause a considerable gain in the precision of the results. On the other hand, for cells with dimension over 2000 m (inflection point), the precision tends to reach a sill. Hence, the ideal size should be closer to 2000 meters, taking as reference the last range value of La and Cr, 2100m and 2400m, respectively in the variograms.

Based on these results, the area was divided into 767 square cells with 39 columns and 34 lines, where each cell has a 2000 m side.

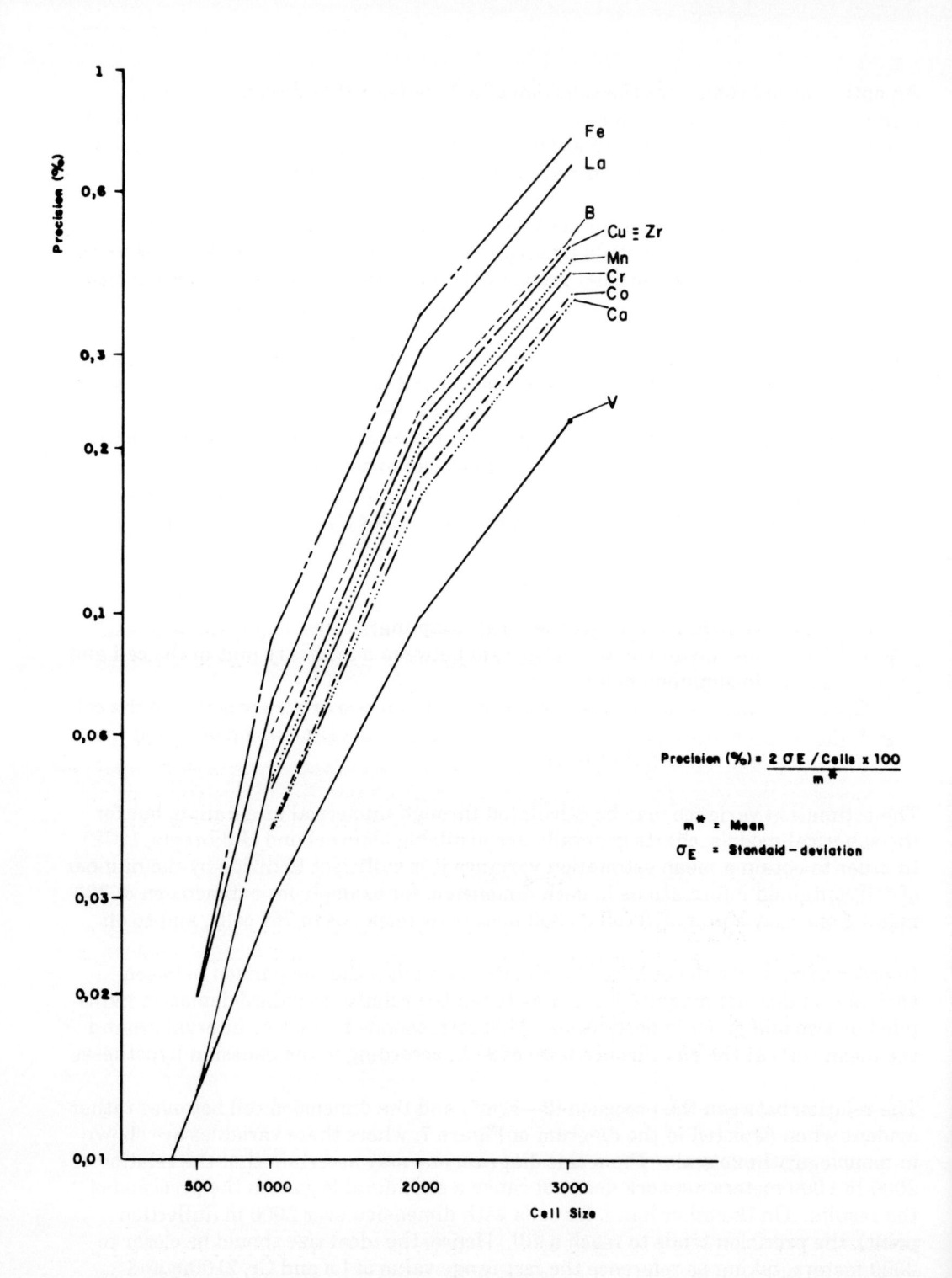

Figure 7. Relationship between precision and cell size.

$$\text{Precision (\%)} = \frac{2\sigma_E / \text{Cells} \times 100}{m^*}$$

m^* = Mean
σ_E = Standard-deviation

CHARACTERISTIC ANALYSIS

After the distribution of the data in the cells, the characteristic analysis was applied, beginning with the transformation into binary form, that is, the value ("1") implies the presence of the variable value in that place, whereas assigning zero ("0") means its absence. The term favorability is defined in local basis, where a variable is considered "prospectable" in a determined site, if his measured value is higher than the neighboring one. McCammon and other (1983) adopted a ternary representation in the following scheme: the value ("1") has the same meaning as in the binary representation, the value ("-1") would indicate a nonfavorable place, and the value ("0") would be reserved to sites not submitted to evaluation. In order to make easier the calculations, we adopted the binary representation.

According to Botbol and others (1978) the characteristic analysis technique implies the knowledge of four concepts: the second derivative of a bidimensional surface, the boolean transformation, the formulation of a prospecting model, and a regional assessment of the cells.

The concept of anomaly and the second derivative

One of the major challenges in the field of geochemical exploration is the question involving the definition of geochemical anomaly. The usual procedure is to separate the data set into two populations: background and anomalous.

Within the characteristic analysis technique, Botbol and other (1978) with the purpose of allowing an adequate partition in these populations, defined as anomalous any value locally higher than the neighboring ones. Hence, the anomaly is a set of values higher then inflection points, which are defined as mathematical surface representing the spatial distribution of a variable. An inflection point is present when a second-order derivative changes sign.

The calculation of the second-order directional derivative for a function $z = f(x,y)$ is resumed to the determination of critical points according to the following expression (Kaplan, 1972):

$$A \cos^2 \alpha + 2B \operatorname{sen} \alpha \cos \alpha + C \operatorname{sen}^2 \alpha$$

where the following abbreviations are used:

$$A = \partial^2 z / \partial x^2 (x,y) \qquad B = \partial^2 z / \partial x \partial y (x,y) \qquad C = \partial^2 z / \partial y^2 (x,y)$$

Let $z = f(x,y)$ be a defined function with continuous partial and second derivatives in a domain D. Let (x_o, y_o) be a point of D where $\partial z / \partial x$ and $\partial z / \partial y$ are nulls. Under these conditions, the following examples are considered:

(a) B - AC < 0 and A + C < 0, relative maximum at (x_o, y_o);
(b) B - AC < 0 and A + C > 0, relative minimum at (x_o, y_o);
(c) B - AC > 0, inflexion point at (x_o, y_o);
(d) B - AC = 0, critical pint of unknown nature.

Points indicating an geochemical anomaly consist of maximum relative and inflexion point. For the calculation of the second-order derivative, and algorithm was conceived, where the value of each cell is computed in function of the eight neighboring cells.

Boolean representation

The boolean representation makes use of the binary logic in the resolution of problems. This representation type is of fundamental importance to the multivariate structure of the characteristic analysis. The binary form shows a certain advantage, because is an amplitude independent measurement, in which the anomaly does not have influence on the area defined as anomalous. Besides, the boolean representation allows an information increase during the modeling, as the variable number capable of being used are enlarged.

Formulation of a model and assessment of the cells

One of the most important steps in the characteristic analysis is the definition of the models and respective reference cells. Mostly of the cells should represent anomalies which would have some link with the mineralization model type. Therefore, it should be noted that the models are a representation of a reality (mostly inaccessible) built from information obtained from several ore deposits, trace-elements dispersion and researcher experience in mineral exploration, which can not be thoroughly quantified.

Figure 8 presents the four mineralizations models formulated for the study area. They contain the element associate with those detected, the typical lithology, and the selected lithotype in the area, the sampling parameters incorporated in the model namely: grain size, basin area, and pluvial conditions.

The degree of association of each cell with the model is calculated through the multiplication of the binary vector which represents the variables of each cell by characteristic vector (root of the product-matrix). The association is expressed in the form of high values.

The mean concentrations in the cell was calculated through the inverse distance function, as this keep practically unaltered the original geochemical relief. The data of pluvial conditions and the basin area also were submitted to the same weighting method.

MODEL	ASSOCIATED CHEMICAL ELEMENTS	TYPICAL LITHOLOGIES	SELECTED LITHOLOGIES	MODEL SAMPLING PARAMETERS
BASIC/ULTRABASIC	Fe-Mg-Mn-Co-Cr-Cu-Ni-Sc-V	basic/ultrabasic rocks	gneiss (GN) migmatites (MG) metabasic rocks S.Roque Group (MB)	grain size (GL)
VOLCANO-SEDIMENT.	Fe-Mn-Cr-Cu-Mo-Ni-Pb-V	volcanics, cherts iron formations	S.Roque Group (MB) lithologies (FI) gneiss (GN) schists (XS)	grain size (GL)
GRANITOIDS	Ti-B-Ba-Cr-Cu-La-Mo-Nb-Pb V-Y-Zr	granitoids rocks pegmatites	granitoids (GR) granodiorite (HG) metadiorite (MD) migmatite (MG)	drainage area (BC) grain size (GL)
BAUXITE	Fe-Ti-Mn-Cr-Ga-Nb-V	aluminium-contained rocks	S.Roque Group (MB) lithologies (FI) gneiss (GN) schists (XS)	grain size (GL) rainfall (PL)

Figure 8. Characteristic analysis models in mineral appraisal.

In the situation of the categorical variables (lithologies) it was utilized the criteria of absence (1) presence (0) for each cell. The criteria of presence of a certain lithology was assumed when this parameter had been recognized in approximately 30% of the cell area, that is, nearly 1,33 km^2.

The weights or contributions of the variables to the four models are given in Table 4. The results of the modeling parameters, which supply the robustness degree of the built models are depicted in Table 5. The degree of the association with the models was generated in bidimensional maps (Fig. 9 to 12).

RESULTS GENERATED BY THE MODELS

The results obtained for the four models are the following:

Model for Basic/Ultrabasic Rocks (Fig. 9) — In decreasing order of importance for the model the significant variables were: Fe, grain size, Mg, Cu, Co, Sc, V, Mn, and gneiss. The participation of the grain size is justified in function of the sampling parameters, as this variable has a great influence on the Fe, Cu, and Co concentrations. In this model seven cell stand out as a result of a high degree of association. The cells along the metabasis rocks of the Sao Roque Group and in the south-southeastern portion of the area are the most significant, as they coincide with the selected areas by the statistical methods, where the Cu, Co, and Ni concentrations showed to be high.

Table 4. Variable weights in models.

VARIABLES	MODELS	BASIC/ ULTRABASIC	VOLCANO- SEDIMENTARY	GRANITOID	BAUXITE
E	Fe	0.597	0.275	*	0.289
L	Mg	0.374	*	*	*
E	Ti	*	*	0.277	0.155
M	Mn	0.159	0.307	*	0.155
E	B	*	*	0.408	*
N	Ba	*	*	0.090	*
T	Co	0.246	*	*	*
	Cr	0.002	0.178	0.365	0.370
	Cu	0.352	0.225	0.277	*
	Ga	*	*	*	0.370
	La	*	*	0.235	*
	Mo	*	0.290	0.164	*
	Nb	*	*	0.164	0.155
	Ni	0.002	0.434	*	*
	Pb	*	0.251	0.303	*
	Sc	0.246	*	*	0.289
	Sr	*	*	*	*
	V	0.246	0.127	0.365	0.155
	Y	*	*	0.211	*
	Zr	*	*	0.182	*
LITHOLOGY					
	FI	*	0.228	*	0.276
	GN	0.128	0.0	*	0.133
	GR	*	*	0.483	*
	HG	*	*	0.094	*
	MB	0.081	0.583	*	0.568
	MD	*	*	0.094	*
	MG	0.014	*	0.119	*
	XS	*	0.031	*	0.078
SAMPLING					
	PL	*	*	*	0.198
	BC	*	*	0.019	*
	GL	0.388	0.061	0.156	*

(*) no included in the model

USE OF CHARACTERISTIC ANALYSIS IN MINERAL RESOURCES APPRAISAL   175

Table 5.   Model cell parameters.

MODEL PARAM.	BASIC/ULTRABASIC	VOLCANO-SEDIMENTARY	GRANITOID	BAUXITE
NUMBER OF CELLS/VARIABLES	6	9	9	6
TOTAL NUMBER OF CELLS	78	108	162	78
EIGENVECTOR (CHARACTERISTIC ROOT)	6.95	21.76	15.56	16.15
TOTAL NUMBER OF 1'S (ANOMALY)	18	38	40	29
CONSISTENCY INDEX (*)	0.38	0.57	0.39	0.56

(*)   CONSISTENCY INDEX = EIGENVECTOR/TOTAL NUMBER OF ANOMALIES
                                        (1'S)

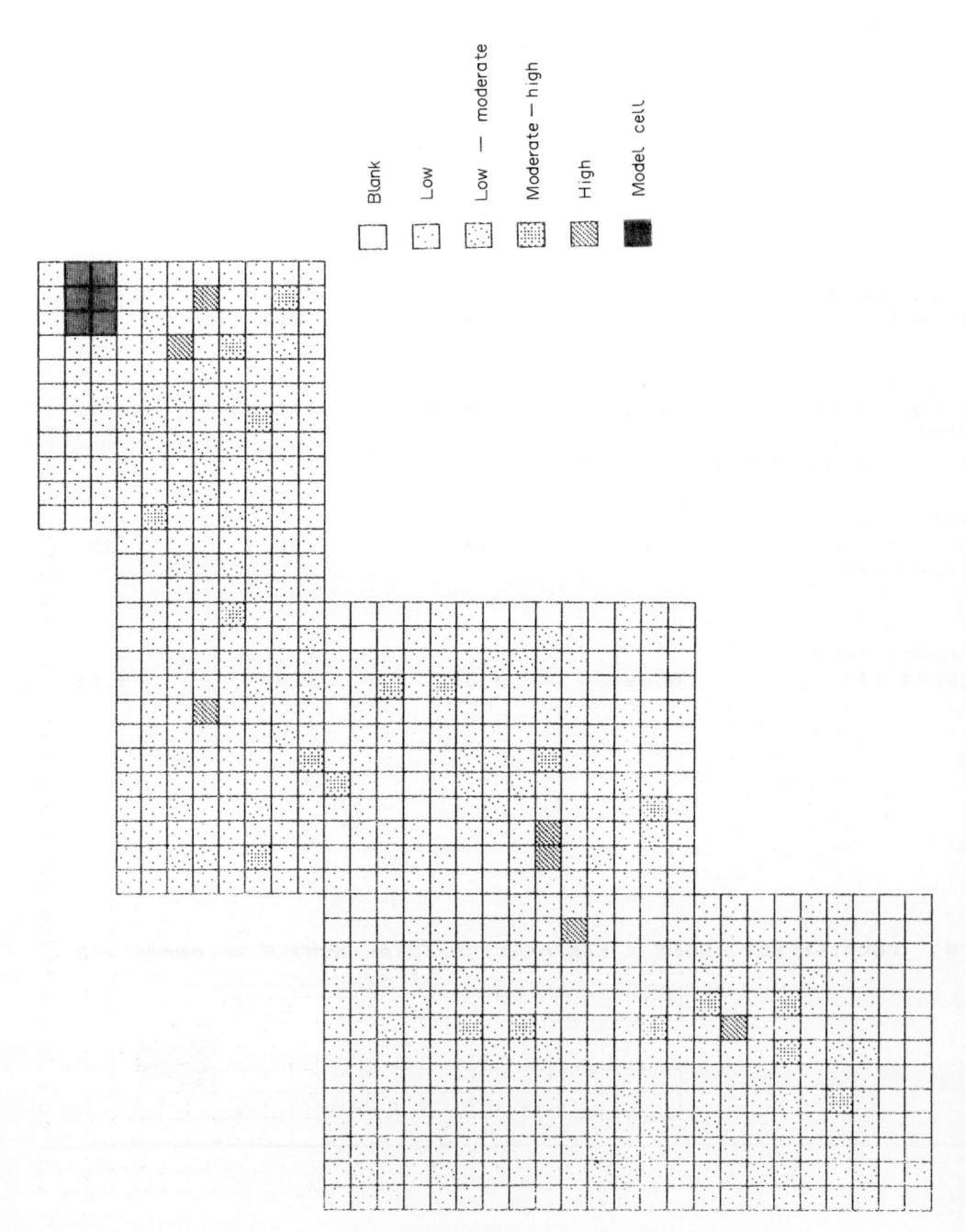

Figure 9. Degree of association map with basic/ultrabasic model.

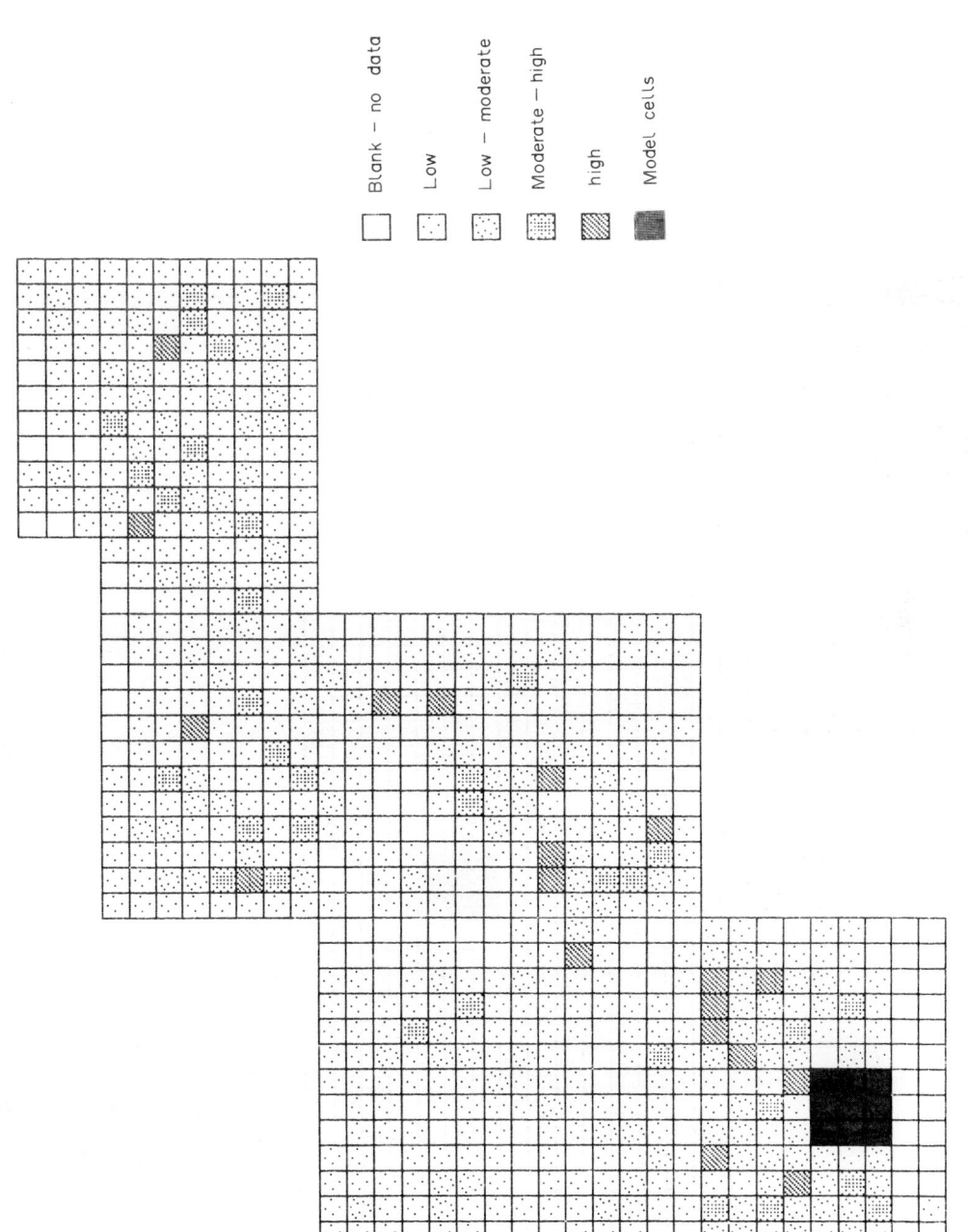

USE OF CHARACTERISTIC ANALYSIS IN MINERAL RESOURCES APPRAISAL 177

Figure 10. Degree of association map with Yollano-sedimentary model.

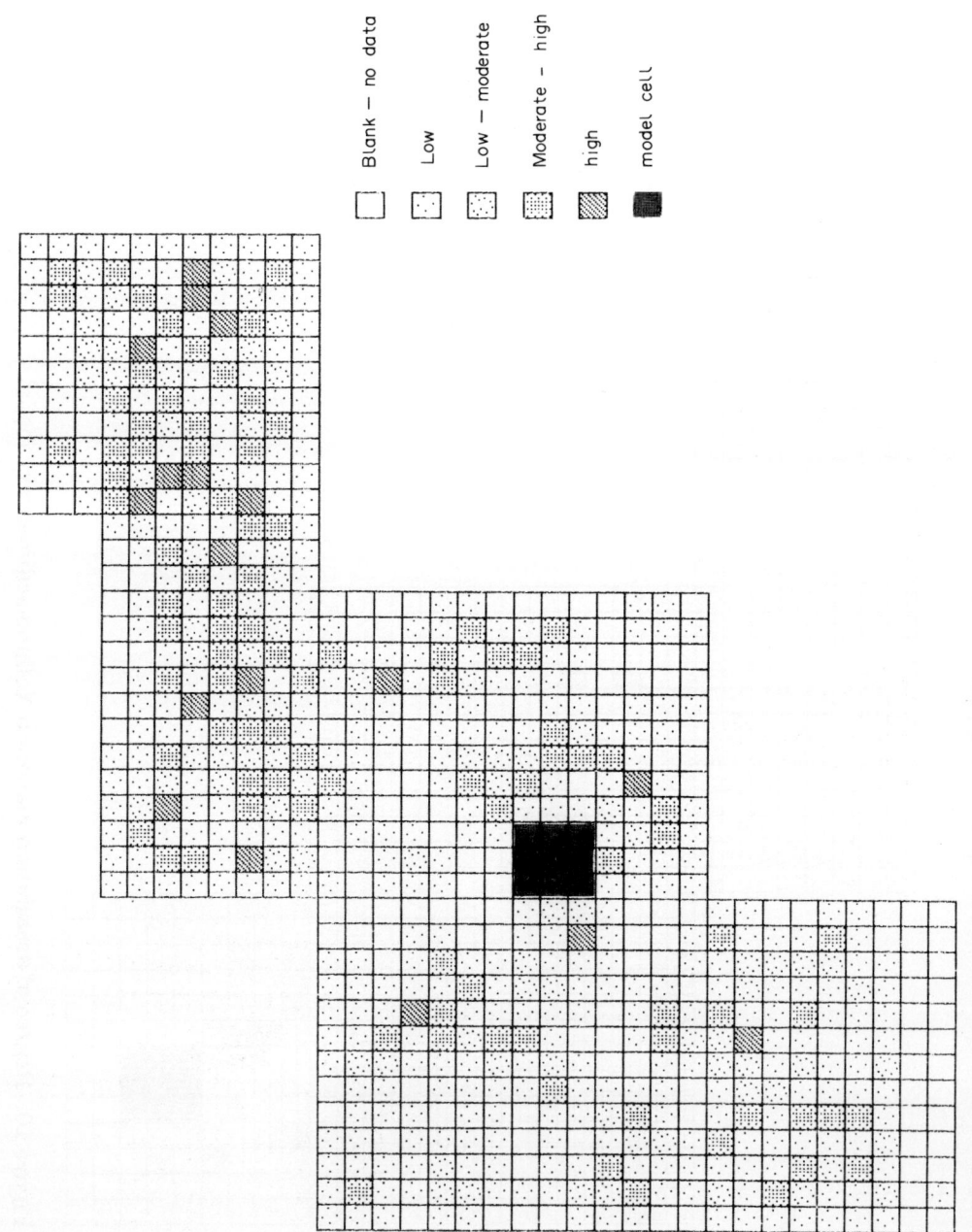

Figure 11. Degree of association map with granoid model.

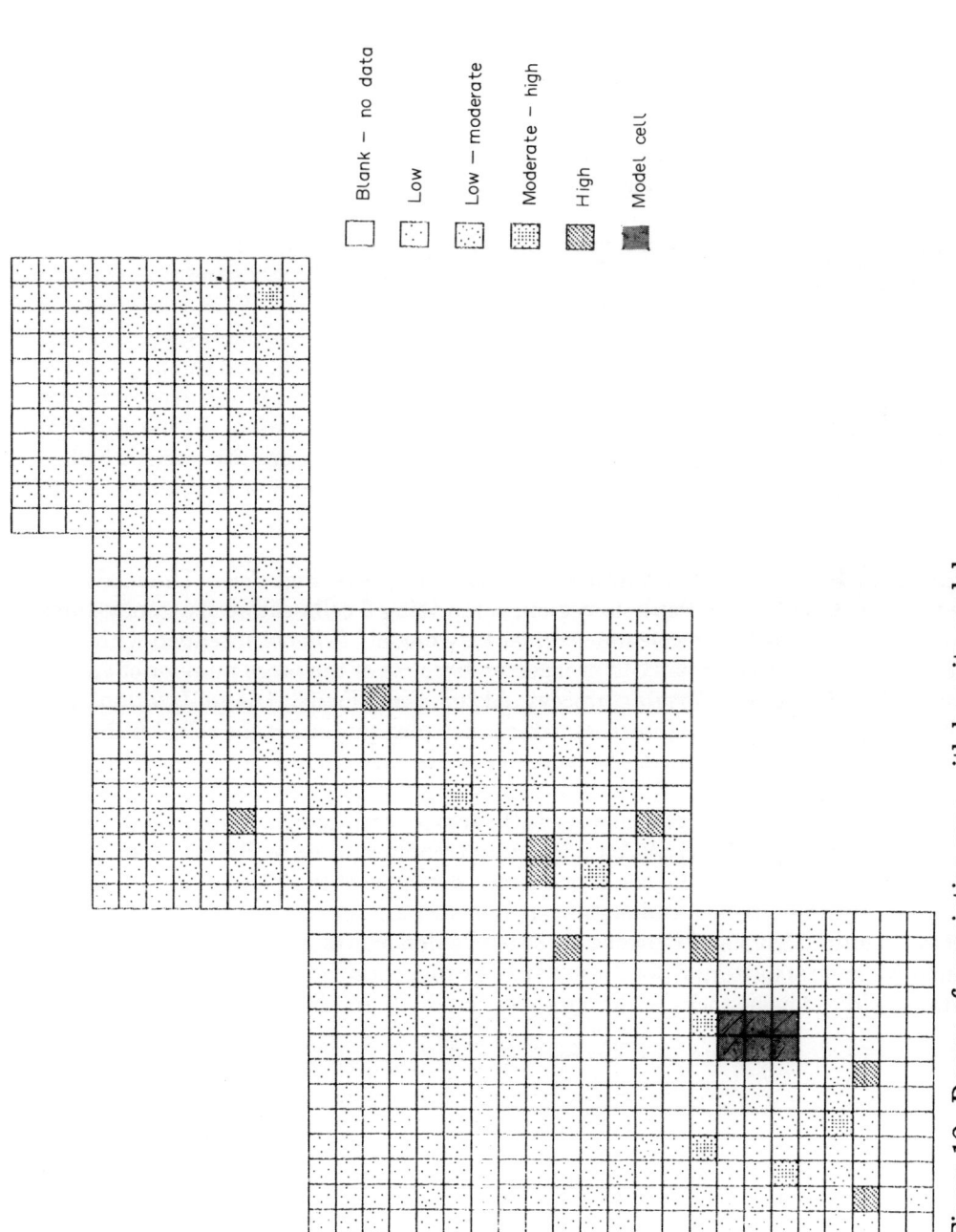

USE OF CHARACTERISTIC ANALYSIS IN MINERAL RESOURCES APPRAISAL 179

Figure 12. Degree of association map with bauxite model.

Volcano-Sedimentary Model (Fig.10) — It comprised nine models cells and reached the highest consistence degree, indicating a good consistency of the model, that is, a suitable equilibrium between the number of model cells and the variables. In this model the variables with a higher weight were lithologies of Sao Roque Group, Ni, Mn, Pb, Mo, Fe, Cu, and V. The model cells lie nearby the Itaberaba region described by Coutinho and others (1982), being considered as a good potential area for volcanogenic exalative deposits.

A trend striking NE is observed in bidimentional representation of cells of this model, which may indicate a continuity of this volcano-sedimentary formation described by the authors. This is confirmed by the concentrations of Co, Cu, Ni, Cr, Mo, and V, where the geochemical reliefs coincide with the cells generated by the model and the contour with Sao Roque Group limits. The NE areas nearby the model cells are being subject of geological exploration by IPT to investigate base metals (Cu, Pb, Zn, Ni, and Co). Primary gold mineralization associated with tuffaceous rocks and metachaert layers within a volcano-sedimentary sequence, also have been recognized.

Model for Granitoid Rocks (Fig. 11) — In this model the variables with higher weights were: granitoid rock, Cr, V, Pb, Ti, La, Y, Zr, Mo, Nb, and the migmatitic rocks. Most of the cells concentrated in the northern portions of the area, indicating a good possibility for the previously proposed metallic association. In the northwestern zone there are occurrences of sulphides associated with granitic rocks which coincide with the domain #6 (Fig. 6), causing an enhancement of anomalies of Cu, Zr, La, and Y.

Model for Bauxite and Associated Mineralizations (Fig. 12) — The model cells lie in deposits of Nazare Paulista region, comprising 6 cells with consistency index equal to 0.56, close to the volcano-sedimentary model. It is not intended with this model to select areas for bauxite prospecton because of the large number of variables involved in the bauxite-forming process (Al_2O_3, SiO_2, Fe_2O_3, Loss of Ignition) and other geomorphological parameters, which also play an important role, but have not been incorporated in the model because of the scarcity of data throughout the study area.

The results of the characteristic were satisfactory, with special emphasis to the NE cell which is along the NE-SW bauxitization trend proposed by Suszczinski (1978). It also is notable in this model the conspicuous presence of the Sao Roque Group lithologies, mainly amphibolitic and metabasis rocks where the bauxite-forming process acted "in situ". Silva and Oliveira (1988) have demonstrated that the REE (Rare-Earth Elements) pattern from these bauxites suggesting deep-sea ophiolitic character of these altered basic rocks with concentration of Ce and Eu.

CONCLUSIONS

The use of characteristic analysis reveals a powerful tool for the treatment and analysis method do not provide adequate answers for spatial variations of geological/geochemical variables. This constraint has serous implications where the geochemical contrasts are low and the information shows a high heterogeneity and different quality levels. The use of second derivatives make it possible toby pass such constraints, because it is reliable technique and it is not necessary to submit to statistical hypothesis restrictions.

The integration with quantitative auxiliary technique through interactive system is a essential requirement for a good performance of characteristic analysis to achieve a high number of alternatives for proposed models. The cell size represents essential parameters that could influence the final results. The geostatistical methods play an important role in furnishing optimum dimension criteria.

The models represent an attempt to define some geochemical and metallogenic patterns in areas with low-geochemical contrast and few mineral occurrences.

The results obtained in this work were not indicative for the selection of favorable areas concerning mineralizations associated with basic and ultrabasic rocks and bauxite deposits. Nevertheless, they have pointed to a trend, striking NE, which includes the favorable areas for volcano-sedimentary deposits and those associated with granite-pegmatite rocks.

ACKNOWLEDGMENTS

The authors are grateful to PRO-MINERIO/SCT-SP for the financial support for this work and CNPq-Conselho Nacional de Desenvolvimento Cientifico e Technologico for the travelling facilities to participate in the COGEDATA Symposium in Helsinki. Dr. J.M. Botbol (USGS) provided some early help in the concepts of characteristic analysis. J.M. Quintanilha helped in the concept of statistical analysis and the implementation of some computer programs. Help in the English version of the manuscript by Professors R.P. Xavier and A. Chouduri is much appreciated.

REFERENCES

Botbol, J.M., 1970, A model way to analyse the mineralogy of base metal mining districts: Soc. Min. Engl., March, p. 56-69.

Botbol, J.M., and others, 1978, A regionalized mutivariate approach to target selection in geochemical exploration: Econ. Geology, v. 73, no. 4, p. 534-546.

Coutinho, J.M.V., and others, 1982, Geology and petrology of volcano-sedimentary sequence of Sao Roque Group in the Serra de Iaberaba, SP., *in* 32nd Congresso Brasileiro de Geologia, Salvador, Salvador, SBG. v. 2, p. 624-640 (in Portuguese).

Figueiredo, B., and Suslick, S.B., 1988, Computer-based files on mineral deposits in Brazil - The IDEM data system in Sao Paulo State. International Symposium on Computer Applications in Resource Exploration: Prediction and Assessment for Petroleum, Metals and Nonmetals, 21-23 July 1988, Espoo (Helsinki) Finland (in press).

Gaál, G., 1984, Evaluation of the mineral resource potential of Central Finish Lapland by statistical analysis of geological, geochemical and geophysical data: Report of the IGCP Program: Geol. Survey Finland, Rept. Invest. 63, 69 p.

Garrett, R.G., 1969, The determination of sampling and analytical errors in exploration: Econ. Geology, v. 64, no. 5, p. 282-283.

Journel, A.G., and Huijbregts, J., 1978, Mining geoestatistics: Academic Press, London, 600 p.

Kaplan, W., 1972, Advanced calculus, v. 1: Edgar Blucher Ed., Sao Paulo, 200 p. (in Portuguese).

Matheron, G., 1970, La theorie des variables regionalisees it ses applications: Les Cahiers du Centre de Morphologie Mathematique de Fontainebleau 5, ENSMP Paris, 212 p.

McCammon, R.B., and others, 1979, Drill-site favorability for concealed porphyry copper prospect, Row Canyon, Nevada, based on characteristic analysis of geochemical anomalies: Soc. Min. Eng. Fall Mtg. (Preprint 79-311) p. 1-7.

McCammon, R.B., and others, 1983, Characteristic analysis, 1981: Final program and a possible discovery: Jour. Math. Geology, v. 15, no. 1, p. 59-83.

PRO-MINERIO, 1985, Regional Geochemical Program. Programa de Desenvolvimento de Recursos Minerais: SICCT Sao Paulo, Brazil, 12 p. (in Portuguese).

Quintanilha, J.A., and Suslick, S.B., 1982, *Precision controls in paired geochemical samples*. An methodological approach *in*: 33rd Congresso Brasileiro de Geologia, Rio de Janeiro, 1984, Anais... Rio de Janeiro, v. 7, p. 1912-1920 (in Portuguese).

Quintanilha, J.A., Suslick, S.B., and Lewis Jr., R.W., 1984, Quantification of sampling parameters in geochemical prospecting, *in*: 33rd Congresso Brasileiro de Geologia, Rio de Janeiro, 1984, Anais...Rio de Janeiro, v. 7, p. 4871-4891 (in Portuguese).

Silva, M.L.M.C., and Oliveira, S.M.B., 1988, Ferruginous bauxite from Nazare Paulista/Sao Paulo - Brazil: Geochemical Evolution. VI International Congress of ICSOBA. Sao Paulo. 1988, in press.

Suslick, S.B., and Quintanilha, J. A., 1982, Programming library of geological data analysis version 2, Sao Paulo: IPT/SMGA, Internal Report, 180 p. (in Portuguese).

Suslick, S.B., Quintanilha, J.A., and Lewis Jr., R.W., 1984, Exploratory data analysis applied to geochemical data, *in*: 33rd Congresso Brasileiro de Geologia, Rio de Janeiro, 1984, Anais... Rio de Janeiro, SBG. p. 4864-4876 (in Portuguese).

Suslick, S.B., 1986, Mineral resources appraisal based in geochemical data: a study applied to Itaquaquecetuba, Piracaia, Igarata, Camanducaia e Monteiro Lobato quadrangles, Sao Paulo: unpubl. doctoral dissertation, IG-USP, 304 p. (in Portuguese).

Suszczynski, E.F., 1978, New bauxite ore deposits in the oriental portion of the Brazilian shield, *in*: 4th International congress of ICSOBA, Athens (Greece) 1978: *Proceeding* ... Athens, National Technical University of Athens/Dept. of Mineralogy, v. 2, p. 837-839.

Thompson, M., and Howarth, R.J., 1978, A new approach to the estimation of analytical precision: Jour. Geochemical Expl., v. 9, no. 1, p. 23-30.

INTERCRAST — The Technology for Prognosis and Quantitative Assessment of Mineralization in Regions of Intrusive Magmatism Based on Numerical Modeling

V. G. Zolotarev
VNII Zarubezhgeollogia, USSR Ministry of Geology, Moscow

ABSTRACT

The objectives of INTERCRAST are to forecast hidden mineralization around and inside intrusive plutons by two-dimensional models reflecting real geological structures of study areas in horizontal sections, computer modeling of thermal processes that accompany magma emplacement, and special techniques of data treatment.

The studies include: a paleoreconstruction of geological structure at a time of magma appearance; the analysis of the dynamics (in space and in time) of magma crystallization, of the development of temperature and energy fields, and zones of thermal fracturing (fluid paths); and calculation of quantitative thermal parameters which influenced metal migration and concentration. Anomalous thermal-energy regimes favorable for the formation of certain mineralization types are revealed and analyzed together with all quantitative and geological-geochemical data. Narrow zones potentially favorable to locate mineralization of corresponding metal (mineral) in them are mapped.

Linear dependencies between thermal energy resources of these zones during the period of metal concentration and metal resources provide the quantitative assessment of possible metal content. The final estimation needs test geochemical sampling.

INTERCRAST was applied in the scales from 1:100 to 1:500,000 to solve geochemical, ore, and metallogenic problems and to forecast U, Th, REE (granite and alkaline magmatism), Mo, Sn, W, Pb, Zn, Cu, Sb, and Au (granite and diorite magmatism) mineralizations of various genetic types related and nonrelated to intrusive events, as well as to study how younger intrusives affect the existing ore bodies.

TYPES OF ORE-LOCALIZING CONDITIONS

A wide spectrum of hydrothermal ore deposits associated with the intrusive magmatism is the evidence in favor of a high variety of the conditions of metal deposition and of the relation between ores and intrusive bodies. Several ore-localizing conditions can be defined on the basis of factual data and modeling of thermal processes in intrusive systems.

The intra-intrusive ore-localizing conditions occur inside cooling plutons when metals separate into fluids in ore-producing magmas. Initial deposition of metals occur under an orthomagmatic regime which is replaced later by a regime of convective fluid filtration. This late stage is characterized by the redeposition of metal concentrations which were formed earlier as well as by new metal associations brought up from other sources.

The extra-intrusive ore-localizing conditions in the roof area (in endocontacts and exocontacts) are determined by the participation of magmatic and meteoric fluids during the early stage of ore formation in a contrasting thermal gradient; the proportions between fluids of different origin may vary significantly. This is influenced by a magma type, its metal and fluid content, by the course of differentiation and crystallization processes. The late stage of ore formation is governed by fluids of a convective system.

The extra-intrusive ore-localizing conditions in the side endocontacts and exocontacts of plutons differ from the previous one by the subordinate role of melt differentiation and, consequently, by the reduced accumulation of metal-bearing fluids and volatiles. Ore formation is related mainly to convective fluid circulation that begins in host rocks and in solidified parts of plutons in parallel to the crystallization of magma chamber and simultaneously with the development of fissure zones. In both extra-intrusive conditions various deposits of Sn, W, Mo, Cu, U, Pb, Zn, etc. of the porphyry, stockwork, or vein type are formed. Mineralization is of greisen, skarn, and hydrothermal genetic types including remobilization; the latter concerns the process of the formation of vein ore (Zn, Pb, W, U, REE) in the thermal field of an intrusion from the existed stratiform disseminated mineralization. Correspondingly, different mechanisms of ore formation are of importance in each particular situation, and various factors of ore location influence the spatial distribution of the deposits.

The outside-intrusive ore-localizing conditions concern the deposits which were formed in nonimmediate proximity to the intrusive contact. These deposits may be regarded by geologists as related to the particular pluton or as having no relation to it. Such definition is characteristic to the traditional geological approach by which the presence of obvious links between the ores and the pluton are demanded; the varieties in the manner of influence of igneous intrusions in sedimentary, volcanic, or metamorphic rocks on the processes of mobilization, transportation, and concentration of metals are not taken into account. That is why many deposits of Au, Pb, and

Zn, Sb, As, etc., which were formed in different temperature ranges, were described as those related to small intrusive bodies, or as those of telethermal type, or as the deposits of "unclear formational belonging". The search for the latter ones might be carried out on the basis of wrong prediction criteria.

A group of volcanogenic massive sulphide-polymetallic deposits, which were affected thermally by younger subvolcanic or hypabyssal intrusions also may be regarded, despite paradoxicality, as those associated with the intrusive magmatism. Such relations are peculiar to some of the deposits of Rudny Altai (USSR), Finland (the Vihanti deposit), and Sweden. The intrusion influence results in metamorphic transformation of the existed ore bodies which can lead up to total dissapearance of primary features and to the formation of remobilized polymetallic veinlet mineralization younger than the intrusive body. In other situations primary ores can be preserved close to the intrusive contacts.

Among thermal regimes that influence the location of ores in intrusive systems are the following.

(1) The crystallization dynamics expressed in the successive movement of meltrock phase boundary in the melt chamber.
(2) The dynamics and quantitative parameters of thermal gradient fields; temperatures, temperature gradients, rates of the movement of isotherms, and the maxima distances at which high temperature isotherms advanced into surrounding rocks. Certain mineral paragenetic associations of ores are formed either in the narrow temperature ranges (temperature optima) or under temperature changes in a wide range but at fixed thermal gradients. The latter is peculiar to certain types of Au mineralization (Buryak, 1983; Shilo, Goncharov, and Sidorov, 1979; Zolotarev, 1984a).
(3) The dynamics of energy fields and the energy provision of metal concentration and ore deposition which determine the amount of metal resources and is related to the energy capacity of magmatic chamber (Smirnov, 1981; Zolotarev 1982).
(4) Development of zones of micro- and macrofissures covering large areas of plutons and of host rocks (Titley and others, 1986) that determine fluid paths and the location of ores. The nature of such fracturing is debatable, and different authors suggest a different genesis. What is certain is its role in ore location and the development during early and late stages of the existence of magma chambers. The morphologies of these zones and the intensities of fracturing in them determine the interaction of circulating fluids with intrusive and host rocks, the movement of fluids along or across isotherms, the effects of pressure changes along fluid paths, and, consequently, physical-chemical changes in the solutions influencing their composition and deposition of minerals.

Thus, a large complex of labor-intensive researches has to be done for a studied region including a group of intrusive bodies, surrounding rocks, and ore deposits located there. It involves petrological, structural, geochemical, micromineralogical, and other studies of rocks and ores, as well as predictive metallogenic analysis. Altogether, much time and financial expenses are needed, but to carry out such researches full time and is impossible for large areas. This limitation in turn will affect the details of the forecast and, thus, will not result in the considerable decrease of expenses on exploration including drilling. Traditionally, the problem is solved by a wider application of methods of exploration geochemistry and geophysics which, nevertheless, do not bring light to the genetic aspects of the development of ore-forming processes and, thus, to their understanding.

THE INTERCRAST TECHNOLOGY

During the last few years the methods of computerized prediction and modeling have been applied widely to forecast and to evaluate metal resources. Among the INTERCRAST* technology, developed in VNII Zarubezhgeologia, has no analogies because of its orientation toward the reconstruction of the course of thermal processes in two-dimensional models which reflect real geological situations. It affords the study and analysis of the spatial temporal dynamics of crystallization of melt chambers, of the distribution of thermal fields, of the development of thermal fracturing that is produced by thermal stresses when they exceed the strength of rocks (up to 6-7 times), of the development of energy fields represented at certain moments of time, and as integrated energy values of units of an area for prolonged periods of time. Anomalous regimes, favorable for the formation of the mineralization of a particular type and genesis, thus are revealed. Comparison of calculated and factual data in well-known areas of the modeled region is a way to check how the factors of ore location which are revealed in the previous studies suit to this particular region and to ore deposits and ore manifestations in it. If these factors are valid (or the new ones are established) then a prognosis of the concrete potential metal-bearing sites can be made with a fine detection (the zones suggested for prospecting or exploration are 5 to 10 times more local than those determined by traditional geological methods, including exploration). The most distinctive features of INTERCRAST are the presentation of: (a) the visual understanding about the development of the processes within the whole region; (b) the possibility to undertake scientific examination of theoretical ideas how the factors of ore genesis worked at the concrete site of the region; (c) the results in a map form; and (d) the possibility to check quickly the predictive recommendations.

INTERCRAST was created from the Yu. I. Diomin's and M. S. Krass' program of conductive heat transfer and thermal balance, which was modified by V. M. Demenchuk and me, and from the special technique of data treatment and of the analysis of calculated data, developed by me. The latter part represents the combined study of all thermal processes in an intrusive system. Despite the fact that the previous studies carried out by the author included as well three-dimensional modeling and two-dimensional modeling of thermal processes in vertical sections of intrusive chambers of different shapes, the modeling aimed to predict ore-bearing sites should be done in horizontal two-dimensional models of the concrete objects; ore deposits, ore fields, and ore regions and provinces. Such an approach affords the

---

* INtrusive TEmperatuRe CRystallization ANistrophy STructure

most reliable geological data input to a model and to verify the results. This modeling is truthful when intrusive contacts are vertical or steep. If contrary (when the side contacts slope gently) then a model should be done either for a deeper horizon or in a vertical section. In the latter situation more local studies can be done, otherwise they should be regraded as the test ones. Numerical modeling requires high-speed computer operations and memory. The program is in ALGOL and PASCAL and is compatible with Soviet BESM-6 and EC computers, as well as with IBM.

The conductive modeling of thermal processes in horizontal two-dimensional models is the only one possible because convection cannot be realized in them. The special analysis and the review of the data on the convective and the conductive heat transfer in vertical models (Cathles, 1977; Norton, 1978, 1979; Parmentier and Schedl, 1981; Kolstad and McGetchin, 1978), on the unpredictable changes in rock permeability under high P-T conditions (Vitovtova and Shmonov, 1982), on the counter effects on cooling rates produced by filtration of fluids and by the decrease in thermal conductivity of rocks with the increase of temperature as well as the comparison of calculated temperatures and real temperatures of mineral-paragenetic associations of ores and hornfels, which showed their close correspondence, have been carried out. Both of them justify such simplification of calculation model. Heat transfer along the vertical axis: the outflow from the given horizon in upward direction and the inflow onto it from the depth both can be neglected. Different models with different initial and boundary conditions which represent magma chambers in vertical and horizontal sections (two-dimensional and three-dimensional models) were compared by their main characteristics (eg., time of crystallization, etc.). One can see that the cooling through the Earth's surface was unimportant during the most significant period for ore formation; the period, when contrasting thermal gradients existed.

The variety of genetic and predictive problems solved by INTERCRAST is illustrated by several examples: the REE-Th-U Bear Lodge deposits (USA), the vein Zn Saint-Salvy deposit (France), the Cu-Mo porphyry Tekhute deposit (USSR), and the multimetallic ore province in the USSR.

The porphyry and successive vein mineralization at the Bear Lodge is located inside the massive part of phonolit-trachyte intrusive body of Tertiary age. It intruded Phanerozoic platform carbonate and clastic rocks and underlying Archean granites and schists. The question why these largest Th-REE deposits of a new porphyry type have been formed there received no solution after the detailed geological-geochemical works carried out by American scientists (Staatz, 1983). It was shown by modeling that the mineralization has been formed because of the successive-simultaneous action of several factors. The area of metal concentration corresponds to the area of residual magma chamber, the rates of crystallization in which were uniform in all the directions and equal to 3.6-5.4 cm/year, in contrast to highly uneven crystallization during the early period after the magma emplacement and the formation of the entire chamber. The same area during both magmatic and post-magmatic stages was under conditions of maxima temperatures and thermal energies and of sharp minimum (along and across the intrusive body) heat flow. The latter probably was the cause why metals were not disseminated from the melt and from the solid rock. The concentration first of rare earths, then of thorium, and later of uranium occurred in a decreasing temperature. During this period the isotherms changed their morphology and the center of maxima temperatures migrated on a 470 m distance. The location of the mineralization of each of the metals is determined by the morphology of isolines of energy fields at corresponding time periods. The resources of REE and Th, calculated for 73 grid blocks (cells), are in direct correlation with the energy resources of these blocks during the periods of concentration of REE and Th. The amount of thermal energy decreased from the earlier to the later period of concentration of Ree, Th, and U together with the decrease of metal content in this row. The gradients of REE resources are correlated with the gradients of energy resources of the period of REE concentration. The vein mineralization, which is mineralogically and geochemically similar to the porphyry one, has been formed by redeposition of metals from the porphyry mineralization by fluids which circulated in a residual thermal field. This statement is proved by the similarity in geochemical zoning in vein mineralization (U/Th, Ce/Y, Yb/La, Gd/Sm) and in orientations of zones of Th depletion in the porphyry mineralization and of isotherms of the residual thermal field. The prediction includes the determination of the main factors of the location of the porphyry ores which can be applied to locate similar deposits, of local sites enriched by a certain element, and of probable concentration of U outside the Bear Lodge intrusive body.

The vein mineralization of the Saint-Salvy deposit was formed, according to French geologists (Foglierini and others, 1980; Barbanson and Geldron, 1983), from the disseminated Cambrian sedimentary Zn mineralization under the influence of Permian Sidobre granite (in its exocontact). It was shown by modeling that the Saint-Salvy fissure-vein structure has been formed inside the zone of thermal fracturing. The isothermal regime 400°C existed at the deposit site during first 75,000 years after the granite emplacement. During this time metals could migrate by diffusion as far as 160m if diffusion coefficient, D, was 2.72×10^{-5} cm^2 s-1 (Zolotarev, 1984; Korzhinsky, 1982). Simultaneously, pore fluids could migrate in opposite direction because of the laws of convection thus limiting the distance of metal dispersion from granites and then they could move along the zone of thermal fracturing. Favorable temperature and structural conditions enabled deposition of metals in the vein structure. The axes of maxima thermal stresses coinside with the compression axes reconstructed from the orientations of conjugated fissures and faults. The prediction includes the sites at which the zone of isothermal regime coinsided with the Saint-Salvy structure, especially with its western flank [this conclusion by Zolotarev (1985) was confirmed later by geochemical exploration studies by Verraes (1987)].

The Tekhute Mo-Cu porphyry deposit was formed (Zolotrev, Rusadze, and Tvalchrelidze, 1987) in two stages: (1) in thermal gradient of quartz diorite Shnokh-Kohb pluton the ore-bearing structure of the entire ore field was installed and the concentration of metal occurred in a background of wide quartz-feldspar alterations; (2) in a thermal gradient of small intrusive bodies (the vein phase of the pluton) new narrow zones of fracturing and related to them quartz-sericite alteration occurred causing the redeposition of metals and their recurrent concentration. Ore location was the result of the development of zones of thermal fracturing and to energy maxima inside them. Energy values are correspondent quantitatively to the resources of metals. By maximum values of correlation coefficients between the energy and metal resources (the former were calculated for a number of periods: from 0-70,000 to 1-300,000 years for the first stage and from 0-300 to 0-1100 years for the second stage) the duration of early and late concentration events was estimated as 200,000 and 625 years, correspondingly. The axes of maxima thermal stresses coincided with the compression axes determined by our predecessors (Kekelia, Chichinadze, and Starostin, 1985) at the Tekhute ore field. Predictive recommendations reduce the area to be explored in 7.6 times only at the ore field.

Modeling of the thermal fields of batholithic granite intrusions in a large polymetallic province lead to understanding the relations between the location of ore-bearing zones and ore fields and the position of zones of anomalous regime: the isothermal (constant temperatures) ones and the stationary thermal gradient (fixed temperature gradients) ones. The latter provided Au concentration, whereas under different temperatures in the former ones Hg and Sb, Pb and Zn, Mo and some other rare metals were accumulated. New possibilities for the metallogenic analysis thus were received, and the size of the areas which were perspective to locate mineralization in them was reduced in more than 5 times in comparison with the previous studies.

In these situations, prediction of the potential ore-bearing sites enables a reduction of exploration works at least at one of the exploration stages and, thus, saving of time and financial expenses. The usage of INTERCRAST leads first of all to a deeper understanding by a field geologist and by a researcher of the course of thermal processes, of the peculiarities of their development in a studied region and in its local sites, and thus to scientific bases of ore-controlling factors in predictive purposes.

SUMMARY

The existence of direct correlations between the resources of thermal energy and the resources of metals allows:
(1) an evaluation of amounts of metals which were concentrated at an already known site of interest;
(2) an evaluation of the metal endowment at the flanks of a deposit if they were insufficiently explored;
(3) an evaluation of potential resources of metals for the predicted sites suggested for exploration by the usage of the INTERCRAST technology. The latter is provided by a diagram of linear dependency between metal resources in ore deposits (those mentioned here and others) and energy resources calculated for them. When energy resources are established for a new site, potential metal resources can be estimated from the diagram. The final assessment needs test geochemical sampling at a number of profiles across the predicted zones. Thus, quantitative assessment by INTERCRAST which is given for a local site can be checked easily. The most important feature is the exact linear correspondence of energy/metal dependency among ore deposits of different genetic and mineral types.

Depending on the scale of the research and the size of a geological object studied by modeling, INTERCRAST makes it possible to solve either purely geochemical (the analysis of the behavior and distribution of metals) and ore (the formation and the location of deposits) problems, or metallogenic ones. It was tested at scales of 1:100, 1:1000, 2:10,000, 1:24,000, 1:50,000, 1:59,000, 1:100,000, and 1:500,000. Besides the experience in solving genetic problems of ore formation and in predicting mineralization of U, Th, REE (granite and alcaline magmatism), Mo, Sn, W, Pb, Zn, Cu, Sb, Au (granite and diorite magmatism), as well as in determining the conditions of preservation of primary massive sulphide ores in thermal fields of younger intrusions and of formation of remobilized mineralization, other types of ore deposits which are known in intrusive systems can be predicted. This is guaranteed by a complete verification of effects of thermal processes on ore formation in the terms of scientific analysis.

INTERCRAST provides a researcher by the exclusive data that cannot be received by geological and geochemical methods, and by a possibility to solve efficiently the problem of the prediction of potential ore-bearing site in regions of intrusive magmatism.

REFERENCES

Barbanson, L., and Geldron, A., 1983, Distribution du germanium, de l'argent et du cadmium entre les schistes et les mineralisation stratiformes et filoniennes a blende-sidérite de la region de Saint-Salvy (Tarn): Chron. rech. miniere, v.51, no. 470, p. 33-420.

Buryak, V.A., 1983, The genetic model of metamorphogenic-hydrothermal ore formation in: Genetic models of endogenous ore associations, v. 2: Nauka Publ., Novosibirsk, p. 139-145 (in Russian).

Cathles, L.M., 1977, An analysis of the cooling of intrusives by ground-water convection which includes boiling: Econ. Geology, v. 72, no.5, p. 804-826.

Fogliérini, F., Beziat, P., Tollon, F. and others, 1980, Le gisement filonien de Noailhac-Saint-Salvy (Tarn): Zn (Ag, Ge, Pb, Cd): Congrès Géol. Intern. (Paris) Publ. 26-e, 43 p.

Kekelia, S.A., Chichinadze, L.L., and Starostin, V.I., 1985, Geological-structural aspects of the location of porphyry-copper mineralization within a massive sulphide bearing province: Geol. Ore Deposits, No. 7, p. 71-79 (in Russian).

Kolstad, C.D., and McGetchin, T.R., 1978, Thermal evolution models for the Valles caldera with reference to a hot-dry-rock geothermal experiment: Jour. Volcan. and Geotherm. Res., No. 1-2, p. 197-218.

Korzhinsky, D.S., 1982, The theory of metasomatic zoning: Nauka Publ. Moscow, 104 p. (in Russian).

Norton, D., 1978, Sourcelines, source regions, and pathlines for fluids in hydrothermal systems related to cooling plutons: Econ. Geology, v. 73, no. 1, p. 21-28.

Norton, D., 1979, Transport phenomena in hydrothermal systems: the redistribution of chemical compents around cooling magmas: Bull. minér., v. 102, no. 5-6, p. 471-486.

Parmentier, E.M., and Schedl, A., 1981, Thermal aureoles of igneous intrusions: some possible indications of hydrothermal convection cooling: Jour. Geology, v. 89, no. 1, p. 1-22.

Shilo, N.A., Goncharov, V.I., and Sidorov, A.A., 1979, Physical-chemical zoning of hydrothermal ore-forming systems in marginal volcanic belts (abst.) : 14th Pacific scientific congress, Khabarovsk, 1979, Section B6, Moscow, p. 57-59 (in Russian).

Smirnov, V.I., 1981, The energy basis of post-magmatic ore formation: Geol. Ore Deposits, No. 1, p. 5-17 (in Russian).

Staatz, M.H., 1983, Geology and description of thorium and rare-earth deposits in the southern Bear Lodge Mountains, northeastern Wyoming: U.S. Geol. Survey Prof. Paper, 1049-D, 52 p.

Titley, S.R., Thompson, R.C., Haynes, F.M., and others, 1986, Evolution of fractures and alteration in the Sierrita-Esperanza hydrothermal system, Pima County, Arizona: Econ. Geology, v. 81, no. 2, p. 343-370.

Verraes, G., 1987, Exemple d'utilisation combinée de la mercurométrie et de la geochimie en roches pour la détection de colonnes minéralisées du filon à Zn-Pb de Saint-Salvy (Tarn), partiellement repouvert par des depots tertiaires: En: 12 IGES - 4 SMGP Resumés. Orléans, p. 111-112.

Vitovtova, V.M., and Shmonov, V.M., 1982, Permeability of rocks under the pressures up to 2000 kg/cm^2 and temperatures up to 600°C: USSR Acad. Sci. Repts., v. 266, no. 5, p. 1244-1248 (in Russian).

Zolotarev, V.G., 1982, Thermal energy conditions and the emplacement dynamics of intrusive massifs in connection with post-magmatic ore formation: Geol. Ore Deposts, No. 6, p 39-51 (in Russian). In English see: Intern. Geol. Review, 1984, v. 26, no. 4, p 445-455.

Zolotarev, V.G., 1984a, The peculiarities in the location of gold mineralization during the progressive stage of the development of thermal fields of granitoids: USSR Acad. Sci. Repts., v. 278, no. 2, p. 421-425 (in Russian).

Zolotarev, V.G., 1984b, Formation of geochemical fields of concentration in the thermal gradient fileds of intrusive bodies: Jour. Geochem. Explor., v. 21, no. 1-3, p. 229-237 (in Russian).

Zolotarev, V.G., 1985, Spatial-temporal dynamics of processes related to granite consolidation and their role in ore formation (an application of numerical modeling): Sci. de la Terre, Inform. Géol., no. 23, p. 27-48.

Zolotarev, V.G., Rusadze, V.G., and Tvalchrelidze, A.G., 1987, Thermal-structural and thermal energy factors in the formation of the Tekhute Mo-Cu porphyry deposit: Geol. Ore Deposits, no. 3, p. 60-70 (in Russian).

DATA INTEGRATION IN MINERAL EXPLORATION BY IMAGE PROCESSING AND OTHER TECHNIQUES

Use of Image Prrocessing and Integrated Analysis in Exploration by Outckumpu Oy, Finland

Jussi Aarnisalo
Outokumpu Oy Exploration, Outokumpu, Finland

ABSTRACT

The digital image-processing and data-integration methods applied by Outokumpu Oy Exploration are aimed at displaying the variations in, and the structure of, the explorational data in the optimal way for final visual interpretations. The integration of different geodata sets is performed either digitally, optically, or manually, aiming at the most efficient and economic way of extracting necessary information. The processed data consist of both low-altitude airborne and ground geophysical data, of satellite imagery, as well as of some topographical, geochemical, and drilling-log data.

Satellite imagery is especially contributing to the revision and detection of regional structures. Again, the properly processed geophysical images reveal generally well both structural and lithological features of the bedrock and the integration of different data sets further enhances the discernibility of individual rock units. Especially principal components analysis of geophysical data, and image-processing and color compositing of the results, has been determined useful in distinguishing targets of differing lithological characteristics. This approach has in many instances substantially promoted the exploration of base and precious metals in Finland. For example, it has been used successfully in the search for ultramafites potential for Ni-Cu-deposits, and in the exploration of Cu-Co-Zn and Zn-S massive sulphide deposits.

INTRODUCTION

The basis for digital image-processing and integrated analysis as an exploration tool in Finland was laid down mainly in the 1970s. Several major remote-sensing research projects were carried out at that time at universities, at the Technical Research Centre of Finland (VTT) and in the exploration organizations. For instance,

Outokumpu Oy Exploration participated in a number of these projects, providing ground survey data and other support necessary for the work. Consequently, since the mid-1970s Outokumpu Oy Exploration has been using remote sensing in regional exploration studies to aid the planning of exploration strategies and to guide the targeting of costly field operations (Aarnisalo and others, 1982).

The digital image-processing of data obtained with nonimaging systems, such as geophysical and geochemical methods, has been a subject of extensive research and development during the past decade (e.g. Anuta, 1977; Missalati, Prelat, and Lyon, 1979; Chavez and others, 1980; Arkimaa and Nikander, 1980; Kuosmanen, Kuivamaki, and Tuominen, 1981; Duval, 1983; Green, 1984; O'Sullivan, 1986; Kowalik and Glenn, 1987). In late 1979 Outokumpu Oy Exploration set up together with VTT and the Geological Survey of Finland a working group to develop and study the digital image-processing of low-altitude airborne geophysical data and its integration with LANDSAT imagery (Aarnisalo and others, 1982) (Fig. 1). The first images digitally processed and plotted on film from low-altitude airborne magnetic data were from the Rautuskyl_ test sitein western Lapland (Plate 1). The images were geologically so encouraging that the digital image-processing techniques were approved rapidly as a method of vizualizing low-altitude airborne geophysical data. Consequently, in 1981 - 1986 Outokumpu Oy and VTT worked together two development projects for the application of digital image-processing and integration methods in mineral exploration (Fig. 1). At the same time, these methods have been used to process various airborne, ground geophysical, geochemical, and drilling-log data, and also topographic data for the exploration of different types of base and precious metal deposits. To date several hundreds of explorational geodata images have been processed from about 100 target areas in Finland and abroad.

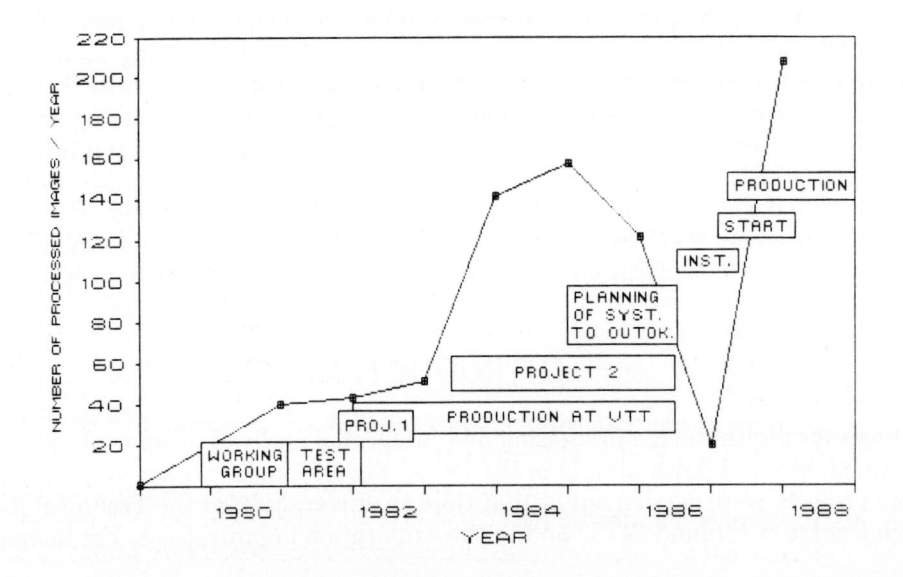

Figure 1. Development phases of digital image processing at Outokumpu Oy Exploration, and number of images processed annually.

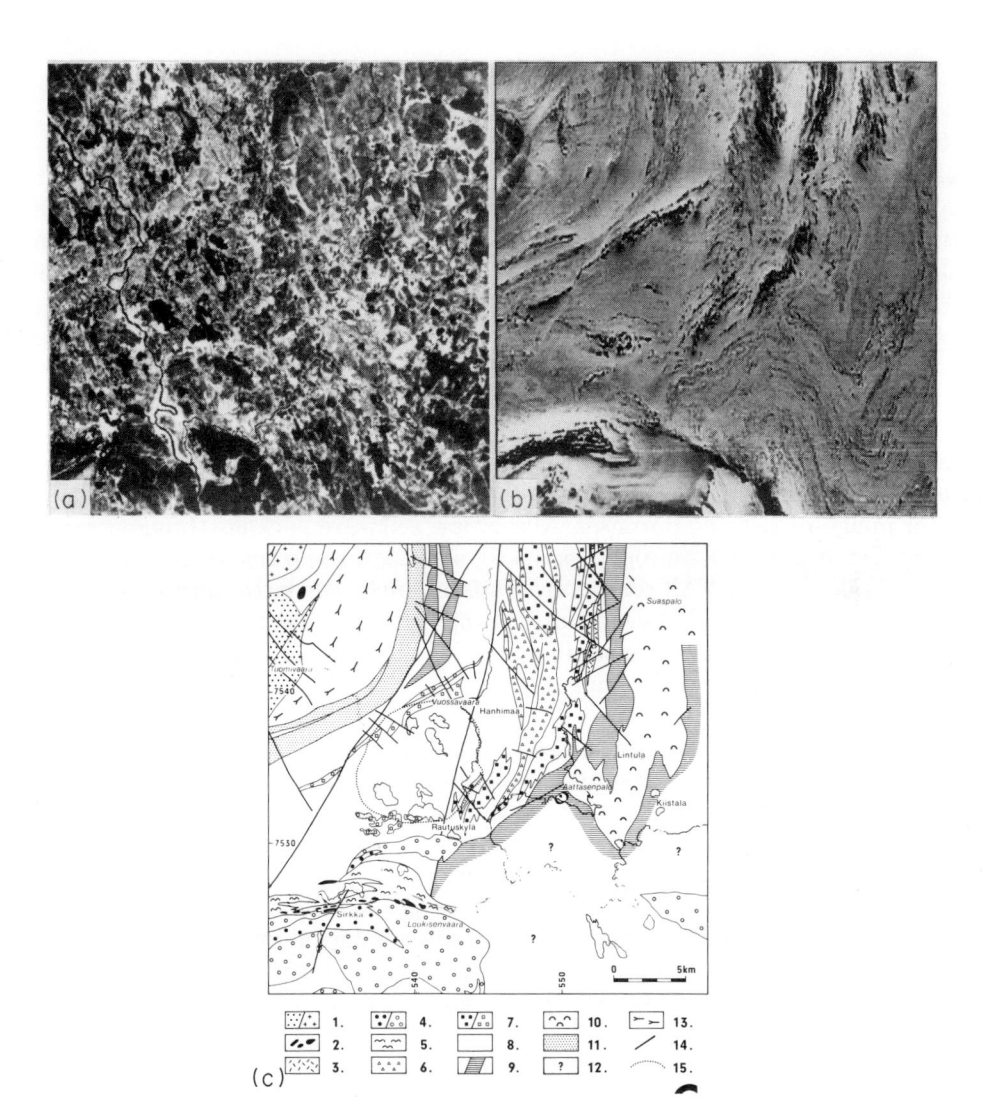

Plate 1. Comparison of LANDSAT and aeromagnetic data on the Rautuskylä area, 30 km x 30 km (Aarnisalo and others, 1982). A, LANDSAT MSS image, band 7 (see text). B, Image of low-altitude airborne magnetic data provided by GSF, digitally enhanced with high-pass convolution filtering and linear stretching. C, General bedrock map of Rautuskylä area, compiled by J. Eeronheimo. Intrusive rocks: 1 Tepasto granitoid complex / other granites; 2 ultramafic rocks; 3 quartz-feldspar porphyries. Jatulian: 4 Kumpu Formation; conglomerate/quartzite. Lapponium: 5 phyllite-tuffite formation; 6 Kellolaki Formation of intermediate pyro-clastics, cherts, and phyllites; 7 basic pyroclastics and jaspilites; 8 upper volcanite formation of basaltic and andesitic rocks; 9 cherts and black schists; 10 lower volcanite formation of basaltic lavas; 11 quartzite-mica schist formation; 12 undefined formations of Kittilä greenstone complex in general. Archaean basement; 13 granite gneiss complex. Structures: 14 fractures and faults; 15 ring structure interpreted from LANDSAT imagery (Aarnisalo and others, 1982, p. 50-53).

PROCESSING SYSTEMS USED

The first studies by Outokumpu Oy Exploration on the digital enhancement of LANDSAT data were conducted in late 1970's using the image-processing facilities - Nova computer and Comtal color display unit - then available at the VTT Laboratory of Land Use. In 1980 VTT installed a SADIE 2.0 digital image-processing system, developed by Thompson and others (1980) at the University of Minnesota, in their CDC CYBER 170 mainframe computer. In the course of the two joint research projects between Outokumpu Oy Exploration and VTT, the SADIE 2.0 system was improved by Franssila (1982), making it better for remote sensing, geophysical and geological exploration studies, and production runs.

In 1984, during the second joint development project, the VTT Laboratory of Land Use acquired a VAX 11/750 computer with Salora color display and B&W display units for image-processing purposes. Also a new software package (Image Processing Software (Imaged/Slip/Disimp), version 3.0), was acquired from CSIRONET, Australia, and installed in the VAX computer. Consequently, the latter part of the second joint development project between Outokumpu Oy Exploration and VTT involved transfer of the existing explorational applications to the new image-processing system, and their further development.

In 1985, Outokumpu Oy Exploration and VTT started planning an image-processing unit to be opened in the new premises in the town of Outokumpu, eastern Finland, when Outokumpu Oy Exploration moved there from Espoo in 1986. At the same time, Outokumpu Oy established a new subsidiary company named Tietokumpu Oy to take care of the increasing needs for computer services in the local organization units in the town of Outokumpu. The hardware purchase part of the original plans was completed finally in the summer of 1986 when a digital image-processing system was acquired for Tietokumpu Oy, and the necessary photographic laboratory for the Geoanalytical Laboratory of Outokumpu Oy Exploration.

The main units of the image-processing system are a VAX 11/750 computer, a Salora IPS-512 color image display unit, a Salora VPT-64 B&W display unit, and a Matrix QCR digital film recorder (Fig. 2). The software consists of the present version of the aforementioned Image Processing Software plus later developments, the MOVIE.BYU Version 4.2 - A General Purpose Computer Graphics System - package acquired from Brigham Young University, USA, and of the driver softwares for the QCR film recorder. Installation and calibration of the system were started in June 1986 and production level image-processing started at the beginning of 1987.

IMAGE PROCESSING AND INTEGRATION METHODS

Since the early 1970s the Finnish exploration organizations and the Geological Survey have conducted airborne geophysical surveys at low altitude (30 - 50 m) with a line spacing of either 125 or 200 m. The data points are placed 20 - 30 m apart except in gammaray spectrometric surveys, where the data are summarized over a

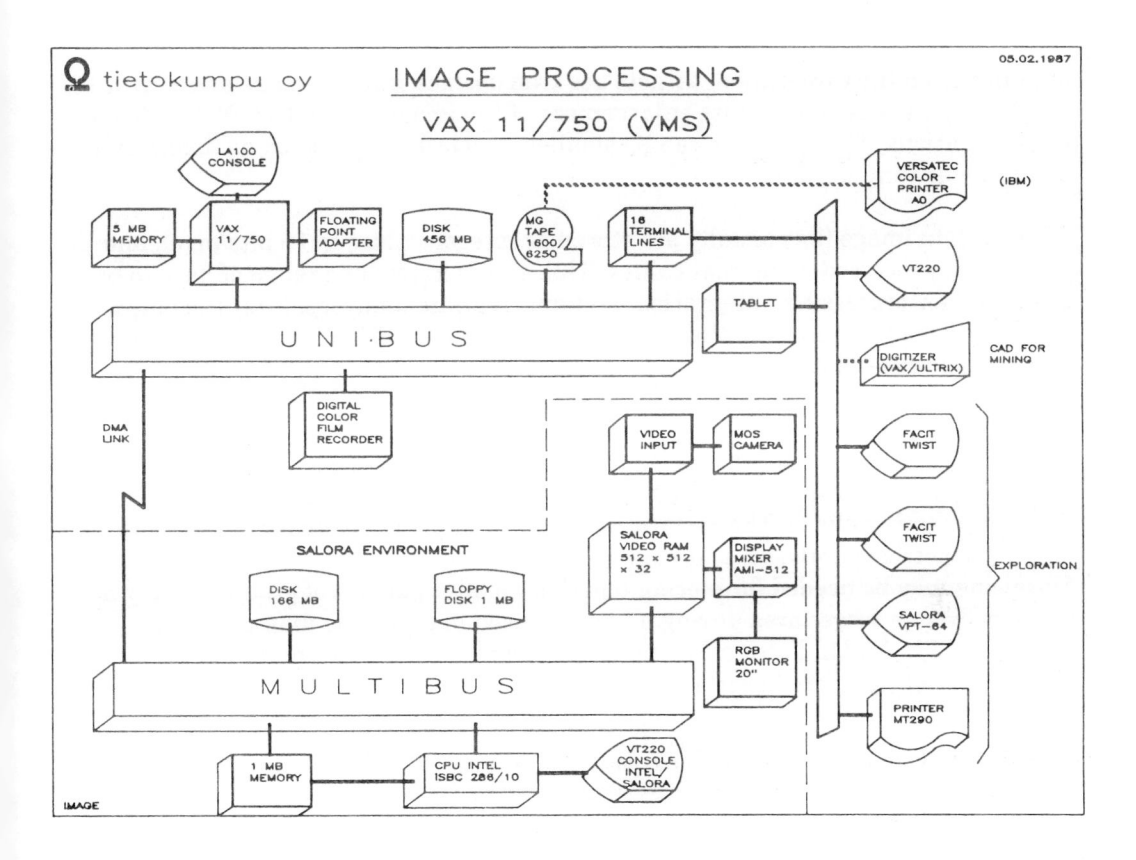

Figure 2. Configuration of image-processing system installed at Tietokumpu Oy, subsidiary of Outokumpu Oy.

distance of 50 -60 m along the traverse. Simultaneous magnetic, electromagnetic, and multichannel gammaray spectrometric measurements are recorded digitally (Lakanen and others, 1978; Multala and Vironmäki, 1981; Peltoniemi 1982) for computer processing. Nowadays even the data collected in systematic magnetic, electromagnetic, gravimetric, and other geophysical ground surveys from the exploration target areas are recorded digitally; earlier, manually collected data are digitized also for computer processing.

At Outokumpu Oy Exploration, the assay results of the geochemical sampling and diamond drilling are all entered in the main frame IBM computer and stored in the appropriate data files for future use. Similarly, the information on ore boulders, mineralized outcrops and explorationally interesting samples sent in by amateurs is stored in the appropriate data files in the computer.

Because all the digitally stored geodata are in map projection, they can be processed easily and plotted as symbol, contour, or color maps, or transformed into an image array for digital image-processing and integration with other data. Usually regional exploration data, such as airborne geophysical and the geographically rectified satellite data, are transformed into a uniform array with 50 m x 50 m pixel size for direct correlations and integrations. The ground geophysical and other detailed data obtained from the exploration targets are transformed into an image array with 10 m x 10 m pixel size to facilitate enlargement of the images up to 1:10,000 scale for field operations. The pixel size can be smaller in detail studies such as image-processing of the drilling-log data.

The geodata image files usually are stored as integer values in 16 bits/pixel, and the images are processed using real values. This allows a dynamic range of data to be preserved for processing, integration, and multivariate analyses. First the output images are transformed into the 8-bit integer mode for CRT display, film plotter or electrostatic color plotter.

Digital image-processing involves several steps depending on the objectives of the study. It is generally divided into four main phases (Fig. 3).The necessary image restorations, and mosaicing and generation of multichannel images are made during the preprocessing phase. The restoration procedures compensate for data errors, noise, geometric distortions, and map-projection transformations in the images. Mosaicing may be needed for piecing together smaller patches of ground survey data collected from a target area during a longer period of time. The map projection system used in Finland also may necessitate coordinate transformations and mosaicing of regional data sets in certain areas.

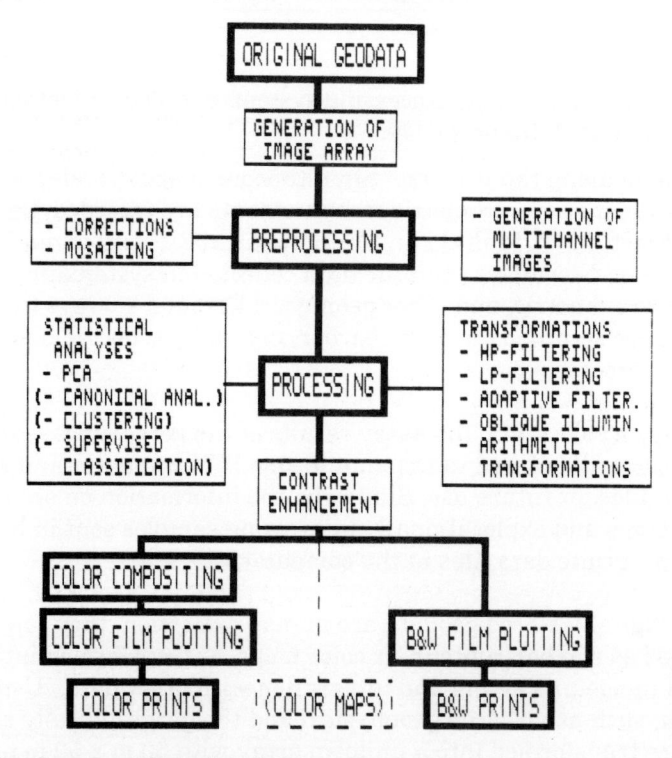

Figure 3. Main phases of digital image processing and some of procedures used by Outokumpu Oy Exploration.

The image-processing software used offers a wide variety of options for spatial filtering, normalizing, rationing, and other arithmetic transformations, oblique illumination, perspective viewing, color coding and compositing, and for multivariate analyses of geodata. Some of these processing methods are used for basic image enhancement, whereas others are applied in special studies for extracting explorationally interesting information from the available geodata.

The skewness of the frequency distributions of most of the geodata sets poses a major problem in the optimal contrast enhancement of the output images. The typical range of variation in the original values may be in the order of 1,000 or even 10,000 units, but the image-display units have only 256 density levels; films, printing papers, and the human eye can distinguish even fewer brightness levels or tones. Thus, several different types of contrast stretching function have been applied in efforts to optimize the output images for further visual interpretation.

In the final phase the images either are plotted on B&W film or interactively integrated into color composites using the color image-display unit and then plotted on color film. Thereafter the images are enlarged to prints of adequate scale for manual combinations and visual interpretations with other variously processed data sets.

The data for mineral exploration may be integrated in many ways besides digital image-processing, from the conventional manual to completely digital methods (e.g. Anuta, 1977; Aarnisalo, 1978; Proskuryakov, Negrutsa, and Yakovleva, 1977; Aarnisalo and others, 1982; Eliason, Donovan, and Chavez, 1983; Fabbri, 1984; Kuosmanen, Arjunaa, and Lindquist,1985). A scheme for the integrated use of explorational data as applied in Outokumpu Oy Exploration is shown in Figure. 4. Because the variability and characteristics of the data available for a certain integrated exploration study differ greatly, it usually is considered more practical and economical to used different types of data analysis and integration method on data sets differing markedly from each other. Figure 4 shows three different data processing and integration levels: digital with ADP and image-processing, photographic B&W and color processing, and manual with maps, photos, transparencies, and overlays. These may be applied in this order to combine data sets and interpretations of different origin. The procedure usually is carried out as teamwork between the geologists responsible for the exploration field work and drilling, and a number of specialists, for example a geophysicist, a geochemist, a remote-sensing and image-processing geologist ,and ADP personnel. The results are displayed as images,thematic maps, and diagrams, and in principle none of the original data used should be at odds with the geological interpretations made.

SATELLITE AND GEOPHYSICAL IMAGERY IN EXPLORATION

The aim of explorational applications of the digital image-processing conducted by Outokumpu Oy has been to display the structure of, and the variations in, the explorational data in the optimal way for final visual interpretations (Aarnisalo and others, 1982; Aarnisalo, 1984). However, other workers in Finland, or instance,

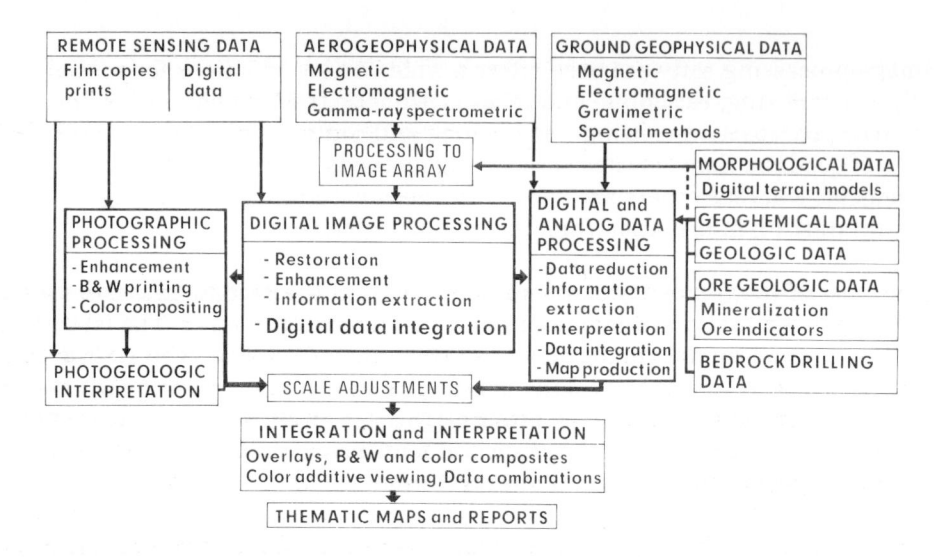

Figure 4. Approach to image processing and integrated analysis of geodata for mineral exploration.

Kilpelä and others, (1978), Arkimaa (1982) and Kuosmanen, Arkimaa, and Lindquist (1985) have determined that automatic interpretation methods, such as unsupervised and supervised classification, also are useful in geological mapping and in predicting favorable areas for base-metal deposits. In all these studies, satellite and geophysical images have been used as basic research material.

Satellite imagery

The applicability of digital image processing to the analysis of satellite and other digital scanner data is well established the world over. The basic processing techniques are widely known and include routines for compensating for data errors, correcting for atmospheric effects, noise and geometric distortions, and for conducting various image enhancement and information extraction procedures (see e.g. Sabins, 1978, Siegal and Gillespie, 1980).

Adequately processed satellite imagery permits the study of regional geological features that were difficult or even impossible to discern in earlier data. In Finland, satellite imagery has contributed to the revision and detection of regional structures, such as major fracture zones, ring structures, and lithologic/tectonic units (Tuominen and Aarnisalo, 1976; Tuominen and Kuosmanen, 1977; Talvitie, 1979; Vuorela, 1982; Aarnisalo and others, 1982; Kuosmanen, Arkimaa, and Lindquist, 1985). The wide and repetitive coverage and the mostly synoptic view of satellite images also has contributed to the study of Pleistocene glacial geomorphology and dynamics, which are important in efforts to trace the origin of ore floats (Punkari, 1982).

In more detailed exploration studies the usefulness of satellite imagery can be improved by interpreting the images together with other explorational data. In some instances this approach has substantially promoted the exploration of base metals in Finland (Aarnisalo and others, 1982).

Geophysical images

During the 1980s the use of digitally processed gray-tone images and color composites of both low-altitude airborne and ground geophysical data has gained wide acceptance among exploration geologists in Finland. Image-type presentation makes better use of the spatial and intensity resolution of geophysical data for visual interpretation than contoured maps. Both the bedrock and structural information are delineated easily from the images, which turn the geophysical data from readings into a more or less "geological image" of the area. In a regional context this image may have a marked resemblance to the geological features discernable on the outcrops in mesoscopic scale. Especially the structural components, such as interfering polyphase folding, shearing, faults and fractures, intrusive structures, and even the overall tectonic styles of the formations of different ages, can all be discerned through close study of properly enhanced magnetic, electromagnetic, and gammaray spectrometric images.

Plate 1 shows the Rautuskylä area in western Finnish Lapland, which was used as a site for testing of the image enhancement of low-altitude airborne magnetic data, when it was first used in Finland (Aarnisalo and others, 1982). The difference between satellite imagery and the gray-tone image of the aeromagnetic data is obvious. Multispectral satellite data provide information directly from the vegetation, water, exposed surface soils, and outcropping rocks in the Finnish environment. On the other hand, the aeromagnetic data mostly contain direct information on the magnetic characteristics of the bedrock, both from the deep seated sources and from close to the surface of the bedrock.

The geographically rectified image of band 7 data of the LANDSAT-2 scene 208/12, acquired on 22 July 1977 (Plate 1A) shows clearly the lakes and rivers, forested hills, open fields, and bogs in the area. The distribution of these features reveals, in a general way, the main lithologic units and their trends (see Plate 1C), but few of the boundaries between them. On the other hand, bedrock structures, such as several distinct lineaments and a ring structure, can be interpreted from the image.

The image of the low-altitude airborne magnetic data (Plate 1B) shows the distribution and trends of the differently magnetized bedrock layers that act as "marker horizons", revealing the polyphase folding of the bedrock. The magnetized Tepasto granitoid complex, with some radial dikelike anomalies extending eastwards from it, can be discerned clearly in the upper left (Plate 1 B and 1 C). The difference in magnetization and tectonic styles between the Jatulian Kumpu Formation and Lapponian Formations also is marked, and there is an apparent discontinuity zone between them. Disruptions and dislocations of magnetic anomaly belts, and narrow, linear minima in the magnetic field indicate several major fault and fracture zones. For the sake of clarity, however, only a few of them are drawn on the generalized bedrock map (Plate 1C). The ring structure visible in the LANDSAT image is shown as an "augen-like" area of rather flat magnetic field in the magnetic image, with major faults transecting and partly bordering on it.

Plate 2 gives another example of the usefulness of low-altitude airborne geophysical images in displaying regional structures. The area (30 km x 37 km) covers a small part of the well-known Suvasvesi Fault in eastern Finland (Väyrynen, 1954; Gaál, 1972; Koistinen, 1987). Comparison of the images derived from various geophysical data sets with each other, indicates that each type of data seems to reveal best a certain type of regional structure.

The distribution and patterns of the magnetized and conductive rock units and layers (Plate 2 A and 2 B) reveal again plastic deformation features such as polyphase folding and ductile shearing. Their discernability for interpretation may be enhanced greatly by using adequate spatial filtering techniques to bring out even the smallest local variations in the data, as shown in Plate 2E.

Brittle deformation features, faults, and fracture zones, are best brought out by electromagnetic out-of-phase (Plate 2C), gammaray spectrometric (Plate 2D) and LANDSAT images (Plate 1A). The gammaray spectrometric images have particular use in detecting fracture and fault patterns relevant for regional geological mapping.

More detailed information on the bedrock and structures of an exploration target may be obtained from the images of geophysical ground survey data. Plate 3 gives an example of this from the environment of the Stormi Ni-Cu mine in southwestern Finland. The folding patterns and fracturing of the bedrock are discerned clearly , and bedrock layers differing in physical properties can be delineated. On the other hand, only some types of the ultramafic rock have densities high enough to be discerned in the image of filtered gravity data (Plate 3C and D).

It is evident that more aspects of both rock types and structural features can be revealed when images of different data types are integrated. For instance, photographic color compositing of magnetic and electromagnetic images assigns different colors to rock types differing in physical properties. In many situations, however, rock types or thinly layered rock units possess both magnetic and conductive properties, causing similar shades in the color composite images and making it difficult to discriminate between different types of anomaly combination. This difficulty may be overcome by using principal components analysis during the integration of various data sets (Plate 3B). This has been determined especially useful in Finland when enhancing geophysical anomalies caused by magnetized ultramafic and mafic bodies potential for Ni-Cu mineralizations or indicative of a favorable geological environment for Cu-Co-Zn mineralizations, as in the Outokumpu ore district (Aarnisalo and others, 1982; Aarnisalo, 1984).

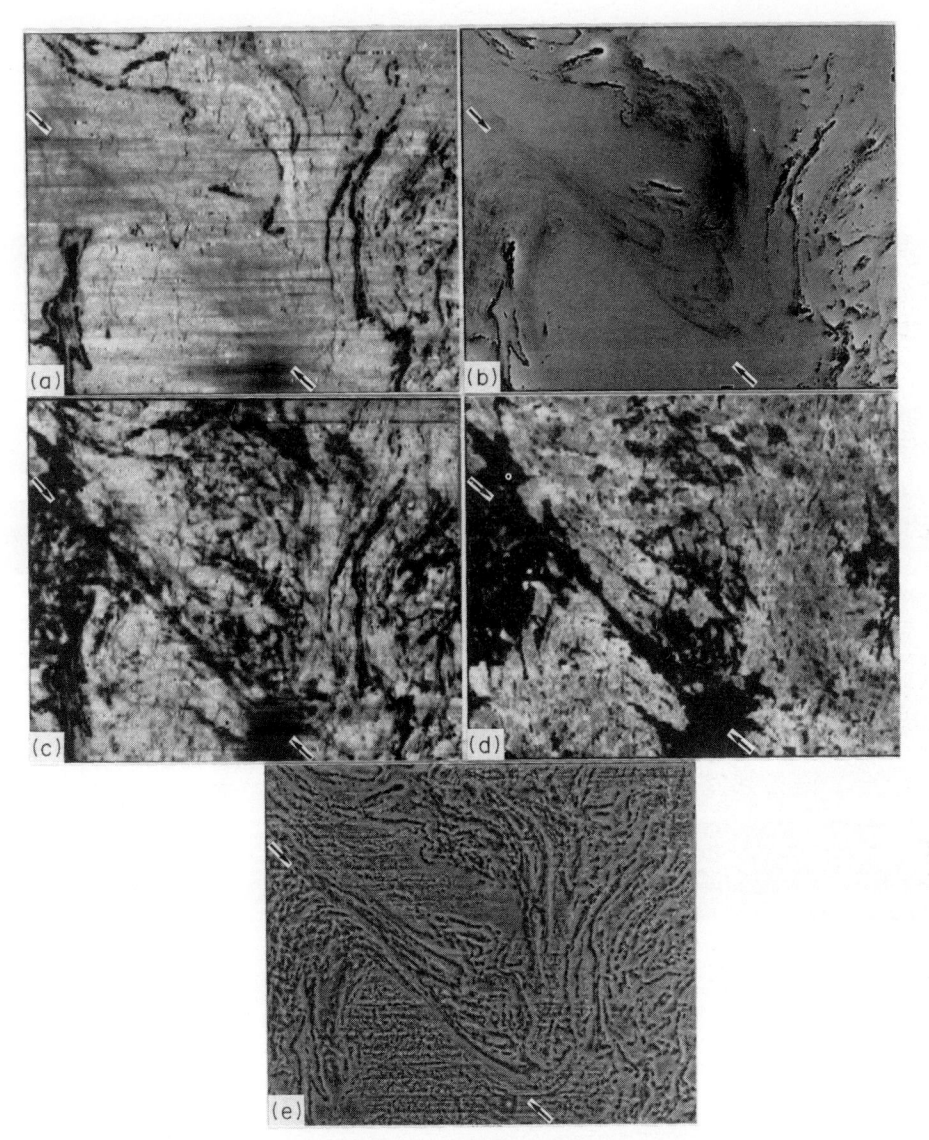

Plate 2. Part of Suvasvesi Fault zone, eastern Finland. Size of the area is 30 km x 37 km. Principal fault zone is marked with arrows. A, Low-altitude airborne magnetic image, enhanced with high-pass filtering (91x91 window mean filter) and edge enhancing with 5x5 window mean filter, linear stretch. B, Low-altitude airborne EM in-phase image, linear stretch. C, Low-altitude airborne EM out-of-phase image, linear stretch. D, Low-altitude airborne gammaray spectrometric total radiation image, linear stretch. E, Low-altitude airborne magnetic data filtered with adaptive high-pass filter (21x21 window), which enhances even smallest local variations in data and reduces large ones. This filtering effectively brings out trends in the magnetized bedrock layers, even when magnetic intensity differs markedly along layers. Thus structural patterns of bedrock are revealed clearly in images of this type.

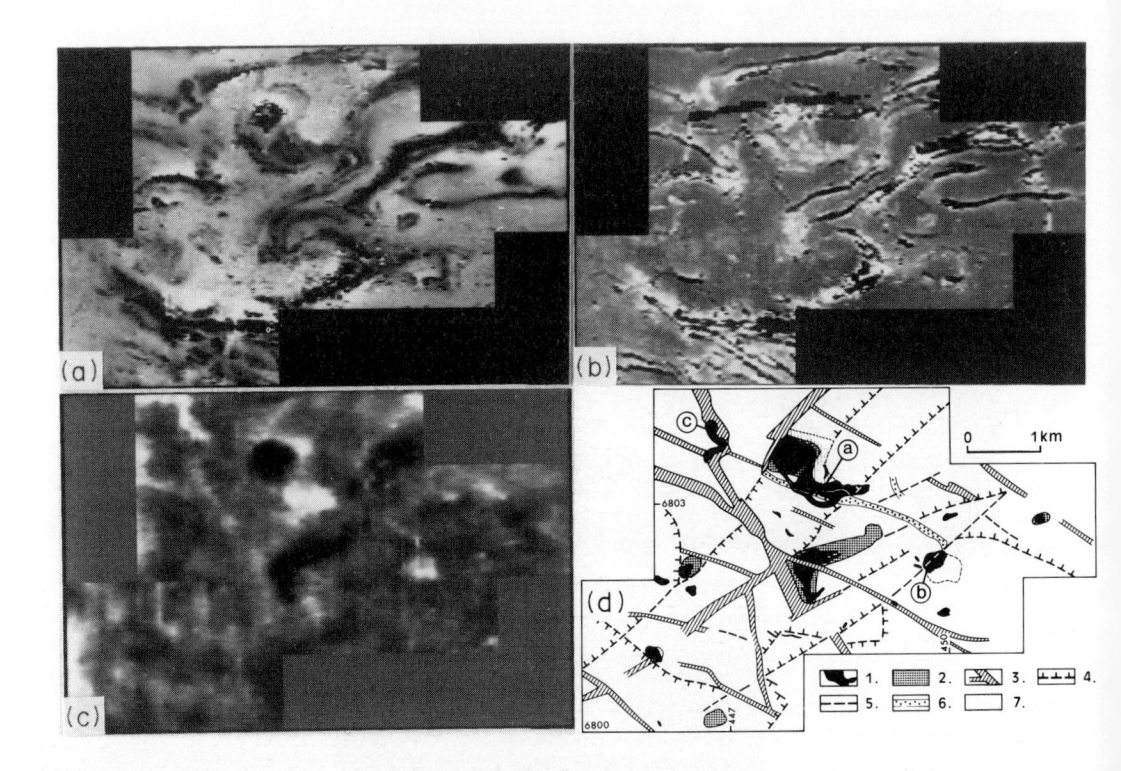

Plate 3. Images processed from ground geophysical data for vicinity of Stormi Ni-Cu mine (Aarnisalo 1984). A, Image of magnetic data, logarithmic stretch. B, Image of second principal component of PCA of magnetic, EM (slingram) in-phase and out-of-phase data, linear stretch. C, High-pass filtered image of gravity data (Bouguer anomalies). Regional anomalies filtered with 91x91 moving average convolution filter, linear stretch. D, Fracture patterns of vicinity of Stormi Ni-Cu mine as interpreted from gravity image (C, 1 ultramafic or mafic body, 2 positive gravity anomaly, 3 fracture valley, 4 fracture line scarp, 5 inferred fracture, 6 river valley filled with soil, possibly also fracture zone, 7 inferred valley caused by weathering of bedrock. Ni-Cu deposits: a Stormi; b Kovero-oja; c Ekojoensuu.

IMAGES OF OTHER DATA SETS

According to Ketola (1987), the discoveries of sulphide and precious metal mineralizations made in Finland in the past few years show that integrated application of geophysical and geochemical methods holds promise for explorationally difficult areas.

Geochemical data have been presented traditionally with symbols or profiles to emphasize the anomalously high-metal contents of explorational interest. Regional variations in the geochemical data may be presented with trend surfaces and their color coding. These methods have developed rapidly during recent years because of the advance in appropriate computer hardware and software technology. This also has led to the use of digital image-processing techniques for integrating geochemical and other geodata sets (e.g. Conradsen, Nilsson, and Thyrsted, 1984; Green, 1984; Forrest and Green, 1986). In Finland, the Geological Survey of Finland in particular has contributed to the development of new display techniques for visualizing geochemical data on maps (Björklund and Gustavsson, 1987).

At Outokumpu Oy, geochemical, drilling-log, and ore geological data usually have been processed separately by computer to produce B&W and colored maps, sections, and diagrams for further manual integration and interpretation with other explorational data (see Fig. 4). However, in some instances, digital image-processing has been used for displaying geochemical data for exploration studies. These data have comprised the results of multielement analyses of detailed and regional till sampling, of a regional organic stream sediment survey and lithogeochemical data from current diamond drilling of ore deposits.

The few examples studied so far have demonstrated that image-processing and color display can be applied successfully to the interactive analysis of regional geochemical patterns or local anomalous concentrations of single elements and multielement data for explorational studies. The interactive stretching of the graytones of geochemical images on the display screen of the image-processing system seem to offer a particularly interesting approach for studying variations in the data. The relations between local anomalies and regional patterns are discerned readily on the display screen, and the images that most efficiently visualize the geochemical data can be processed quickly for further integration with other explorational data. The color compositing of images of three elements reveals the relative variations in the contents of the elements as color changes, thus enabling ready pick-up of the multielement anomalies. Thus it should be possible, for instance, to visualize lithogeochemical aureoles around ore deposits in the drilling-log data. Even more information can be extracted from the geochemical data when multicomponent statistical methods are applied together with the interactive image-processing and display techniques.

Besides satellite imagery and geophysical and geochemical data sets, topographic data or the digital-terrain model has been used as an explorational geodata set in a case study in the Vammala region, southwestern Finland (Aarnisalo, 1984). The study demonstrated clearly that the digital-terrain models will eventually provide a new, as yet almost untapped source of data for lithological, structural, and even glaciomorphological studies in Finland.

DISCUSSION

The experience gained by Outokumpu Oy during the processing of several hundreds of geodata images differing in type has shown that digital image-processing offers the exploration community a powerful tool for displaying and integrating explorational data. At Outokumpu Oy, it also has promoted substantially the study of geological features, and stimulated new ideas relevant to the exploration of base and precious metals, although only a few examples of these have been described briefly in the literature.

The combined application of principal components analysis (PCA), image-processing, and color compositing in the integrated analysis of low-altitude airborne geophysical data, followed up by field sampling, has revealed a number of previously unknown ultramafic to mafic bodies potential for Ni-Cu mineralizations in the Vammala region in southwestern Finland and in the environment of the Hitura Ni mine in western Finland (Aarnisalo, 1984). In eastern Finland, too, in southeastern parts of the Kotalahti Ni-belt, Outokumpu Oy Exploration has used successfully digital image-processing and the PCA of airborne geophysical data, integrated with gravimetric interpretations and the information on mineralized erratic boulders, in the search for Ni-Cu bearing ultramafic bodies (Pietilä and Ahokas, 1985).

Even in the search for massive sulphide deposits, multivariate analysis and image-processing of geophysical data has revealed favorable lithologies and facilitate the modeling of the geology of an exploration area. This has been the situation in the Vihanti region, for instance, where Outokumpu Oy has carried out exploration projects to locate new ZnS deposits (U. Kuronen, 1988, pers.comm.). The same approach has been used in the Outokumpu region to discern the geophysical anomalies caused by the rocks of the Outokumpu association, particularly the serpentinites close to which the Outokumpu type Cu-Co-Zn orebodies are known to occur (Aarnisalo, 1984; Häkli, 1987).

Much new valuable information, structural as well as geological, has been obtained from the integration of airborne geophysical images with LANDSAT images, and with geological and other data. Study of the images of low-altitude airborne magnetic data alone immediately prompted new ideas on how to interpret the regional geology and structures of the Outokumpu area, and thus decisively furthered understanding of the geological entity that includes the Outokumpu Cu-Co-Zn ore deposits (Mäkelä, 1983; Rekola and Ahokas, 1987; Koistinen, 1987). The images of ground geophysical data also have been used in a comprehensive structural analysis of the surroundings of the Kylylahti ore deposit, the latest discovery in the area (Koistinen 1986).

Similar image-processing and integration studies also have been conducted to promote the search for precious metal deposits, both PGE and Au mineralizations, in Finland. Low-altitude airborne geophysical data images are in great demand nowadays whenever a new regional exploration campaign is launched. Even the ground survey explorational data are processed into images for detailed field work and drilling.

It should be noted that the same image-processing steps are not used in each situation, that is, a "production routine" should not be applied; instead, the data should be studied thoroughly and processed in order to obtain an optimal display. If these principles are followed, digitally processed geophysical images are superior in revealing structures and lithological features relevant to the geological mapping of exploration areas. The integration of different data sets, their multivariate analysis, and the color compositing of the output images further enhances the interpretability of the data. Thus, when applied not only to geophysical but also to LANDSAT, geochemical, topographic ,and eventually to other types of data, these methods will provide effective tools for mineral exploration in glaciated Precambrian terrains similar to those in the Fennoscandian Shield.

ACKNOWLEDGMENTS

The author is indebted to Outokumpu Oy for permission to publish this paper. Mrs. Gillian Häkli kindly corrected the English of the manuscript. To her and the staff of Outokumpu Oy, who contributed to the preparation of this paper, the author expresses his appreciation.

REFERENCES

Aarnisalo, J., 1978, Use of satellite pictures for detecting major shield fractures relevant for ore prospecting, northern Finland: Geol. Surv. Finland, Rept. Invest., No. 21, 59 p.

Aarnisalo, J., 1984, Image processing and integration of geophysical, Landsat and other data as a tool for mineral exploration in glaciated Precambrian terrain: ERIM Proc. Intern. Symp. on Remote Sensing Env., Third Them. Conf., Remote Sensing for Explor. Geol., Colorado Springs, Colorado, v.1, p. 107-128.

Aarnisalo, J., Franssila, E., Eeronheimo, J., Lakanen, E., and Pehkonen, E., 1982, On the integrated use of Landsat, geophysical, and other data in exploration in the Baltic Shield, Finland: Photogr. Jour. Finland, v. 9, no. 1, p. 48-64.

Anuta, P.E., 1977, Computer-assisted analysis techniques for remote sensing data interpretation: Geophysics, v. 42, no. 3, p. 468-481.

Arkimaa, H., 1982, Some examples of multi-channel analyses of Landsat and geophysical data: Photogr. Jour. Finland, v. 9, no. 1, p. 38-47.

Arkimaa, H., and Nikander, J., 1980, Monimuuttuja-analyysi Kangasjärven geofysikaaliselle mittausaineistolle: *in* Kaukokartoitusprojekti, toimintakertomus vuodelta 1980: unpubl. Rept., Arch. Univ. Oulu, p. 16-33, (in Finnish).

Björklund, A., and Gustavsson, N., 1987, Visualization of geochemical data on maps: new options: Jour. Geochem. Explor., v. 29, p. 89-103.

Chavez, P.S., Edwards, K.B., Swann, G.A., Termain, P.A., and Watson, R.D., 1980, Digital image processing applied to multispectral, radar, geophysical and geological map data - A tool for illustration and interpretation (abst.): 6th W.T. Pecora Symposium, Sioux Falls, South Dakota, Program Abstracts, p. 83.

Conradsen, K., Nilsson, G., and Thyrsted, T., 1984, Application of integrated Landsat, geochemical and geophysical data in mineral exploration: ERIM Proc. Intern. Symp. on Remote Sensing Env., Third Them. Conf., Remote Sensing for Explor. Geol., Colorado Springs, Colorado, v. 2, p. 499-511.

Duval, J.S., 1983, Composite color images of aerial gamma-ray spectrometric data: Geophysics, v. 48, no. 6, p. 722-735.

Eliason, P.T., Donovan, T.J., and Chavez, P.S., 1983, Integration of geologic, geochemical and geophysical data of the Cement oil field, Oklahoma, using spatial array processing: Geophysics, v. 48, no. 10, p. 1305-1317.

Fabbri, A.G., 1984, Image processing of geological data: Van Nostrand Reinhold Company, New York, 244 p.

Forrest, M.D., and Green, P.M., 1986, Mineral exploration using digital image processing - the GISA service: Mining Mag., v. 155, no. 2, p. 125-129.

Franssila, E., 1982, Digital image processing system documentation, SADIE-RS 1.0, 1/7/82: unpubl. Rept., Arch. Lab. Land Use, Techn. Res. C. Finland, Espoo, 120 p.

Gaál, G., 1972, Tectonic control of some Ni-Cu deposits in Finland: Intern. Geol. Congr., 4th, v. 4, p. 215- 224.

Green, P.M., 1984, Digital image processing of integrated geochemical and geological information: Jour. Geol. Society, London, v. 141, no. 5, p. 941-949.

Häkli, T.A., 1987, The Outokumpu project, in Häkli, T. A., ed., Otto Trüstedt symposium in Finland on June 3-5,1985: Geol. Surv. Finland Spec. Paper 1, p. 51-58.

Ketola, M., 1987, The role of geophysics and geochemistry in sulphide and precious metal exploration in the light of some recent ore discoveries in Finland: Exploration'87, Third Decen. Intern. Conf. on Geophys. Geochem. Explor. for Minerals and Groundw., Toronto, Canada, (Sept. 27 - Oct. 1): Abstr. Vol., p. 24. Manuscr. Arch. Outokumpu Oy Explor., Outokumpu, 24 p.

Kilpelä, E., Jaakkola, S., Kuittinen, R., and Talvitie, J., 1978, Automated Earth resources surveys using satellite and aircraft scanner data. A Finnish approach: Techn. Res. C. Finland, Build. Techn. Commun. Dev., Publ. 15, 174 p.

Koistinen, T., 1986, Kylylahden profiilimalli : unpubl. rept., Arch. Outokumpu Oy Explor., Outokumpu, 23 p., (In Finnish).

Koistinen, T., 1987, Review of the geology of North Karelia, *in* Häkli, T. A., ed., Otto Trüstedt symposium in Finland on June 3-5, 1985: Geol. Surv. Finland Spec. Paper 1, p. 35-40.

Kowalik, W.S., and Glenn, W.E., 1987, Image processing of aeromagnetic data and integration with Landsat images for improved structural interpretation: Geophysics, v. 52, no. 7, p. 875-884.

Kuosmanen, V., Kuivamäki, A., and Tuominen, H., 1981, Structural distribution of sulphide ore deposits in the Lake Ladoga-Bothnian Bay Zone, Laatokan - Perämeren rakennetutkimus: unpubl. Final Rept., Arch. Dept. Geol. Univ. Helsinki, Helsinki, 38 p.

Kuosmanen, V., Arkimaa, H., and Lindqvist, E., 1985, Structural positions of predicted and known Cu-Zn ore indications: ERIM Proc. Intern. Symp. Remote Sensing Env., Fourth Them. Conf., Remote Sensing for Explor. Geol., San Francisco, Calfornia, v. 1, p. 57-71.

Lakanen, E., Ahokas, T., Rekola, T.P., and Katajarinne, T.V., 1978, Geologisen tutkimuslaitoksen ja Suomen Malmi Oy:n matalalentomittaustulosten vertailua Outokumpu Oy:n Malminetsinnän saaman kokemuksen perusteella: unpubl. Rept., Arch. Outokumpu Oy Explor., Outokumpu. (in Finnish).

Missallati, A., Prelat, A.E., and Lyon, P.J., 1979, Simultaneous use of geological, geophysical, and Landsat digital data in uranium exploration: Remote Sensing Env., v. 8, p.189-210.

Multala, J., and Vironmäki, J., 1981, The new aerogeophysical equipment of GSF: Meet. Nordic Ass. Appl. Geoph., Lulea, Sweden (Jan. 1981): Arch. Geol. Surv. Finland, Espoo, 7 p.

Mäkelä, M., 1983, Outokumpu-jakso ennen ja nyt: Vuoriteollisuus- Berghanteringen, v. 41, no. 1, p. 18-22. (English summary: Outokumpu zone before and now).

O'Sullivan, K.N., 1986, Computer enhancement of Landsat, magnetic, and other regional data: Publ. 13th CMMI Congr., Singapore. V. 2, Geology and Exploration, p. 1-10.

Peltoniemi, M., 1982, Characteristics and results of an airborne electromagnetic method of geophysical surveying: Geol. Surv. Finland Bull. 321, 229 p.

Pietilä, R., and Ahokas, T., 1985, Geophysical methods for mapping nickel-bearing structures in eastern Finland and the discovery of the Laukunkangas ore body, (abst.), *in* Abstract Papers Subm. 15th Meet. Nordic Assoc. Appl. Geoph., Espoo, Finland (Jan. 1985): Geoexploration, v. 23, no. 3, p. 417.

Proskuryakov, V., Negrutsa, V., and Yakovleva, V., 1977, The study and mapping of faults in the eastern part of the Baltic Shield, *in* Kortman, C., ed., Fault tectonics in the eastern part of the Baltic Shield: Proc. Finnish-Soviet Symp. held Finl. (20th-24th Sept., 1976): The Committee for Scient. and Techn. Co-oper. between Finland and the Soviet Union, Helsinki, p. 105-111.

Punkari, M., 1982, Glacial geomorphology and dynamics in the eastern parts of the Baltic Shield interpreted using Landsat imagery: Photogram. Jour. Finland, v. 9, no. 1, p. 77-93.

Rekola, T., and Ahokas, T., 1987, On the latest geophysical surveys in the Outokumpu zone, Finland (abstr): Techn. Progr. and Abstract Papers., European Assoc. Explor. Geophysicists., 49th Meeting, The Sava Congress Centre, Belgrade, Yugoslavia (9-12 June 1987), p. 106.

Sabins, Jr., F., 1978, Remote sensing, principles and interpretation: W.H. Freeman and Co., San Francisco, 426 p.

Siegal, B.S., and Gillespie, A.R., 1980, Remote sensing in geology: John Wiley & Sons, Inc., New York, 702 p.

Talvitie, J., 1979, Seismotectonics in Finland: Geol. Fören. Stockholm Förh., v. 100, p. 247-253.

Thompson, W., Hunt, B., Lemche, C., and Kopet, T., 1980, SADIE-2.0, 3/4/80, Documentation: unpubl. Rept., Arch. Univ. Minnesota, Minneapolis, 100 p.

Tuominen, H.V., and Aarnisalo, J., 1976, Major crustal fractures in Baltic Shield: Final Report of Program No. SR 580-03, NASA E77-10027. NTIS, U.S. Dept. of Commerce, Springfield, 91 p.

Tuominen, H.V., and Kuosmanen, V., 1977, Investigation of Landsat imagery on correlation between ore deposits and major shield structures in Finland: Final Report of program 28600, NASA, NTIS, U.S. Dept. of Commerce, Springfield, 53 p.

Vuorela, P., 1982, Crustal fractures indicated by lineament density, Finland: Photogram. Jour. Finland, v. 9, no. 1, p. 21-37.

Väyrynen, H., 1954, Suomen kallioperä, sen synty ja geologinen kehitys: Tiedekirjasto N:o 27. Otava, Helsinki, 260 p. (In Finnish).

Mappable Data Integration Techniques in Mineral Exploration

D. Bonnefoy and A. Guillen
Bureau de Recherches Géologiques et Minières, Orléans, France

ABSTRACT

Increasingly numerous and diversified mapping data are available to earth scientists (geologists, town and country planners,hydrogeologists, etc.). These technically skilled experts combine these data and, depending on their needs, draw thematic or predictive maps, which are powerful tools to improve decision-making.

A geographic information system (GIS) MARICA has been built to assist decision-makers: it takes into account and integrates data of various sources and natures. The diversity of these integrated data requires the use of various processing facilities. For example, particular remote-sensing data are illustrated by images and other information, such as geochemical data, will be represented by grids or vectors. MARICA also takes into account the expert's knowledge as production rules and built-up objects. MARICA was created by BRGM, and runs with UNIX or VMS operating systems. Standard languages (FORTRAN 77, C) are used, as well as standard graphic libraries (GKS, X-WINDOWS version 11).

MARICA consists of three main parts :

(1) Part I comprises standard processing functions of remote-sensing images (developed by the 2AI Company and BRGM) with a graphic editor to visualize the interpretation of the processed images.

(2) Part II, which accepts as input results obtained from interactive graphic interpretations and from the classifying functions performed by part I, also takes into account and processes grid data or vectors. This module has two interfaces to input outside data: interpolation of scattered data points and digitizing of documents. Numerous processing functions allow diversified operations: production of maps following a combination written in natural language, multivariate statistics, classification, signal processing, graphic processing, graphic correlations, display and reading-out of information.

(3) Part III makes use of expert-system techniques and also takes into account and applies the expert's technical skill in the field under study by a simple method using an expert system generator interfaced with the GIS. The expert-system makes it possible to express a view in a particularly interesting field by means of triplets: "zone subject-ed to diagnosis", "database", and "knowledge base".

There is a wide field of applications for such techniques, which are particularly useful for geoscientists involved in mineral exploration.

INTRODUCTION

Geoscientists and particularly exploration geologists generally are faced with decision processes with heavy possible economic consequences. The development and the results of mineral-exploration programs depend on the soundness of the decisions.

The decision process is based on extremely various data. The confrontation of data from satellite imagery with geological, geochemical, geophysical, and topographic data with a reliable selection (in terms of economic potential) of exploration targets will minimized risks of error.

The numerous data are of extremely various nature. The volume of remote-sensing data is large. Geochemical exploration data, if multielement analysis is involved, may easily represent 0.5 million single analytical data for a regional exploration program. A computer-processed geological map contains hundreds of elementary arcs.

Because of the various sources of information, the knowledge necessary for interpretation covers a considerably wide range which generally is not in hands of a single person. It therefore is difficult to locate geologists skilled enough to conduct accurately an exploration program with all its facets.

To help the decision-makers and the exploration geologists focus their efforts on an area where the chances of discovering a deposit are good and the risk of error is the lowest possible, we have created a Geographic Information System (GIS) MARICA. MARICA takes into account, processes and compiles extremely various data. MARICA also takes into account the knowledge of one or several experts who are specialists in their respect-ive field.

DESCRIPTION AND CAPABILITIES OF MARICA SYSTEM

MARICA comprises three main parts: part I provides facilities for processing of remote-sensing data ; part II takes into account and combines data of various sources; and part III enables the integration of an expert's knowledge in a specific field of the decision process (Fig. 1).

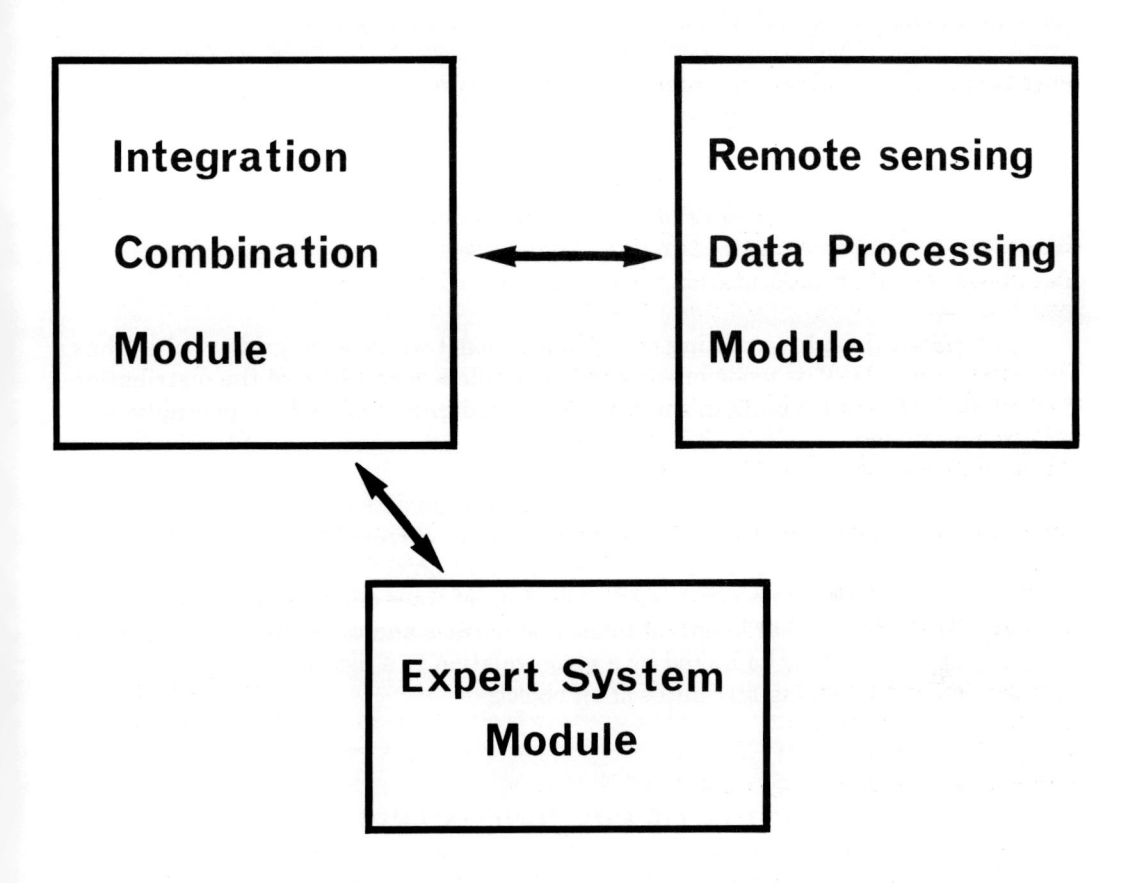

Figure 1. GIS design of MARICA.

The first part uses the conventional single and multichannel processing techniques of remote-sensing data. Principal component factor analysis and classification functions are available. Specific facilities such as a graphic editor are made available to the user enabling interactive interpret-ation of satellite or processed image. Such interpretation may lead to the identification of spectral discontinuities or even to draw up facies considered to be homogeneous. A vector/point/surface file is created which may serve as an entry to part II of the so-called combinational system.

Part II of MARICA allows the combination of data from various sources leading to the definition of different data representation modes.

DATA REPRESENTATION

Two modes of data representation have been selected (Fig. 2):

(1) A classical mode consisting of a gridded model where each grid or "pixel" has a value or a code. This mode provides for example a description of the distribution of trace-element contents in soil samples, or a digital model of a topographic or geological map.

(2) The second mode is termed vector mode and consists in subdividing a level of information into elementary graphic objects that correspond either to one or several points, a line, or a surface. Each graphic object may possess a set of variables each having a value such as content of trace elements in isolated samples, orientation or length of lines, and surface and coefficient of elongation. Each object also may be linked by a given relation to a list of other objects, relation and list being attributes of an object.

CREATION OF SPECIFIC INTERFACES

An outside interface enables the gathering of data of scattered data in independent database such as the GDM system used at BRGM. Interpolation algorithms generate a gridded model of the phenomenon: classical interpolation with weighting by inverse of distance, interpolation by the least-square method, or estimation by linear kriging without drift. Mineral-exploration data which may be entered from the GMI-PACK system (Gravity and Magnetic Data Interpretaion Software), FIESTA (Geochemical Exploration Data Interpretation Software), or from any conventional data-processing system which may produce a text file of the X, Y, and Z format. An interface with field digital models is provided by the Institut Géographique National of France. Another way of entering into the system is by capturing the vector data from the CAD system such as the Intergraph system set up at BRGM. This enables the geocodation of geological or interpretative maps for comparison with other data.

GRID MODE

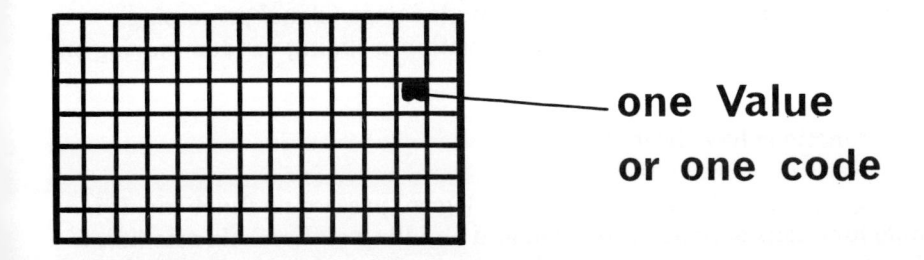

— one Value
or one code

VECTOR MODE

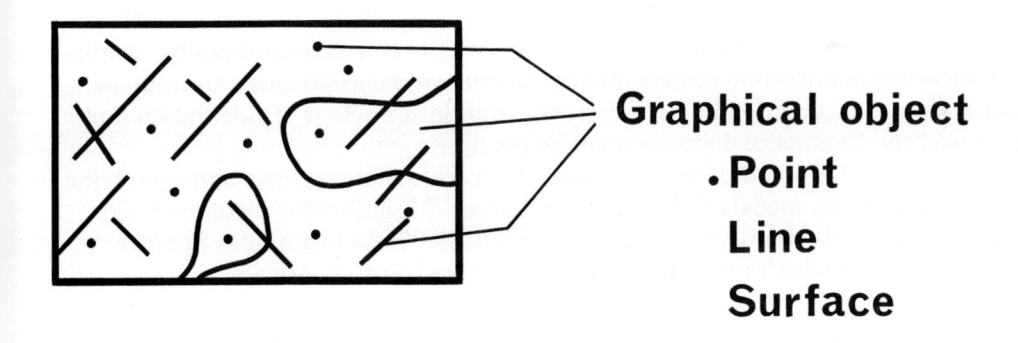

Graphical object
- Point
  Line
  Surface

Figure 2. Data representation.

The interface with the vector files produced by the interpretation of remote-sensing data is a preferential entry in the combinational system.

MARICA has a wide range of processing functions of various levels of information.

Raster processing

This type of processing is concerned with data represented on a regular grid and the aim is to combine several images or to process a single image (signal processing). These combinations are carried out pixel by pixel, within a sliding window moving on the entire map (concept of neighboring).

Several groups of functions have been developed:

(1) A first group contains the arithmetical and logical processing functions. It includes simple functions of image threshold and handling of threshold sets (intersection, disjunction). A most interesting function is the evaluation function which enables the user to formulate a combination expression using operators +, -, /, and x, basic mathematical functions (SIN, COS, MIN, MAX, square root, and logarithm) as well as conditional operators "IF condition THEN action X ELSE action Y".
Condition and action are arithmetical expressions concerned with images. The expression is evaluated pixel by pixel and the result is saved as a new printout image.
(2) A second group of functions aims at processing two or more images by statistical algorithms including among others the following classification techniques:
- supervised classification by definition of a standard nucleus of calculation and storage of the Euclidean distance pixel by pixel;
- multicriteria classification by definition of several training areas and search for similarity with the models defined in these areas (Mahalanobis distance or discriminant analysis) (Faugeras,and Berthod, 1979). These techniques produce geocoded images which represent the geographic extent of each class.
The statistical functions also provide facilities for the drawing of histograms, comparison of populations defined by coded images, drawing of two-dimensional histograms with extraction of groups of pixels, and mapping of selected subpopulations. A few statistical functions such as calculation of correlation coefficients, standard deviation and coefficient of variation, performing in the vicinity allow the user to study the data.

(3) A third group of functions provides facilities for the study of geometrical relationships between one or several threshold images.

(4) A last group of functions enables the application of signal-processing algorithms on an image: filtering and convolution.

Vector processing

Some data cannot be processed in raster mode. For example, a representation of a gridded model is not suitable for establishing therelationships between a discontinuity (e.g., a fault) and some points (e.g., mineralized occurrences). The optimal description of a polygon formed by n elementary arcs using a pixel representation may generate a number of incompatible data. This is the reason why another mode of representation, the vector mode, has been selected. It is different from the gridded model mode in that it introduces the concept of a vector image formed by n graphic objects.

A graphic object may be a point, a line or a polygon. Thus it is possible to represent discontinuities resulting from interactive processing of remote-sensing images, mineralized occurrences (points), or geological units (surfaces).

Processing these images consists in studying the relationships between these objects. The neighboring relationship between points and discontinuities may be studied. For example it is possible to know whether an object is included in another object or whether it belongs to another object, to classify the objects or to know whether two lines are roughly concordant.

The vector image is a special graphic file which can be modified interactively and superimposed on a raster or another vector image. The file may be created from the interpretation of images successively appearing on the screen.

Each object is described by an almost unlimited number of variables which may be used to save the results of the relationship study. These variables enable the user to select objects using the management functions.

Intelligent classification

GIS also provides capacities for establishing a diagnosis on the interest of a given region through the utilization of specific knowledge bases fed by one or several experts (Fig. 3). This part of GIS uses expert-system techniques and the CLASSIC software created by INRIA (Institut National de Recherche en Informatique et en Automatique). The typical problem set up is as follows: the geologist is asked to express his views on a potentially valuable area on the basis of collected data that describe and define this area and also by using knowledge directly related to the expert field.

After having defined on the screen the potentially interesting area and the descriptive parameters, an attribute file representing the initial database may be sent to the expert system. The engine starts its inference cycle and makes its diagnosis if the initial input data are sufficient, otherwise additional data are requested.

EXAMPLES

Raster mode processing - geochemical exploration of gold-bearing shear zones

The soils developed on the French Hercynian basement are residual so that they reflect the geochemical signature of underlying rocks and mineralizations, if any. Multielement exploration geochemistry enables to distinguish specific geochemical associations of the various types of gold-bearing or barren shear zones (Bonnemaison, and Braux, 1987).

In this example, GIS is used to combine various parameters and namely monoelementary maps of trace or major elements contents. The result of the classification is the discrimination of distinct units, based on their geochemical characteristics. The geochemical signatures include 3 categories of elements:

(1) Major elements initially present in the host rocks, but whose concentrations have been modified during introduction of the mineralization and associated hydrothermal alterations.

(2) Elements introduced by mineralizing processes and characterizing the mineralized structure on a regional scale. These elements generally are associated with the early stage of the formation of the shear zone, responsible for disseminated, generally subeconomic, gold mineralization.

(3) Elements specific of rich mineralized zones, so-called pathfinders, associated with later stages of the evolution of the shear zone.

The supervised classifying functions are used to charaterize each of the described phenomena. Three standards were defined in a multivariate space (As, Sb, Pb, Au) characteristics of the three evolution stages of the gold-bearing shear zone. In a next step, the Euclidean distance between each pixel and each standard is calculated, so that "isosimilarity maps" may be drawn, referring to the 3 metallogenic processes, including the one responsible for the main gold mineralization. Such predictive maps where actually used to rank by order of priority the exploration targets in that region.

Vector mode processing

The gold-bearing lodes of the Saint-Yrieix district (Haute-Vienne deparment, French Massif Central) occur for several kilometers along mylonitic and cataclastic shear zones trending NE-SW to ENE-WSW.

A remote-sensing study (Image LANDSAT Thematic Mapper) was carried out in the district to identify spectral discontinuities which, after calibration, were shown to be related with the sought mineralizations. (Bouchot, Dutartre, and Gros, 1988).

one interesting sector
a data base
a knowledge
gives a diagnostic on an area

area

n levels
of information

knowledge base

DIAGNOSTIC

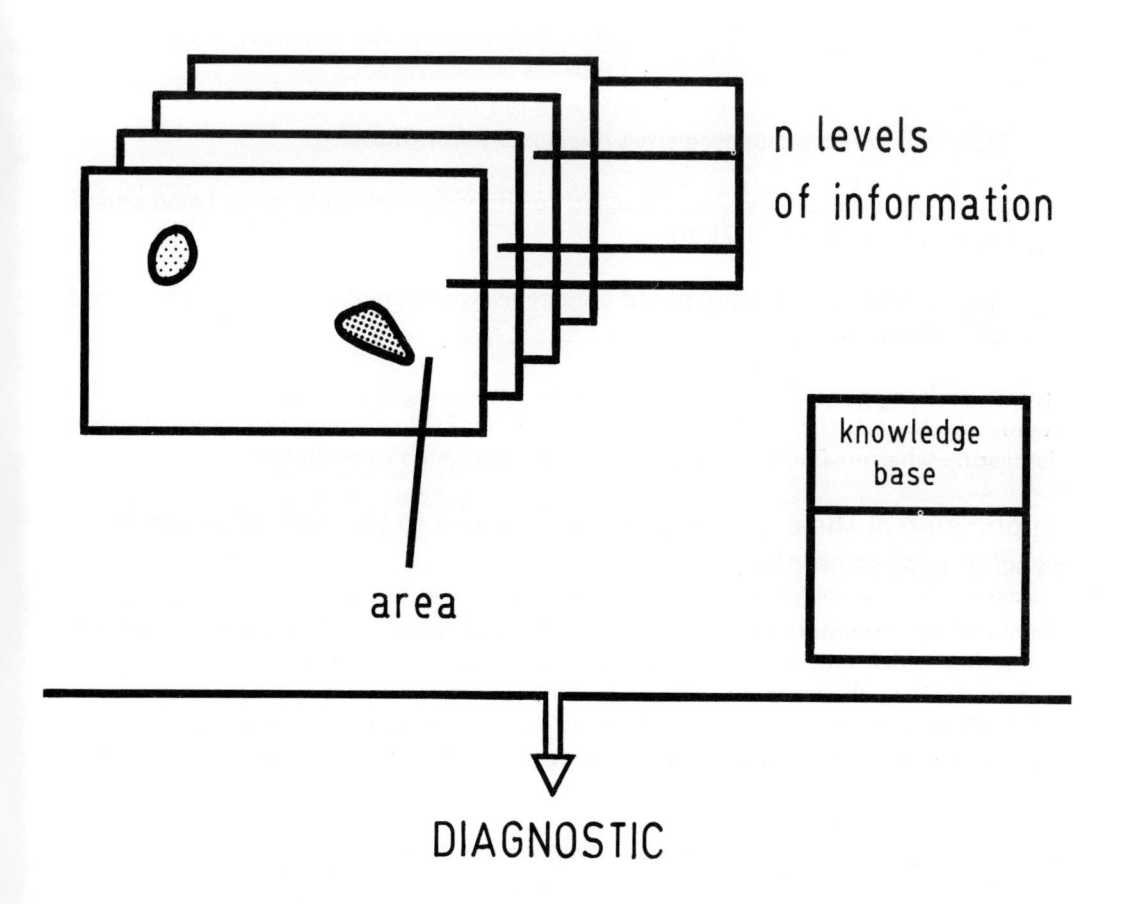

Figure 3. Intelligent classification.

The selection of potentially gold-bearing discontinuities is based on:

(1) Topographic criteria: crest lines are favorable parameters in-so-far as they are related to silicified faults.

(2) Geometrical criteria: for instance, NW-SE oriented discontinuities are not taken into consideration because no significant gold-bearing structures are known along these fracture zones.

(3) Metallogenic criteria: mineral occurrences (stored in the databank) present in the neighborhood of discontinuities may increase their potential.

The processing functions we have developed investigate the geometric relationships between objects, for instance:

(1) Search for superimposed crest lines and discontinuities.

(2) Search for and selection of discontinuities with a given average trend and of limited extent (0.5 to 2.5 km).

(3) Search for known mineralized occurrences or ancient workings in the vicinity of discontinuities.

Iterative filtering is achieved through vector processing functions that measure the geometric characteristics of each discontinuity, make a selection on these criteria and determine whether two lines (e.g. crest and lineaments) are similar.

Application of the expert-system techniques to the classification of geochemical anomalies

Among the preoccupations of the exploration geochemist is the selection of significant anomalies which may reveal a deposit. The necessity to select mineral target as early as possible, has shown the need for reliable selection procedures for geochemical anomalies, taking account of highly varied nonnumerical factors, such as sample type, morphoclimatic and geological settings, and anthropic context (Bonnefoy and others, 1986).

In this situation, the area subjected to diagnosis is a geochemical anomaly. MARICA provides capacities for the enhancement of the attributes of the area in the form of files. This file forms the initial database of the expert system whose role is to give a value to the "final diagnosis" attribute. A specific knowledge basis is used to process the classification of geochemical anomalies.

This knowledge is expressed as production rules and is divided into large sets including:

- a description of the geographic environment, for example topography, drainage characteristics, proximity of a village or a town;

- a description of the geochemical anomalies : area, number of anomalous sampling sites, etc...;

- information about the geological environment ;

-a catalog of the geochemical signatures (pathfinder elements) characteristic of the main exploration targets for the area under study.

After semiautomatic feeding from GIS, the expert system starts making deductions. To obtain a diagnosis on the quality of the anomaly investigated, the operator only has to answer complementary questions asked by the system.

CONCLUSION

MARICA is a geographic information system which enables a true comparison between various data such as remote-sensing, geophysical, and geochemical data. Because of the modes of data representation used, raster and vector modes, the system is a precious aid to decision-makers and is used at BRGM for optimization of decisions. When coupled with the CLASSIC expert-system, it may classify and make diagnostics on geological objects by specific knowledge bases.

REFERENCES

Bonnemaison, M., and Braux, C. 1987, Les zones de cisaillement aurifères du socle hercynien francais: Chron. rech. min. n° 488, p. 29-40.

Bonnefoy, D., Jebrak, M., Rousset , M.C., and Zeegers, H., 1989, Serge: .an expert system to recognize geochemical anomalies: Jour. Geochemical Exploration, 12th IGES, 1987, Orléans, v. 32, no. 1-3, p. 343-344.

Bouchot , V., Dutartre, P., and Gros ,Y. 1988, Contribution de l'imagerie satellite Landsat thematic-mapper à la caractérisation et à la carto-graphie des zones de cisaillement aurifères du district de Saint-Yrieix (Haute-Vienne, Limousin, France): Rap. B.R.G.M. 88 DAM 008 DEX/TED, 32 p.

Faugeras, O., and Berthod, M. 1979, Scene labeling: an optimization approach: IEEE proc., PRIP Conf., p. 318-326.

APPENDIX

Configuration of hardware and software

The data system described in this paper is termed MARICA (MARiage de l'Image et de la CArte, "combination of image and map"). Its configuration is made up of:

- a workstation with 4-8 megabytes memory
- a virtual memory
- a 256-color screen, about 1000 x 1000
- a disk of 200 megabytes
- A UNIX or VMS operating system
- GKS software
- X-Windows 11 software
- FORTRAN 77
- C

The CLASSIC (ILOG company) and LE-LISP software from INRIA are required for the Expert System Module. This system is being installed on VAX series computer, SUN workstation.

The Use of Digital Elavation Models Computed from Spot Stereopairs for Uranium Exploration

Pierre Leymarie
CNRS, URA Géodynamique, Valbonne
JacquesDardel, and
CEA - DAMN, Paris
Laurent Renouard
ISTAR, Valbonne

ABSTRACT

The area selected for study was the Lodeve Basin (Herault, France). The computation of a Digital Elevation Model (DEM) from a SPOT stereopair yields to results which fall well within 1:50000 mapping specifications. Using this DEM makes possible a better registration of SPOT images to reference maps and to geological data, and also the production of simulated perspective views. Faults, contacts, and key beds which are not apparent on SPOT images may be detected by mathematical filters applied to the DEM. SPOT derived DEMs also offer the posibility of ascertaining whether a line detected on a satellite image is the trace of a plane on the ground surface and, if the test is positive, to compute the strike and dip of this plane. Finally, this paper provides the results of ground controls achieved on the examples studied on DEMs and SPOT imagery.

INTRODUCTION

The examples which we intend to present here apply to the Lodeve Basin where COGEMA (Compagnie Generale des Matieres Nucleaires) works, at Saint Martin du Bosc (Herault), a major uranium mine (Fig. 1).

The SPOT satellite makes it possible to obtain pairs of stereo images thanks to inclined viewing directions during revisits over the district under study. Figure 2 illustrates the right-side image of the pair which we have used. The method employed in the computation of a Digital Elevation Model (DEM) from a SPOT stereopair already has been described in a more detailed way by Renouard (1988).

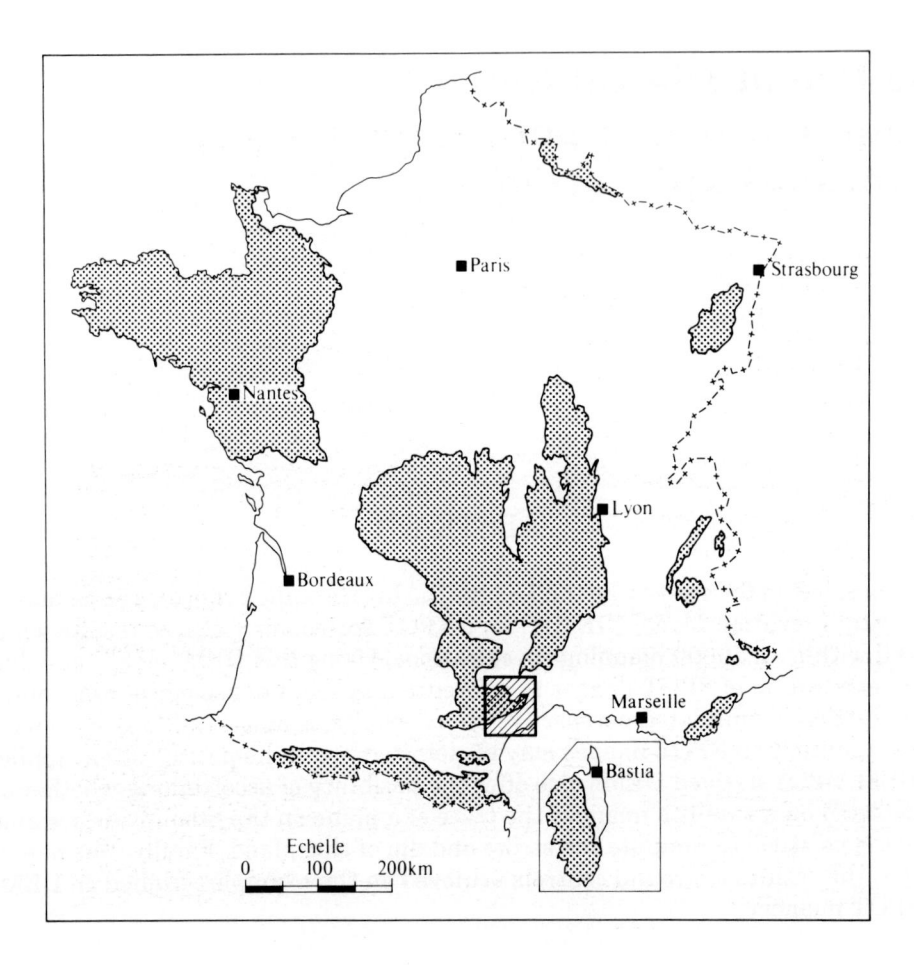

Figure 1. Scope area under study.

PROGRAM DESCRIPTION

The first step of the computation of the DEM is fully automatic. Matching of the two stereopairs is accomplished through a hierarchical method followed by a subpixel correlation. In areas where the correlation is unsatisfactory, elevations are interpolated between neigboring values. In the second step of the method, we detect the points where the altitude differs significantly from that of their neighborhood. If they are located neither on a crest nor on a valley, they are replaced by interpolated values, as in the preceeding step. We obtain in this manner what we will hereafter refer to as the raw DEM, which then is smoothed by a method which minimizes simultaneously both the difference in altitude between the raw and the smoothed DEM and the gradient of the smoothed DEM. We will be using only smoothed DEMs in our study here. In the last step of our method, the DEM and its computed orthophoto are projected into the user-specified standard reference system.

USE OF DIGITAL ELEVATION MODELS FOR URANIUM EXPLORATION 227

Figure 2. Right-side image SPOT-1 panchromatic HRV1 stereopair, preprocessing level 1A, sampled 1 pixel out of 8, scene (KJ) 046-262 (central point N 43° 40', E 003° 34'). Right-side image of 04 April 86, 11h 09mn 49s UT, viewing angle E 29°, left-side image of 15 March 87, 10h 34mn 19s UT, viewing angle W 20°. Images which follow usually contain 800 lines x 1000 columns.

The program, if so desired, can process the totality of the area shared by the two stereopair images, about 3800 square kilometers. The orthophoto (Fig 3) and the elevation contours (Fig 4) obtained in this way make it possible to prepare maps at a scale of 1:50 000 which, if necessary, can be enlarged to 1:25 000.

RESULTS

The horizontal accuracy and the standard deviation of *absolute* errors for altitude are under 10 m. Altitudes are given in decimeters.

DEMs obtained in this manner are considerably less expensive than those obtained through other techniques. Once a good-quality stereopair is available for use, the required time limit to develop a DEM is normally about three weeks and can even be reduced to a few days in urgent situations.

SOME USES FOR SPOT-DERIVED DEMS IN GEOLOGICAL STUDIES

Analysis of landforms

SPOT-derived DEMs are more detailed in spatial high-frequency (HF) components than DEMs obtained through the digitization of topographical maps which always simplify to some extent the landforms. Details resulting from tree-cover, forest paths, matching errors, etc..., which have no geological significance whatsoever, are eliminated by low pass filtering. But other details which may be absent in traditional DEMs may correspond to faults, contacts, or particular rock types. Because of this possibility, care must be taken to ensure that filtering does not eliminate them.

We isolated the HF components of the SPOT-derived DEM, in order to identify their origin, by substracting the smoothed DEM from the raw DEM (Fig 5). As expected, in open terrain the HF components usually are positive and caused by trees, shrubs, etc... In forested land they usually are negative and caused by roads, clearings, etc...

It should be noted that the first pair of stereoimages on which we worked had been taken at a one-year interval. We therefore were working under conditions which were considerably unfavorable to matching and which certainly were responsible for the appearance of numerous insignificant details. Whatever the situation may be, the normal filtering methods (Adam de Villiers, 1985; Adam de Villiers and Leymarie, 1984), applied to the smoothed SPOT-derived DEMs, provide results comparable to those obtained on the DEMs resulting from the digitization of a topographical map. We shall see next how contacts and faults can be detected on images derived from the mathematical processing of the SPOT DEM.

USE OF DIGITAL ELEVATION MODELS FOR URANIUM EXPLORATION 229

Figure 3. Orthophoto, left-side image, with Lambert grid and elevation contours spaced at 25m. 1 pixel = 10m.

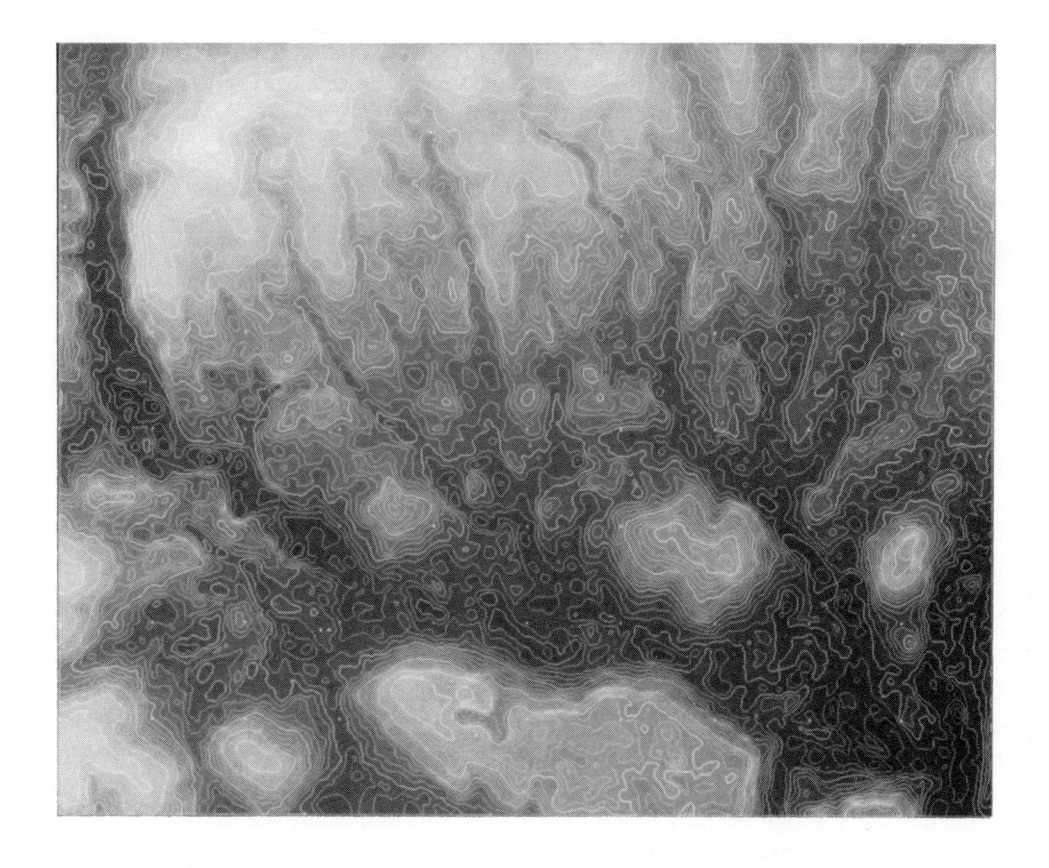

Figure 4. DEM represented in gray scale and in elevation contours spaced at 10m. 1 pixel = 10m.

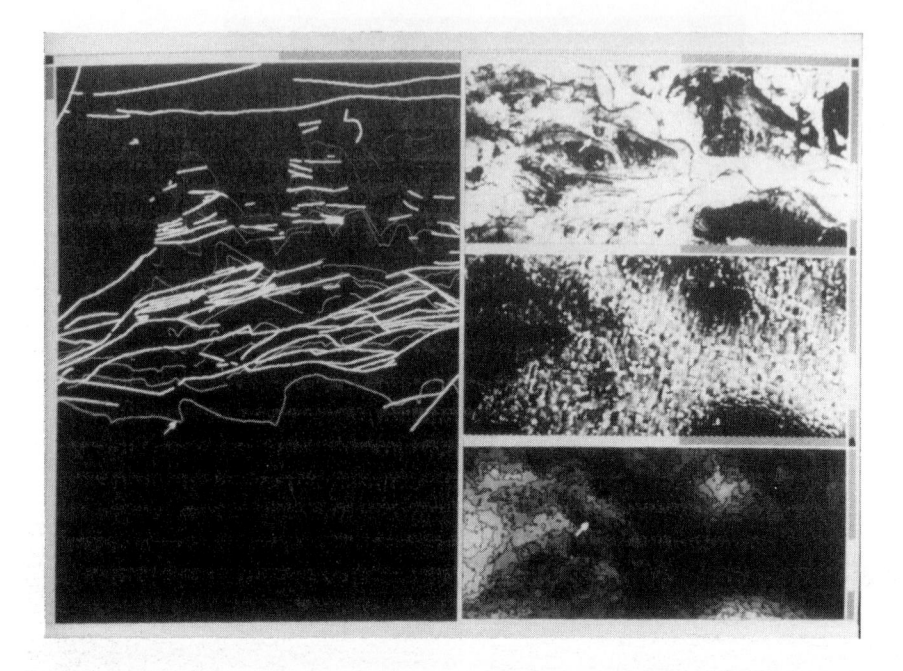

Figure 5. On left, map of faults and key beds. On right and from top to bottom, (1) orthophoto, (2) HF components of DEM, and (3) raw DEM. Cursor designates Saxonian conglomerate.

Comparison of geological knowledge and SPOT data

One of our major objectives was to demonstrate the advantages of SPOT images in the prospecting of uranium districts and provinces. At the outset, the idea was to compare the SPOT data to present geological knowledge concerning a few mining districts, in particular knowledge on the Lodeve Basin, for which we possess a geological map scaled to 1:50 000 of the BRGM and a geological map scaled to 1:20 000 and worked out by COGEMA. Both of these maps have been digitized.

DEMs derived from a SPOT stereopair make possible a better coregistration of satellite images to reference maps. Because horizontal accuracy of the orthophotos is around 10 m., the comparison of geological data and SPOT data becomes precise enough to enable us to affirm without too much risk that a given linear feature does in fact correspond to a given contact or a given fault. On the other hand, when a DEM is not available, it may happen that the positioning errors prevent us from concluding with certainty that the known geological structures are visible on satellite images (Leymarie and others, 1985). In order to take advantage of this improvement, we nevertheless were obliged to develop a program better adapted to precise comparison that methods which employ the flickering of two images on the display screen or which use graphic overlays (Allain, 1988). Comparison of digitized geological docu-

USE OF DIGITAL ELEVATION MODELS FOR URANIUM EXPLORATION 231

ments and SPOT data is facilitated by a cursor designating the same pixel on all of the images displayed on the screen, whatever their scale. These images can be scrolled in the display windows. In the left-hand window, two reference images can be switched as desired, here (Fig 5) the faults and key beds with the geological contacts.

Figures 5 to 7 illustrate how it thus is possible to record the contacts and key beds visible on SPOT images. Figures 8 to 12 show that the key beds or the faults invisible on these images may manifest themselves as topographical high frequencies which can be detected through the use of appropriate filters applied to the SPOT-derived DEM. We at first thought that a layer of hard, white cinerite some decimeter thick would be visible on the SPOT images, because its presence lights up the ground within a width of several dozen yards. Such was not the situation, but study of the slope angle enabled us to localize this layer of cinerite as shown in Figure 8.

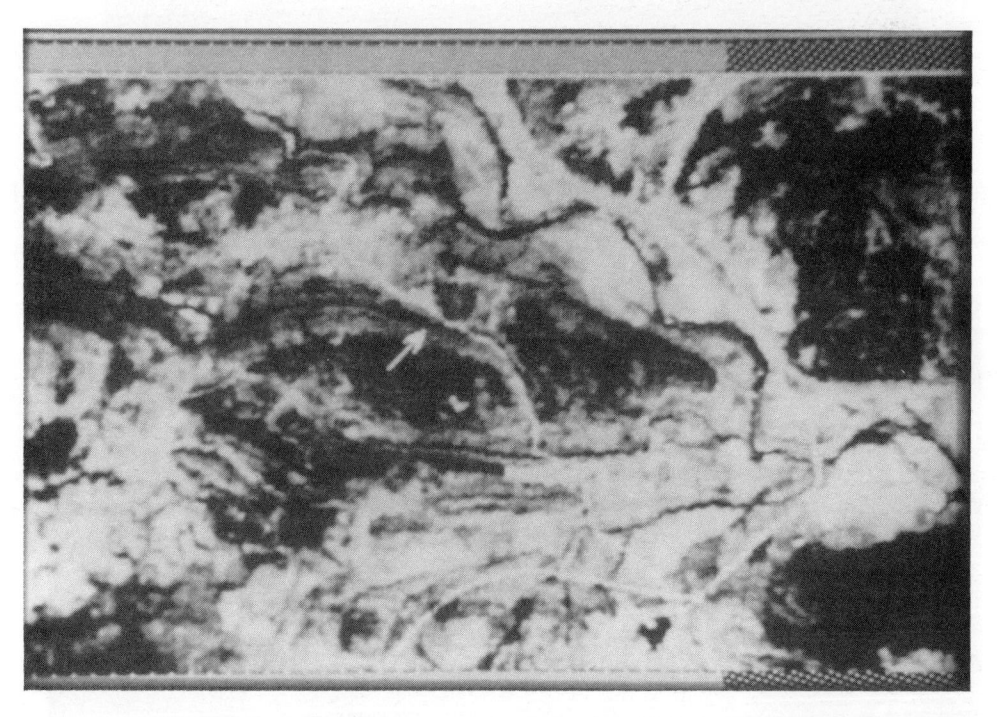

Figure 6. Enlargement of Figure 5 - 1. Saxonian conglomerate designated by cursor is clearly visible.

Figure 7. Exposure of Saxonian conglomerate. Dark line on SPOT image is result of shadow of cliff and to vegetation line above.

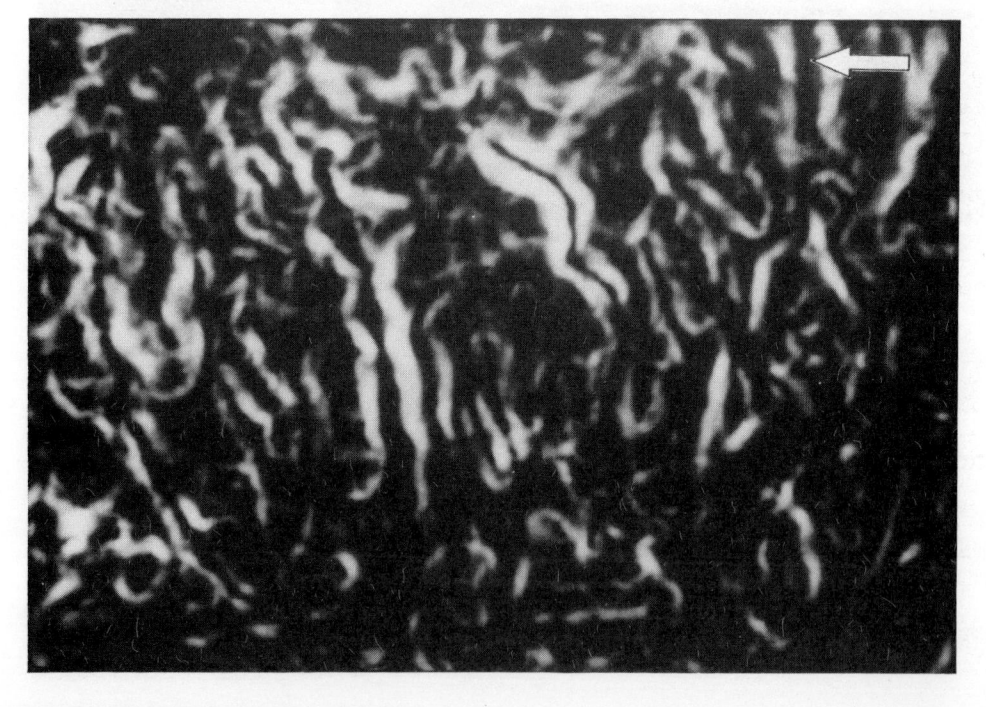

Figure 8. Key bed X, designated by cursor, appears clearly on map of slope angles deduced from SPOT DEM.

USE OF DIGITAL ELEVATION MODELS FOR URANIUM EXPLORATION 233

Figure 9. Photo of hill surrounded by outcrops of key bed X.

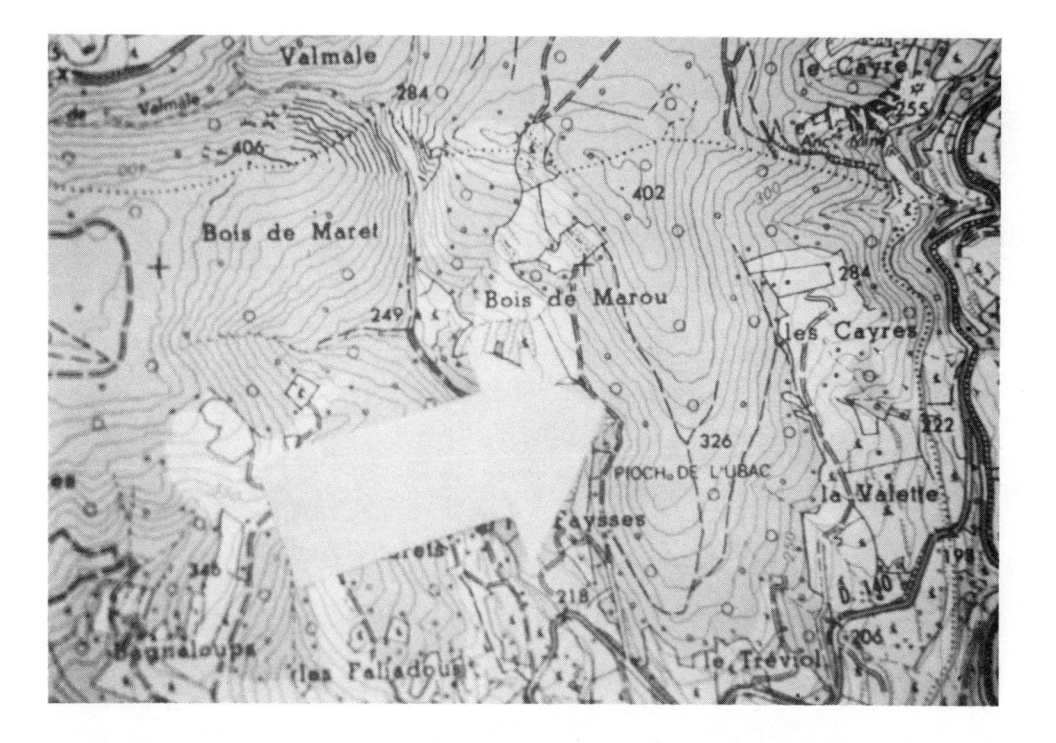

Figure 10. Close-up on topographic map (1:25000), corresponding to Figures 8 and 9.

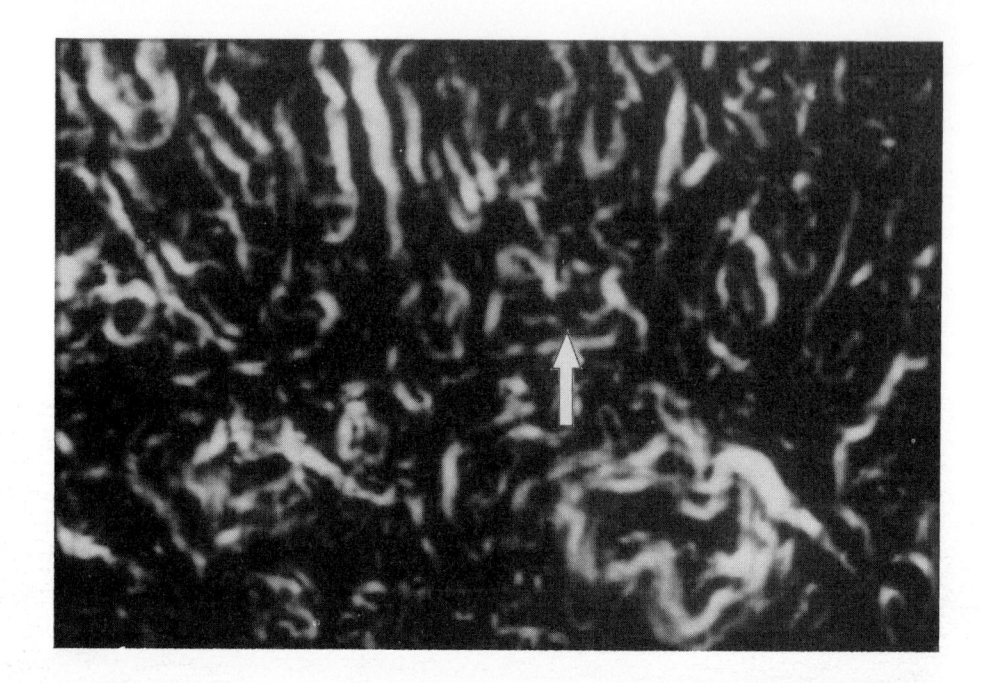

Figure 11. EW fault picked up by cursor appears clearly on map of slope angles deduced from SPOT DEM.

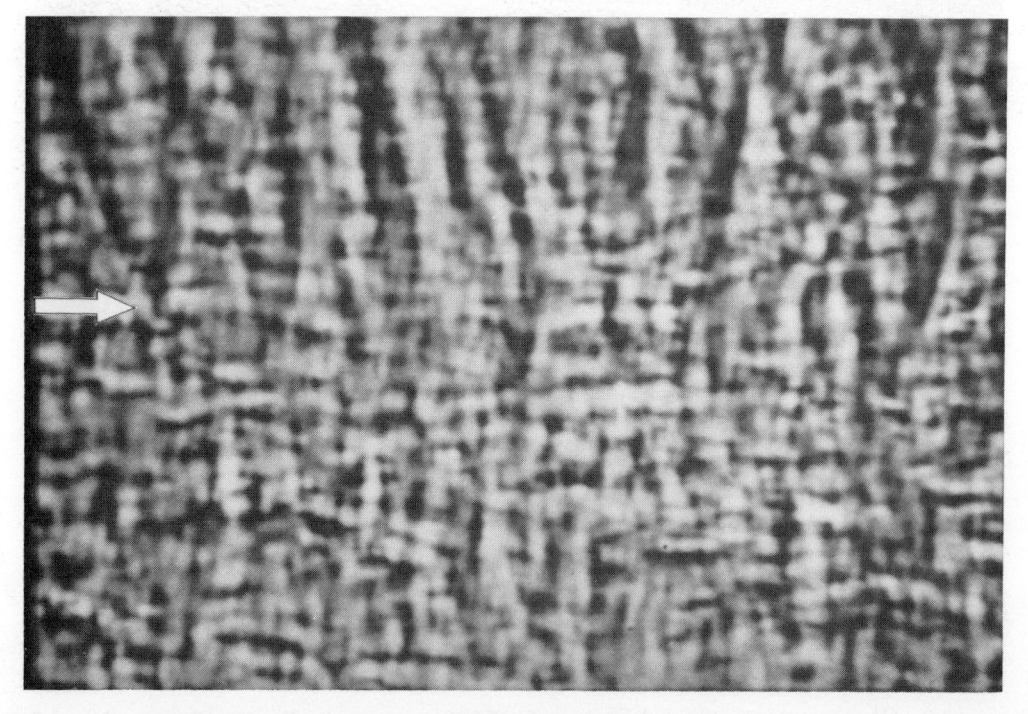

Figure 12. EW fault picked up by cursor appears clearly after processing SPOT DEM by filter which amplifies structures EW and NS (image of $\partial^2 Z/\partial X.\partial Y$).

Determining Strikes and Dips

SPOT-derived DEMs also offer the possibility of ascertaining whether a line detected on a satellite image is the trace of a plane on the ground surface. If the linear element under examination does not fit a plane, it is rejected, but if it does, the program computes the strike and dip of the plane and allows for their illustration (Fig. 13) on the orthoimage. This program currently is operative on linear features digitized in interactive mode; in the future it will enable us to process in the same manner automatically detected linear features.

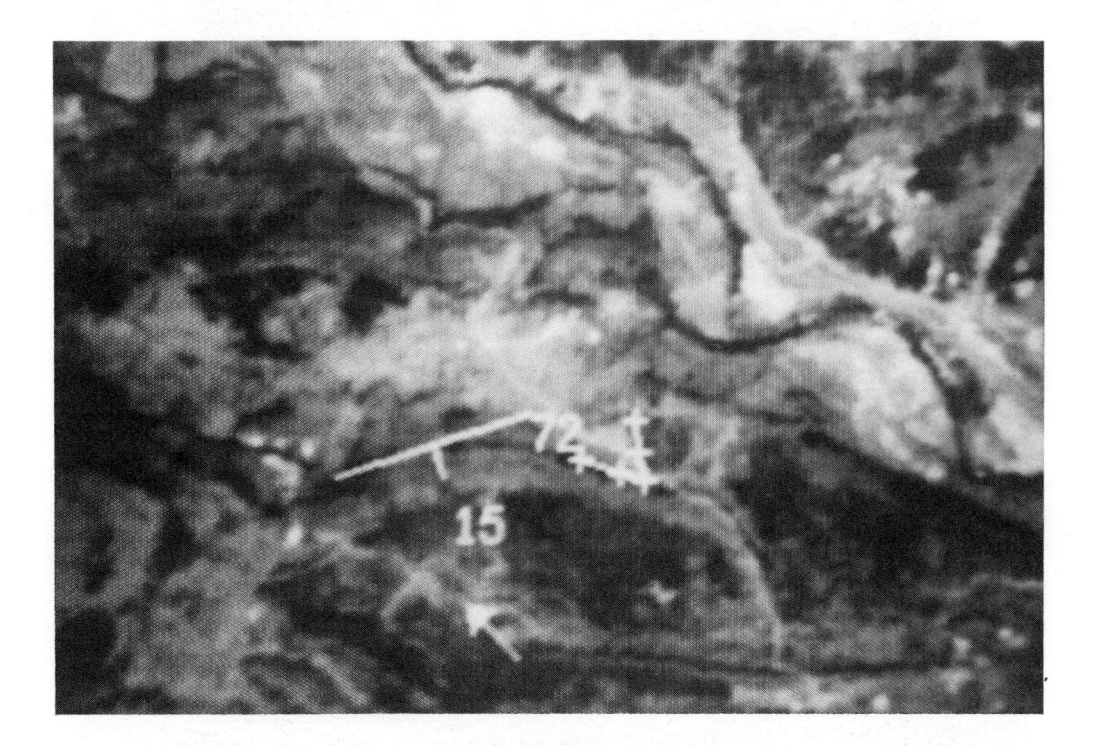

Figure 13. Line acquisition in interactive mode, and automatic determination of strike and dip of plane, where test shows that such line is tracing of plane. Geological map indicates for this layer: strike N80°E, dip 15°S.

Simulated perspective views ("3-D views")

Appropriate software makes it possible to indicate easily the parameters of the 3-D view which we want to work out, based on the DEM and the orthophoto: a focal point in the center of the image, look angle and elevation, focal distance, vertical exaggeration of land forms, and limits of the represented scope area. Figure 14 provides an example of aerial views obtained in this manner, which can be used to create a flight simulation (INRIA-SPOTIMAGE, 1987).

Perspectives also can be worked out from view points accessible to ground controls, and with the contacts, or faults and key beds, in graphic overlay (Figs. 15 to 17).

Figure 14. Simulated perspective view. Focal length 100 mm, view elevation 40°, view direction NNW, distance to image center 20 km.

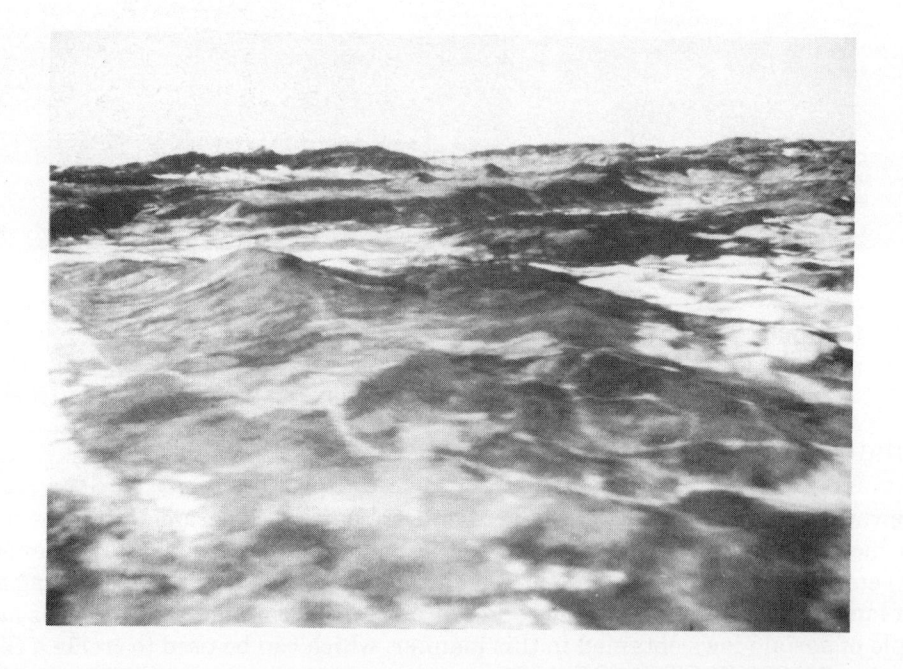

Figure 15. Photo of district under study taken from elevated point located at X Lambert = 690.4, Y Lambert = 3157.6, Z = 522 m. (Mean elevation in scope area is Z = 200 m.)

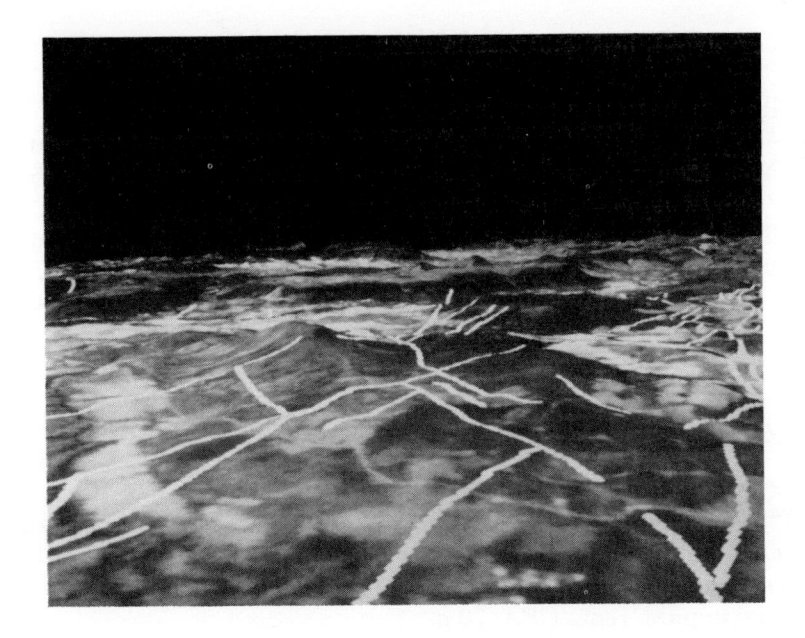

USE OF DIGITAL ELEVATION MODELS FOR URANIUM EXPLORATION 237

Figure 16. Simulated perspective view with graphic overlay of faults, as seen from same viewing point as Figure 15.

Figure 17. Simulated perspective view of rock types, as seen from same viewing point as Figure 15.

REFERENCES

Adam de Villiers, C., 1985, Utilisation géologique du traitement d'images appliqué aux modèles numériques de terrain. Exemples du Nord Limousin et du plateau de Millevaches: thèse , Ecole Nationale Supérieure des Mines de Paris, 150 p.

Adam de Villiers, C., and Leymarie, P., 1984, Cartographie automatique des situations topographiques en vue d'utilisations géologiques, *in* : JJ. Royer, ed., Computers in earth sciences for natural resources characterization, p. 333-350.

Allain, E., 1988, Doublevue, manuel d'utilisation: unpubl. rept., INRIA, 124 p.

INRIA - SPOTIMAGE, 1987, Voyage dans une photo satellite, Vidéocassette, duree 6 mn.

Leymarie, P., Durandau, A., Adam de Villiers, C., Rousselin, T., and Lerouge, G., 1985, Etablissement de règles de prospection de l'uranium basées sur l'explication et la précision des relations entre minéralisations uranifères et linéaments structuraux: unpubl. rept., CEA, 70 p.

Renouard, L., 1988, Création automatique de MNT à partir de couples d'images SPOT: SPOT 1 utilisation des images, bilan, résultats, p. 1347 - 1356.

APPLICATIONS IN PETROLEUM EXPLORATION

Conditional Simulation in Oil Exploration

H. Burger, M. Eder, A. Mannaa, W. Skala
Freie Universitaet Berlin, Berlin
Bundesanstalt fuer Geowissenschaften
und Rohstoffe (BGR), Hanover

ABSTRACT

Oil exploration is of probabilistic nature and includes large elements of uncertainty and risk. Two main sources of uncertainty which influence decision-making can be defined by (a) geologic uncertainty and (b) oil-occurrence uncertainty. Geologic uncertainty can be related to the distances of drillholes and to the complexity of the geologic structures. Unfortunately, drilling is most expensive and groups of closely spaced wells for detailed structural analysis seldom do exist. Even in that situation an uncertainty about the exact location, size, and shape of the structure would remain yet.

Interpolation methods (kriging, etc.) are able to detect local anomalies which are significantly larger than borehole distances. Trend surfaces and their respective residuals are highly dependent on the spatial distribution of the data points and are unstable at the border of the field.

However, conditional simulation techniques are suitable to model small-scale variations of a regionalized variable and to indicate local anomalies by conditioning the simulated data to control points. This approach takes into account the spatial structure of the stratigraphic layer which can be analyzed using data of the entire oil field.

In a well-known part of an oil field in the Rhein Graben attempts have been made to correlate local geologic structures indicated by conditional simulation with known oil occurrences.

INTRODUCTION

Oil exploration is of probabilistic nature and includes large elements of uncertainty and risk. Two main sources of uncertainty which influence decision making can be defined by (a) geologic uncertainty and (b) oil occurrence uncertainty. Geologic uncertainty can be related to the distances of drillholes and to the complexity of the geologic structures. Unfortunately, drilling is most expensive and groups of closely spaced wells for detailed structural analysis seldom do exist. Even in that situation an uncertainty about the exact location, size and shape of the structure would remain.

Various attempts have been made in order to map potential oil traps (Harbaugh, Doveton, and Davis, 1977; Olea and Davis, 1977; Robinson, Merriam, and Burroughs, 1978). All methods focus in delineating structural anomalies which seem to be favorable for oil accumulation. If a potential oil-bearing layer is dissected by faults (with significant uplift on one side) then the well-proven methods such as trend surface analysis or universal kriging must be improved by taking into account these discontinuities.

Interpolation methods (kriging, etc.) are able to detect local anomalies which are significantly larger than borehole distances. Trend surfaces and their respective residuals are dependent on the spatial distribution of the data points and are unstable at the border of the field.

However, conditional simulation techniques are suitable to model small-scale variations of a regionalized variable and to indicate local anomalies by conditioning the simulated data to control points. This approach takes into account the spatial structure of the stratigraphic layer which can be analyzed using data of the entire oil field. It seems to be possible to localize and to estimate shapes and sizes of structural anomalies. Each realization of a regionalized variable defines areas of positive and negative closure in an isoline map. These areas may be random and disappear in the following simulation run or they may be stable in almost all simulation runs resulting from the conditioning process.

In a well-known part of an oil field in the Rhein Graben attempts have been made to correlate local geologic structures indicated by conditional simulation with known oil occurrences.

GEOLOGICAL BACKGROUND

The sedimentary fill of the Alsatian part of the Rhein Graben in the area of Pechelbronn (Fig. 1) consists of around 650 m of Cenozoic, and around 1000 m of Mesozoic; top basement - thus granite - is at 1600 m depth.

In the graben area of Pechelbronn we distinguish two types of faults or fault systems (Schnaebele, Haas, and Hoffman, 1948):

- Antithetic faults or anormal faults. They are striking parallel to the main fault of the Rhein Graben and are dipping with 50° to the West. Partially synsedimentary movements can be observed.

- Synthetic faults or normal faults are less abundant in the Pechelbronn Field.

Besides these major faults there are numerous smaller faults with smaller (<20 m) throw (see Fig. 2).

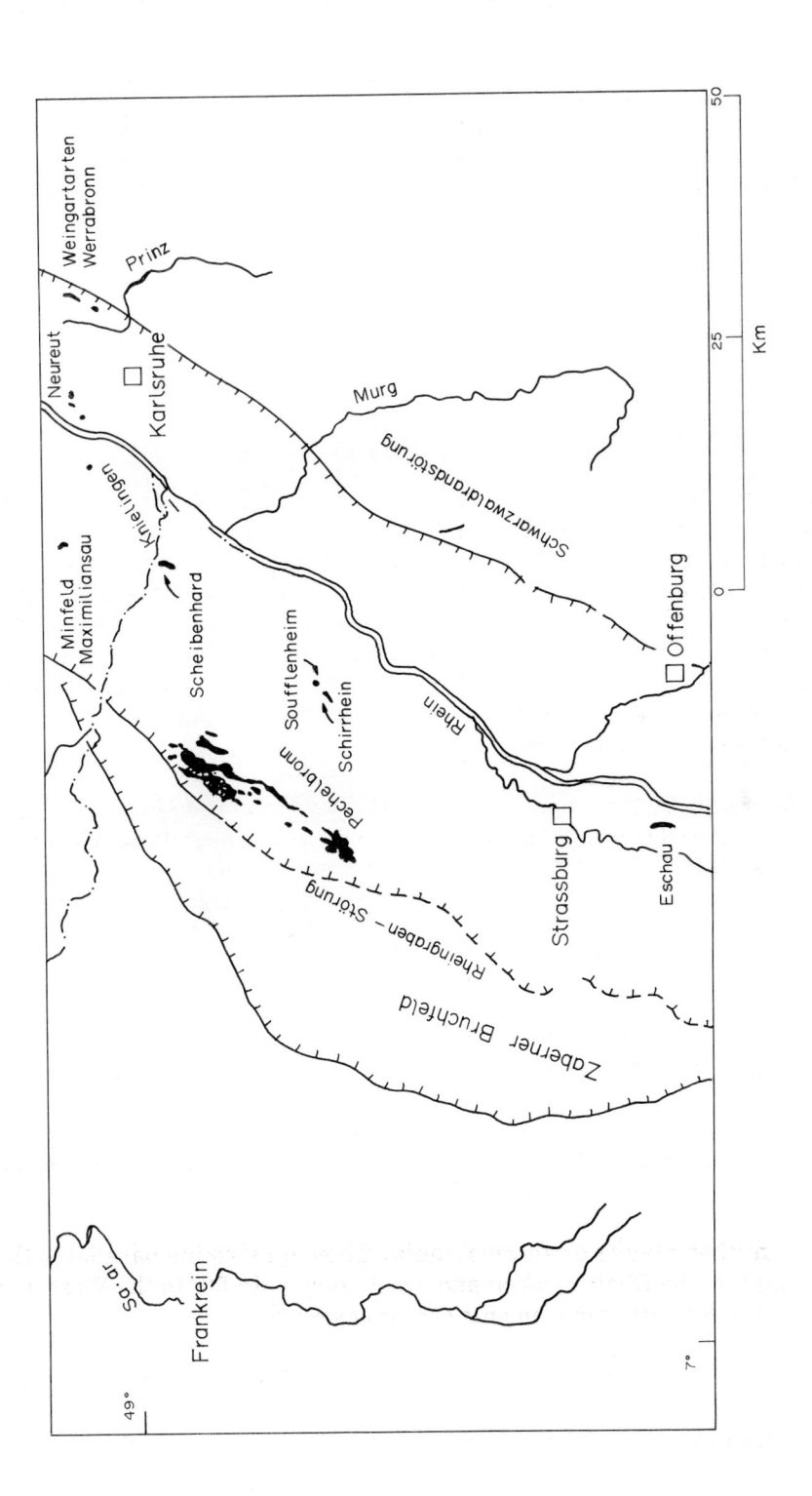

Figure 1. Location map of test area.

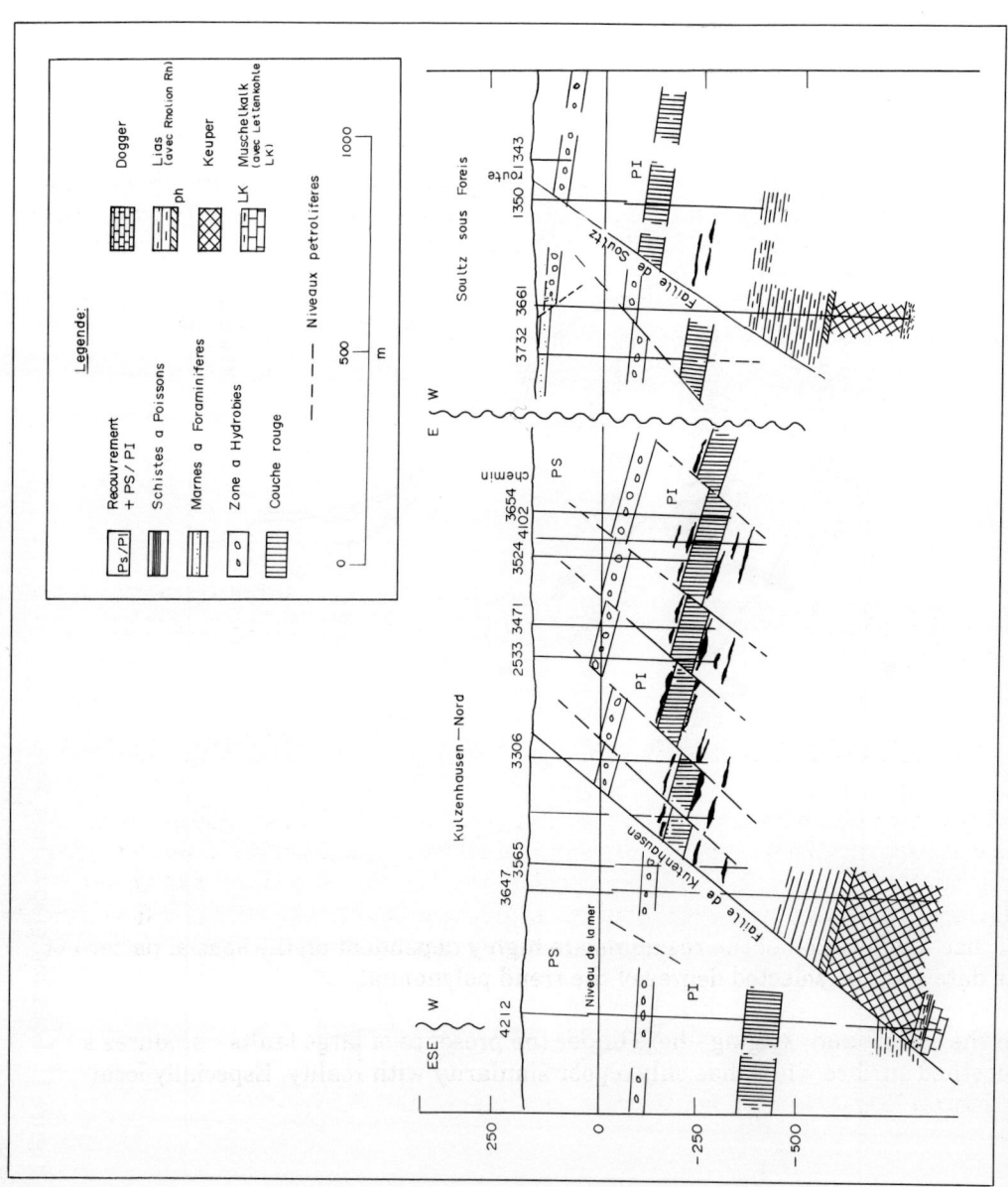

Figure 2. Fault system near Pechelbronn (from Schneabele, Haas, and Hoffman, 1948)

Downthrown since the Middle Eocene - 45 million years ago - the rift valley is cut into a Paleozoic Shield with a Mesozoic cover. Nearly 1000 m of this cover were preserved at Pechelbronn and are source rocks. The oil fields in the graben are generally linked with structural highs along faults. The accumulation of these fields on the western side of the graben axis is a result of the asymetric shape of the graben, the ascent of the layers to the west with its larger drainage area of source-rock sediments.

It can be postulated that the oil migration into Pechelbronn Field is related to deep groundwater flow. Lateral migration occurred mainly along permeable layers and vertical movement along faults.

DATA DESCRIPTION

The data used for this study have been extracted from internal drilling reports stored on microfiche (supplied by the BGR and SGAL). Main problem was the inhomogeneity of the available data set starting with first drillings in 1920! Oil occurrences were described using qualitative items ("barren", "traces", "fluent," etc), usually without exact information about the depth of the oil-bearing layer.

The oil occurrences were classified according to the following scheme:

　　　　0: rocks (sterile, no hydrocarbon traces)
　　　　1: traces of hydrocarbon (stained)
　　　　2: separate phase in water
　　　　3: saturated
　　　　4: oil (productive).

Finally 48 wells were selected with an almost complete set of information (location of borehole, depth of lithological units, hydrocarbon class) covering an area of about 12 sq km (Fig. 3).

CONDITIONAL SIMULATION OF STRUCTURAL SURFACES

The test area is subdivided by two major fault systems: the Kutzenhausen Fault in the west and the Soultz Fault in the east (Fig. 3). Trend-surface analysis and especially the mapping of residuals indicate clearly the presence of structural anomalies (Harbaugh and Merriam, 1968; Harbaugh, Doveton, and Davis, 1977). Unfortunately the size and location of the residuals are highly dependent on the spatial pattern of the data and the selected degree of the trend polynomial.

On the other hand, kriging - here under the presence of large faults - produces a smoothed surface which has only a poor similarity with reality. Especially local structural "highs" and "lows" are suppressed (Olea and Davis, 1977).

Conditional spatial simulation, introduced by Journel and Huijbregts (1978), combines the advantages of both methods:

- It takes into account the spatial variability of the data (instead of using an arbitrarily selected trend polynomial)

- the conditioning process reveals local "highs" indicated by neighboring data.

Variograms can be calculated and modeled in areas with sufficient data points. In our situation the central block which seems to be more or less undisturbed was used for a detailed variography. The experimental variograms confirm the continuous behavior of all relevant strata (Fig. 4). These models also are used for adjacent areas with sparse data, that is the assumption is made that the blocks are up- or down-lifted along faults without internal structural deformation.

Conditional simulation was performed for the elevation of each oil-bearing layer separately. Local anomalies are indicated on the contour line plots by circular or elliptic "closed" areas. The mapping of these closed areas on each contoured realization of the layer and the intersection of all these areas indicate the optimal location of a structural "high". The nature of these structures generally is not known (faulting, folding, etc.) but can be inferred from the local tectonic setting.

The number of intersecting "closures" and the variability of their location allows a ranking of favorable areas for drilling (Fig. 5).

The volume of the indicated anomaly which is an important economic factor, can be estimated by simple numerical integration (Harbaugh, Doveton, and Davis, 1977).

RELATIONSHIP BETWEEN SIMULATED "CLOSURES" AND OIL OCCURRENCES

Kriging of the elevation data of the Lower Pechelbronner Schichten (PBI) give no clear hint for local structural anomalies (Fig. 6). Compared to this the simulation runs (Fig. 7) reveal a more detailed image of the PBI - surface. The problem to solve is to decide whether these structural highs are real or only a result of the simulation procedure.

The superposition of all closed areas (maxima only) is shown in Figure 8. A comparison with the kriging result for the hydrocarbon classes (Fig. 9) shows that the main occurrences of oil coincide with the location of closures. A more detailed comparison with the original oil classes (Fig. 3) shows that the correlation between the productive class 4 and predicted highs is rather good. It may be interesting that there exists a promising structure in the western part of the area with no drillhole in its neighborhood.

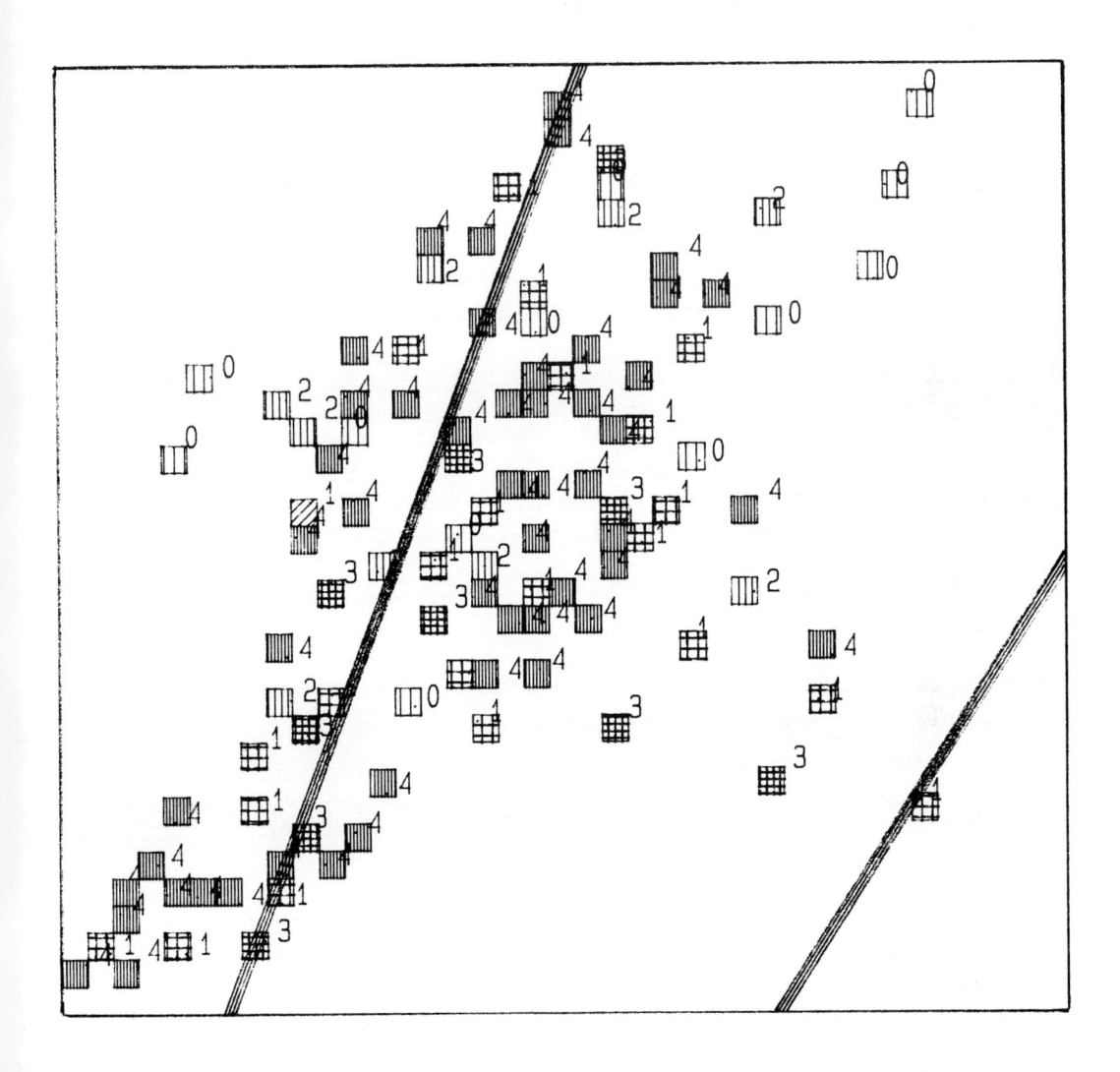

Figure 3. Distribution of hydrocarbon classes in Lower Pechelbronner Schichten (PBI, for details see text).

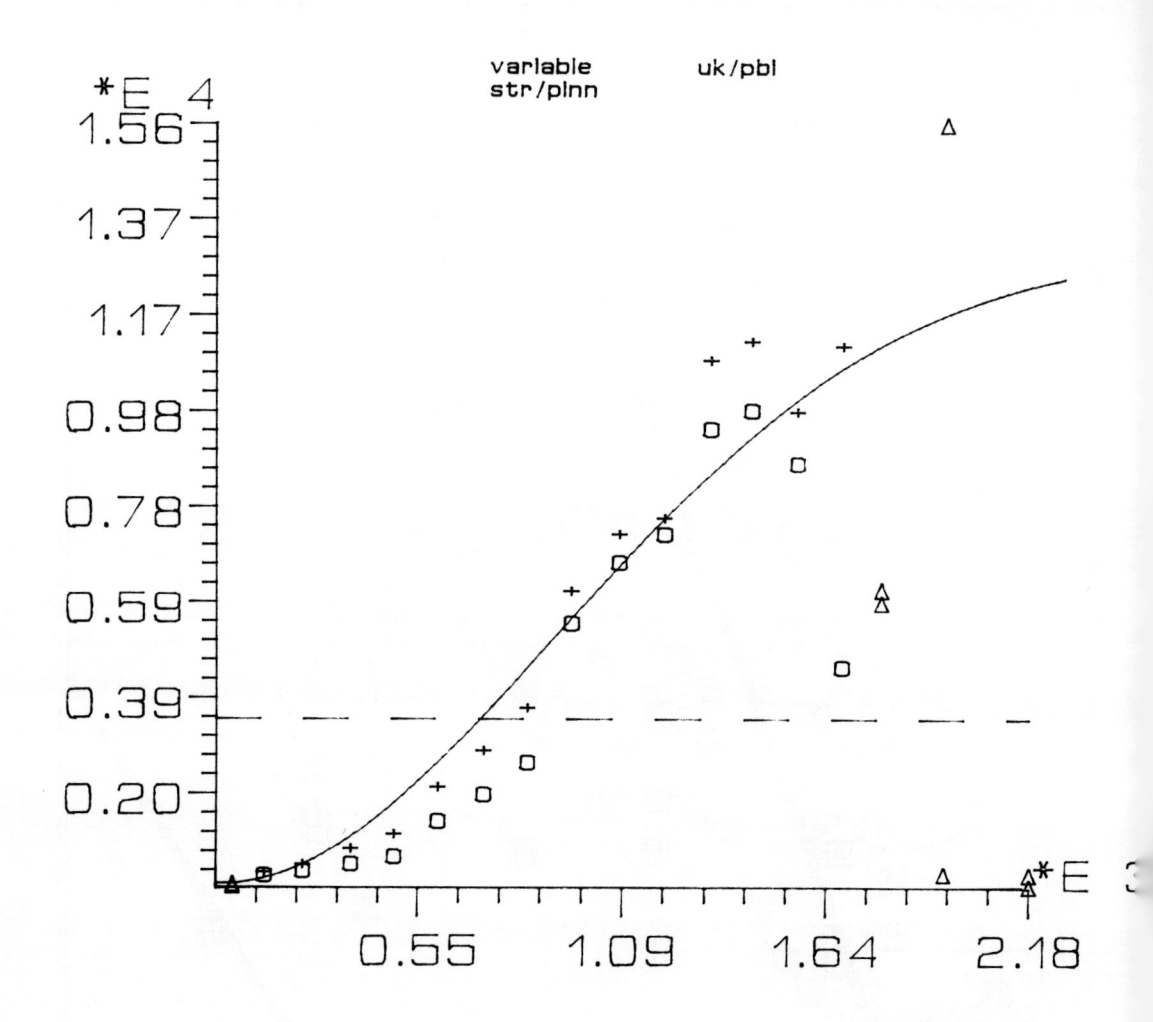

Figure 4. Variogram of PBI elevation data (data only from central part of area).

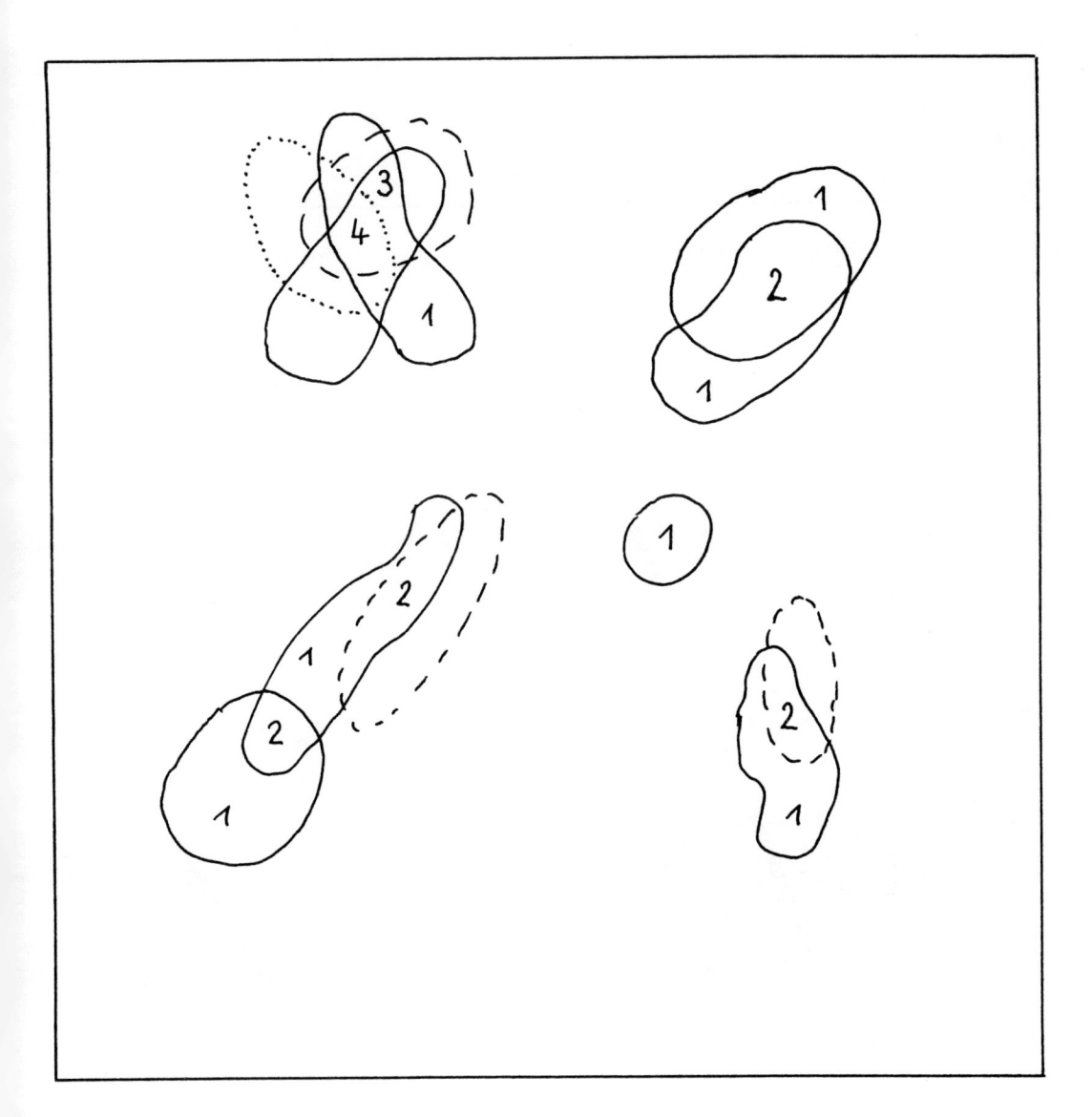

Figure 5. Intersection of hypothetical "closures" for producing probability map for structural "highs".

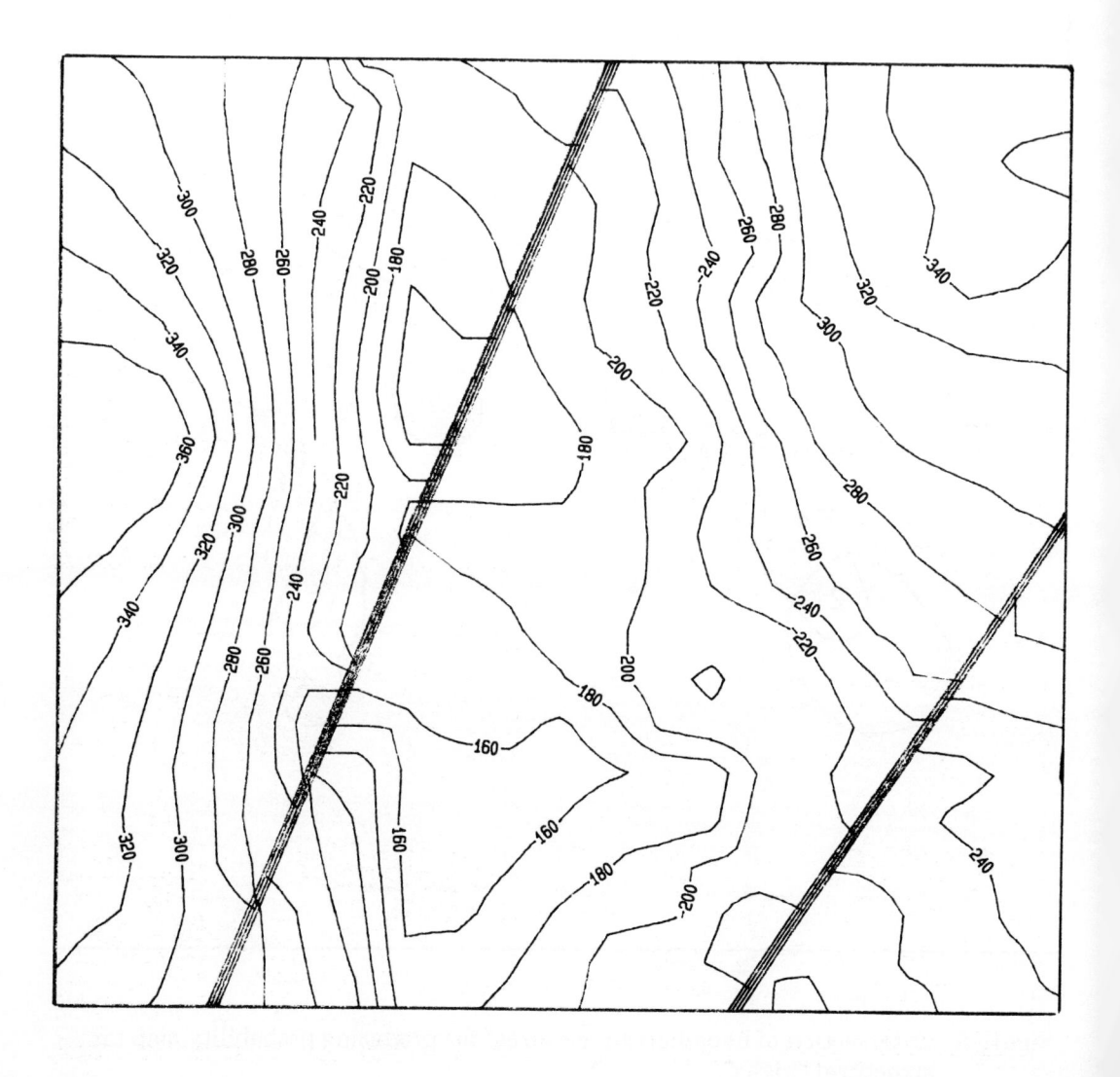

Figure 6. Contour map of kriging results for elevation values of PBI - structural surface. Smoothing effect hides local anomalies.

Figure 7. Four conditional simulations of elevation values of PBI - surface. Local "highs" are hatched.

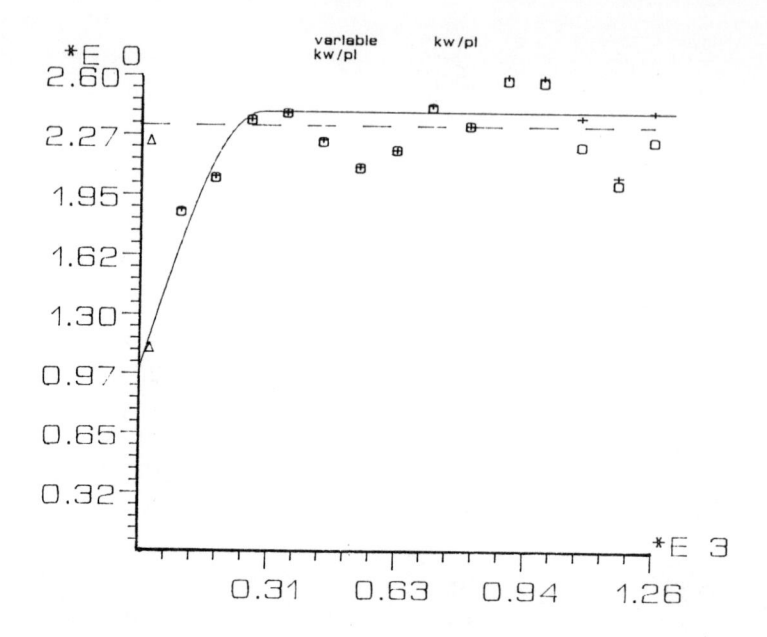

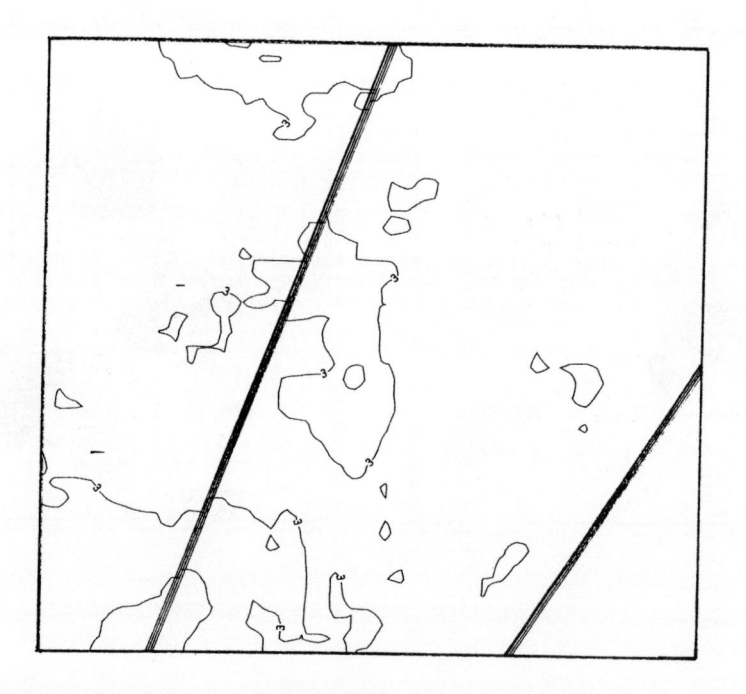

Figure 8. Superposition of 10 conditional simulations of PBI-surface. Outline shows areas with more than 3 intersections of local structural "closures", hatched areas with more than 5 intersections.

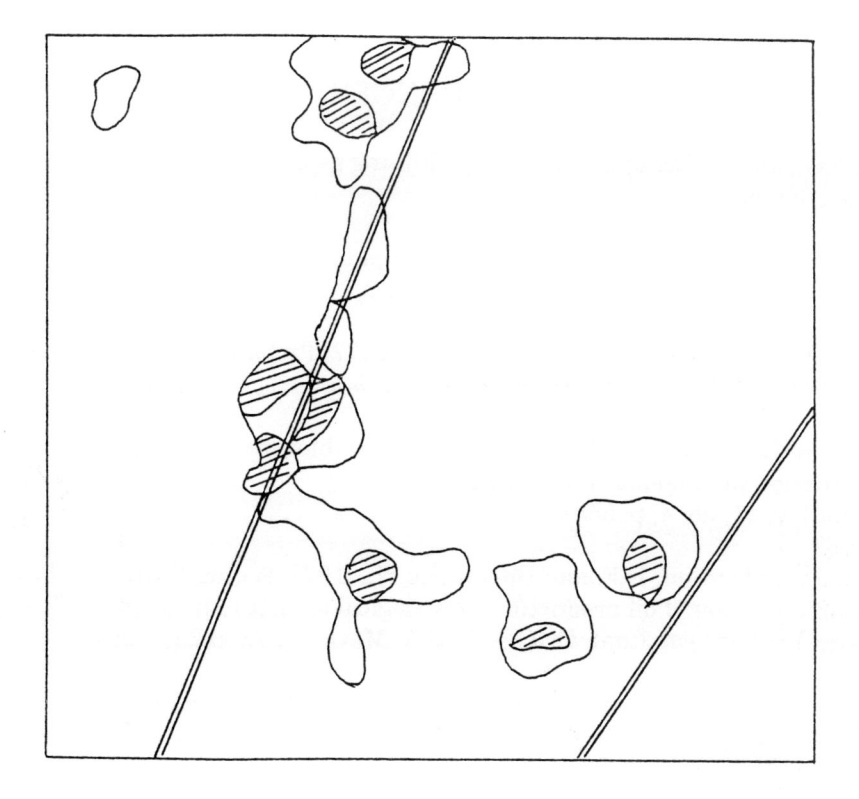

Figure 9. Variogram and kriging result of hydrocarbon classes of PBI-oil-bearing layer (see Fig. 3).

CONCLUSION

Conditional simulation seems to be a valuable tool for estimating the presence of structural highs. It is necessary that the spatial variability of the surface (described by the variogram) can be estimated from neighboring parts of the oil field. A more detailed case study (with production data for feedback) is planned to obtain more quantitative results for the correlation between simulated "closures", oil occurrences, volume estimates, and production rates. Therefore an additional objective of this project can be the development of more flexible strategies in grid drilling programs (Singer, 1972).

REFERENCES

Journel, A.G., and Huijbregts, Ch.J., 1978, Mining geostatistics: Academic Press, London, 600 p.

Harbaugh, J.W., and Merriam, D.F., 1968, Computer applications in stratigraphic analysis: John Wiley & Sons, New York, 282 p.

Harbaugh, J.W., Doveton, J.H., and Davis,J.C. 1977, Probability methods in oil exploration: Wiley Interscience, New York, 269 p.

Olea, R.A., and Davis,J.C., 1977, Regionalized variables for evaluation of petroleum accumulation in Magellan Basin, South America: Am. Assoc. Petroleum Geologists Bull.,v. 61, no. 4, p. 558-572.

Robinson, J.E., Merriam, D.F. and Burroughs,W., 1978, A quantitative technique to determine relation of oil production to geological structure in Graham County, Kansas: Proc. Intern. Conf. COGEODATA, Mexico 1978, p.256 - 264.

Schnaebele, R., Haas, J.O., and Hoffmann, C.R., 1948, Monographie geologique du champ petrolifere de Pechelbronn: Mem. du Service de la Carte Geophysique d'Alsace et de Lorraine, No. 7, 244 p.

Singer, D.A.,1972, Elipgrid, a FORTRAN IV program for calculatingting the probability of success in locating elliptical targets with square, rectangular and hexagonal grids: Geocom Programs 4, 16 p.

Computer-Assisted Estimation of Discovery and Production of Crude Oil from Undiscovered Accumulations

D. J. Forman and A. L. Hinde
Bureau of Mineral Resources, Canberra

ABSTRACT

SEAPUP is a computer program designed to simulate repeated drilling of petroleum traps in onshore Australia and to estimate discovery and production of crude oil during 1987 to 2000. During each iteration, the program simulates drilling a range of petroleum traps and when discoveries are simulated it estimates their sizes, the years of discovery, their economic viability, the lead times from discovery to production, and the annual production. Output from the program includes estimates of the number of wells drilled, the number of oil discoveries, the amount of oil discovered, success rates, discovery rates, number of accumulations brought into production, and oil production, both for each year and for the 14 year period, as well as estimates of the accumulation sizes in their simulated order of discovery. The program also simulates drilling all of the petroleum traps and provides an assessment of undiscovered crude oil resources.

INTRODUCTION

The basic principles for building a model for assessing undiscovered petroleum resources were outlined by Kaufman (1975). These same principles were used by Eckbo, Jacoby, and Smith (1978) and Band (1987) to build models of future discovery and production in the North Sea which incorporated a number of steps in the oil exploration, discovery, and production process. The model that we describe in this paper also contains a number of steps, but it differs from the earlier models, mainly in that it incorporates the method used by Forman, Hinde, and Cadman (1988) for assessing undiscovered petroleum resources and in that it covers all of the petroleum provinces throughout onshore Australia.

The SEAPUP model is made up of six submodels (Fig. 1) each of which is defined by one or more stochastic equations relating the input variables and the required output variable. Because exact values cannot be estimated for the variables in each equa-

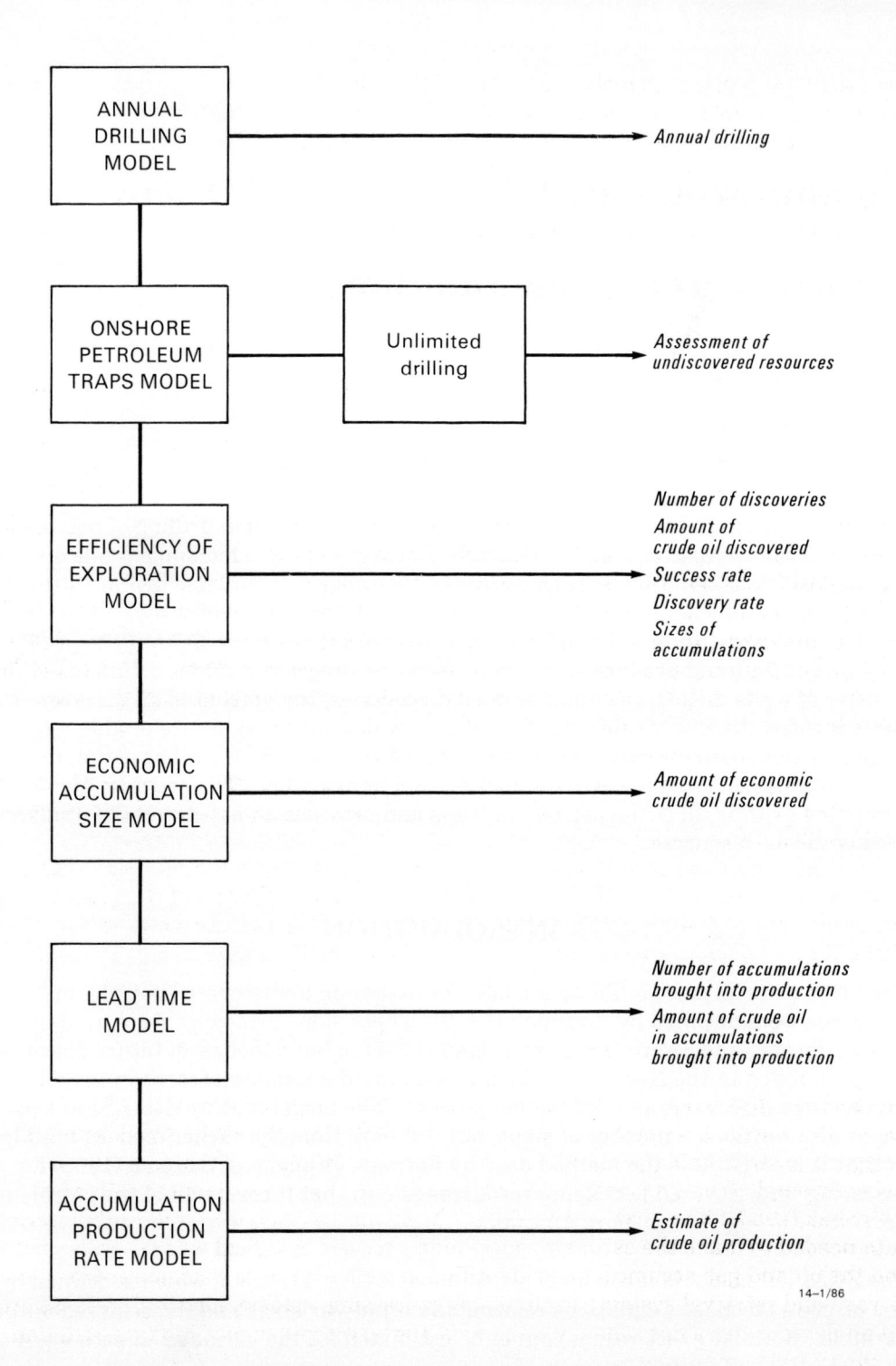

Figure 1. Diagramatic illustration of steps involved in SEAPUP program and information that is estimated.

tion, they are given as distributions of possible values. The input distributions were determined by groups of colleagues, referred to later in this paper as expert committees, who used their specialist knowledge and historic data to make subjective judgements of the values likely to apply to each factor during 1987 to 2000. This paper outlines the relationships that were used in each model, shows how the necessary distributions were determined, and gives examples of the output distributions obtained. It draws on unpublished work by Forman and Hinde (1987, 1988) and Hinde and Forman (1988).

THE SEAPUP MODEL

SEAPUP (Simulated Exploration And Production of Undiscovered Petroleum) is a computer program made up of a number of submodels and designed to estimate future discovery and production of crude oil in onshore Australia from 1987 to 2000. Simulated drilling of all the traps within the onshore petroleum traps model (Fig. 1) simulates the discovery of all the undiscovered crude oil accumulations and provides a probabilistic assessment of undiscovered oil resources. Drilling a defined number of traps, equal to the estimated number of new-field wildcat wells drilled from 1987 to 2000, provides the basis for estimating future rates of discovery and production of crude oil from the undiscovered oil accumulations within this period, using a number of other assumptions regarding efficiency of exploration, economic accumulation sizes, lead times, and production rates.

The results of several thousand iterations, covering a wide range of assumptions, are presented as cumulative probability distributions showing estimates of the number of wells drilled, the number of discoveries made, success and discovery rates, the number of accumulations brought into production, and crude oil production rates, both for each year and for the 14 year period, as well as estimates of the sizes of the accumulations in their simulated order of discovery. The estimates can be made either for any play or combination of plays or for all of Australia.

PREPARATION FOR THE ESTIMATES

The initial preparation for estimating crude oil discovery and production from undiscovered resources in onshore Australia was to divide the petroleum traps within the prospective sedimentary basins into a number of superplays. Each superplay consists of a single trap type within an independent petroleum system (as defined by Ulmishek, 1986). The next step in the estimate was to compile all relevant historic data needed to make the estimate, particularly for the new-field wildcat wells drilled and the oil and gas accumulations identified in each play, into a computer-based data storage and retrieval system. Preliminary processing, sorting, plotting, and statistical analysis of these data then were carried out by computer ready for examination by the expert committees.

ANNUAL DRILLING MODEL

The number of new-field wildcat wells drilled each year from 1960 to 1986 (Fig. 2) has been characterized by uptrends and downtrends with a high degree of autocorrelation from one year to the next. Our preferred drilling model provides an estimate of the number of new-field wildcat wells that could be drilled onshore Australia for each year from 1987 to 2000, assuming that uptrends and downtrends, similar to those of the past, will occur in the future. It includes the possibility of low drilling in early years and high drilling in later years and vice versa. It also takes into account the limitations on increasing well numbers from one year to the next and the number of years for which an uptrend or downtrend is likely to last.

The number of wells (N_i) to be drilled in a particular year (i) during an iteration is given by the equation:
$$N_i = N_{i-1} + n_i, \quad 1 < i < y_t \quad (1)$$

where N_{i-1} is the number of new-field wildcat wells drilled in the previous year, n_i is the increase (or decrease) in the number of wells drilled during year i, and y_t is the number of years for which the uptrend (or downtrend) continues before changing to the opposite trend. Some modification to the basic equation is needed to determine the length and duration of the initial downtrend that began in 1985.

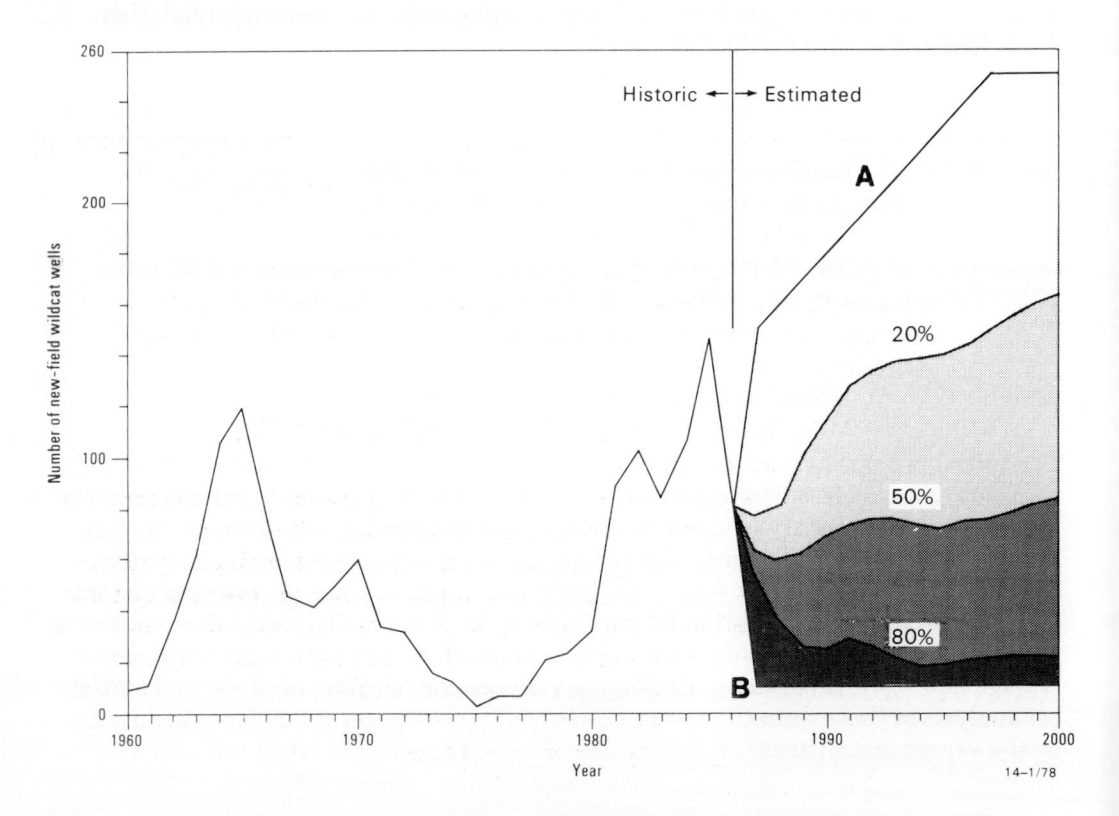

Figure 2. Annual new-field wildcat drilling onshore Australia from 1960 to 1986 together with upper (A) and lower (B) and 20, 50, and 80 percent probability estimates of drilling from 1987 to 2000.

The input data for the model include distributions for the values of n_i and y_t (Fig. 3), derived directly from the data in Figure 2, and upper and lower limits to the likely level of annual drilling (N_j) which were determined subjectively by an expert committee. The model provides a drilling profile for each iteration of the program by substituting values for n_i and y_t, selected by random selections from the probability distributions, into the equation and by repeating the process for each succeeding downtrend or uptrend until the year 2000 is reached. The model also provides cumulative probability distributions showing the estimate of annual new-field wildcat drilling for each year from 1987 to 2000 (Fig. 2) and for the whole 14 years period.

Alternative drilling models may be used in the program. For instance the number of wells to be drilled in each year of the estimate may be specified directly as a range of values.

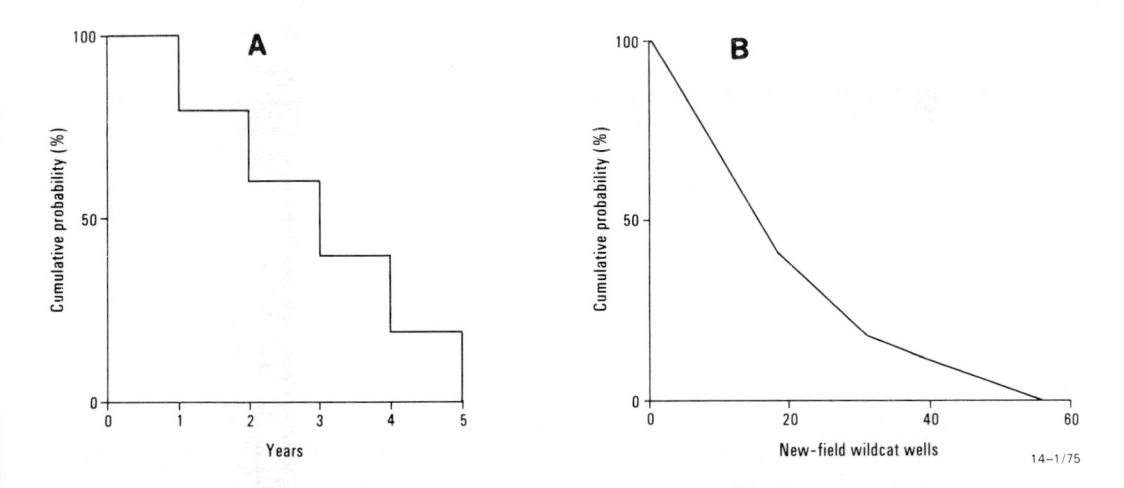

Figure 3. A, Cumulative probability distribution showing duration of past uptrends and downtrends in annual drilling. B, Cumulative probability distribution showing how number of new-field wildcat wells drilled has changed from one year to next.

ONSHORE PETROLEUM TRAPS MODEL

The model of the onshore petroleum traps contains a variety of information for each play concerning the number of traps, the amount of petroleum that each trap could contain, the success rate for future drilling, the proportion of oil to oil plus gas, the smallest size of accumulation to be included as a resource, and the probability that petroleum occurs in at least one of the traps. One of the main uses of the model is to estimate the risked average amount of petroleum that each petroleum trap could contain; referred to as the risked average trap size (R.A.T.S)

$$(R.A.T.S)_j = [\sum_{i=1}^{N_{iter}} V_{ij} d_{ij} c_{ij}] E_r / N_{iter} \qquad (2)$$

where N_{iter} is the number of iterations, V_{ij} is the simulated volume of recoverable oil and oil-equivalent gas in trap j during iteration i, d_{ij} is 0 or 1 depending on whether a random number is greater or less than a random value selected for iteration i from the distribution for the success rate, c_{ij} is either 0 or 1 depending on whether V_{ij} is smaller or larger than the random value selected for iteration i from the distribution for the smallest size of accumulation to be included as a resource, and E_r is the estimated probability that hydrocarbon resources occur in at least one trap in the play.

Another major use of the model is to calculate the amount of recoverable oil (V) that each discovery could contain using the equation

$$V = A_{clos} V_A P_o \qquad (3)$$

where A_{clos} is the area of closure of the trap, V_A is the volume of recoverable oil and oil-equivalent gas per unit area of closure, and P_o is the proportion of oil to oil plus gas. The method is based on the concept that A_{clos} is a rough measure of the drainage area of the trap and V_A is a measure of the vertical height of oil and oil-equivalent gas that can be caught and retained in the trap (Forman and Hinde, 1986; Forman, 1986; Bureau of Mineral Resources, 1988; Forman, Hinde, and Cadman,1988). Loglinear models are used to provide input distributions of A_{clos} and V_A for each discovery and a triangular distribution is used to provide input values for P_o.

Determination of area of closure

Forman and Hinde (1985) and Forman, Hinde, and Cadman,(1988) showed that computer simulations of a successive sampling model, incorporating (1) undrilled petroleum traps whose closure areas (A_{clos}) form a lognormal distribution truncated at each end and with a comparatively low standard deviation and (2) an order of drilling equivalent to random sampling without replacement and proportional to A_{clos} raised to a constant power lambda (λ), will yield plots in which the average values of log A_{clos} versus drilling sequence number are approximately linear, particularly when the value of lambda is low. In these plots, the distribution of possible values for log A_{clos} for each drilling sequence number is close to normal and has a similar standard deviation. The simulations also showed that there is a dependent relationship among lambda, the slope and intercept of the fitted line, and the standard deviation of the residuals which is given by the equation

$$B = B_{max}(1 - e^{-s\lambda}), \quad \lambda > 0 \qquad (4)$$

where B is the average slope for a given value of lambda (λ), B_{max} is the slope of the fitted line when lambda equals infinity, and s is the standard deviation of the logarithms of the areas of closure.

The loglinear model is used to input the areas of closure and the likely order of drilling of the undrilled traps in each play. Depending on how the data are compiled and sampled, the model forms the basis of two methods, one statistical and one geological, for determining the areas of closure (A_{clos}) of the undrilled traps in the play. The statistical method may underestimate the areas of closure (Forman and Hinde, 1988) whereas, because of the more subjective approach, the geological method could overestimate the areas of closure. It is advisable, therefore, to use both methods as a cross-check on consistency.

In the statistical method, the linear model is derived from a plot of the areas of closure of the traps that already have been drilled versus their drilling sequence numbers (Fig. 4). In the situation of a geological assessment, the linear model is derived from a plot of the areas of closure of the traps that remain to be drilled in their likely order of drilling. In areas where information on the areas of closure is incomplete, the linear models are constructed using the additional types of information, such as density of structuring and analogy with other plays, that are outlined in Baker and others (1986). The parameters of these models are the slope and intercept of the straightline fitted to the plot by the method of least squares, the standard deviation of the residuals to the fitted line, and the mean and standard deviation of the normal distribution of lambda.

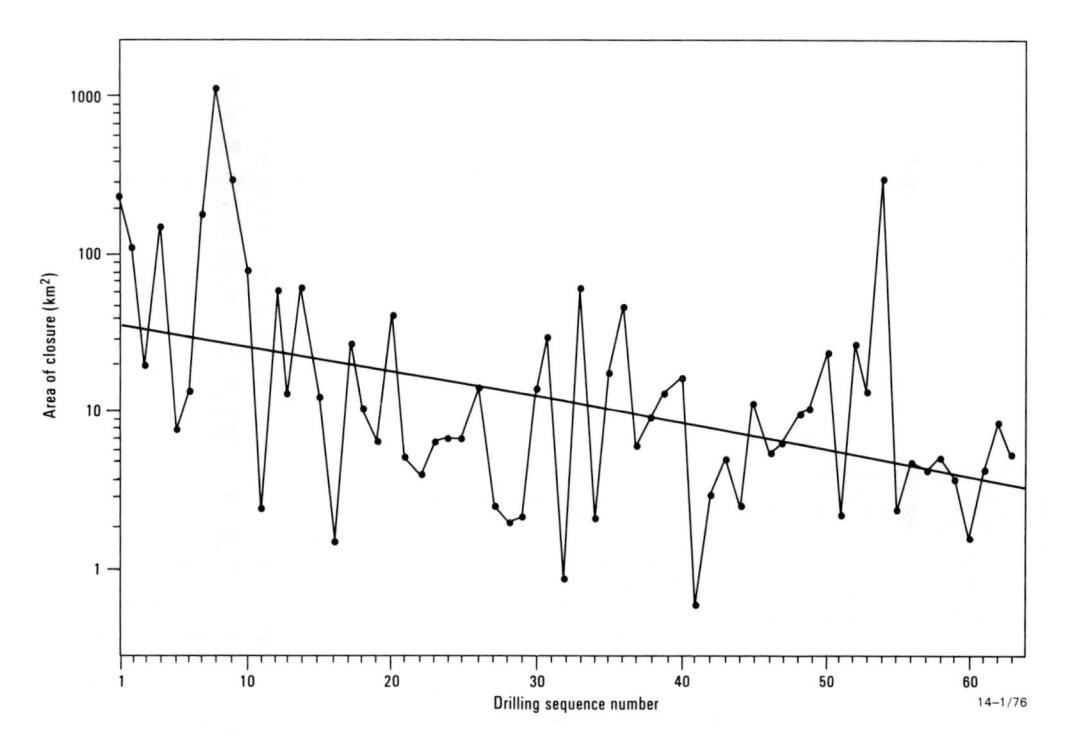

Figure 4. Areas of closure (A_{clos}) of anticlinal traps within a superplay (part of Eromanga Basin) plotted against their drilling sequence numbers.

Determination of resources per unit area

Resources per unit area for the undiscovered accumulations in each play are determined (Forman, Hinde, and Cadman 1988) by sampling or projection of linear models of the logarithms of the resources per unit area (log V_A) for the undiscovered accumulations in their likely order of discovery. The models are selected after interpretation of plots of log V_A versus discovery sequence number for the identified accumulations in each trap type in each independent petroleum system (Fig. 5). For a play that has been tested by only one or two wells, a distribution of V_A may be determined by quantitative geochemical and geological modeling if both the time of formation of the trap with respect to generation, the relationship between the closure area and the drainage area of the trap, the likely recovery factor, and the sealing capacity of the trap are known. Alternatively, a distribution of V_A or a linear model of log V_A versus discovery sequence number may be selected from better explored plays with analogous source, reservoir, and cap rocks, and drainage efficiency.

EFFICIENCY OF EXPLORATION MODEL

The efficiency of exploration model is designed to ensure that the values of lambda for the simulated order of discovery of the undiscovered fields fall within a range established by an expert committee after examination of the historic trends visible in the plot of log accumulation size (log V) versus discovery sequence number (Fig. 6) and consideration of likely future trends. The task is complicated firstly because the expert committee cannot assess the values of lambda directly and secondly because the program does not use field size and lambda but instead determines the probability (P_j) of drilling a trap next using the equation

$$P_j = \{[\sum_{i=1}^{N_{iter}} V_{ij} d_{ij} c_{ij}] E_r / N_{iter}\}^{\propto} \text{constant} \quad (5)$$

where the first factor on the right hand side of the equation is the risked average trap size, obtained from the petroleum traps model, raised to the power alpha ($\propto$). Alpha is an adjustable parameter of the model which gives the program a tendency to simulate early drilling of the most prospective traps.

During the ordering process, the existence risk (E_r) is changed to one if a play is successfully risked, thus automatically changing the order of drilling for all subsequent traps. The year in which each trap may be drilled then is estimated by matching up the traps in their drilling order with the number of new-field wildcat wells drilled for each year. The program then simulates drilling the traps, and when discoveries are simulated it estimates their sizes and the years of discovery.

DETERMINATION OF THE ALPHA EXPONENT

An expert committee examined the historic trends visible in a plot of log accumulation size versus discovery sequence number for the oil accumulations that already have been discovered onshore Australia (Fig. 6) and decided on a range of trends that could apply in the future. The values of lambda for these trends in the accumulation size then were calculated. The SEAPUP program was run a number of times, each time using a different value of alpha, so that the value of lambda for the simulated oil discoveries could be calculated and the relationship between lambda and alpha could be established. Once this had been done, a distribution of values for alpha was selected such that SEAPUP would simulate discovery of oil accumulations according to the selected range of values for lambda.

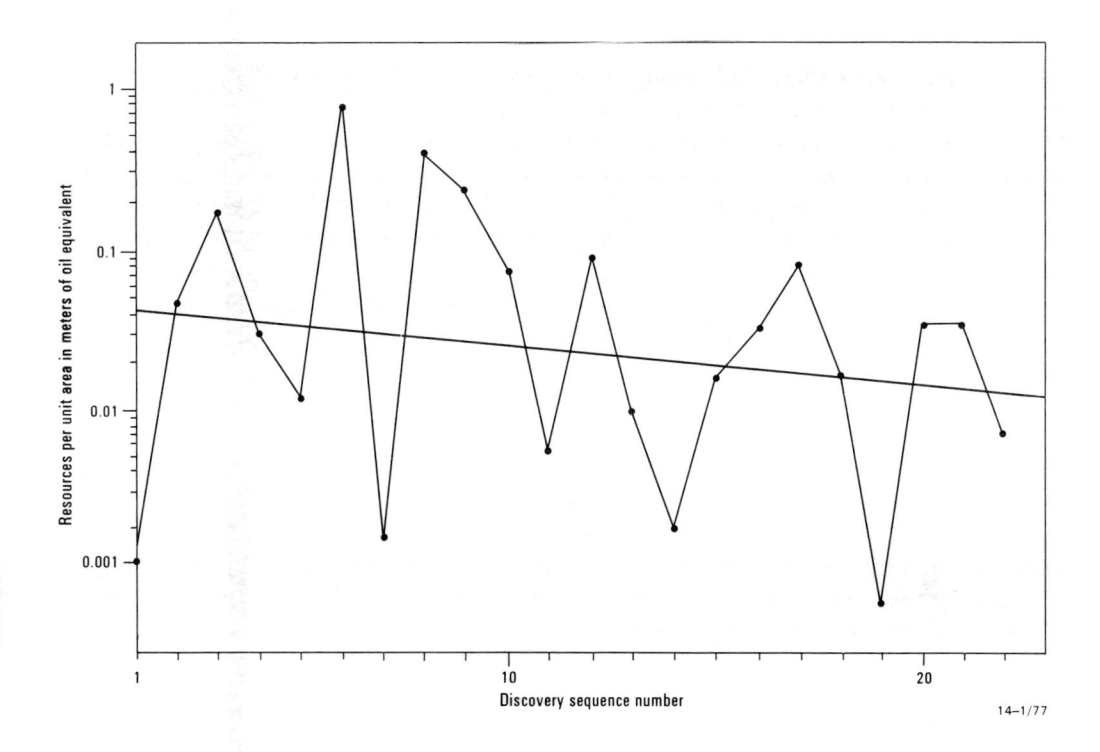

Figure 5. Resources per unit area (V_A) within anticlinal traps within superplay (part of Eromanga Basin) plotted against their discovery sequence numbers.

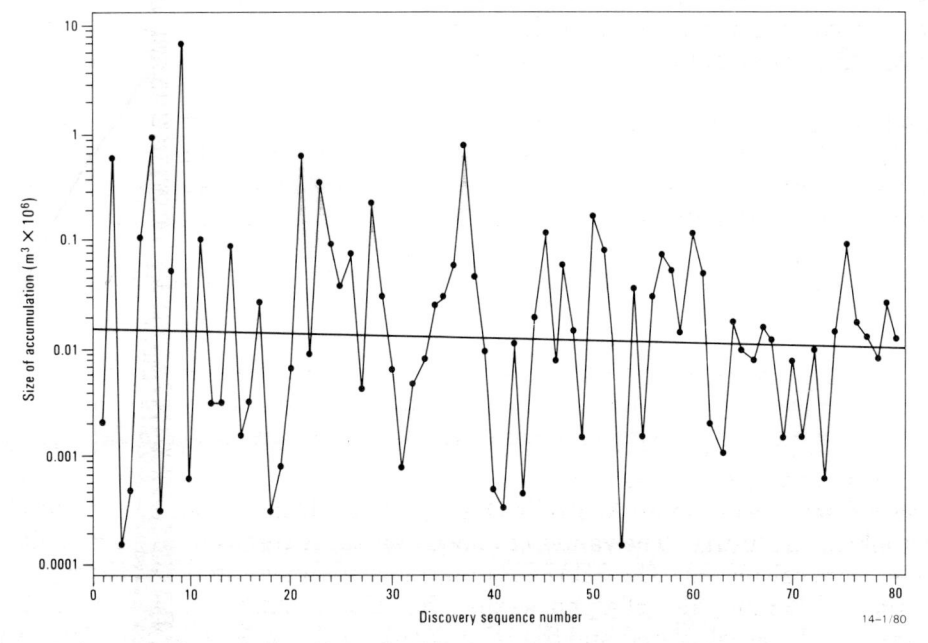

Figure 6. Logarithm of oil accumulation size plotted against discovery sequence number for Australia's onshore oil discoveries. Straightline has been fitted by method of least squares.

ECONOMIC ACCUMULATION SIZE MODEL

The likelihood of bringing an accumulation into economic production is determined by comparing the estimated size of the accumulation with a size selected at random from a distribution showing the range in the smallest size of accumulation that it is thought could be brought into economic production during 1987 to 2000. A distribution, which is consistent with experience of recent developments in Australia and with oil prices within the range of $14-28 per barrel (1986 US dollars), has been prepared for each onshore region by expert committee. These distributions together make up the economic accumulation size model.

LEAD TIME MODEL

The lead time between simulated discovery and production of each oil accumulation is determined by random selection from a probability distribution of likely lead times. The distribution has been selected by an expert committee after consideration of historic lead times (Fig. 7) and subjective opinion concerning the likely distribution of future lead times. For instance, although it has taken as long as 21 years to bring an oil accumulation into production in the past, comparison of the price of oil and the infrastructure that existed in the 1960's with those expected in 1987 to 2000 suggests that such long lead times are unlikely in the future. Hence, a histogram of future lead times has been obtained by truncating, contracting, and smoothing the histogram of historic lead times.

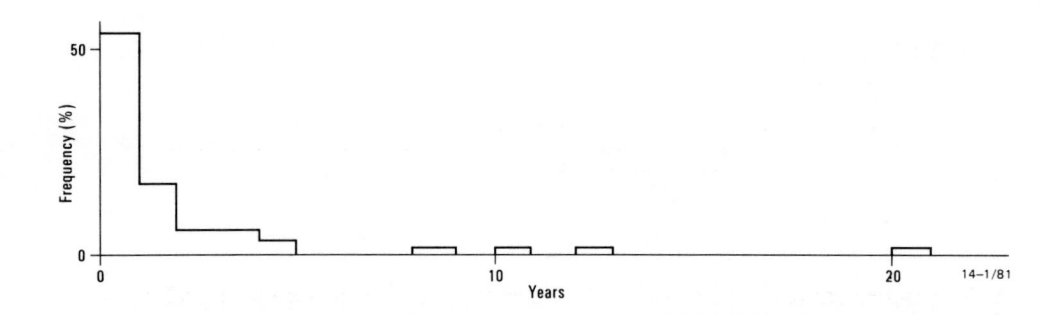

Figure 7. Frequency histogram showing lead times between discovery and production for all producing oil accumulations onshore Australia to October 1985.

ACCUMULATION PRODUCTION RATE MODEL

Consideration of the production profiles for Australia's producing onshore fields by an expert committee suggests that the fields typically conform to the pattern shown in Figure 8 and that oil production (P_i) in each year, i, can be determined using the equations

$$P_i = \begin{cases} ib/a, & 0 < i < a \\ b, & a < i < a+c \\ b(1-d)^{i-(a+c)}, & a+c < i \end{cases} \quad (6)$$

where a is the number of years taken to reach maximum production, b is the rate of maximum production (expressed as a fraction of the total resources), c is the number of years for which maximum production is maintained, and d is the proportion of the previous years production by which production declines from the level of maximum production.

The program estimates a production profile for each simulated oil discovery by substituting random values, selected from probability distributions for a, b, c, and d, into Equations (6), (7), and (8). The four distributions, which have been compiled by an expert committee assuming that future production rates will fall into the same broad range as past production rates, are: (a) a uniform distribution providing input for the period, from 0 to two years, that production may take to build up to a plateau; (b) a uniform distribution providing input for the period, from three to nine years, that production will remain at plateau levels; (c) a uniform distribution providing input for the percentage of the resources, from seven to 15, produced during each year of plateau production; and (d) a uniform distribution providing input for the percentage fall in production rate, from five to 20, during each year of the production decline.

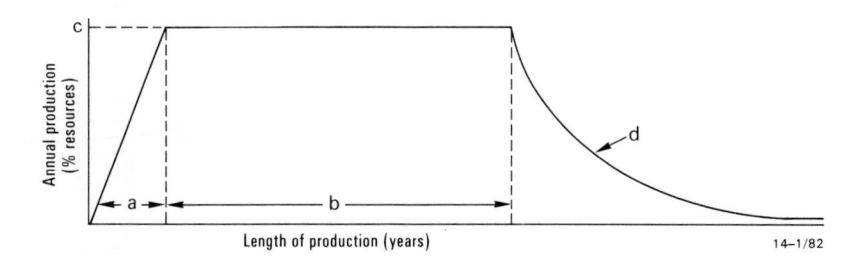

Figure 8. Diagrammatic representation of typical crude oil production profile, onshore Australia: (a) period of production build-up; (b) period of plateau production; (c) percent of recoverable resources in accumulation that are produced in each year of production plateau; and (d) production decline, typified by regular decline in production rate.

OUTPUT FROM THE SEAPUP PROGRAM

The main types of output obtained routinely from the SEAPUP program are shown diagrammatically in Figure 1. Initially, the program obtains an assessment of undiscovered crude oil resources by simulated drilling of all of the traps within the Australian onshore petroleum traps model. The assessment includes a cumulative probability distribution (Fig. 9) indicating a potential for further discoveries of somewhere between about 20 and 150 million m³ of crude oil, with an average of about 60 million m³ (380 million barrels). The assessment also includes an estimate of the risked average undiscovered crude oil resources within each play. These may be summed to give a summary estimate of the undiscovered crude oil resources of each basin (Fig. 10), which suggests that the Canning, Eromanga, and Otway Basins are the most prospective.

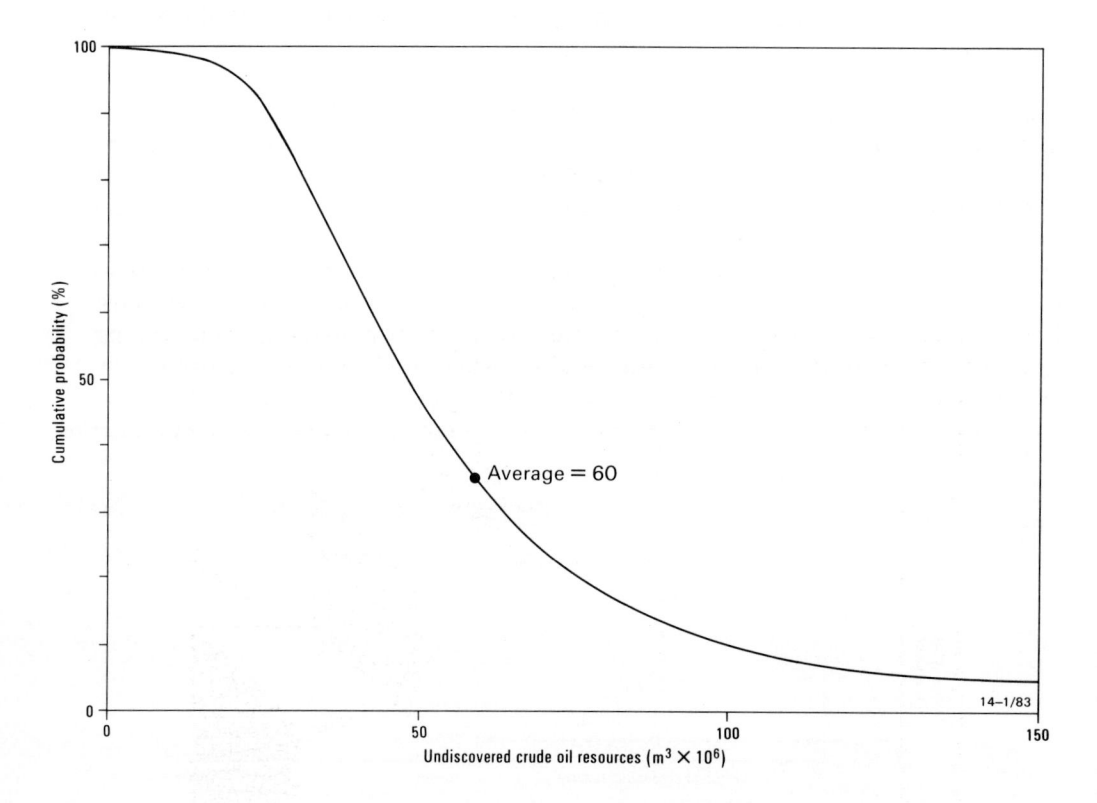

Figure 9. Cumulative probability distribution showing assessment of undiscovered crude oil resources in onshore Australia, as of May 1986.

The program also provides estimates of the number of wells drilled, number of discoveries, the amount of crude oil discovered, success and discovery rates, number of accumulations brought into production, and crude oil-production rates resulting from the range of assumptions made in each of the models. Figure 11 shows the estimates that were prepared for each year from 1987 to 2000 assuming that annual drilling would lie in a range of 50 to 150 new-field wildcat wells, with a most likely value of 100 (Figure 11A). Corresponding estimates also are prepared for the whole period from 1987 to 2000 (Table 1). In addition, the program provides estimates of the amounts of economic and uneconomic oil to be discovered and the amount of economic oil to be discovered in fields that may be brought into production during 1987 to 2000.

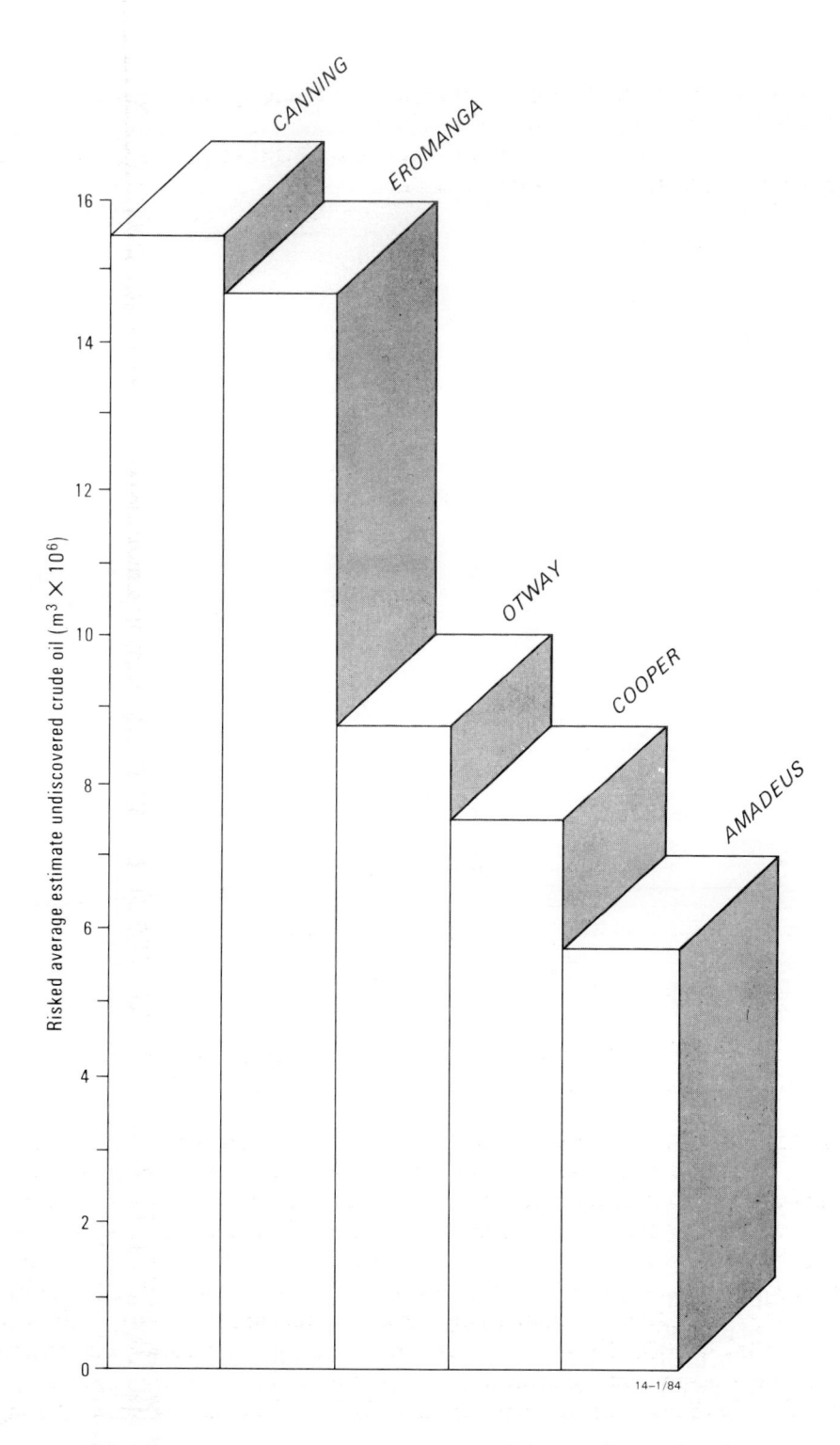

Figure 10. Five of most prospective sedimentary basins onshore Australia ranked in order of risked average assessment of their undiscovered crude oil resources.

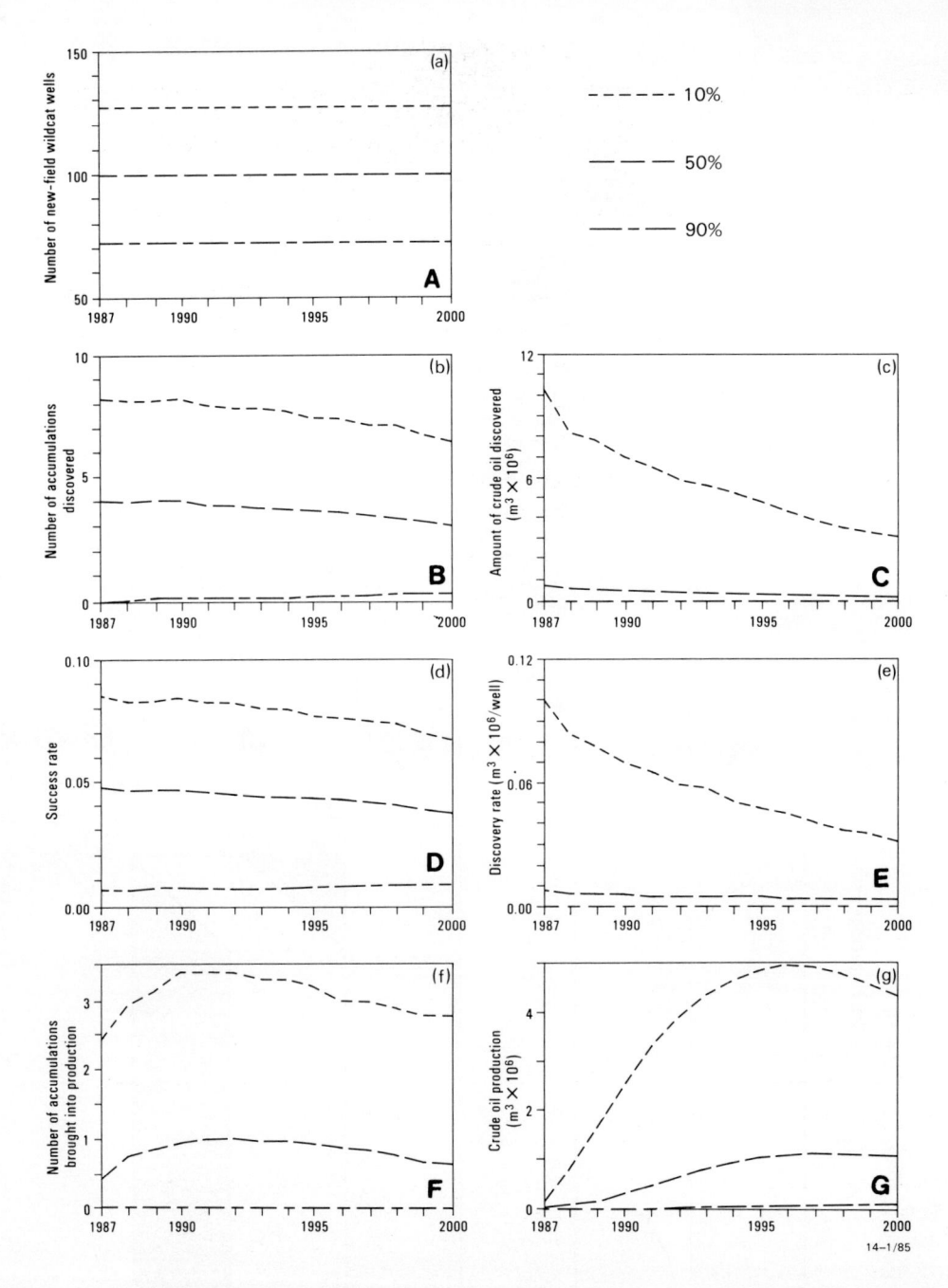

Figure 11. Annual estimates for onshore Australia at 10, 50, and 90 percent cumulative probability: A, number of new-field wildcat wells drilled (using alternative drilling model); B, number of accumulations discovered; C, amount of crude oil discovered in million m^3; D, success rate (number of oil accumulations discovered per new-field wildcat well drilled); E, discovery rate (million m^3 of crude oil discovered per well drilled); F, number of accumulations brought into production; and G, crude oil production in million m^3.

The size of each simulated discovery is recorded in a histogram according to its discovery sequence number and these are converted into cumulative probability distributions from which the information shown in Figure 12 has been obtained.

The identical range of information may be obtained for any of the plays or for any sedimentary basin onshore Australia.

Table 1. Estimates at 90,50, and 10 percent probability levels of drilling, discovery, and production for onshore Australia during period 1987 to 2000.

	Probability of exceeding value shown		
	90%	50%	10%
Number of new-field wildcat wells	1310	1400	1490
Number of accumulations discovered	23	66	86
Amount of crude oil discovered (million m^3)	4	27	80
Success rate (discoveries / well drilled)	0.018	0.05	0.06
Discovery rate (m^3 / well drilled)	60 000	20 000	3000
Number of accumulations brought into production	6	23	33
Amount of oil in accumulations brought into production (million m^3)	2.5	22	75

LIMITATIONS OF THE MODEL

The estimates refer only to undiscovered conventional crude oil resources; that is conventional resources contained in traps that are not presently known to contain either oil or gas. Hence, they do not include any allowance for new pool discoveries in identified oil accumulations or for enhanced recovery of crude oil from identified or undiscovered accumulations.

The main deficiency in the estimates is a lack of knowledge of all the circumstances in which oil may occur throughout Australia. We cannot assess oil that we cannot conceive. Hence, it is possible that oil occurs in well-known play types in as yet unknown sedimentary sequences or in unknown play types in well-known sedimentary sequences. A major problem is that for many sedimentary basins we have insufficient information on which to base a reliable estimate of their undiscovered

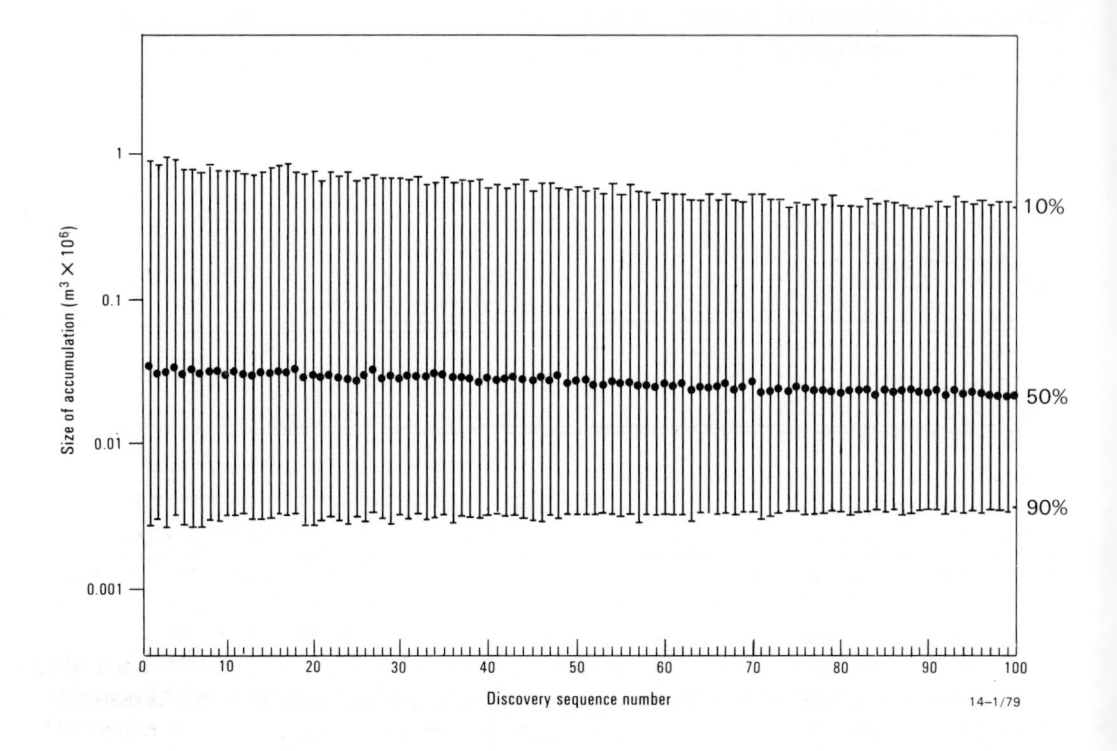

Figure 12. Estimates at 10, 50, and 90 percent cumulative probability of sizes of first 100 oil accumulations, onshore Australia, in their simulated order of discovery.

crude oil resources. Even in well-known plays in well-known sedimentary sequences, our perception of the number of petroleum traps and their size distribution will increase with the quality and density of the seismic surveys that have been undertaken (Forman, Hinde, and Cadman, 1988). Also, our perception of the amount of petroleum contained in each identified accumulation and hence in the undiscovered accumulations tends to increase as a result of continuing revisions to reserves.

Modeling of the rate at which the resources may be discovered and brought into production is limited by the historic information available and by our inability to know all the factors involved, such as oil prices, production costs, and the level of future drilling. Consequently, the range of values given in the estimates may be conservative and they must be used with caution.

CHECKING THE ESTIMATES AND THE MODEL

An obvious technique for diagnostic checking of the model is to run the computer program using data appropriate to some time in the past, such as 1976, to see if it predicts subsequent events with reasonable precision. This procedure is applied to predictive models, but SEAPUP contains both the experience of the events that occurred from 1976 to 1986 and subjective expert opinion and it can not be converted into a model appropriate to 1976 with any reliability. For instance, it would be particularly difficult to attempt to reproduce the model of the petroleum traps as it may have been conceived in 1976. Such diagnostic testing as is to be carried out, therefore, is best performed by checking the accuracy of the estimates against the actual outcome.

Because of the lead time between discovery and production, especially in offshore areas, comparison of the actual number of wells drilled, number of discoveries, the size of the discoveries, the amount of oil discovered, success rate, and discovery rate with those estimated should give advance warning of whether the estimate of crude oil production will remain within the predicted ranges. For instance, because of the advent of drilling funds in Australia, about 157 new-field wildcat wells now are known to have been drilled onshore in 1987, just exceeding the maximum limit of 150 wells built into both drilling models. The wells drilled discovered about 12 new oil accumulations containing demonstrated resources probably less than 1.6 million m^3 (10 million barrels) of crude oil (Laws, 1988). The number of oil discoveries is high and the amount of oil discovered is low compared to the median values of the estimates (Fig. 11); this is attributed partly to the high level of drilling and partly to a strategy of drilling in established oil provinces where success rate has been comparatively high and discovery rate low. No oil accumulations were brought into production during the year.

These results illustrate the importance of the input data. The record number of wells drilled in 1987 exceeded the number input to the drilling model and as a result more accumulations were discovered than estimated. Although all the other exploration results for 1987 seem to have been satisfactorily predicted by the estimates, it is evident that new input should be provided to the drilling model. Nevertheless, even close matching of all the estimates for onshore Australia with the observed outcomes need not imply that the model is correct; for onshore Australia contains a large number of superplays and there could be compensating errors. The model itself should be tested, therefore, by comparing the estimates made for each play with the actual outcomes for each play.

Each estimate is given as a cumulative probability distribution against which the outcome is easily compared. There are no widely accepted standards, however, by which differences between individual estimates, given as ranges of values, and their outcomes may be judged when concerned with highly skewed distributions. It seems better that acceptability of an estimate be left to the judgement of the experts familiar with the region and the underlying reasons for the trends. Hence, monitoring of the results of exploration in individual plays and comparison with the estimated results by a group of expert geoscientists are the best ways of checking the model and deciding if either the model or the input data require alteration to reflect the new experience.

In many situations, alterations will be made to areas of the program or data that are relatively insensitive to change. For instance, sensitivity testing carried out by differing prospectiveness, drilling rate, efficiency of exploration, economic cut-off of accumulation sizes, and the lead time from discovery to production demonstrates that the assumed prospectiveness and the assumed rate of drilling of new-field wildcat wells are the most important factors. Assumptions about the efficiency of exploration and the lead time from discovery to production have a significant impact on results obtained, but are of lesser importance. Assumptions about the economic cut-off values for accumulations in the different regions seem to have least effect on the average estimate of total crude oil production. Hence, many changes will result only in fine-tuning of the estimate.

ACKNOWLEDGMENTS

We thank P.L. McFadden, A.P. Radlinski, and C.S. Robertson of the Bureau of Mineral Resources (BMR) for critically reviewing the manuscript and the BMR drafting office for drafting the figures. The paper is published with the permission of the Director, BMR.

REFERENCES

Band, G.C., 1987, U.K. North Sea production prospects to the year 2000: Jour. Petroleum Technology, January 1987, p. 64-70.

Baker, R.A., Gehman, H.M., James, W.R., and White, D.A., 1986, Geologic field number and size assessments of oil and gas plays, *in* Rice, D.D., ed., Oil and gas assessment - methods and Applications: Am. Assoc. Petroleum Geologists Studies in Geology 21, p. 25-31.

Bureau of Mineral Resources, 1988, Australia's petroleum potential, *in* Petroleum in Australia: The first century: Australian Petroleum Exploration Association Ltd., p. 48-90.

Eckbo, P.L., Jacoby, H.D., and Smith, J.L., 1978, Oil supply forecasting: a disaggregated process approach: The Bell Jour. Economics, v. 9, no. 1, p. 218-235.

Forman, D. J., 1986, Australia's potential for further petroleum discoveries (from May 1986): Bureau of Mineral Resources, Australia, Record 1986/34 (unpubl.), 6 p.

Forman, D.J., and Hinde, A.L., 1985, Improved statistical method for assessment of undiscovered petroleum resources: Am. Assoc. Petroleum Geologists Bull., v. 69, no. 1, p. 106-118.

Forman, D.J., and Hinde, A.L., 1986, Examination of the creaming methods of assessment applied to the Gippsland Basin, offshore Australia, *in* Rice, D.D., ed., Oil and gas assessment - methods and applications: Am. Assoc. Petroleum Geologists Studies in Geology 21, p. 101-110.

Forman, D.J., and Hinde, A.L., 1987, Factors affecting the availability of Australia's undiscovered crude oil resources. Speaking notes and figures: Bureau of Mineral Resources, Australia, Record 1987/17 (unpubl.), 24 p.

Forman, D.J., and Hinde, A.L., 1988, Australian crude oil production from new discoveries to year 2000. Speaking notes and slides: Bureau of Mineral Resources, Australia, Record 1988/11 (unpubl.), 24 p.

Forman, D.J., Hinde, A.L., and Cadman, S.J., 1988 Use of closure area and resources per unit area for assessing undiscovered petroleum resources in part of the Cooper Basin, South Australia, *in* Sinding-Larsen, R., and others, eds, Proceeding of IUGS Research Conference on Petroleeum Resouurce Modeling and Forecasting, Loen, Norway, 1987: IUGS Special Publication Series, in press.

Hinde, A.L., and Forman, D.J., 1988, SEAPUP: a computer program for estimating future production of crude oil from undiscovered resources: Bureau of Mineral Resources, Australia, Record 1988/4 (unpubl.), 14 p.

Kaufman, G.M., 1975, Models and methods for estimating undiscovered oil and gas— what they do and do not do, *in* Grenon, M., ed., Methods and models for assessing energy resources: Intern. Inst. Applied Systems Analysis Proc. Ser. v. 5, Pergamon Press, p. 173-185.

Laws, R.A., 1988, Geological significance of recent discoveries and developments in Australia and Papua New Guinea: The APEA Journal, v. 28, no. 2, p. 59-66.

Ulmishek, G., 1986, Stratigraphic aspects of petroleum resource assessment, *in* Rice, D.D., ed., Oil and gas assessment - methods and applications:Am. Assoc. Petroleum Geologists Studies in Geology 21, p. 59-68.

Geostatistical Characterization of Selected Oil-Shale and Phosphate Deposits in Israel

D. Gill
Geological Survey of Israel, Jerusalem

ABSTRACT

Strata of Upper Senonian age in southern Israel exhibit a variegated lithologic and mineralogic section containing economic phosphate deposits, chert beds, and organic-rich, bituminous marls ("oil shales"). The phosphorites occur in the Mishash Formation and the oil shales in the overlying Ghareb Formation. The occurrence of economic deposits is controlled structurally; they are confined to long and narrow synclinal morphotectonic valleys.

The phosphate deposits consist of several thin (1 to 2 m thick) economic layers with phosphorus pentoxide (P_2O_5) concentrations of 20-32%, which are separated by layers of waste material. The oil-shale deposits (a 10% concentration of organic matter is considered as ore-grade threshold) consist of a single layer in which the concentrations of organic matter and the calorific value of the ore rise gradually with depth to 25% and 1600 cal/g, respectively.

Geostatistical methods were used to evaluate certain aspects of the Rotem oil shale and two phosphate deposits. The variables which were investigated include the thicknesses and grades of the ore-bearing strata, the thicknesses of the upper overburden and internal waste layers, waste/ore ratios, and structural elevation. Typically, the sequence of operations included variogram modeling, block kriging, assessment of waste volume, and grade and volume of ore, computations of the distribution of reserves according to different cutoff grades and waste-to-ore ratios, and investigations of questions concerning optimal sampling networks.

On the basis of these preliminary studies, some common characteristics were noticed. With reference to spatial variability, two situations are discerned. One involves shallow stratigraphic units (the overburden and the two uppermost phosphorite beds) whose thicknesses were affected by surficial physical erosion. Here variability is relatively high, a large portion of the total variance is random, and the distance of interdependence is short. The second involves ore concentrations (in all horizons) and thicknesses of deeper strata. These variables have a low variability, the share of the random component is small, and the continuities persist for longer ranges.

A drilling density of one borehole per square kilometer is sufficient to estimate the global means of most variables with a precision of 20% or better (at the 95% confidence level). Thus, for feasibility and reconnaissance surveys, it usually is sufficient to drill on a square grid with a spacing of 1000 m. The number of required drillholes increases significantly with variability and as the size of the evaluated area gets smaller.

The geostatistical reserve estimates, which are based on block kriging, are in good agreement with estimates obtained from area-of-influence polygons, even when the former utilize data from fewer drillholes.

OBJECTIVES AND PROCEDURES

In Israel the use of geostatistical methods is rather new; only during the past five years or so have these methods been utilized to analyze local problems. The studies were concerned with stratified sedimentary deposits of phosphates and oil shales. The present report reviews the main case studies conducted thus far and highlights some of the lessons that may be drawn from these preliminary attempts.

The main practical issues dealt with by geostatistical methods can be grouped under the following general headings: (1) three-dimensional deposit modeling; (2) volumetric resource appraisal; (3) quantitative deposit characterization; (4) evaluation of estimation error; and (5) optimization of exploration and development drilling programs.

The data processing was carried out by several computer programs from the MINEVAL (Control Data Corp., 1979) and the GEOSTAT (Geostat Systems International, 1981) geostatistical software packages. The procedure adopted in these investigations followed the traditional standard practices established for similar applications. The main components and stages of analysis are portrayed diagramatically in Figure 1. The first stage entails the preparation of experimental variograms in several directions for the variables of interest, detection of anisotropies, and the fitting of appropriate theoretical models that faithfully represents the spatial variability of each variable. Given a variogram, it is possible to proceed along two avenues, block kriging and its various derivatives, and estimation error analysis.

In geostatistics, spatial problems are handled by discretizing the orebody into small rectangular blocks. The attributes of blocks are estimated by kriging. The results of block kriging serve as a database for a diverse array of analyses. They facilitate the evaluation of many aspects of the deposit under externally prescribed constraints which are relevant to the economics of the deposit, such as different cutoff grades, different waste-to-ore ratios (WOR), and so on. Among other things, block kriging also serves as a gridding algorithm for contour mapping. The chart (Fig. 1) lists the principal applications for which block kriging may be useful. All of these are ways for providing a concise summary of a certain aspect of the deposit on the basis of which concrete operational decisions can be made.

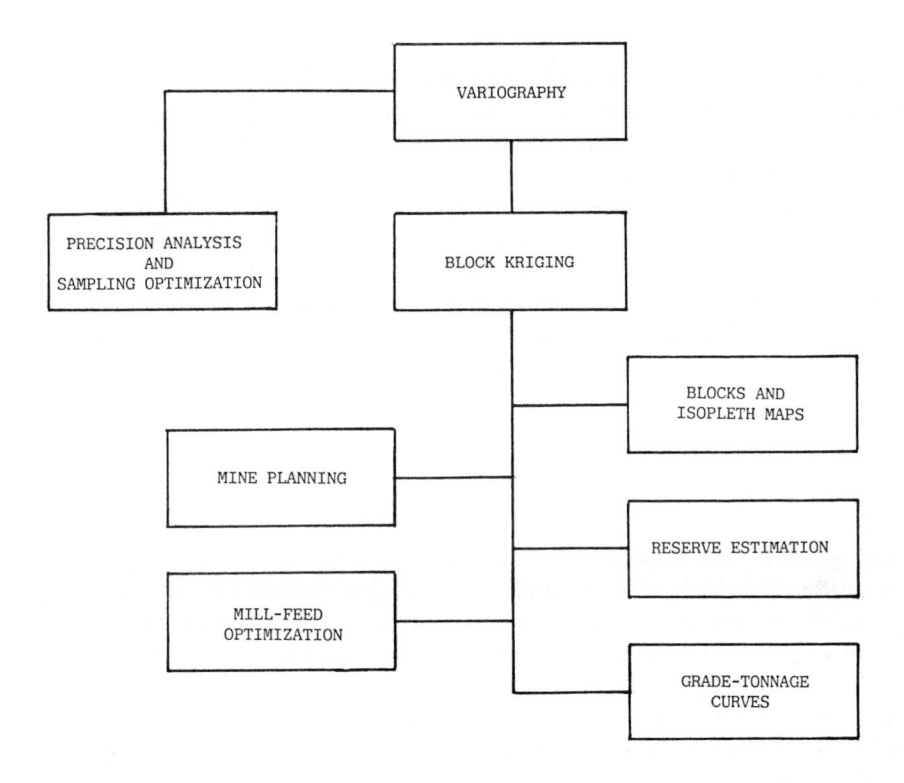

Figure 1. Schematic overview of main procedural steps and analytical options in geostatistical ore-reserve valuation.

The precision of the estimates are a function of the number and distribution of the information control points (i.e., the drillholes). The estimation errors can be reduced by additional drilling. A second direction of analysis pertains to the evaluation of the estimation errors and to the formulation of a cost-benefit function that suggests an optimal way to reduce the errors.

GEOLOGICAL SETTING AND GENERAL BACKGROUND INFORMATION

Strata of Late Senonian age in southern Israel exhibit a heterogeneous mineralogic and petrographic succession containing, in stratigraphic order, chert beds, phosphorites, and organic-rich, bituminous marls ("oil shales"). The same natural association of biogenic-derived constituents, entailing the cooccurrence of the elements phosphorus, silica, and carbon (usually referred to as "P-Si-C trinity"), is known to recur at different geologic times in geographically widely separated areas throughout the world. The association of phosphorus and carbon, two of the principal nutrients, with silica (derived from sliceous tests), is indicative of high biomass concentrations in depositional environments marked by high organic productivity (see Kolodny, 1980; Reiss, 1988, and references therein). The Senonian mineral-bearing sequence in Israel is part of a much larger belt of similar facies which encompasses strata of Senonian to Eocene age in shelf areas of the Mesozoic Tethys Ocean that encircled the northwestern edges of the Arabo-Nubian Massif. Today, these areas form a belt along the eastern and southern shores of the Mediterranean Sea, from Turkey to Morocco. This belt is referred to as the "Mediterranean" or the "North African - Middle Eastern" phosphogenic province.

In the Negev (southern Israel) the phosphate-bearing rocks (phosphorites) occur in the upper part of the Mishash Formation (of Late Campanian age). The oil shales occur immediately above the phosphorites, in the lower part of the Ghareb Formation (Maastrichtian age). Recent drilling encountered bitumen-rich marls also within the Mishash Formation (Minster and others, 1986). The deposition of the high-grade facies of these commodities was controlled by geologic structure. The phosphatic and the organic-rich facies developed only in synclinal morphotectonic depositional basins. These basins were formed during the early stages of the Late Cretaceous to Eocene "Levantine" folding phase which is recognized throughout the Levant (see Garfunkel, 1978; and Horowitz, 1979, for recent reviews). In the northern Negev, this folding phase formed a series of essentially parallel, NE-SW trending, long and narrow mild folds which are arranged in an echelon pattern (Fig. 2). The present-day landscape reflects the underlying geological structure faithfully - the anticlines form prominent ridges whereas low-lying valleys developed over synclinal depressions. The synclinal valleys may be less than 5 km wide. The folds are asymmetric; the anticlines have steep eastern flanks and gently inclined western flanks. In the synclines the asymmetry is reversed (Fig. 3).

The economic concentrations of phosphate and oil shale are confined to a narrow zone along the axis and gentle flank of the synclines. Within the synclinal valleys, the richer concentrations are segregated into more or less discrete bodies, because of small-scale structural irregularities that create local depressions which are separated from one another by small structural highs.

Because the commodities in question are economic only with open-pit mining, the thickness of the overburden is one of the most important parameters in evaluating the economic aspects of a deposit. In the synclines, the Ghareb Formation is overlain by a thin veneer of younger formations (usually shales of the Paleogene Taqiye Formation and conglomerates of the Neogene Hazeva Formation) and by alluvial sediments. The Ghareb itself constitutes part of the overburden of the phosphates. Therefore, a thick oil shale layer, in a sense, is detrimental to the mineability and attractiveness of the underlying phosphates. Furthermore, from an economic standpoint, oil shales did not become an interesting commodity until the oil crisis of the 1970's, and, save for special circumstances (see next), they are not an economic commodity yet. For these reasons, in the prospection for phosphates, the strategy was to avoid areas deliberately where the Ghareb Formation was expected to be thick. Consequently, the information about oil shales is more limited.

Thus, the geological guidelines for locating potential deposits are fairly well understood and "discovering" them does not present a difficult problem. Most of the more promising locations for phosphates have already been tested by at least some reconnaissance drilling. Hitherto, some 37 separate phosphate deposits have been outlined (Anonymous, 1977; Nachmias and Shiloni, 1985). Of these, 18 are located in

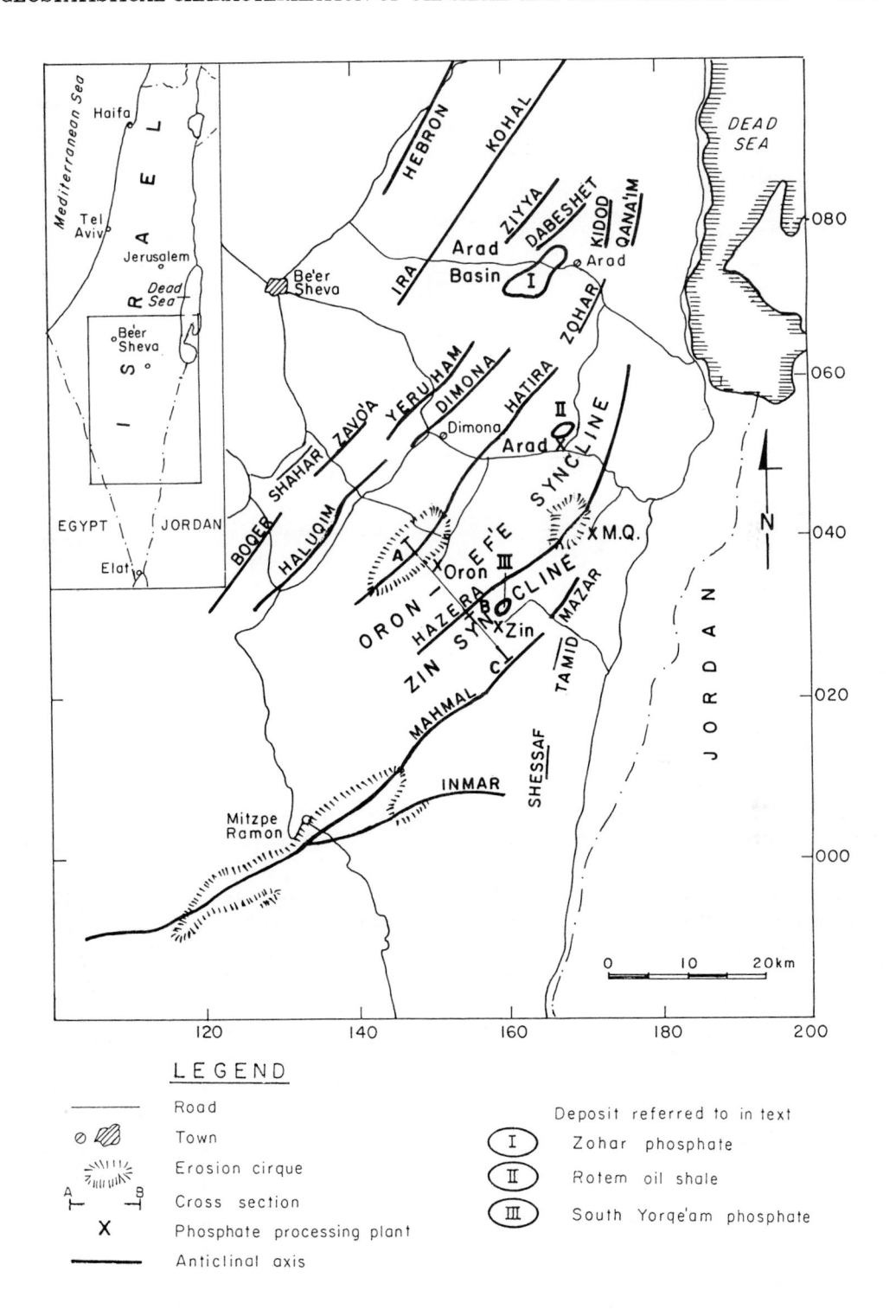

Figure 2. Structural sketch map of northern Negev showing main folds and location of ore deposits referred to in text.

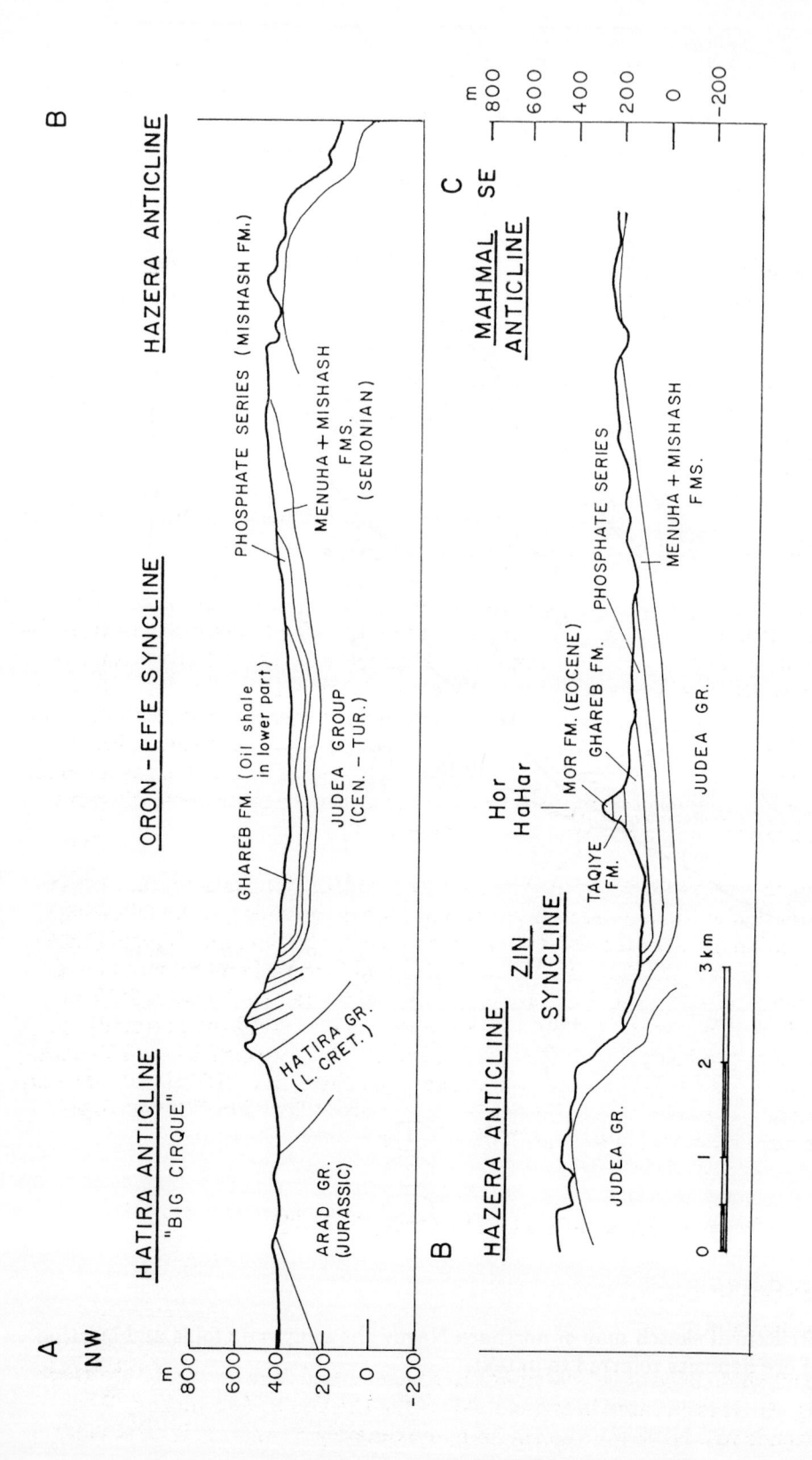

Figure 3. Regional geological cross section A-B-C (see Fig. 2 for location) showing stratigraphic and structural context of phosphorites and oil shales (modified from Roded, 1982; certain components of overburden are not shown).

the central Negev and the Arava Valley (Soudry and Mor, 1985). Because of their less favorable geographic location and infrastructural circumstances, compounded by inferior economic attributes, the latter, at present, are only of secondary interest. The bulk of the operational reserves occur in two of the largest synclines in the northern Negev, the Oron-Ef'e Syncline (four deposits) and the Zin Syncline (ten deposits), and in the Arad Basin (Fig. 2).

Phosphates are one of the principal mineral commodities produced in Israel; in terms of cumulative value of past production, phosphates have been the second most valuable (see reviews in Gill and Griffiths, 1984; Gill, 1984). The phosphates are mined and processed by Negev Phosphates Ltd. (a subsidiary of Israel Chemical Ltd.). Hitherto, processing plants were built adjacent to four of the principal deposits. In historical order these include the plants in Oron (1951), Makhtesh Qatan (1966), Mishor Rotem (the Arad plant in 1970, and the Rotem Fertilizers plant in 1981), and Zin (1977). The first two already are defunct because of the exhaustion of the nearby reserves. The most acute concern today is to locate new operational reserves in order to prolong the operation of the other two plants. In this respect, the deposits of the Arad Basin become particularly significant (see next).

Israel possesses large reserves of oil-shale material. Seventeen deposits thus far have been identified. The Rotem deposit (also known as "Ef'e"), discussed next, is the most throughly explored. Several others, including the deposits of Nebi Musa, En Boqeq, Hartuv, Oron, and Mishor Yamin, have been explored in some depth, but for the remaining ones, only scanty data are available at the present time. Minster and Sharav (1984) indicate proven reserves of 11.5 billion tons in 10 deposits, with an energy value equivalent to 5 billion barrels of oil. The total resource base is undoubtedly larger; additional indicated and probable reserves of the same order of magnitude as the proven ones already have been delineated.

An extensive literature exists on the Israeli phosphate and oil-shale deposits. Additional information on various aspects of the subject are contained in the following recent publications, and in the references cited in them: Nathan and Shiloni (1989) for a general overview; Nathan and others (1979) and Soudry, Nathan, and Roded (1985) on geological setting and inorganic geochemistry; Bein and Amit (1982) on organic geochemistry; Soudry (1987) and Soudry and Lewy (1988) on petrography; Reiss and others (1985) and Reiss (1988) on fauna and biostratigraphy; and Vengosh, Kolodny, and Tepperberg (1987) on stable isotope geochemistry. Additional information and references to research on Israeli oil shales are included in Gill and David (1984), Minster and others (1986), and Shirav (1987).

REVIEW OF CASE STUDIES

Rotem Oil-Shale Deposit
General background

The Rotem oil-shale deposit is located about 20 km east of Dimona and 30 km south of Arad (Fig. 2). It is situated in the central part of the Oron-Ef'e Syncline. This syncline extends in a NE-SW direction for approximately 60 kms. It is subdivided into several elongated depressions, namely (from southwest to northeast), Oron, Yamin, Rotem, Zef'a-Ef'e, and Hatrurim. The deposit (at times referred to as the "Ef'e" deposit after the geographical name of the valley in which is is located) was discovered in 1962 in the course of prospecting for phosphates in the northeastern part of the Oron-Ef'e Syncline (Shahar, 1965). The area is known as "Mishor Rotem" (the "Rotem Plain"); in it are present today several industrial facilities, including: (a) Arad Phosphates plant - processes phosphates from nearby deposits; (b) Rotem Fertilizers Ltd. - produces phosphoric acid; (c) Dead Sea Periclase Ltd. - produces periclase (M_gO) from the end-brine of the evaporation ponds in the Dead Sea, which are transported to the plant by pipe as a slurry; (d) the terminal of the Dead Sea Works conveyor belt which transports potash from the plant at Sedom to the railroad terminal in the Rotem Plain; and (e) Energy Resources Development Ltd. ((PAMA), the phonetic transliteration of the acronym of the firm's name in Hebrew, the name usually used in making reference to the company). PAMA (formerly "Energy 2000 Ltd.") was established by the Government of Israel in 1977 to explore the various aspects of exploiting the large oil-shale resources of the country. Following the successful results of an extensive research and development program, including experimental production runs in a pilot plant of semiindustrial scale, the company is currently in the process of building a full-scale power station which will utilize oil shales for the generation of steam and electrical power (Yerushalmi, 1984; PAMA, 1987). The energy-generation process is by direct combustion, using fluidized-bed technology. The power station will have a capacity of 12 megawatts. Of these, eight will be used by Arad Phosphates and four will be taken up by the state-wide electrical network. The station is scheduled to become operational by the end of 1989. The operation entails burning 300,000 to 400,000 tons of ore per yuear. The ore will be obtained from an open-pit mine in Nahal Havarbar. The dimensions of the pit are 100m by 200m. This site was selected because of its proximity (only 4 km) to the plant, and because in this area the overburden is the thinnest. It is estimated that the price of energy (including only mining and processing costs and not counting any other previous investments in the project) will be commercially competitive. The economic viability of this enterprise hinges to a large extent on the fact that the entire operation is geographically localized. Both the raw material and the principal consumer are located in immediate proximity to the plant.

In the northern Negev the Ghareb Formation, the source of the oil shales, consists mostly of marls and some chalks. From a lithological standpoint, the use of the term "oil shale" for the organic-rich rocks of the Ghareb is inappropriate because there are no genuine shales in the formation. The formation is divisible into a lower bituminous part, and an upper, bitumen-free, part. The boundary between the two is marked by a distinct change from dark gray to yellow. For technoeconomic reasons only material with more than 10% organic matter (OM) is considered of ore-grade rank. This concentration boundary passes within the uppermost part of the gray zone. Thus, as far as mining is concerned, the upper part of the gray zone and the entire upper (yellow) part of the formation constitute an overburden of waste material. Within the organic-rich part the concentration of organic matter increases more or less systematically downwards. The concentrations range between 10% (the technologically determined cutoff) and 25% and the richest part is near the base of the formation (see data in Shahar, 1965; Shirav, 1978). The formation is poorly stratified; it consists essentially of a single massive layer. The lack of internal waste imparts a significant economic advantage as it obviates selective mining. The lower boundary with the phosphorites of the Mishash Formation is marked by a change in lithology, a significant decrease in the OM content, and an increase in the concentration of P_2O_5. The variations in the OM content in the Ghareb (and the Mishash) Formation have been attributed by Shiloni (1988a; 1988b) to late diagenetic changes. According to this interpretation, the Ghareb was originally bituminous throughout its entire vertical extent, and the present lack (or reduction) of OM in its upper part is the result of late (Neogene to Recent) oxidation, induced by subaerial exposure and circulating groundwater.

Additional data on some material properties of the Rotem oil shales, compiled from several sources (Shahar, 1965; Shahar and Wurzburer, 1967; Shirav and Ginzburg, 1978, 1983; Issahari, 1982; Yerushalmi, 1984) are summarized here. The average OM contents of the ore-grade material (> 10% OM) is 14%. Calorific values range between 700 and 1570 cal/gr; the average for the deposit is about 1050 cal/gr. Extractable liquid hydrocarbons constitute about 44% of the organic matter. Fischer assay values (which express the relative volume of extractable liquid hydrocarbons as a percentage of the dry sample) range between 5% to 10%. For the representative ore material with 14% OM, the value is about 6%. These figures translate to oil yields of between 10 and 28 gal/ton with an average of 16 gal/ton (equivalent to about 0.4 barrels/ton). The specific gravity of the extracted crude is 0.954-0.974.

Extensive additional analytical and tchnological data have been compiled in numerous proprietary reports by PAMA. Additional data on the organic and inorganic geochemistry of the oil shale material are in Spiro (1980), Bein and Amit (1982), Tannenbaum (1983), Spiro and others (1983), Shirav and Ginsburg (1983), and Shirav (1987).

It is of interest to note that the average values of the main material-quality properties of the oil shales of the Rotem deposit, that were determined at an early stage of their exploration on the basis of data from only nine drillholes (Shahar and Wurzberger, 1967), are valid today, some 60 drillholes later. This is a result of the low spatial variability of the material attributes. This observation should serve as an important lesson for the design of sampling schemes for similar deposits in the future.

Geostatistical analysis and reserve estimates

The exploration and evaluation of the deposit progressed in several stages. As a general rule, as exploratory drilling was extended laterally along the axis of the syncline, more and more reserves were confirmed. Since its discovery in 1962, some 70 boreholes have been drilled in Rotem and about 100 others to the southwest of it, along the Oron-Ef'e Syncline. Hitherto four deposits have been recognized (from southwest to northeast): the Biq' at Zin, Oron, Mishor Yamin, and Rotem. Their areal extents are 40, 13, 35, and 22 sq. km, respectively, and their estimated reserves 3.0, 1.0, 4.0, and 2.5 billion tons, respectively (Shirav and Minster, 1984; Minster and Shirav, 1984). The individual deposits are separated from one another by shallow structural highs which, in places, elevate the Mishash cherts to the surface. PAMA employed geostatistical methods to evaluate several aspects of the deposit. Some preliminary results were published by Slotky and others (1983) and by Padan and Slotky (1988). However, the bulk of the results remain unpublished and proprietary. A geostatistical analysis of the thickness of the Rotem deposit was carried out by Gill and David (1984). Their study was based on data from 29 drillholes for which information was available at that time. The main results of this study are presented next.

In order to detect the anisotropies, variograms were constructed in several directions. It was suspected initially that spatial continuity might be better in the strike direction. Thus, experimental variograms were prepared along the strike (NE-SW), along the dip (NW-SE), and in the E-W and N-S directions as well. The variograms are shown in Figure 4. The directional variograms are similar, which indicates that the variations in thickness do not have any directional preference. Because the thickness seems to be isotropic, the more representative average variogram was used for the kriging computations. The spatial continuity of the thickness of the deposit can be approximated by a spherical or a Gaussian variogram without local random variations, with a sill (variance) of 19.8 m, and a range (zone of influence of a sample) of 1125 m. A spherical model was used for the kriging; its parameters are listed in Table 1. The kriging was computed for 237.5 m long square blocks. The results of the block kriging are shown in Figure 5. Within each block, the upper number is the estimated average thickness of the oil-shale beds and the lower number is the relative standard deviation of the thickness estimate for the block. An estimated average thickness (T) with a relative standard deviation (Sd) means that there is a 95% probability that the true average thickness of the block is in the range T+/-(2x2dx100)/T. In blocks that contain drillholes, the standard deviations are small, mostly between 0.07 and 0.15. On the other hand, along the periphery and in areas with sparse drilling, the standard errors are high, around 0.4, which indicates that the true value may lie anywhere within +/-80% of the estimate. Figure 6 is an isopachous map of the Rotem deposit. The map is based on the kriged block data shown in Figure 5. The largest oil shale thicknesses of 65 m and 69 m occur along the axis of the syncline, or slightly to the east of it. The map is similar to the hand-contoured map of the same data (Shirav, 1978, Fig. 4).

The estimation variance also can be regarded as a regionalized variable which can be expressed as a continuous surface. The kriging standard errors in Figure 5 were contoured to produce the estimation error map shown in Figure 7. South of coordinate-line North 053 there are only two drillholes and the precision is poor. North of it, the precision surface appears as a fairly regular dome. Most of the area lies within the 0.2 isopleth, implying that within this area the precision of the estimated thicknesses is 40% or better (at the 95% confidence level).

With the number and distribution of prospection holes drilled until 1984, the average thickness of the deposit (37.7m) was determined with a precision of about 10% (i.e., the margin of error was about 4 m). The holes were not drilled on a regular geometric grid. The geostatistical evaluation revealed that almost as good a precision could have been obtained from half as many holes drilled on a 1000 m square grid. Based on the assumption that Rotem is fairly typical of the Ghareb oil shale occurrences in

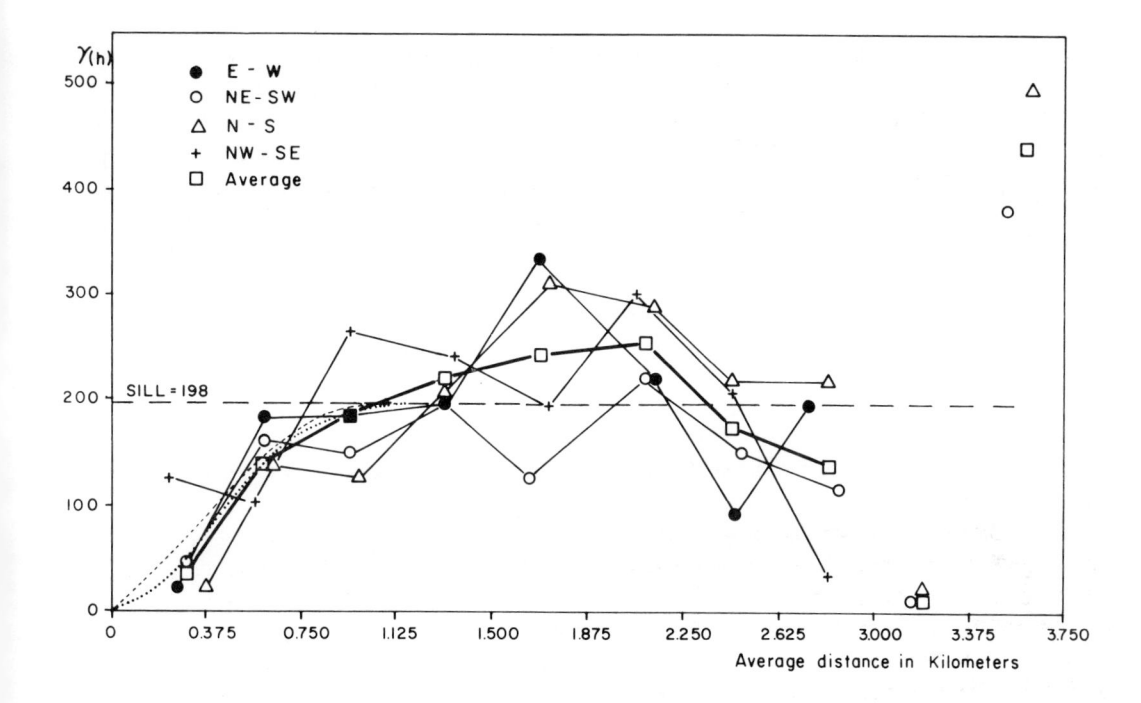

Figure 4. Rotem oil-shale deposit, directional and average variograms of thickness of section containing more than 10% organic matter. Dashed line - Gaussian model; dotted line - spherical model.

the northern Negev synclines, it was suggested that the reconnaissance exploration of the nearby deposits be conducted on a square grid with a spacing of 1000m. In Rotem, only a marginal improvement in precision can be gained from additional drillholes at wide spacing. It will be more profitable to drill a small part of the deposit along a dense grid, on the order of 100 to 200 m, in order to learn more about small-scale variations which might be present.

South Yorqe'am Phosphate Deposit

The South Yorqe'am phosphate deposit (SYD) is one of several phosphate deposits located along the axis of the Zin Syncline. This syncline extends in a NE-SW direction for approximately 50 km. It is bounded on the west by the Hazera Anticline and on the east by the Mahmal and Mazar Anticlines (Fig. 2). Some ten individual phosphate deposits have been delineated within this syncline (Roded and others, 1972). Its total proven and indicated reserves are large, on the order of 200 million tons (Nachmias and Shiloni, 1985). The mining and processing of phosphates commenced in 1977 with the completion of the Zin plant and the construction of a railroad line to the site.

The SYD is located to the southwest of Nahal Yorqe'am, about 3 km to the north of the Zin plant. It extends for about 2.5 km along the synclinal axis, and is about 1.25 km wide. Structurally, the deposit is positioned on a low amplitude synclinal undula-

Table 1. Descriptive statistics and variogram parameters for variables of different deposits in text.

Deposit and variable	area (sq km)	holes	density	variance	range (m)	nugget	n/var	range	mean	sd	sd/m
Rotem oil shale deposit	15	29	2	19.80	1125	0.0	.00	17.0-69.0	37.70	4.45	.12
Oil shale thickness											
South Yorke'am Phos. deposit	1.5	46	15								
Overburden thickness				30.21	700	0.0	.00	5.0-30.8	17.90	5.50	.31
P4 thickness				0.01	700	0.0	.00	1.4- 2.0	1.71	0.12	.07
P4 grade				2.04	600	1.0	.50	23.2-29.8	27.63	1.43	.05
Zohar phosphate deposit	77	124	1.6								
Overburden thickness				171.00	300	110.00	.64	5.0-30.0	13.07	13.07	.50
Upper phos. thickness				0.34	600	0.12	.35		1.15	0.58	.41
Lower phos. thickness				0.35	1000	0.05	.14		1.43	0.59	.47
Interchert waste thick.				3.80	1800	1.80	.26		4.10	1.95	.39
Interchert phos. thick.				0.60	1800	0.35	.58		2.00	0.77	
Upper phos. grade				8.00	750	0.00	.00	17.0-33.0	25.60	2.80	.11
Lower phos. grade				3.50	1800	0.80	.23	23.0-34.0	28.90	1.87	.06
Interchert phos. grade				4.50	900	0.50	.11	22.0-31.0	26.20	2.12	.08

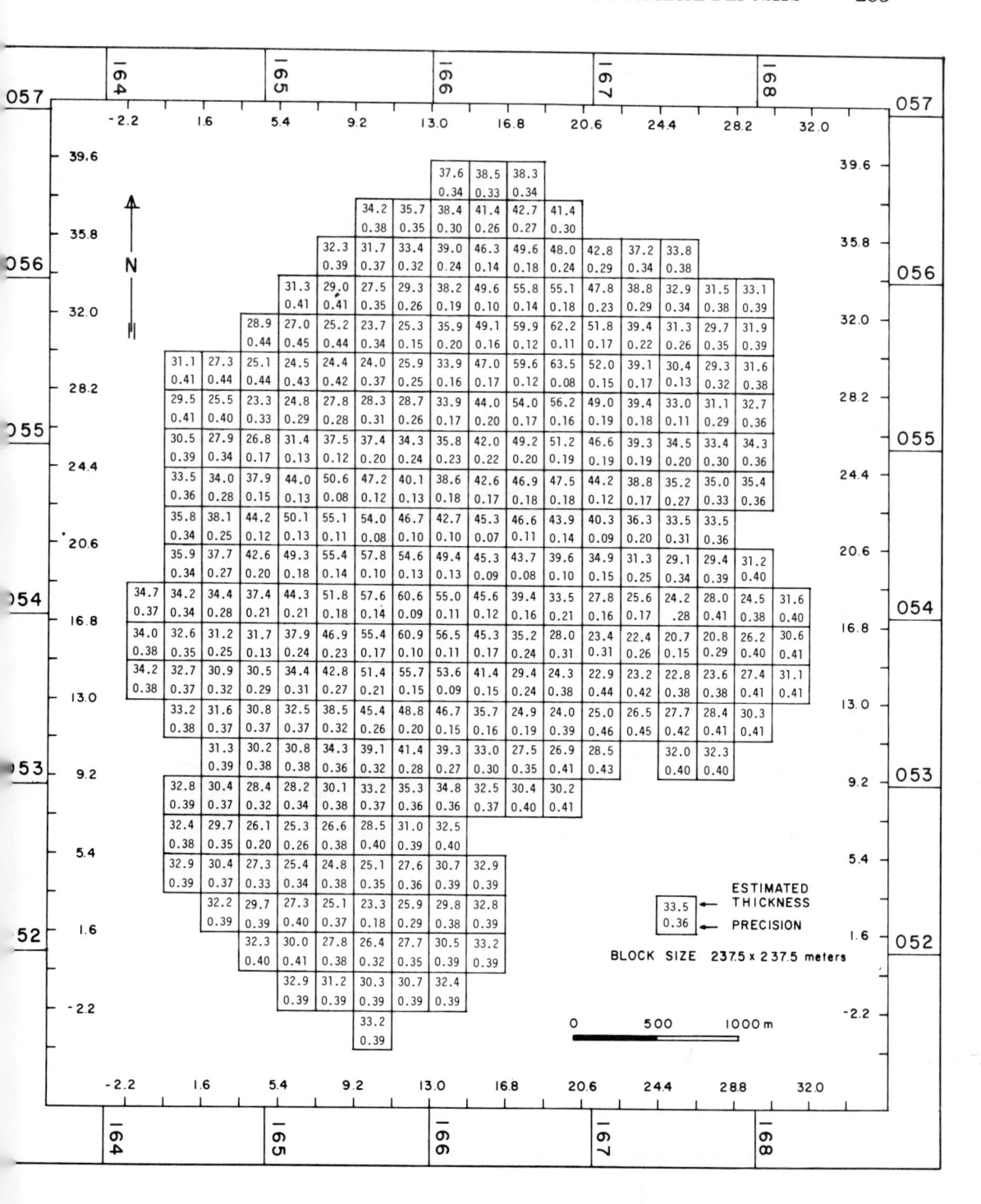

GEOSTATISTICAL CHARACTERIZATION OF OIL-SHALE AND PHOSPHATE DEPOSITS 285

Figure 5. Rotem oil-shale deposit, kriging thickness estimate (upper number inside blocks) and kriging standard error (lower number) block map.

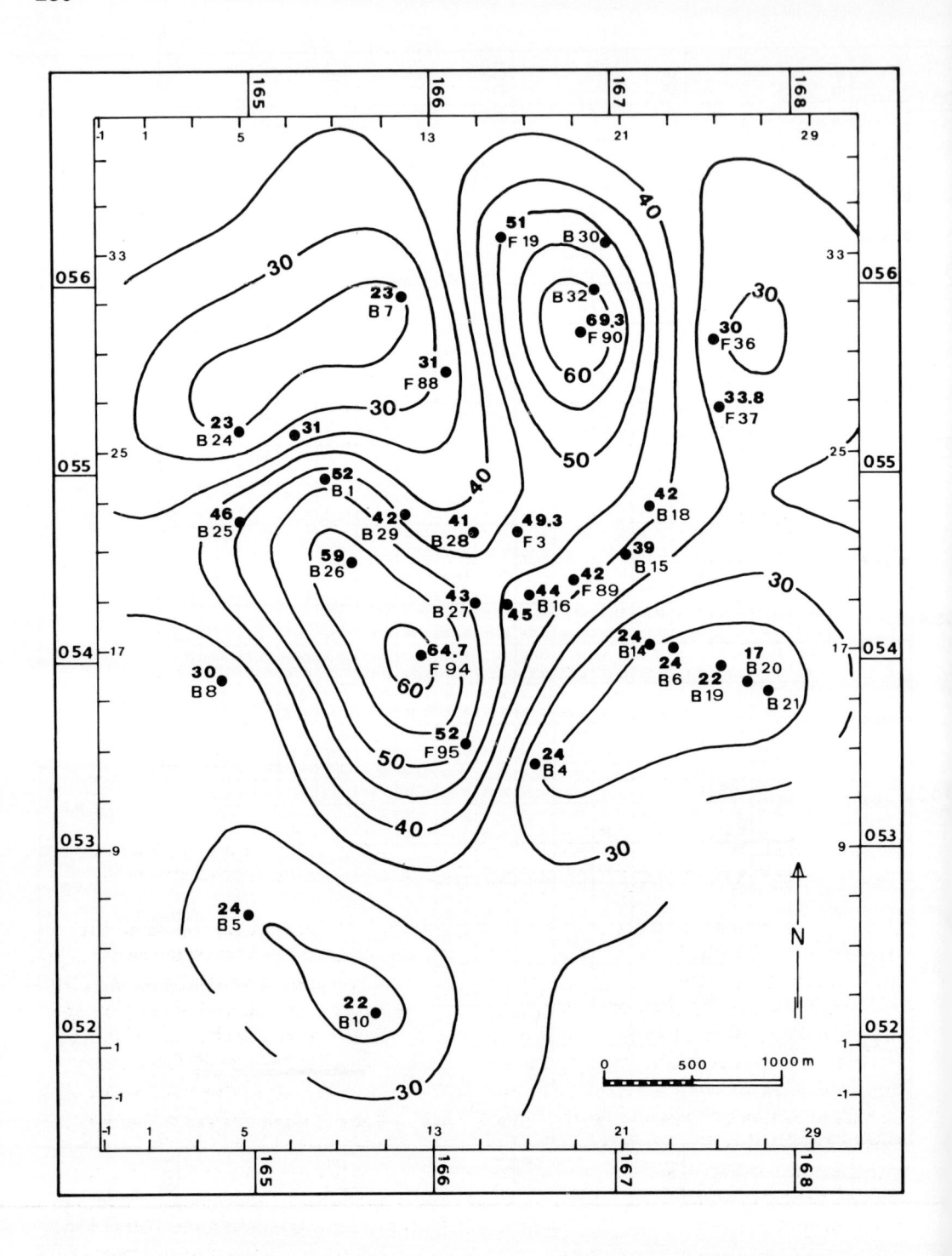

Figure 6. Isopachous map of Rotem oil-shale deposit based on kriged grid of Figure 5.

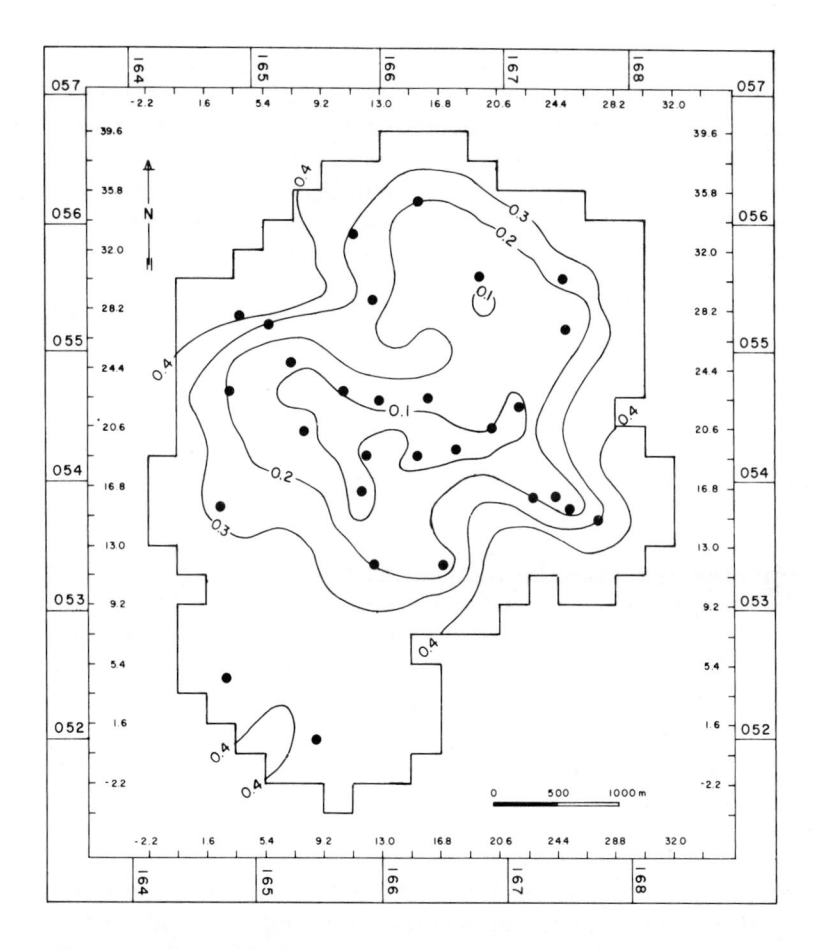

Figure 7. Rotem oil-shale deposit, contour map of kriging estimation errors detailed in Figure 5. Isopleths connect points of equal relative standard error.

tion, with a structural relief of about 20 m, between the steep eastern edge of the Hazera monocline and the main axis of the Zin syncline. The phosphates occur in the uppermost 20 m of the Mishash Formation. The overburden consists of marls of the Senonian Ghareb Formation and the Pliocene Mazar Formation, and of conglomerates of Recent alluvial sediments. The phosphate-bearing part of the the Mishash Formation contains eight thin phosphate beds, separated by layers of marl, chalk, and flint. The average P2O5 content in the various beds ranges between 24 and 28%. The interbedding of ore and waste necessitates selective mining which adds an extra cost component to the final product. The fourth phosphate bed (P4), is the thickest, richest and most widespread ore bed in the SYD. It contains most of the reserves and its properties are most consequential to the economics of the deposit. Some of these properties were evaluated by geostatistical methods by Miller and Gill (1986). The main objectives of the analysis were: (1) to estimate the mineable reserves of P4 and map the distribution of its thickness and grades; (2) to map the thickness of the overburden and estimate its volume; and (3) to evaluate the precision of the estimates and formulate guidelines for sampling optimization. The main results of their study are quoted here.

About 150 drillholes have been drilled thus far in the course of the exploration and development of this deposit. Ninety holes penetrated P4, and of these 46 were selected as the database for the analysis. The selected drillholes are approximately evenly distributed throughout the deposit, with between-hole distances ranging from 100 to 400 m. The variables which were analyzed were the thickness and average grade of the P4 bed and the thickness of the overburden.

Variograms were constructed for two principal directions, along the strike (N4OE) and perpendicular to it. In a strict sense, all variables were anisotropic in that they possess a slightly better continuity along the strike. These geometric anisotropies, however, are minor and they can be disregarded. Spherical models were fitted to the average variograms. The variograms are shown in Figure 8; the variogram parameters and some additional basic statistics concerning the variables are summarized in Table 1.

The variogram that describes the areal variability of the phosphate grade (Fig. 8A) indicates that the total variance of this variable can be attributed to two sources, a structurally continuous component and a random one (nugget effect), each accounting for about 50% of the total variance. The nugget effect seems to be exceptionally large and was suspected actually to be the result of inadequacies in the sample recovery procedure. To test this hypothesis, three new holes were drilled within 1 m of existing ones. Whereas in the old reference holes samples had been collected with a rather primitive device, sample recovery in the new test holes was by an improved dust-collector. The difference in the vertical variability of the grades between corresponding pairs of neighboring holes, which, by and large, can be attributed to the different sampling devices, was evaluated by comparing the respective vertical variograms. It was determined that whereas the ranges remained the same, the variance in the (new) test holes was reduced to about 60% of the variance in the (old) reference holes, and the nugget effect in the test holes was reduced to zero. These results confirm that the large nugget effect in the areal grade variogram to a large extent may be the result of inadequacies in the sample collector that was used during the drilling of the exploration and development holes.

The variogram of the thickness of the overburden (Fig. 8C) has a peculiar shape. Near the origin, for short distances between samples, the curve rises and bends gradually towards the sill in a 'normal' spherical fashion. After reaching the sill and following it for a short distance, the curve rises again along a parabolic trajectory. Thus beyond the range, the variance rises without bound. This is the result of a nesting of two components of structured variability which are of different scales. The spherical part reflects the short-distance continuity of the variable and the parabolically rising part reflects the existence of a longer range trend of variations (David, 1977, p. 267; Journel and Huijbregts, 1978, p. 44). This trend is apparent in the isopachous map of the overburden which shows that the unit thickens in a regular fashion towards the northwest (Miller and Gill, 1986, fig. 10).

The block kriging was used for preparation of contour maps (Miller and Gill, 1986, figs. 8-10), for global reserve and overburden-volume estimations, and for deposit characterization in the form of a graph depicting the cumulative tonnage, and its

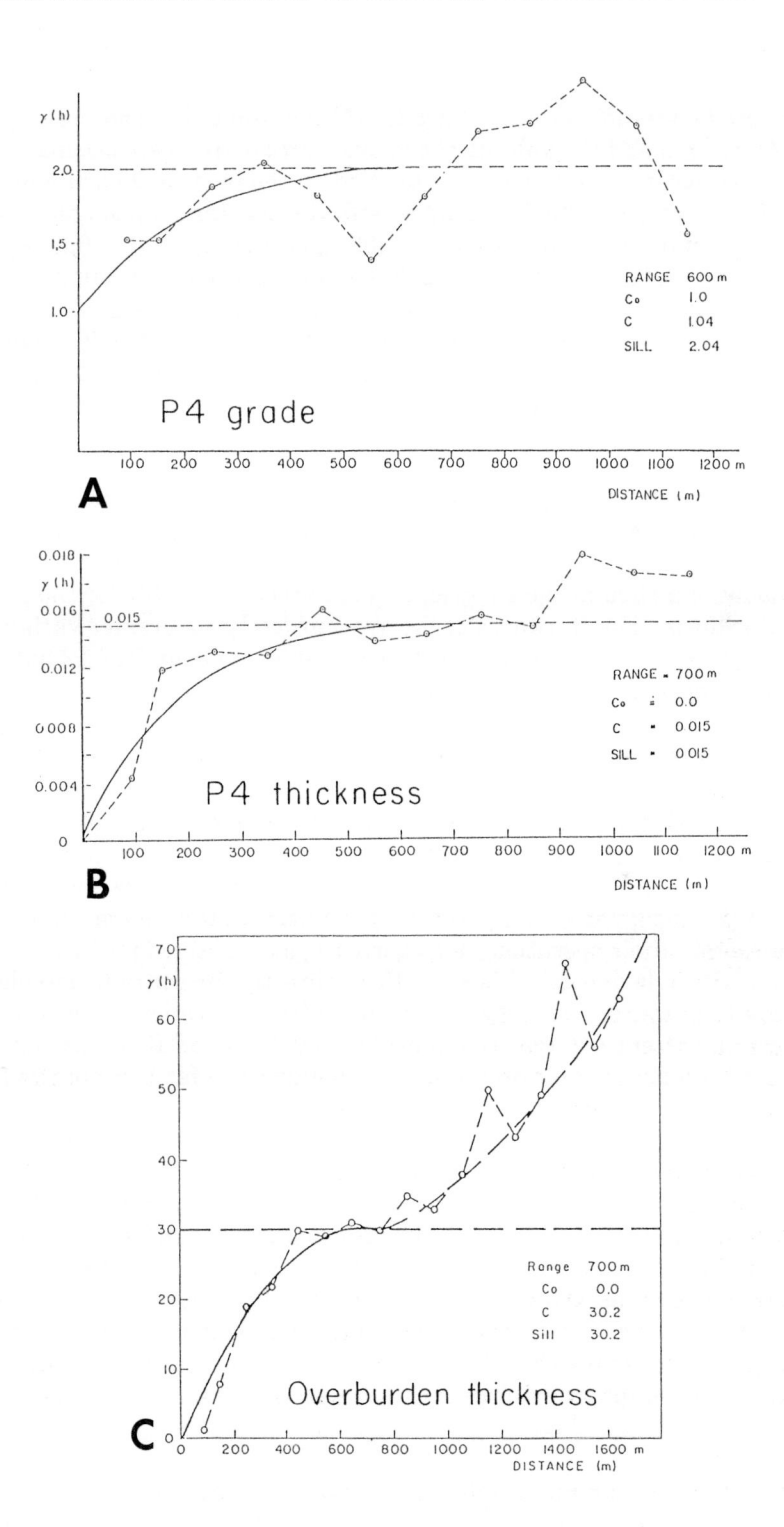

Figure 8. South Yorqe'am phosphate deposit, average experimental variograms (dashed line) and fitted spherical model (solid line) of (A) P4 grade; (B) P4 thickness; and (C) overburden thickness.

average grade, for different cutoff grades (Miller and Gill, 1986, fig. 11). The reserves of the P4 horizon and the volume of the overburden also were estimated by the polygon interpolation and extension method on the basis of 90 drillholes, and it is interesting to compare these estimates with those obtained from block kriging. The estimates from block kriging are 4,669,000 tons of ore and 24,986,000 cu m of overburden, versus 4,478,000 and 22,461,000 obtained by polygon computations. The estimates obtained by the different methods are close. It is significant to note that the kriging-based estimates were derived on the basis of one-half the number of drillholes employed by the alternative approach. Furthermore, it is important to note that actual data from the parts of the deposit that were mined already indicate that the accuracy of the geostatistical predictions is satisfactory.

The spatial variability of the grade and thickness of the P4 horizon is small (see Table 1). Therefore, for an estimation of the global average of these variables, high precisions can be attained with a small number of holes. In this situation three holes at the nodes of a 1000 m square grid, furnish precisions of 3% for the grade and 5% for the thickness. The thickness of the overburden has a greater variance and to attain a similar level of precision for this variable (of about 9%), 12 holes, spaced 500 m apart, are needed.

Phosphates in the Arad Basin

The phosphate deposits in the vicinity of the Arad plant have been exploited since 1970. The reserves of raw material in this area are dwindling rapidly. Therefore, the delineation of additional economic reserves which are close enough geographically to the existing phosphate processing facilities in the Rotem Plain is crucial in order to sustain their operation. Furthermore, as explained in the section concerning the Rotem oil-shale deposit, this also affects directly the continued exploitation of the latter as a local source of energy. Thus, apart from its value as a commodity in its own right, the recently discovered phosphates of the Arad Basin assume added importance because of their proximity to the industrial facilities in the Rotem Plain.

The Arad Basin comprises several low-lying synclinal valleys along the eastern parts of the area referred to as the Beer-Sheva Valley. It extends from the Kidod-Zohar anticlinal chain in the east to the Ira Mountains Anticline in the west, encompassing an area of about 200 sq km (Fig. 2). The prospection for phosphates in this area started in 1980. Most of the exploratory drilling was concentrated in the more favorable synclinal land tracts within the basin. The appraisal of the phosphate resources of this large region is based on data from 190 drillholes. Of these, 170 holes encountered mineable beds. A comprehensive account on the geology and the results of the exploratory drilling in this region is presented in Zohar and Shiloni (1987).

The stratigraphy of the Phosphate Series and its overburden in the Arad Basin is similar to that in the (nearby) Zef'a-Ef'e section of the Oron-Ef'e Syncline. Also here the deposits consist of three principal phosphate beds (referred to, from bottom up, as "interchert", "lower" and "upper" beds), with intervening waste layers, and the overburden consists of the Ghareb, Taqiye, and Hazeva Formations with a top cover of young alluvial deposits and soils. The exploration drilling reveals that the economic attractiveness of the deposit, as expressed principally by the grades and by the ore-to-waste ratios, decreases westward. Therefore, interest is concentrated in the eastern part of the basin (in general terms, east of UTM-coordinate-line east 160). The area in question covers approximately 80 sq km and contains 120 drillholes. The variograms of the variables pertinent to the geostatistical analysis of the deposit are based on these drillholes.

The Arad Basin phosporites also are significant as a potential source for uranium (Shiloni and Gill, 1988; see also Avital, Starinsky, and Kolodny, 1983; and Shiloni, 1984). Preliminary assessments indicate that the Phosphate Series in the Arad Basin contains, on average, about 75 ppm uranium per ton, and that the total amount of uranium in the area is on the order of 60,000 tons. There is a significant positive correlation between the concentrations of phosphate and uranium. The uranium concentration in the richer phosphate ore material is almost twice as high as the mean value, on the order of 145 ppm. It is estimated that if the Arad Basin phosphorites eventually become the raw material for the Arad plant, the amount of uranium passing through the production cycle will be on the order of 150 tons per year.

Opening mines in this area poses difficult environmental problems because of the proximity of the higher grade reserves to the town of Arad. Therefore, one of the principal tasks of the exploration program is to delineate economic reserves in "acceptable" areas that are far "enough" from the town and, at the same time, as close as possible to the Arad and Rotem Fertilizers plants. In this vein, within the more prospective area of the deposit as described, further interest was focussed on its southeastern part, which is closer to the Rotem Plain. This part, referred to as the "Zohar deposit" (Nachmias, 1986a, fig. 1), covers an area of about 45 sq km and 74 of the drillholes are located within it.

Thus, in light of these constraints, at the present stage of the exploration program the objectives of the geostatistical analyses were mainly (1) a reconnaissance appraisal of the overall resource base of the Zohar deposit, and (2) to partition the deposit into smaller mining units, to assess their reserves, and to determine the amount of additional drilling needed to refine the estimates for these smaller areas. The geostatistical investigations were carried out by Negev Phosphates Ltd. Some of the results were published (Rosenberg, 1984; Nachmias 1986b; Nachmias, Braester, and Rholich, 1986; and Nachmias and Shiloni, 1986), and others were summarized in internal company reports (Nachmias and Rosenberg, 1985; Nachmias 1986a; and Nachmias, 1987). Some of the main results of these studies are reviewed next.

The variables relevant to the analysis are the thickness and grade of the phosphate beds, the thickness of the upper overburden and the intervening waste layers, and the overall WOR at the drillhole site. Experimental variograms were prepared for these variables in four principal directions (two sets of strike-dip combinations). All the variables were more or less isotropic. Therefore, an average variogram was used in all situations. For all the variables the adapted theoretical models are simple spherical functions (Nachmias and Rosenberg, 1985). The parameters of the variograms are listed in Table 1. The upper overburden and the upper phosphate bed were affected by surface erosion. As a result, the variability of their thicknesses is the largest. This is expressed by the high ratio of the standard deviation to the mean, by the relatively high share of the random component in the total variance, and by the shorter lateral distances of interdependence between samples. The variogram-range of the thickness of the different layers increases with depth from 300m to 1800m. This reflects the better continuity of the deeper layers.

The appraisal of the overall reserves of the Zohar deposit, and the evaluation of aspects of its mining economics, primarily that of the WOR, are based on the results of block kriging (Nachmias and Shiloni, 1986; Nachmias, 1987). The kriging was computed for 400m by 400m square blocks. The computations yield a wealth of information which can be tabulated and displayed in a variety of contour and block maps, and in the form of grade-tonnage graphs. One effective way to utilize these results is illustrated next.

The results of the block kriging of individual variables can be integrated and organized in a tabulated form, as shown in Table 2. The variable of interest, in this situation the WOR, is divided into size-classes, and the identity and number of the blocks in each class is registered. Using the kriging results of other variables, it is possible to incorporate in the table other attributes of the specific blocks thus assigned to the different classes, such as average grade, average thickness of ore beds, ore tonnage, and the like. This "database" then can be manipulated to compute those quantities and parameters which provide a useful characterization of the deposit. Pairs of parameters then can be cross-plotted in various combinations to furnish a concise graphic summary of the results. Such graphs are easier to comprehend and they are instrumental in facilitating comparisons between deposits, or between alternative subareas within the same deposit. Likewise, in a deposit which consists of several stratigraphically separate ore beds, they can be utilized to evaluate the economics of selective mining.

In scanning the numbers in Table 2, the following statistics about the Zohar deposit emerge. The total reserves amount to 271 million tons. The minimal cutoff grade for this amount is 20%. The average grade of this tonnage is 26.85%, and the average WOR is 3.12. A more comprehensive portrait of the joint variations of the respective quantities and parameters is facilitated by plots of the type shown in Figure 9. In this graph columns (6) and (7) of Table 2 are plotted together, with the axes annotated and graded appropriately, as shown. Thus, for example, to determine how many tons of ore are available for a WOR cutoff of 5.0, start at the upper x-axis, go down to the cumulative tonnage curve, and read on the right-hand y-axis the corresponding amount of 240 million tons. To determine what the average WOR will be for a maximal WOR of 3.0, start at the lower x-axis, go up to the average ratio curve, and read on the left-hand y-axis the corresponding value of 2.05. In the same fashion it can be seen that increasing the constraint to, say, a WOR of 2.0 reduces the quantities to 60 million tons, with an average WOR of 1.5.

The grade-tonnage distribution of the reserves in the three ore layers is summarized in Figure 10. For the minimal grade cutoff of 20% the quantities for the upper,

Table 2. Zohar phosphate deposit, distribution of various deposit attributes according to different waste-to-ore ratio classes (data from Nachmias, 1987).

WOR CLASS	2	3	4	5	6	CUM TONS	8	9
0.50 - 0.75	3	0.70	4.48	3591.0	0.70	3591.0	27.76	27.76
0.75 - 1.00	7	0.88	4.02	7513.6	0.82	11174.6	28.11	28.00
1.00 - 1.25	12	1.13	3.96	12705.9	0.99	23810.6	28.18	28.10
1.25 - 1.50	12	1.34	4.08	13074.1	1.11	36884.6	27.57	27.91
1.50 - 1.75	13	1.64	3.58	12435.0	1.24	49319.7	28.01	27.94
1.75 - 2.00	19	1.86	3.80	19317.1	1.42	68636.8	27.57	27.83
2.00 - 2.25	14	2.11	3.79	14186.1	1.54	82822.9	27.13	27.71
2.25 - 2.50	16	2.35	3.58	15322.1	1.66	98145.0	27.50	27.68
2.50 - 2.75	20	2.65	3.71	19850.9	1.63	117995.8	26.59	27.50
2.75 - 3.00	21	2.87	3.92	21970.7	1.99	139966.6	25.89	27.24
3.00 - 3.25	20	3.12	3.57	19053.8	2.13	159020.3	26.33	27.13
3.25 - 3.50	12	3.37	3.79	12146.7	2.22	171167.0	25.26	27.00
3.50 - 3.75	19	3.61	3.50	17755.8	2.35	198922.9	25.93	26.90
3.75 - 4.00	12	3.86	3.52	11277.1	2.43	200200.0	26.30	26.87
4.00 - 4.25	15	4.12	3.13	12540.6	2.53	212740.6	26.39	26.84
4.25 - 4.50	13	4.36	3.38	11735.5	2.63	224476.2	25.76	26.78
4.50 - 4.75	11	4.64	3.47	10191.5	2.72	234667.7	26.79	26.78
4.75 - 5.00	9	4.83	3.26	7829.6	2.78	242497.3	26.14	26.76
5.00 - 5.25	11	5.09	3.48	10216.0	2.88	252713.3	27.53	26.79
5.25 - 5.50	4	5.36	3.73	3984.3	2.92	256697.6	26.81	26.79
5.50 - 5.75	2	5.56	2.97	1586.4	2.93	258284.0	27.35	26.80
5.57 - 6.00	1	5.92	3.32	886.1	2.94	259170.1	30.23	26.81
6.00 - 6.25	2	6.12	2.95	1576.0	2.96	260746.1	27.39	26.81
6.25 - 6.50	2	6.30	4.15	2216.0	2.99	262962.1	29.89	26.84
6.50 - 6.75	2	6.59	4.43	2367.0	3.02	265329.1	27.84	26.85
6.75 - 7.00	2	6.85	2.82	1505.6	3.04	266834.7	25.82	26.84
7.00 - 7.25	2	7.14	3.75	2004.5	3.07	268839.2	27.87	26.85
7.25 - 7.50	2	7.40	3.33	1780.3	3.10	270619.5	27.26	26.85
7.50 - 7.75	1	7.52	2.96	792.0	3.12	271411.5	26.01	26.85

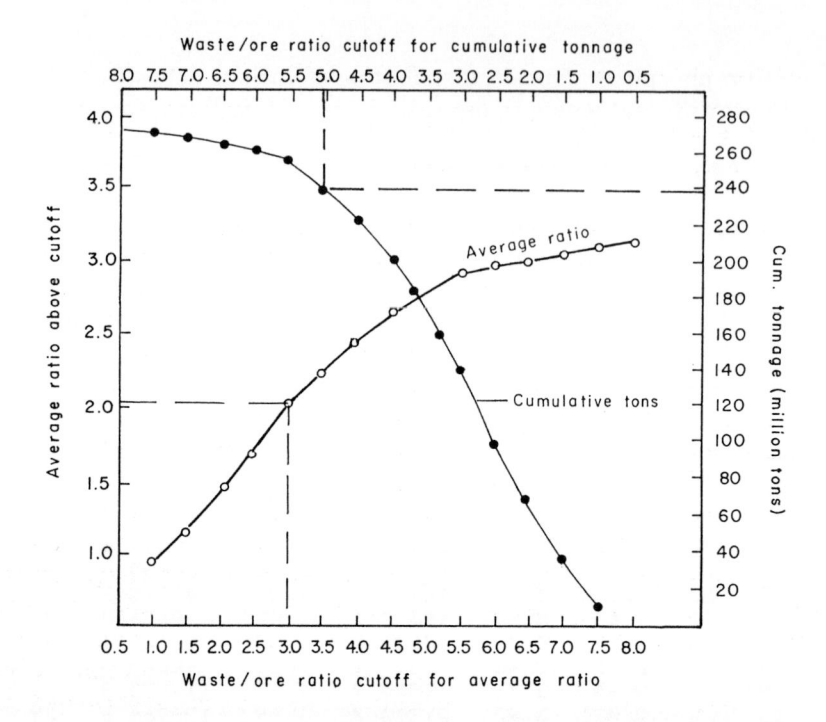

Figure 9. Zohar phosphate deposit, distribution of cumulative ore tonnage according to different waste-to-ore ratios (data from Nachmias, 1987).

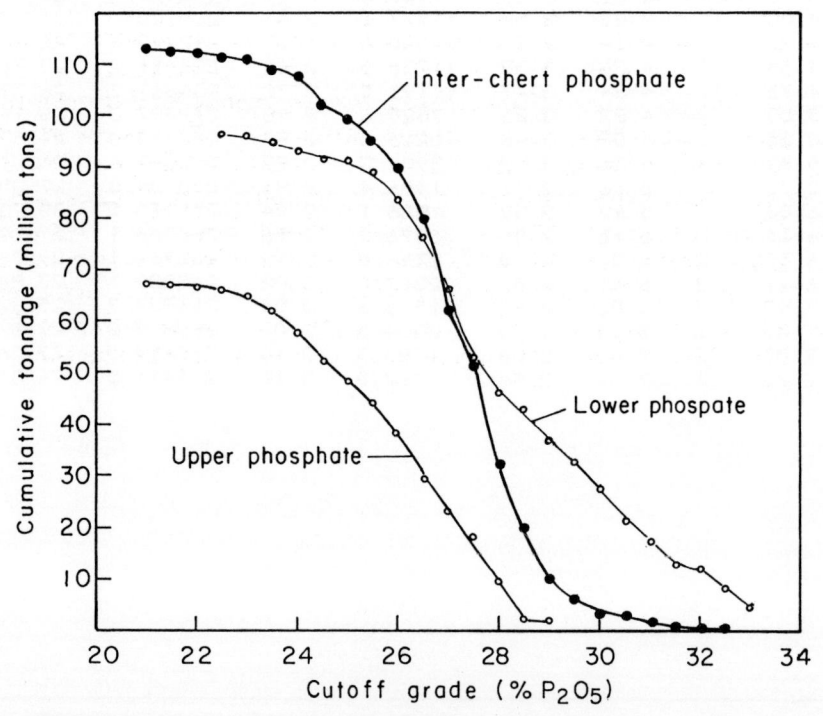

Figure 10. Zohar phosphate deposit, distribution of reserves according to different grade cutoffs, by layer (data from Nachmias, 1987).

lower, and interchart beds, are 65, 111, and 95 million tons, respectively. Raising the cutoff to 26% reduces the quantities to 30, 75, and 80 million tons, respectively. The lower phosphate bed is clearly the richest. Practically all of the highest grade ore (above 30% phosphate) occurs in this layer.

Another important aspect concerns the precision of an estimate and the form of the cost-benefit function for increasing it, or in other words, determining to what extent, and at what cost, the estimation variance of the variables can be reduced by additional drilling. The computations are based solely on the variogram functions. The evaluation can be done with respect to specific new "candidate" drilling sites, or with reference to a regular square network of potentially new sites with a prespecified grid spacing. The results of this exercise for the entire Zohar deposit, using the latter option, are presented in Table 3 (Nachmias, 1986a, 1986b). The computation were carried out by CDC's Drillhole Site Optimization Program (Control Data Corp., 1983). It can be seen that with the present density of drilling, which is on the order of 1.5 holes per square kilometer, the precisions of the average thicknesses of the various units are all better than 20%. This can be regarded as adequate for global, first-cut economic feasibility assessments. However, for more detailed assessments, the situation is drastically different. For detailed mine-planning one has to consider blocks that are no larger than 1 sq km. For this block size the present drilling density is far too sparse; the computed relative errors (precisions) range between 56 and 123%. The right-hand section of Table 3 provides an idea about how much additional drilling is required to attain more acceptable levels of precision, on the order of 30 and 20%, for mining blocks of 1 sq km size. Evidently, to attain these levels will be rather costly. The larger the variability of a given variable, the larger the expenditure. This is clearly the situation with the overburden layer which has the largest variability. Instead of improving the precision of its thickness estimates by additional drilling, which is prohibitively expensive, an alternative way might be to construct a structural map for the base of the overburden (the top of the Phosphate Series) and to subtract this surface from the topographic map. Because the relative variance of this structural surface can be expected to be lower than that of the thickness of the overburden, this procedure will save a substantial amount of drilling. The amount of drilling which will satisfy the precision requirements for the thicknesses will be more than adequate for the grades, because their variability is much smaller. For this reason there was no need to evaluate the precision of the grade estimates.

Table 3. Zohar phosphate deposit, present status of estimation error (with 74 drillholes in 45 sq km area) and additional drilling needed to attain indicated precisions for 1 sq km blocks (data from Nachmias, 1986a).

| variable | Present status ||||||||| Additional drilling required for 1 sq km blocks to attain: ||||
| | For entire deposit |||| For 1 sq km blocks |||| 30% precision || 20% precision ||
	mean thick.	kriging sd.dev.	absol. error	precis. %	kriging sd.dev.	absol. error	precis. %	holes per km	spacing m	holes per km	spacing m
Overburden	17.00	1.64	3.28	19	10.48	21.00	123	25	200	57	130
Upper phos.	1.00	0.07	0.14	14	0.47	0.94	94	15	260	33	170
Lower phos.	1.49	0.07	0.14	9	0.43	0.87	56	6	400	13	280
Interch. waste	4.40	0.20	0.41	9	1.23	4.20	56	5	450	12	290
Interch. phos.	1.65	0.09	0.19	11	0.57	1.14	69	8	350	18	240

CONCLUSIONS

The geostatistical treatments furnish a wealth of quantitative information about the deposits which is not available otherwise. Orebodies involve many variables which are interrelated in a complex manner in time and space. Such situations are difficult to comprehend and manipulate mentally without some controlled simplifications. Block kriging discretization provides a methodology to establish a simple quantitative model which represents the investigated phenomenon faithfully, and that can be manipulated numerically. Such a representation facilitates the comparison and evaluation of the merits of different operational alternatives, be they sites, cutoff grades, or waste-to-ore cutoff ratios.

The studies provide important insight into the spatial variability patterns of the deposits. In this respect, two groups of variables can be distinguished (see Table 1). One includes the thicknesses of shallow stratigraphic units (the overburden and the two uppermost phosphorite beds) which were affected by surficial physical erosion. Here variability is relatively large, as indicated by high ratios of the standard deviation to the mean (0.3-0.5), a large portion of the total variance is random (up to 64% in one instance), and the range of influence (distance of dependence, or "spatial continuity") is short (less than 700 m). The second group includes ore grades (in all horizons), and the thicknesses of some of the deeper strata. These variables have a low variability (standard deviation to mean ratios in the range 0.05-0.10), the share of the random component is small, and the range of influence is 1000 to 1800 m. The spatial variabilities have immediate ramifications concerning the design of the most cost-beneficial sampling network for each deposit.

For the investigated situations, it seems that a drilling density of one borehole per square kilometer is sufficient to estimate the global means of most variables with a precision of 20% or better (at the 95% confidence level). Thus, for preliminary feasibility and reconnaissance surveys, it usually is sufficient to drill on a square grid with a spacing of 1000 m. The number of required drillholes increases significantly with variability and as the size of the evaluated area gets smaller.

In the SYD, where the comparison was made, it was determined that the geostatistical estimates, which are based on block kriging, are in good agreement with estimates obtained from area-of-influence polygons, even when the former utilize data from fewer (in this situation, only one-half as many) drillholes.

ACKNOWLEDGMENTS

Thanks are extended to S. Levi and A. Pe'er for drafting the figures, and to Mrs. B. Katz for helpful editorial suggestions.

REFERENCES

Anonymous, 1977, The industrial and metallic minerals of Israel: Geol. Surv. Israel, Rept. MP/570/77, 60p. (in Hebrew).

Avital, Y., Starinsky, A., and Kolodny, Y., 1983, Uranium geochemistry and fission track mapping of phosphorites, Zefa Field, Israel: Econ. Geology, v. 78, no. 1, p. 121-131.

Bein, A., and Amit, O., 1982, Depositional environments of the Senonian chert, phosphorite and oil shale sequence in Israel as deduced from their organic matter composition: Sedimentology, v. 19, no. 1, p. 81-90.

Control Data Corp., 1979, MIN-VAL, mineral deposit evaluation system—user manual: CDC, Minneapolis, Publ. No. 10127001).

Control Data Corp., 1983, Drillhole site optimization program, general information manual: CDC and Computing Associates International, Publ. No. 41621185, Tuscon, Arizona.

David, M., 1977, Geostatistical ore reserve estimation: Elsevier, Amsterdam, 364p.

Garfunkel, Z., 1978, The Negev - regional synthesis of sedimentary basins: 10th Intern. Congr. Sedimentology, Guidebook, Part I, p. 34-110.

Geostat Systems International, Inc., 1981, The Geostat Package, A computer program library to perform geostatistical ore reserve estimation, version 1.1: Geostat Systems International, Inc., Montreal, Canada.

Gill, D., 1984, Production of mineral commodities in Israel, 1925-1977: Geol. Surv. Israel, Current Research 1983-1984, p. 97-108.

Gill, D., and David, M., 1984, Geostatistical principles illustrated by a thickness analysis of the Ef'e oil shale deposit: Israel Jour. Earth Sciences, v. 33, no. 1-2, p. 48-62.

Gill, D., and Griffiths, J.C.., 1984, Areal value assessment of the mineral resources endowment of Israel: Jour. Math. Geology, v. 16, no. 1, p. 37-89.

Horowitz, A., 1979, The Quarternary of Israel: Academic Press, New York, 394 p.

Issahary, D., 1982, Measured and technological characteristics of oil shales: Proc. 6th Israeli Conf. Min. Eng. p. 84-88 (in Hebrew).

Journel, A.G., and Huijbregts, Ch.J., 1978, Mining geostatistics: Academic Press, New York, 600p.

Kolodyny, Y., 1980, Carbon isotypes and depositional environments of a high productivity sedimentary sequence - The case of the Mishash-Ghareb Formations, Israel: Israel Jour. Earth Sciences, v. 29, no. 1-2, p. 147-156.

Miller, E., and Gill, D., 1986, Geostatistical ore reserve estimation of south Yorke'am phosphate deposit, Zin Valley, Southern Israel: Trans. Inst. Min. Metall. Sect. A: Min. Indurstry, v. 95, p. A1-A7.

Minster, T., and Shirav, M., (Scwartz), 1984, National oil shale survey 1983-84, findings from the northern Negev and geological guidelines for future prospection: Proc. 7th Israel Conf. Min. Eng., p. 112-116 (in Hebrew).

Minster, T., and others, 1986, Oil shale-phosphorite prospects in the east Mediterranean: Industrial Minerals, March 1986, p. 47-60.

Nachmias, Y., 1986a, Planning a drillhole sampling network for the Zohar phosphate deposit with the aid of geostatistics: Negev Phosphates Ltd., Rept. no. MPM/2/86, 14 p. (in Hebrew).

Nachmais, Y., 1986b, Sampling network optimization for mine development by means of geostatistics, the Zohar phosphate deposit, a case study: Proc. 8th Israel Conf. Min. Eng., p. 144-149 (in Hebrew).

Nachmias, Y., 1987, Geostatistical reserve estimation and mine planning for the Zohar phosphate deposit: Negev Phosphates Ltd., Rept. No. MPM/4/87, 6p.

Shirav (Schwartz), M.,1987, Pathway of some major and trace elements during fluidized bed combustion of Israeli oil shale: unpubl doctoral dissertation, Technion, Israel Institute of Technology, Haifa, 198p. (in Hebrew).

Shirav (Schwartz), M., and Ginzburg, D., 1978, A guidebook to the oil shale deposits of Israel: Geol. Surv. Israel, Mineral Resources Division, 20p.

Shirav (Schwartz), M., and Ginzburg, D., 1983, Geochemistry of Israeli oil shales, *in* Miknis F.P., and J.F. McKay, eds., Geochemistry and chemistry of oil shales: Am. Chem. Soc. Symposium 230, p. 85-89.

Shirav (Schwartz), M.,and Minster, T., eds., 1984, Oil shales in Israel; I. Collected reprints - publications of GSI research (1982-84); II. Resources as of April 1984: Geol. Surv. Israel, Rept. No. GSI/24/84.

Slotky, D., and others, 1983, Rotem oil shale deposit - computerizes geological model (CDC's "MINEVAL') versus conventional geological model: Geol. Surv. Israel, Current Research 1982, p. 72-75.

Soudry, D.,, 1987, Ultra-fine structure and genesis of the Campanian Negev high-grade phosphorites (southern Israel): Sedimentology, v. 34, no. 4, p. 641-660.

Soudry, D., and Mor, U., 1985, Phosphorite survey along the northern Arava margins: Geol. Surv. Israel, Rept. No. GSI/19/85, 46 p. (in Hebrew).

Soudry, D., Nathan, Y., and Roded, R., 1985, The Ashosh-Haroz facies and their significance for the Mishash paleogeography and phosphorite accumulation in the northern and central Negev, Southern Israel: Israel Jour. Earth Sciences, v. 34, no. 4, p. 211-220.

Soudry, D., and Lewy, Z., 1988, Microbially influenced formation of phosphate nodules and megafossil moulds (Negev, southern Israel): Palaeogeog. Paleoclim. Paleoecol., v. 64, no. 1, p. 15-34.

Spiro, B., 1980, Geochemistry and mineralogy of bituminous rocks in Israel: unpubl. doctoral dissertation, Hebrew Univ. of Jerusalem, 152 p. (in Hebrew).

Sprio, B., and others, 1983, Asphalts, oils, and bituminous rocks from the Dead Sea area - a geochemical correlation study: Am. Assoc. Petroleum Geologists Bull., v. 67, no. 7, p. 1163-1175.

Tannenbaum, E., 1983, Researches into the geochemistry of oils and asphalts in the Dead Sea area: unpubl. doctoral dissertation, Hebrew Univ. of Jerusalem, 117 p. (in Hebrew).

Vengosh, A., Kolodny, Y., and Tepperberg, M., 1987, Multi-phase oxygen isotopic analysis as a tracer of diagenesis: the example of the Mishash Formation, Cretaceous of Israel: Chem. Geology, v. 65, no. 4, p. 235-253.

Yerushalmi, Y., 1984, A program for utilizing oil shales in Israel - an update: Proc. 7th Israel Conf. Min. Eng., p. 94-103 (in Hebrew).

Zohar, E., and Shiloni, Y., 1987, The Arad Basin phosphate deposit: Geol. Surv. Israel Rept. No. GSI/19/87, 50p. (in Hebrew).

Pore Geometry Evaluation by Petrographic Image Analysis

S.M. Habesch
Poroperm-Geochem Ltd.,Chester

ABSTRACT

There is an increasing demand among production reservoir engineers and petrophysicists for more detailed information on the porosity networks within reservoir lithologies. Variations in measured permeability and reservoir quality cannot be modeled always by a single porosity variable, for example the helium porosity measurement. The situation usually is far more complex, reflecting the interaction of several geometrical parameters.

The two-dimensional pore geometrical structure is well represented by resin impregnated thin sections viewed by back scattered electron microscopy. The problem until recently has been to assess quantitatively the geometrical parameters of thousands of pores on a routine, rapid basis. The introduction of high-powered image analyzers with large storage capacity and array processing facilities, has allowed the rapid digitization of pore network images to a high-resolution pixel matrix. In digitized form after suitable calibration, processing and gray-level thresholding, the geometrical parameters of pores contained within these images can be measured.

The initial or first-order pore parameters include pore area, diameter, perimeter, orientation, and vertical and horizontal connectivity. Using these data second-order pore parameters such as porosity, pore density, specific surface area, pore shape, aspect, and tortuosity also may be calculated. The first- and second-order parameters can be correlated against permeability to model the importance of geometrical networks on reservoir quality. Higher order or reservoir parameters can be calculated using the second-order parameters. These include permeability determination with the porosity and specific surface area data and capillary pressure using the relationship between pore area and cumulative porosity. The second-order parameters - porosity, pore density, specific surface area, and pore shape - are particularly useful in pore geometry classification schemes and the recognition of extreme 'end-member' pore types. These classifications can be used to recognize dominant pore geometry structures within lithological sequences.

This type of analysis is applied routinely to well studies in the North Sea (UK) sector and has been successful in not only identifying pore geometrical controls on reservoir quality on a quantitative basis but also in modeling the changes on pore structure caused by compactional and diagenetic processes.

INTRODUCTION

One of the major controlling factors for the migration of hydrocarbons through reservoir lithologies is the geometrical form of the pore structure. The form of three-dimensional pore networks usually is inferred from laboratory measurement on isolated core plugs (conventional or special core analysis) following helium porosity (%) - (volumetric porosity under ambient or overburden conditions) or air permeability (ability of a fluid phase to move through a lithology) determinations. Permeability is calculated from the Darcy equation (e.g. Archer and Wall, 1986).

$$Q = \frac{K.A. (P_1^2 - P_2^2)}{2\mu.l}$$

where $(P_1^1 - P_2^2)$ is the pressure differential across the plug
l is the length of the core plug
A is cross-sectional area
μ is viscosity of the fluid
Q is rate of flow
K is the permeability (mD)

For oil reservoirs the quality scale illustrated in Figure 1 usually is recognized. Porosity and especially permeability parameters are useful but they do not provide any physical, quantitative data on the size, shape, or connectivity of the pores. The classification of clastic reservoir lithologies in terms of 'porofacies' usually is made using permeability/porosity data (Fig. 1). However, permeability may not be modeled by a single porosity trend on these diagrams and considerable amounts of scatter are observed. This implies that permeability is not controlled necessarily by a single porosity variable - the helium porosity measurement. The situation usually is far more complex, reflecting an interaction of several geometrical parameters.

Pore throat radii (r) can be calculated from capillary pressure (Pc) measurements following mercury injection of core plug samples, using the following equation (e.g. Archer and Wall, 1986).

$$r = \frac{2\sigma \cos\theta}{Pc}$$

where σ is the interfacial tension between the wetting and nonwetting phases, and
θ is the angle between the wetting phase and the pore wall.

Mercury injection techniques however tend to be expensive, time consuming, and sample destructive and the data provided is limited to individual pore throats, not accounting for the pore volume or the network patterns.

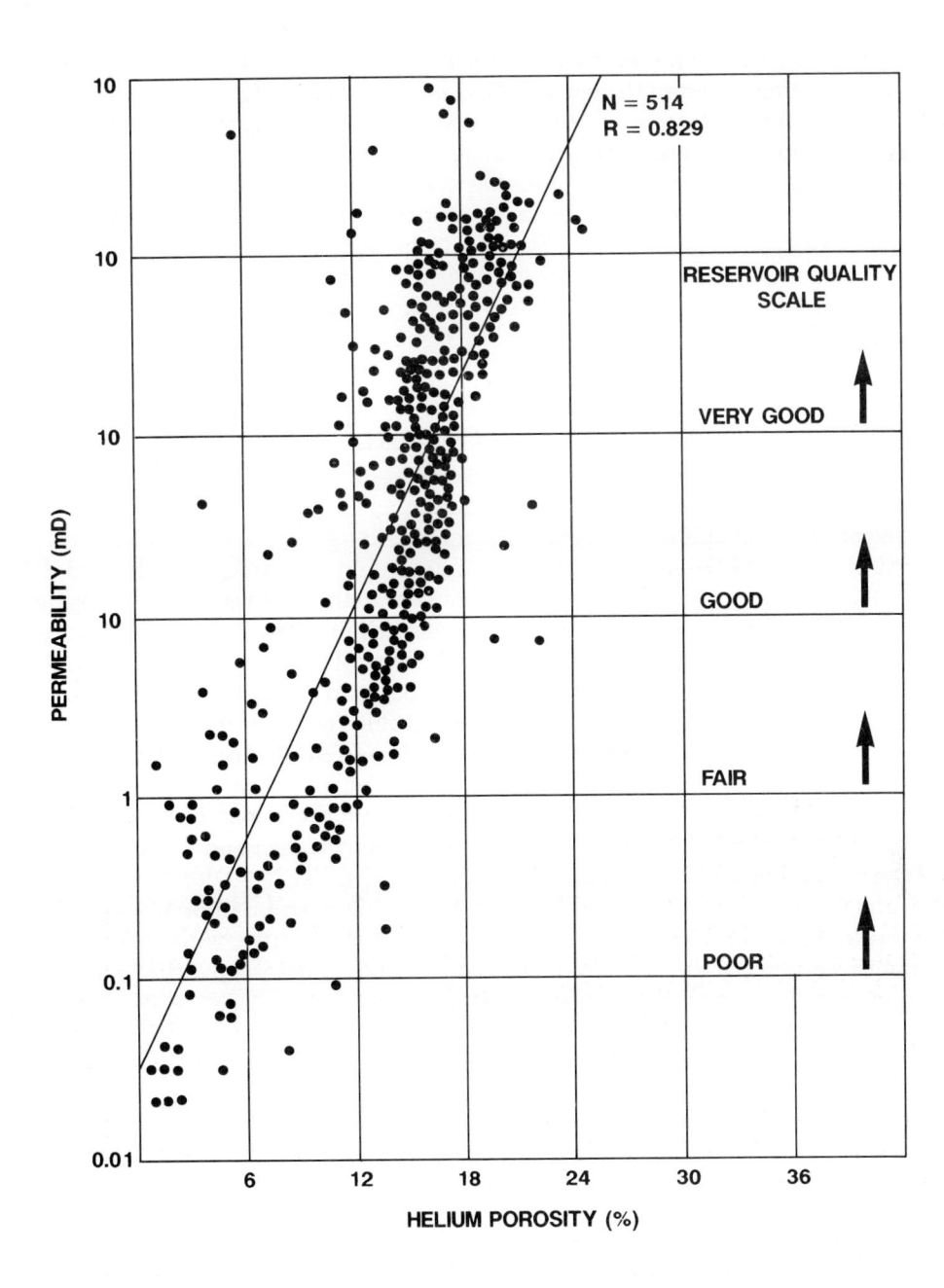

Figure 1. Permeability/porosity crossplot used in reservoir descriptions as 'porofacies' classification diagram. Note trend of data points is not uniform in spite of significant correlation and large scatter in some data points. This suggests that reservoir quality cannot be modeled by single pore network parameter (volumetric porosity) across wide range of lithological types

Porosity and pore network structures in reservoir lithologies are effectively represented in resin impregnated thin or polished 2-dimensional (Fig. 2A) sections. The traditional petrographic approach has been qualitative with the subdivision of porosity into primary intergranular porosity, secondary dissolution porosity (reflecting grain and cement corrosion), and ineffective microporosity and semiquantitative assessment of porosity percentage by point counting. With suitable image processing it is possible to digitize petrographic images, isolate the pore structure (Fig. 2B) by gray-level thresholding, and make measurements on individual pores. This article describes an analytical procedure (POROS) for the quantitative parameterization of pore networks observed in 2D lithological sections by petrographical image analysis.

POROS - A PROCEDURE FOR PORE GEOMETRY EVALUATION

The general field of microscopic image processing and analysis is relatively mature with the general principles, strategies, and applications within different descriptions now well documented (Duda and Hart, 1973; Rosenfeld and Kak, 1976; Fabbri, 1980; Serra, 1982; Rosenfeld, 1984). Image processing and analysis techniques have been applied to pore geometry assessment in reservoir lithologies previously (Rink, 1976; Rink and Schopper, 1978; Ehrlich and others, 1984; Dilkes and Graham 1985; Berryman and Blair, 1986; Ruzyla, 1986). Individual pores in reservoir lithologies are treated as a compositional phase which (with suitable resin impregnation) will possess a specific gray-level intensity which is detected by electron imaging and light microscopy systems. Much early work was carried out using transmitted light microscopy systems (e.g. Ehrlich and others, 1984) and there may be potential applications (although limited) with fluorescence microscopy (Crisp and Williams 1971; Jongerius and others, 1972; Soeder, 1987). However, it has been determined (Dilkes, Parks, and Graham, 1984; Pye and Kinsley, 1984; Dilkes and Graham, 1985; Huggett, 1984 and Ruzyla, 1986) that the highest resolution images, with optimum signal/noise ratios, are obtained from scanning electron microscopy (SEM) with a back scattered electron (BSE) detection system.

POROS (Table 1) is an analytical procedure for the quantitative parameterization of pore networks using BSE images (Fig. 2A) and can be incorporated into the majority of image processing hardware/software (including low cost PC systems) systems presently available on a commercial basis. The necessary requirements for such a system (apart from access to a SEM) are a video scanning SEM interface, a digitizer which converts the monochrome analog signal to digital form, and a microcomputer with sufficient memory to store and process all the gray-level data in the image. Image-processing software and a high-speed data statistics package also are required. An outline of POROS is provided in Table 1. The procedure is fast (a few seconds to analyze and store data from a single image), fully automated, but manual input is possible at critical stages, such as porosity segmentation, and works effectively for a large range of lithological types. Required samples are resin impregnated, polished blocks or sections with a thin coating of evaporated carbon. The samples are orientated initially with respect to primary lamination or structures so that the pore orientation and connectivity data have a suitable reference point. In

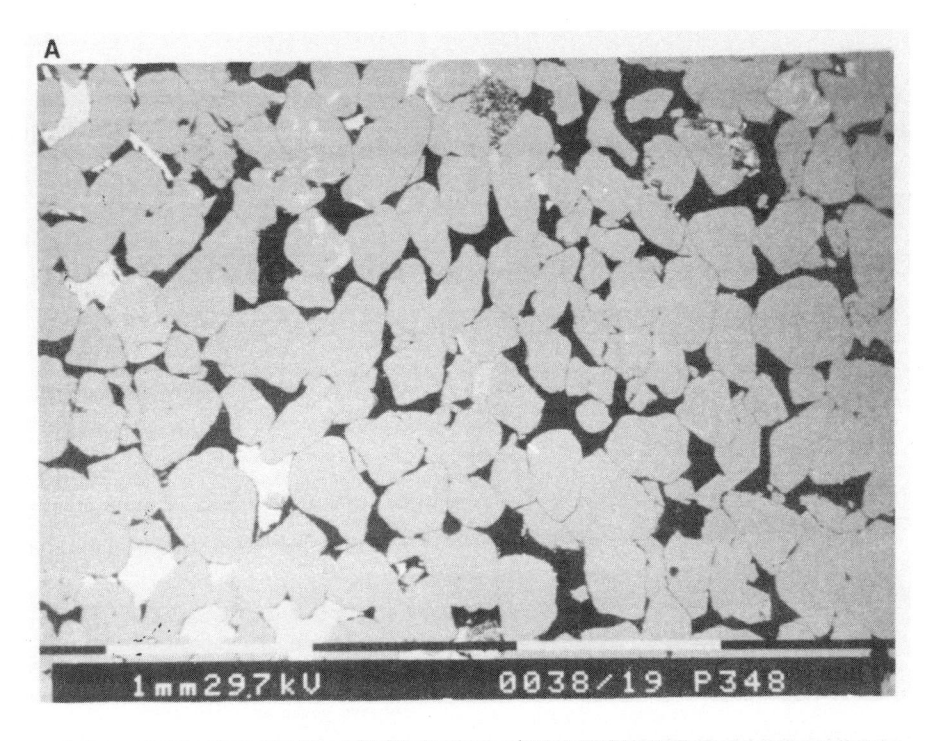

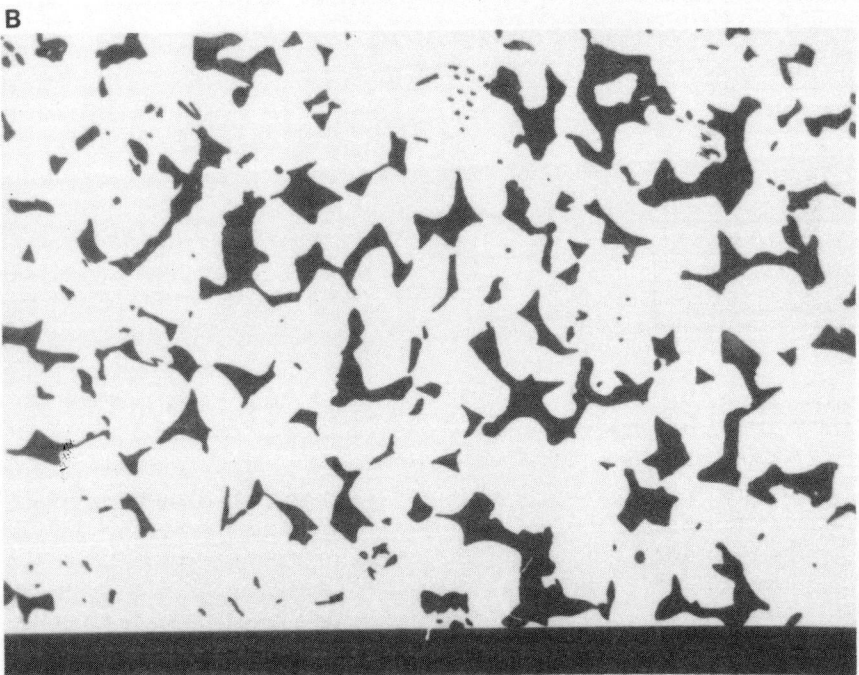

Figure 2. A, Back scattered electron image of typical clastic lithology from Brent (Tarbert) units from North Sea sector. Preserved porosity is black, reflecting injected resin medium; B, Segmented binary image of pore network by gray-level thresholding of (A).

Table 1. Image-processing sequence for pore geometry parametization by POROS.

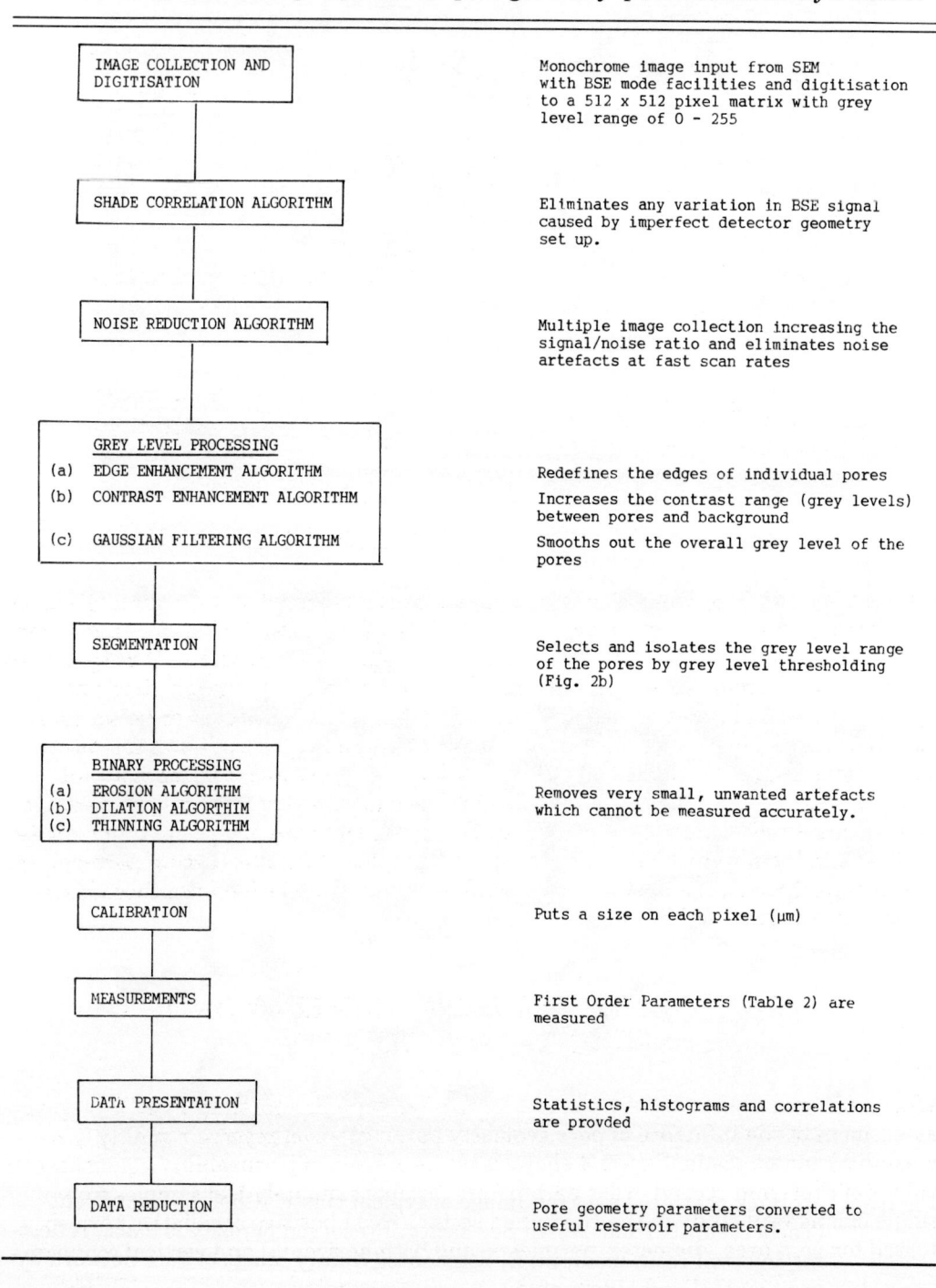

IMAGE COLLECTION AND DIGITISATION	Monochrome image input from SEM with BSE mode facilities and digitisation to a 512 x 512 pixel matrix with grey level range of 0 - 255
SHADE CORRELATION ALGORITHM	Eliminates any variation in BSE signal caused by imperfect detector geometry set up.
NOISE REDUCTION ALGORITHM	Multiple image collection increasing the signal/noise ratio and eliminates noise artefacts at fast scan rates
GREY LEVEL PROCESSING (a) EDGE ENHANCEMENT ALGORITHM (b) CONTRAST ENHANCEMENT ALGORITHM (c) GAUSSIAN FILTERING ALGORITHM	Redefines the edges of individual pores Increases the contrast range (grey levels) between pores and background Smooths out the overall grey level of the pores
SEGMENTATION	Selects and isolates the grey level range of the pores by grey level thresholding (Fig. 2b)
BINARY PROCESSING (a) EROSION ALGORITHM (b) DILATION ALGORTHIM (c) THINNING ALGORITHM	Removes very small, unwanted artefacts which cannot be measured accurately.
CALIBRATION	Puts a size on each pixel (μm)
MEASUREMENTS	First Order Parameters (Table 2) are measured
DATA PRESENTATION	Statistics, histograms and correlations are provded
DATA REDUCTION	Pore geometry parameters converted to useful reservoir parameters.

BSE images (Fig. 2A), the resin with the lowest net atomic number will have the lowest range of gray levels allowing easy segmentation from the background framework grains and authigenic cements (Fig. 2B) with the minimum amount of gray level or binary processing. As several BSE images are collected for a single sample, it is essential that the SEM/BSE working parameters (Kv, working distance, detector geometry, detector gain levels) remain constant. The number of BSE images analyzed and the working magnification will differ between lithological types, reflecting the size, and number of pores to be analyzed. However, it is essential to work at a constant magnification and image number where images are compared directly against each other and related to independent variables, for example permeability. Although the initial BSE images require only minimal secondary processing and more of the computing time/power can be directed towards analysis, a certain amount of gray level (presegmentation) and binary (postsegmentation) processing is required especially in clay-rich lithologies.

With the processing complete, each of the isolated pores are counted and measurements are made. These data are stored in data files which then are analyzed using a traditional statistics package.

PORE GEOMETRY DATA GENERATED BY POROS

A typical output from POROS would involve measurements of several parameters on up to 2000 pores. The parameters are nested into three levels (Table 2). First-order parameters are the raw parameters measured on each isolated pore in the segmented image. Second-order parameters are calculated from first-order parameters and apply to individual pores or characterize the entire network. Higher order (or reservoir) parameters involve calculations using combinations of second- and first-order parameters. As an example, two different lithological pore images with relevant first- and second-order data are provided in Figure 3, showing frequency percentage histograms for pore area, pore specific surface area, and pore horizontal connectivity to illustrate the difference between these two samples. Model statistical parameters, that is mean, median, and standard deviation values are selected from these histograms, and can be used in correlation plots.

CORRELATION OF PORE GEOMETRY DATA
AGAINST PERMEABILITY

Where a large number of litholgical samples have been processed by POROS, a good assessment of the influence of pore geometry parameters on reservoir quality (permeability) can be made. Figure 4 shows a series of plots of permeability against mean values of first- and second-order parameters for approximately 100 samples from a single clastic well study in the North Sea sector. Good positive correlations are observed for pore area, diameter, perimeter, and both horizontal and vertical connectivity. A negative correlation is observed between permeability and specific surface area. These correlations suggest that reservoir quality can be modeled effectively by use of these parameters through a wide range in permeability (0.1 - 10^4 mD).

Table 2. Pore geometry parameters.

FIRST ORDER PARAMETERS	Pore Area (BA)	The number of detected pixels forming the pore's interior.
	Pore Perimeter (BP)	For each point $X(i)$, $Y(i)$ on the pore's boundary, the distance to previous and succeeding points are calculated. The distance between the mid points of these vectors is computed and the perimeter is the sum of the distances.
	Pore Length (BL)	The centre of gravity is calculated as a base point. A point on the boundary $(X1, Y1)$ furthest from base is iteratively selected which will be the maximum chord length.
	Pore breadth (BB)	Using the maximum chord length, the maximum normal distances of all boundary points is computed. Pore breadth is the sum of the two maximum distances - maximum projection normal to length.
	Pore Orientation (BO)	Pore orientation is the angle between maximum chord length and vertical (Y) axis.
	Pore horizontal connectivity (BW)	Pore width - the normal projection onto the horizontal X axis.
	Pore vertical connectivity (BH)	Pore height - the normal projection onto the vertical Y axis.
SECOND ORDER PARAMETERS	Pore Density (pores/mm^2)	The number of pores per unit area.
	Porosity (%)	Total detected pore area divided by analysed field area.
	Field Specific Surface Area* (μm^{-1})	Calculated by $(BP/BA) * (4/\pi) = Ss$.
	Field Shape Factor*	Calculated by $1/3 * Ss/Km$; $Km = 2N/2BP$ (N = Number of pores).
	Pore Aspect	Pore length/breadth ratio (BL/BB)
	Vertical/Horizontal Connectivity Ratio	Pore width/height ratio (BW/BH)
HIGHER ORDER RESERVOIR PARAMETERS	Permeability Estimates (Kia)	Calculated from porosity and field specific surface parameters $Kia = \emptyset^3/5(1-\emptyset^2).Ss^2$
	Cumulative Porosity/Pore Area Relationships +	Pore area frequency histograms are recalculated against measured % porosity (see text for examples)

* Shape factor and specific surface area is also calculated for individual pores

+ Capillary pressure calculations can be made from these relationships

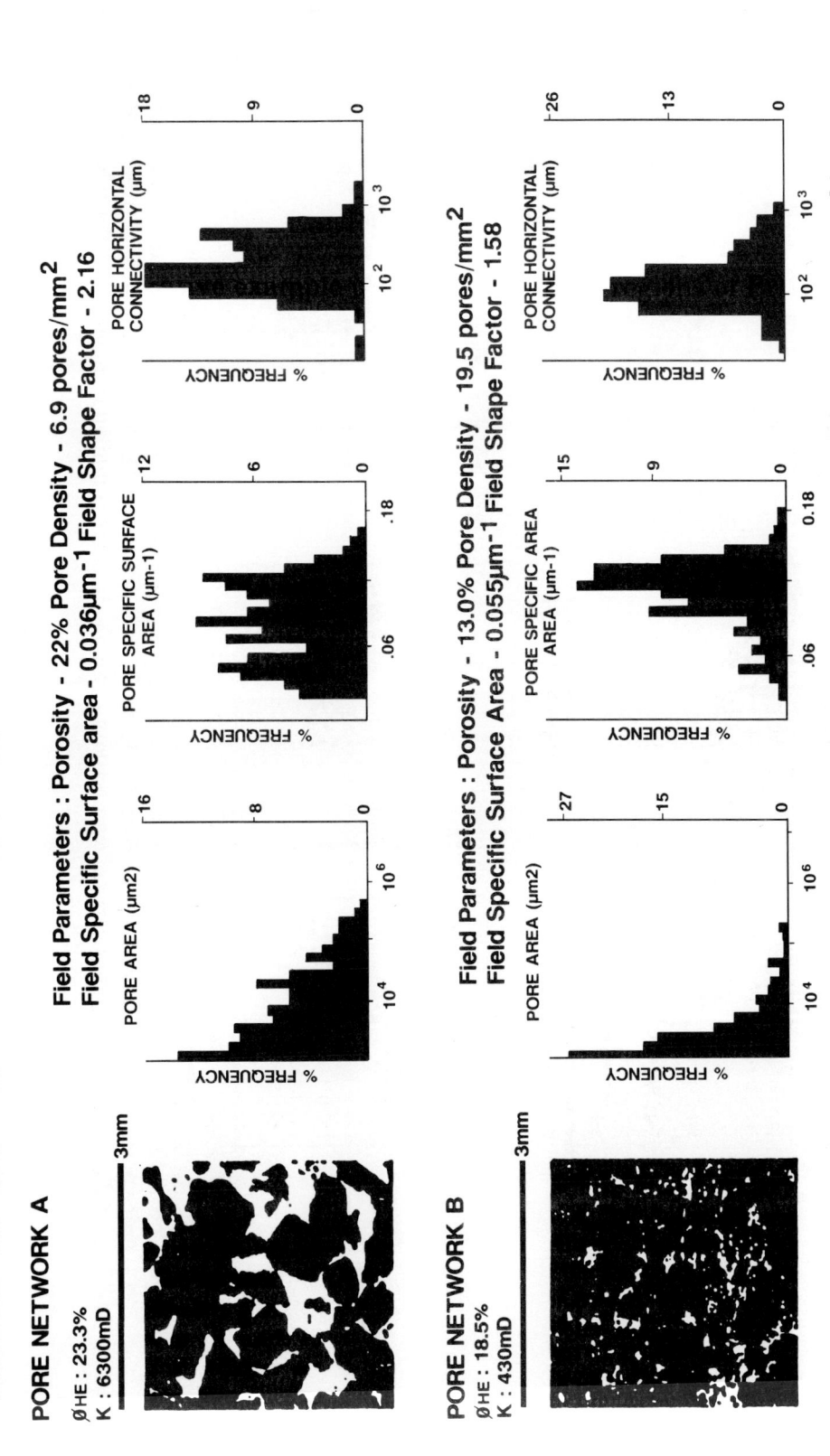

PORE GEOMETRY EVALUATION BY PETROGRAPHIC IMAGE ANALYSIS 309

Figure 3. First- and second-order parameter data for two extreme lithological types. Percentage distribution histograms are provided for pore area, specific surface area, and horizontal connectivity. Binary images are processed, segmented images of effective pore structure; a 3mm scale bar indicates size. Core analysis (helium porosity, air permeability) data, second-order field parameters and Kia estimates (see text) are provided. These two examples are clastic lithologies selected from North Sea well.

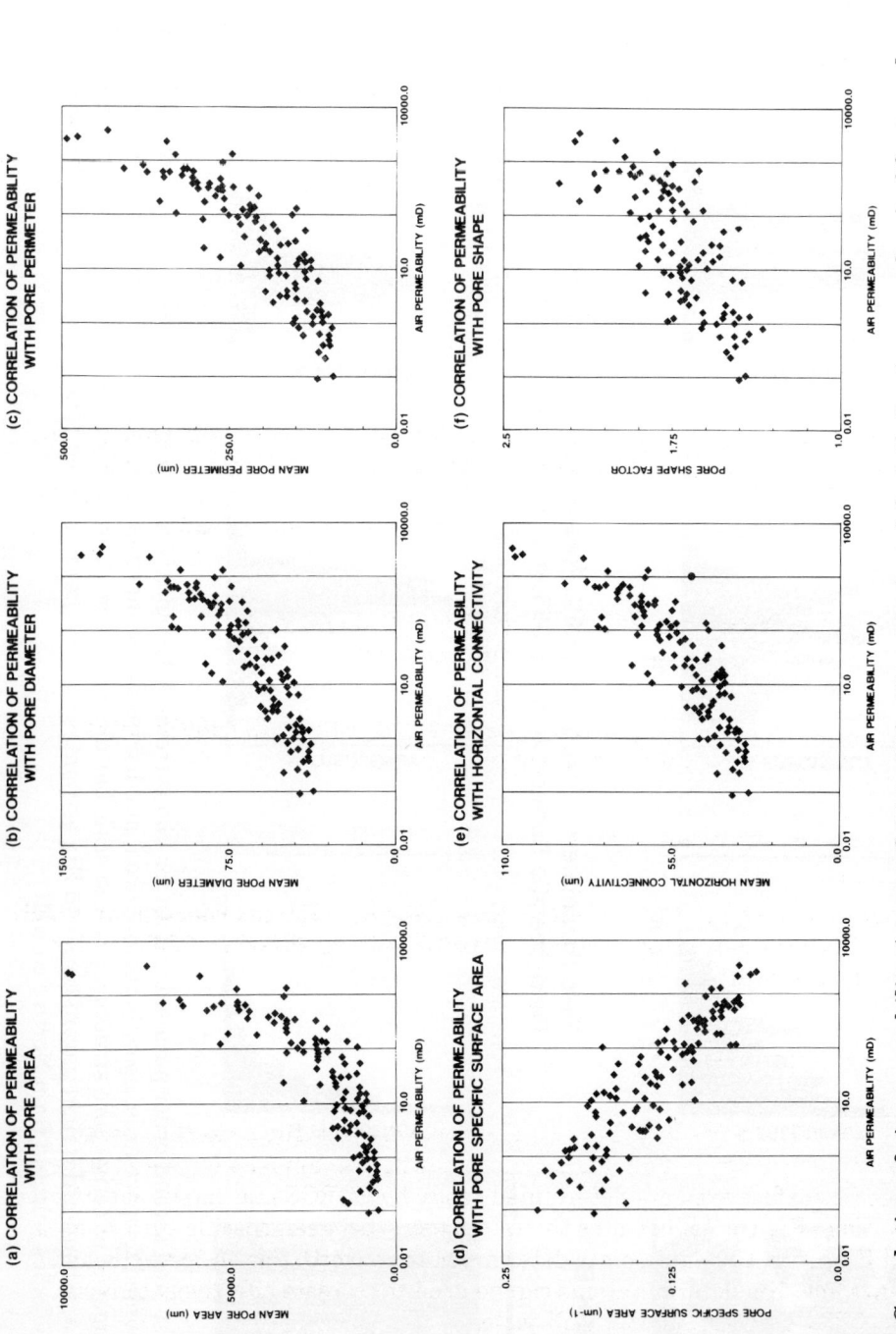

Figure 4. Correlation of air permeability (range of 0.1 to 10^{4}mD) against model statistics (mean values) of first- and second-order parameters. A, Pore area; B, Pore diameter; C, Pore perimeter; D, Pore specific surface area; E, Pore horizontal connectivity; F, Pore vertical connectivity. Sample data are compiled from single well study from North Sea sector.

PORE GEOMETRY CLASSIFICATION

The second-order (Field) parameters (porosity, pore density, specific surface area, and shape - Table 2) can be used in pore network classification schemes for clastic reservoirs. An example is provided in Figure 5 consisting of approximately 130 data points taken from a single well study in the North Sea sector and covering a permeability range of 0.1 - 4000mD. The scatter in the diagrams suggests that considerable variation in pore geometry is observed between these samples and representative illustrative binary images are provided for potential end members. These diagrams also can be contoured for other variables (Fig. 6), for example permeability (A), capillary pressure (B), facies type (C), and authigenic cement (%) (D) to model relationships between pore geometrical style and other important reservoir data. A major advance would be to combine these parameters (with additional pore size data - see cumulative porosity curves) in multivariate classification schemes using principal component analyses (PCA) or factor analysis. This work is currently ongoing.

The purpose of these types of diagrams is to classify reservoir lithologies from a pore network view point, rather, than the traditional grain framework or authigenic cement basis and to identify extreme end-member 'porotype' lithologies which can be calibrated by engineering data (i.e. permeability, capillary pressure, etc.).

PERMEABILITY CALCULATIONS USING PORE GEOMETRY DATA

The Carmen-Kozeny equation (e.g. Collins, 1961; Archer and Wall, 1986) relates permeability to the geometrical structure of pore space;

$$Kck = \frac{\emptyset^3}{Ko(Le/L)^2 \cdot (1-\emptyset^2) \cdot Ss^2}$$

where $\emptyset$ = porosity fraction; Ss = specific surface area; Ko = Kozeny constant and Le/L = tortuosity. It generally is assumed that $Ko (Le/L)^2 = 5$ (e.g. Ruzyla, 1986) and the equation may be rewritten as:

$$Kia = Kck \frac{\emptyset^3}{5(1-\emptyset)^2 \cdot Ss^2}$$

This model implies the porosity to be equivalent to a conduit, the cross section of which has a complex shape but an averaged constant area. Porosity (or pore fraction) and field specific surface area are determined easily by POROS and can be substituted into the equation. Figure 7 illustrates the relationship between specific surface area, porosity and Kia, with 100 data points determined from North Sea sector wells and Figure 8 shows how Kia determinations can be used to recreate effectively conventional permeability assessment on a well scale.

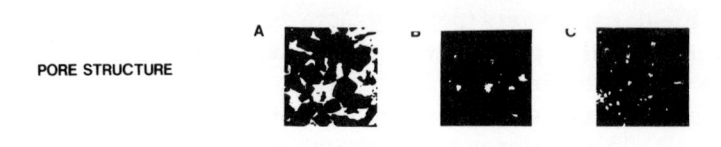

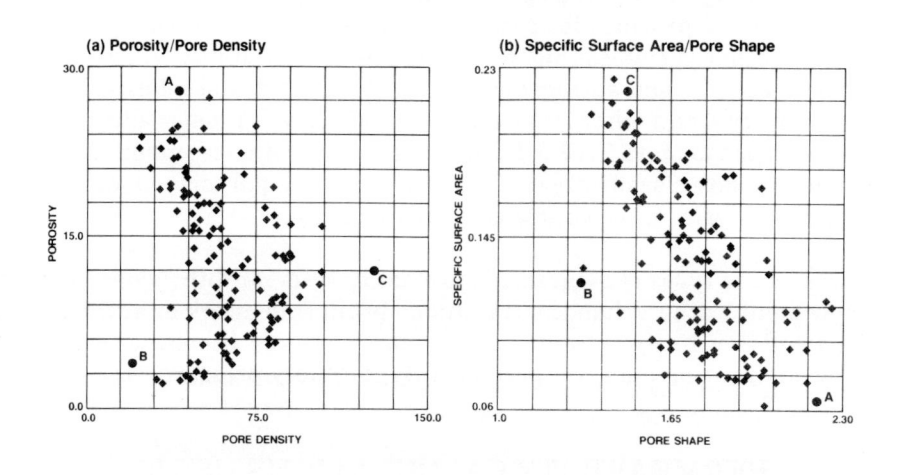

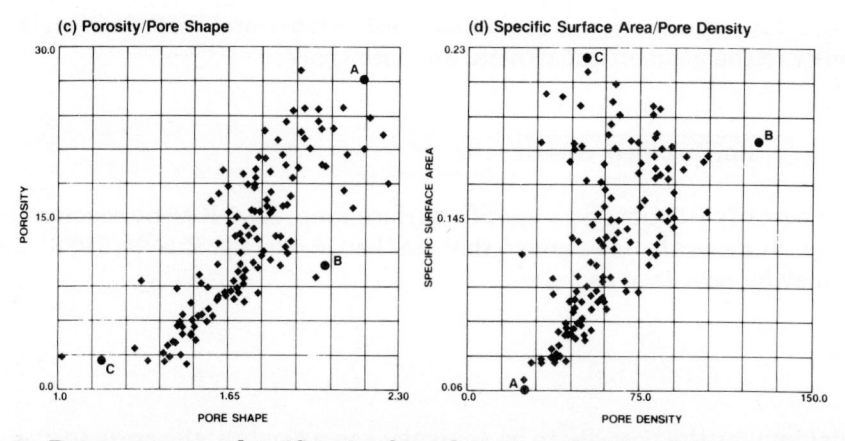

Figure 5. Pore geometry classification based on second-order field parameters representative binary images of extreme, end-member pore networks are included.
(A) Porosity and pore density
(B) Specific surface area and pore shape
(C) Porosity and pore shape
(D) Specific surface area and pore density
Sample data are compiled from single well study from North Sea sector.

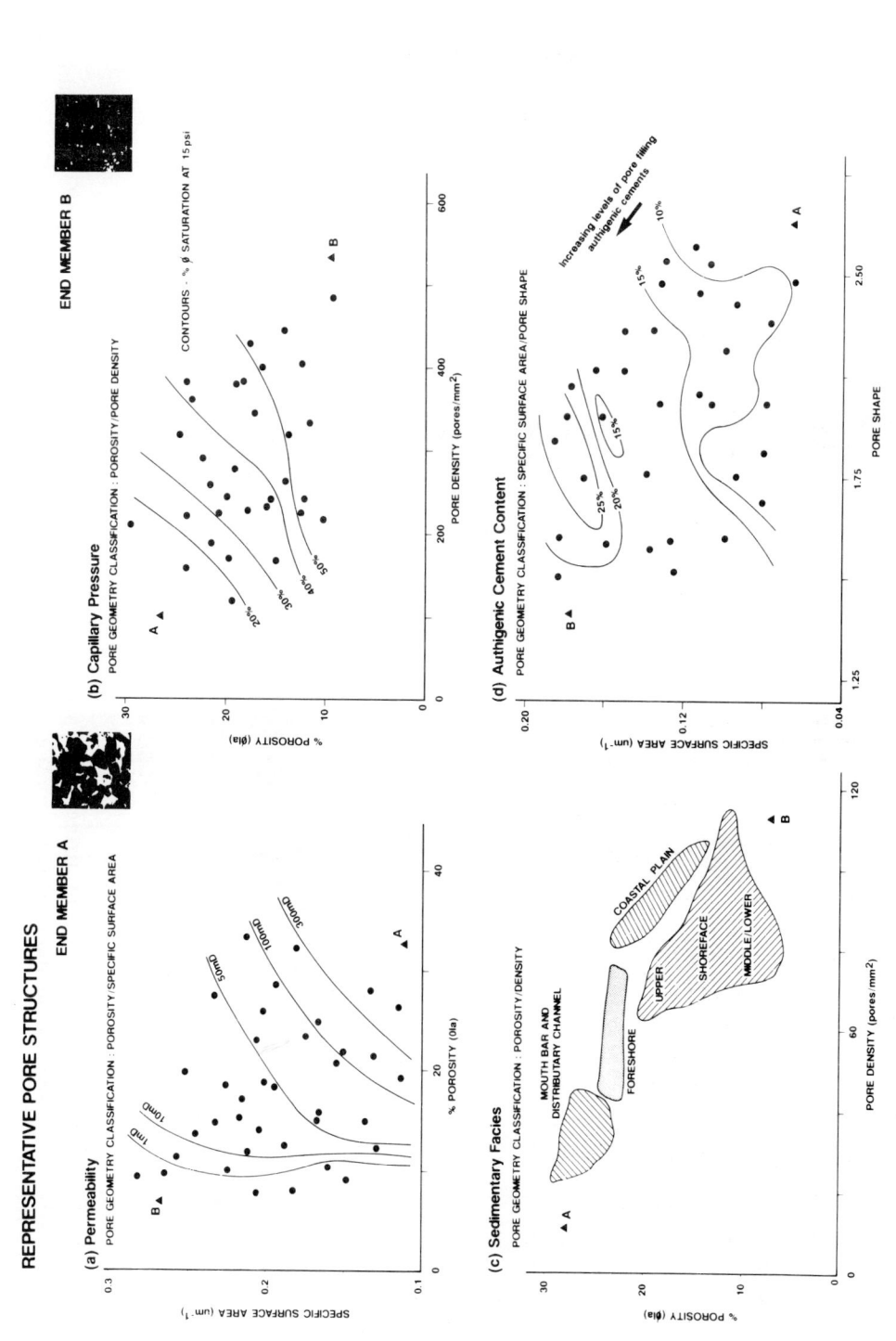

Figure 6. Use of pore geometry classification diagrams for modeling of other reservoir and petrological data. A, Permeability; B, Capillary pressure; C, Sedimentological facies; D, Authigenic cement %. Examples selected from Jurassic Brent lithologies of North Sea sector.

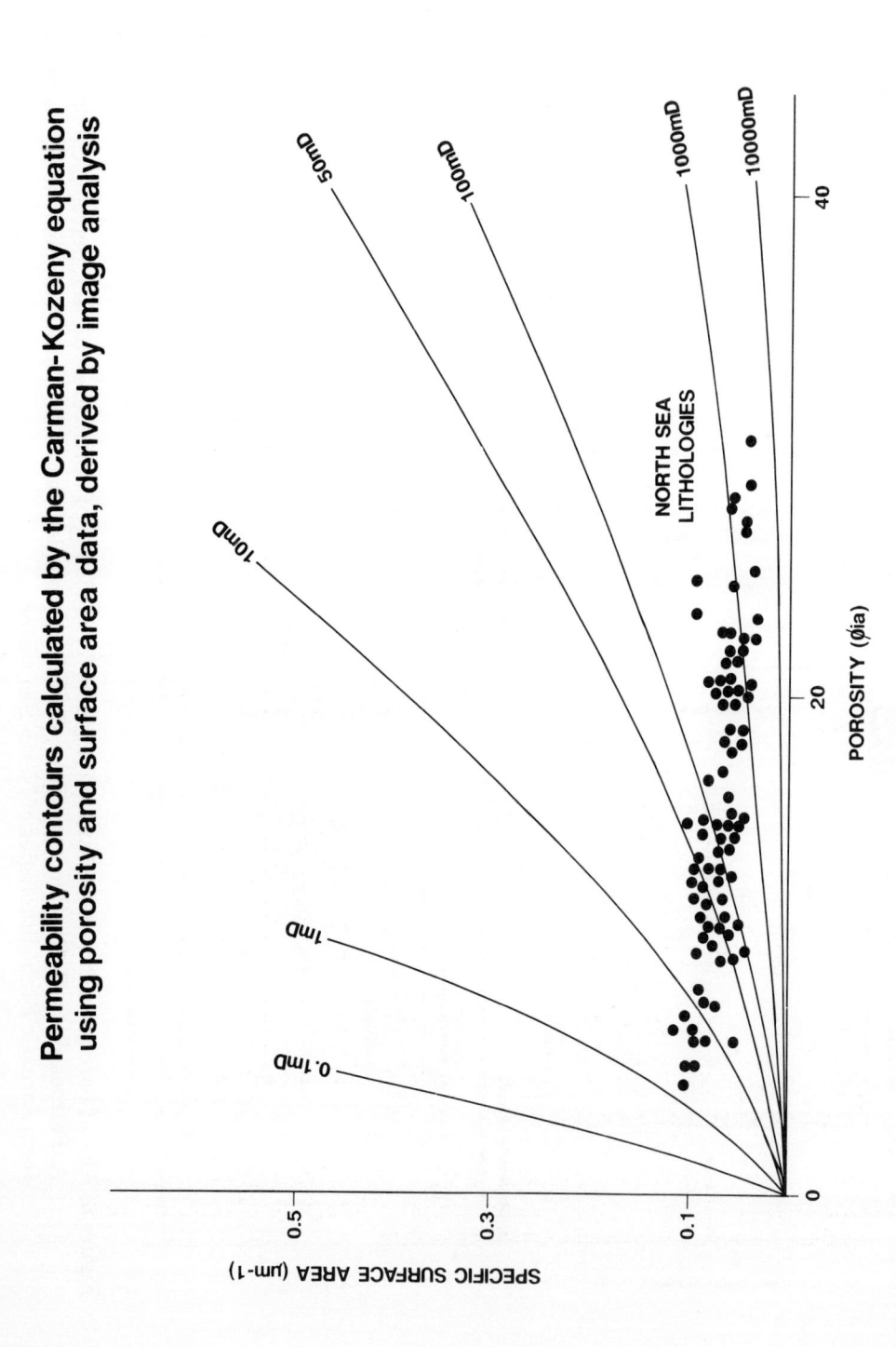

Figure 7. Relationship between porosity, specific surface area (second-order field parameters determined by POROS) and permeability calculated by Karman-Cozeny equation. Approximately 100 samples from North Sea well are used as typical data set.

PORE GEOMETRY EVALUATION BY PETROGRAPHIC IMAGE ANALYSIS

COMPARISON OF PERMEABILITY ESTIMATIONS BY PETROGRAPHIC IMAGE ANALYSIS AND CONVENTIONAL CORE ANALYSIS

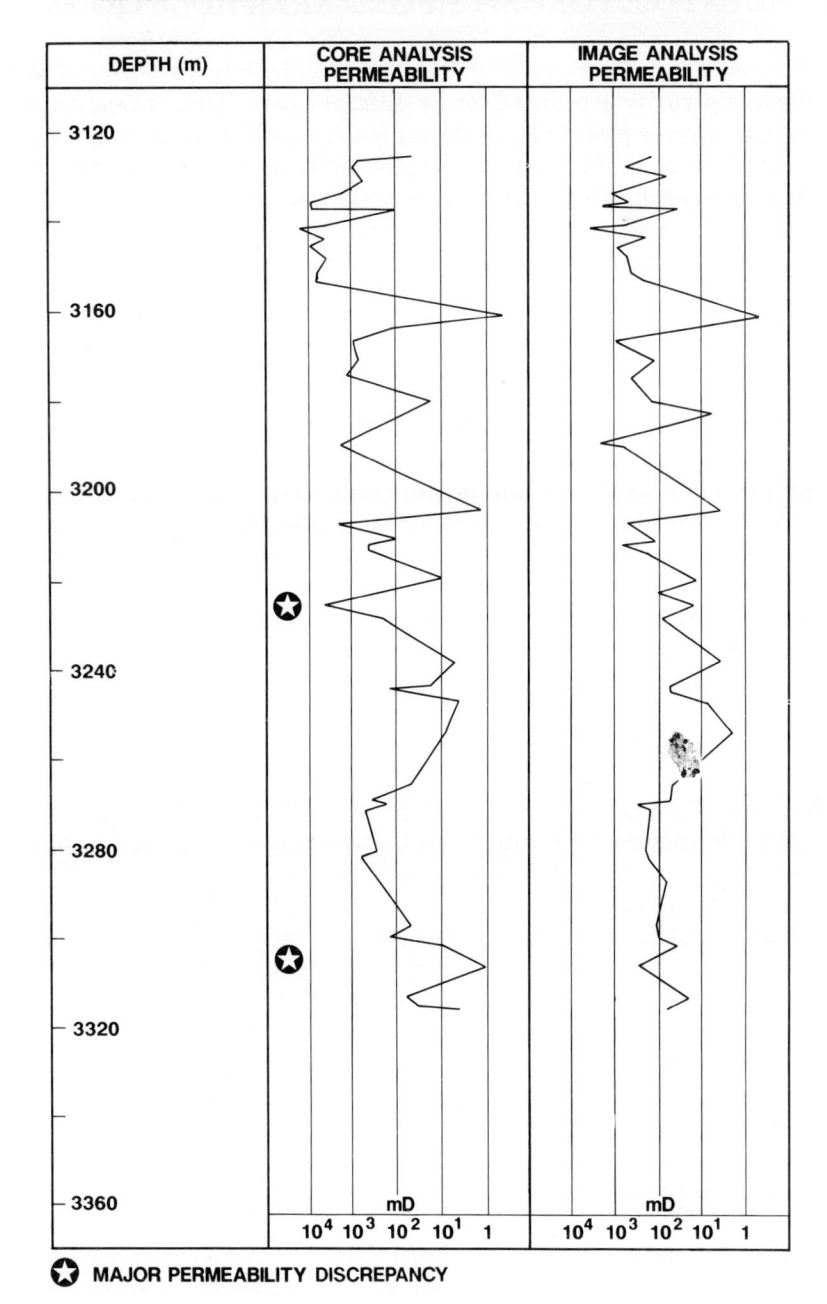

✪ MAJOR PERMEABILITY DISCREPANCY

Figure 8. Comparison of measured and calculated permeability values from single North Sea Well study (depth range - 200m). Note general coincidence of two data sets and only two anomalous points.

CUMULATIVE POROSITY/PORE AREA RELATIONSHIPS

Pore area percentage histograms (Fig. 3) can be recalculated in terms of the proportions of effective porosity accounted for by different pore sizes. These data then are replotted as cumulative plots (Fig. 9) through a range of pore sizes ($1\text{-}10^7\,\mu m^2$). The shape and distribution of these curves is reflected in a variation in sample permeability with the higher permeability lithologies deflected towards the right, illustrating the strong influence pore size distributions have upon effective permeability in reservoir sandstones. This form of data also can be converted to a 'pseudo' capillary pressure curve, with the calculation of capillary pressure (Pc) from the size data as explained in the given equation.

DISCUSSION

POROS can provide a wealth of data on pore size, shape, and structure as envisaged in 2-dimensional slices in reservoir lithologies. These data are an invaluable aid to engineers and geologists concerned with modeling fluid flow in reservoirs.

(A)   In reservoir studies, it is not possible always to take suitable core plugs for conventional core-analysis measurements. However, porosity and permeability determinations can be made from pore-geometry data generated from smaller samples.
(B)   Volumetric porosity data are not sufficient always for the needs of reservoir engineers who are involved in modeling injection and enhanced recovery techniques. Pore geometry data will provide additional information on pore size, specific surface area (roughness), pore shape (coordination - number of throats per pore), and network connectivity.
(C)   Pore-geometry parameterization will allow reservoir description, classification, and discrimination in terms of pore networks rather than by the traditional approach of framework and authigenic mineralogy.
(D)   Quantitative pore-geometry data can be correlated directly against:
   (i)   Mineralogical data (e.g. % clay content; % blocky authigenic cements) so that changes in pore shape and connectivity patterns can be modeled quantitatively in terms of authigenic events.
   (ii) Physical reservoir data (e.g. permeability, capillary pressure, and compressibility). Variations in these data can be modeled quantitatively by pore geometry data.

During the past 12 months POROS has been used successfully to establish detailed pore geometry databases from clastic lithologies throughout the North Sea sector. Table 3 summarizes the extent of data presently available for gas and oil fields in this sector and it is expected that this work will continue in the future.

Cumulative porosity/pore area control on permeability

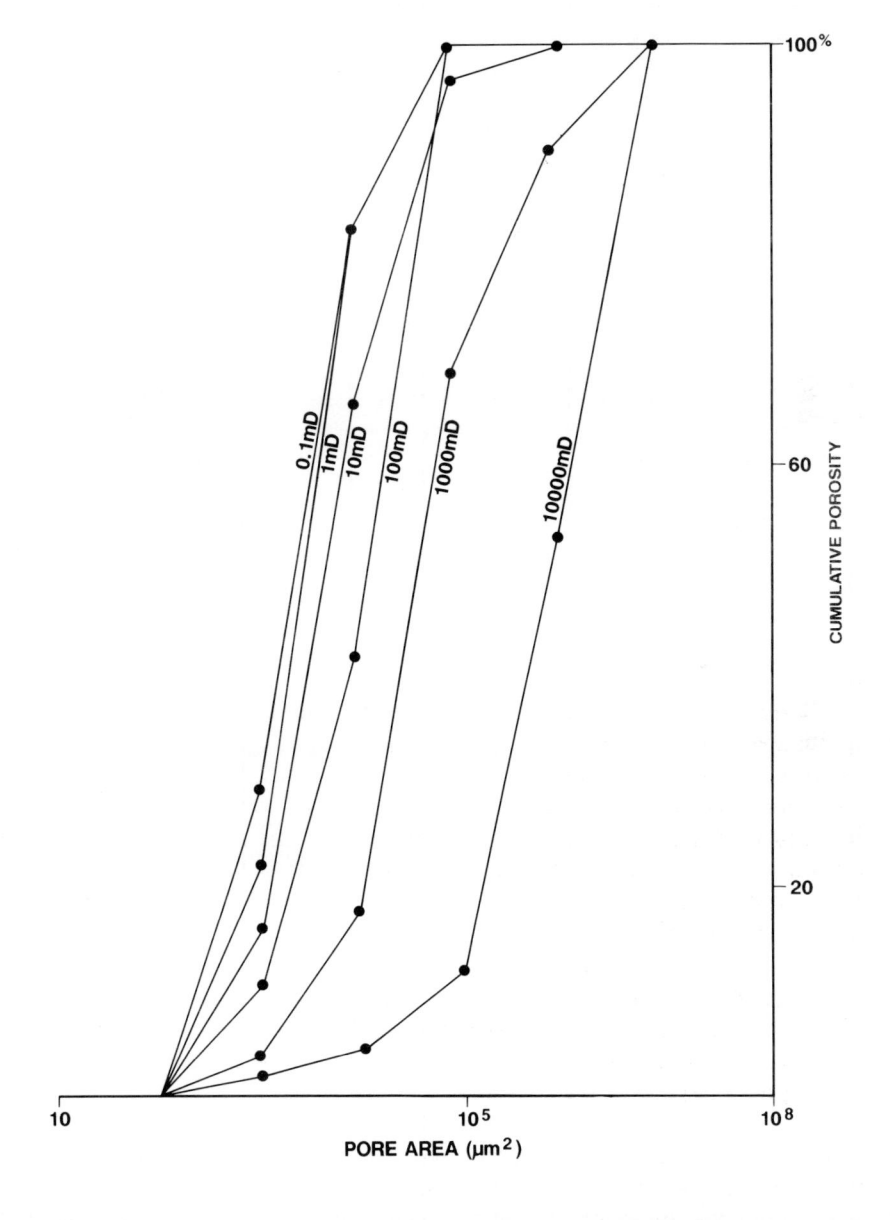

Figure 9. Pore area/cumulative porosity relationship curves compiled from pore area frequency histograms. Note how samples with different measured permeabilities (Kair) are isolated easily with highly permeable samples being deflected to right - illustrating importance of pore-size distributions on reservoir quality and performance.

Table 3. Available pore geometry data base from North Sea Hydrocarbon fields.

	NORTHERN NORTH SEA	CENTRAL NORTH SEA	SOUTHERN NORTH SEA
JURASSIC BRENT	GULLFAKS SØR SNØRRE		
JURASSIC STATFJORD	GULLFAKS SØR		
JURASSIC FULMAR SAND		CLYDE	
PERMO-TRIAS ROTLIEGEND			AMETHYS VULCAN/VALIANT DEIRDRE/DOTTY

REFERENCES

Archer, J.S., and Wall, C.G., 1986, Petroleum engineering; principles and practise: Graham and Trotman Ltd., London, 362 p.

Berryman, J.G., and Blair, S.C., 1986, Use of digital image analysis to estimate fluid permeability of porous materials. 1. Applications of two point correlation functions: Jour. Appl. Physics, v. 60, no. 6, p. 1930-1938.

Collins, R.E., 1961, Flow of fluids through porous materials: Petroleum Publ. Co., Tulsa, 274p.

Crisp, D.J., and Williams, R., 1971, Direct measurement of pore size distribution on artificial and natural deposits and prediction of pore space accessible to interstitial organisms: Marine Biology, v. 10, no. 3, p. 214-26.

Dilkes, A., and Graham, S.C., 1985, Quantitative mineralogical characterisation of sandstones by backscattered electron image analyses: Jour. Sed. Pet., v. 55, no. 3, p. 347-355.

Dilkes, A., Parks, D., and Graham, S.C., 1984, Characterisation of sandstones and their component minerals by quantitative EPMA point counting in the SEM, *in* Roming, A.D., and Goldstein, J.I., eds. Microbeam analysis: San Francisco Press, p. 139-142.

Duda, R., and Hart, P., 1973, Pattern classification and screen analysis: John Wiley & Sons, New York, 482p.

Ehrlich, R., Kennedy, S.K., Crabtree, S.J., and Cannon, R.L., 1984, Petrographic image analysis. I. analysis of reservoir pore complexes: Jour. Sed. Pet., v. 54, no. 4, p. 1365-1378.

Fabbri, A.B., 1980, GIAPP: Geological image analysis program package for estimating geometrical probabilities: Computers & Geosciences, v. 6, no. 2, p. 153-161.

Huggett, J.M., 1984, An SEM study of phyllosilicates in Westphalian coal measures sandstone using backscattered electron imaging and wavelength dispersive spectral analyses: Sedimentary Geology, v. 40, no. 2, p. 233-247.

Jongerius, A., and others, 1971, Electro-optical soil property investigations by means quantimet-B equipment: Geoderma, v. 7, no. 3, p. 177-98.

Pye, K., and Kinsley, D., 1984, Petrographic examination of sedimentary rocks in SEM using backscattered electron detectors: Jour. Sed. Pet., v. 53, no. 3, p. 877-888.

Rink, M., 1976, A computerised quantitative image analysis procedure for investigating features and an adopted image process: Jour. Microscopy, v. 107, no. 8, p. 267-386.

Rink, M., and Schopper, J.R., 1978, On the application of image analyses to formation evaluation: Log Analyst, Jan-Feb. 1978., p. 12-22.

Rosenfeld, A., 1984, Picture processing 1983. Survey; Computer vision: Graphics and Image Processing, v. 26, no. 3, p. 347-393.

Rosenfeld, A., and Kak, A.C., 1976, Digital picture processing: Academic Press, New York, 457p.

Ruzyla, K., 1986, Characteristics of pore space by quantitative image analysis: SPE Formation Evaluations, v. 1, no. 4, p. 389-398.

Serra, J., 1982, Image analysis and mathematical morphology: Academic Press, New York, 610 p.

Soeder, D.J. 1988, Applications of fluorescence microscopy to the study of pores in tight rocks (abst.): Am. Assoc. Petroleum Geology Bull., v. 71, no. 5, p. 616.

Space Modeling and Multivariate Techniques for Prognosis of Hydrocarbons

J. Harff, J. Springer, B. Lewerenz,
Akademie der Wissenschaften der DDR, Potsdam
and W. Eiserbeck
VEB Kombinat Erdöl-Erdgas, Stammbetrieb Gommern

ABSTRACT

For the prognosis of oil and gas two main tasks are distinguished:

- Structural modeling of sedimentary sequences for the locatization of traps on the basis of drilling data and seismic information, and
- Identification of prospective areas by regionalization of potential reservoir beds on the basis of multivariate parameters from layers in drilling profiles.

For the first task 3D-space models are used showing the structural formation of paleosurfaces of lithostratigraphic units during the processs of basin subsidence. These 3D-models are consturcted by connection of 1D-models of sedimentation and compaction using interpolation algorithms.

The regionalization is carried out as the solution of an interpolation task which is multivariate in relation to a complex of reservoir parameters. Here, at first the optimal type of reservoir rocks in the area of investigation is determined by application of numerical classification methods. In a second step the distribution of this type in the plane is analyzed. The boundaries of the searched optimal rock type are determined by regionalization of the plane.

Software elaborated for a 16 bit-PC can be applied for the prognosis and exploration of oil and gas.

A case study demonstrates the appplication of the method to sedimentary Rotliegend rocks of a structural height in the western part of the GDR territory.

INTRODUCTION

The drilling process for the prognosis, search, and exploration of hydrocarbons becomes more and more expensive, but the processing of information by modern computers becomes cheaper. Therefore it is advisable to use computers for extracting as much information as possible from existing data about a region of investigation (well data, seismic data, geological data, etc.). This way an improved geological model can be constructed which leads to a better search strategy.

This paper describes some possibilities for use of personal computers for space modeling and prediction of hydrocarbon deposits. The essential characteristic of the modeling process is the interactive mode by which the geologist constructs his model step by step as the basis for a search and exploration strategy. Two main tasks are discussed:

- Paleospace modeling and modeling of the temporal formation of the structure of sedimentary sequences during the subsidence of a basin, and
- Identification of favorable areas by regionalization of potential reservoir beds in relation to the existing multivariate exploration data.

MATHEMATICAL MODELS AND NUMERICAL METHODS

Decompaction and paleospace modeling

The thickness of a sedimentary layer decreases during subsidence because of the weight of overlying sediments (compaction). Decompaction is the computation of a paleothickness of a layer, that is the reconstruction of the thickness of a layer at a certain time in its burial history. It is assumed here that the changes of thickness and porosity are caused *only* by compaction and that a relation

$$p = p(x)$$

exists between the porosity p ($0 < p < 1$) and the maximum burial depth x and that this relation is known for the considered lithotype. The exponential relation may be

$$p(x) = p_0 e^{-bx}$$

with the initial porosity p_0 and the compaction constant b (Athy, 1930).

The theoretical thickness of a layer if all the pore volume would be removed is termed "solid height". The solid height of a layer is constant in time and can be computed by integrating the solidity $S = 1 - p$ over the depth interval of the layer. This fact leads to the following algorithm for decompaction (see also Perrier and Quiblier, 1974).

If a layer in a drilling profile occurs between the depths x_1 and x_2 ($x_1 < x_2$) and never was buried deeper, its solid height is

$$h_s = \int_{x_1}^{x_2} (1 - p(x))\, dx$$

and can be computed directly by analytical or numerical integration. For every depth position of the upper boundary x_1^0 of the layer during its subsidence history the paleothickness $d = x_2^0 - x_1^0$ can be computed from the equation

$$h_s = \int_{x_1^0}^{x_2^0} (1 - p(x)) \, dx$$

by a suitable iteration technique.

This method of decompaction can be used for improved paleospace modeling, that is the reconstruction of the structure of a sedimentary sequence at different times. For this purpose such times in the history of subsidence are selected, for which the sediment surface in the region of investigation can be approximated by a horizontal plane (after equalizing sedimentation). In this situation the depth $X_i(r,t)$ of a sedimentary unit i at the time t in the point r of the region R of investigation can be determined by summing up the paleo thicknesses d_j of all the units j lying above unit i:

$$X_i(r,t) = \sum_{j=1}^{i-1} d_j(r,t)$$

The paleothicknesses d_j are computed from the top to the bottom by the decompaction method described here.

This procedure can be carried out only at some single points in the region of investigation, where the recent depths, thicknesses, and lithotypes of the layers are known, for example at drilling points. For the construction of a space model the depth $X_i(r,t)$ is considered as a random field

$$X_i(r,t) = m_i(r,t) + Y_i(r,t) \quad \text{for all } r \in R$$

with the expected value function $m_i(r,t)$ and the fluctuation function $Y_i(r,t)$. Geostatistical methods can be used for interpolation between the known values at some points r_k (k = 1,...,K) and lead to the wanted space models in form of isoline maps for paleothicknesses and paleodepths, cross sections through the sedimentary sequence, and three-dimensional representation of reliefs; all this for different times of the burial history.

To investigate the relative vertical movement of the subsurface of a sedimentary unit during a time interval [t,t'] the depths at the times t and t' must be compared. This is achieved by computing a type of "normalized difference".

First for each of the two times, t and t', the experimental fluctuation function Y_i^* is computed as the difference between the depth X_i and the properly estimated expected value function m_i^*. In the simplest situation $m_i^*(r,t)$ and $m_i^*(r,t')$ are constant mean values for the depth of unit i in the region of investigation at the times t and t', respectively. The difference

$$D_i(r,t,t') = Y_i^*(r,t) - Y_i^*(r,t')$$

of the fluctuations represents the relative vertical movement besides the general trend of subsidence. The isoline maps of D_i show areas of relative uplifting ($D_i > 0$) and areas of relative subsidence ($D_i < 0$).

Regionalization by multivariate exploration data

To divide the region of investigation into different homogeneous parts by geostatistical methods, the geologist selects a set of geological features according to the task at hand (David, 1977). In the situation of hydrocarbon prognosis such features can be thickness and depth of the source rock, porosity, contents of methane, water saturation, and others. The selected features $X_1(r), X_2(r), \ldots, X_n(r)$ are suitably scaled and collected in a n-dimensional feature vector

$$X(r) = (X_1(r), \ldots, X_n(r)),$$

where r denotes a point in the region R of investigation. This feature vector is considered as a random vector function

$$X(r) = m(r) + Y(r)$$

with the vector

$$m(r) = (m_1(r), \ldots, m_n(r)) = E[X(r)]$$

of the expected value functions. The vector function m(r) is assumed to be discontinuous with constant values in the subregions $R_j \subset R$, $j = 1,\ldots,J$:

$$m(r) = m^{(j)} = (m_1^{(j)}, \ldots, m_n^{(j)}) \quad \text{for } r \in R_j.$$

It is assumed that

$$m^{(j)} = m^{(k)} \quad \text{for } j = k,$$

but each subregion may consist of several homogeneous regions with the same expected value vector.

For any point r of R we define the j-variances

$$V_j(r) = \sum_{i=1}^{n} E[(X_i(r) - m_i^{(j)})^2], \quad j = 1,\ldots,J$$

as a measure of distance in the feature space between the random feature vector X(r) and the expected value vector $m^{(j)}$. The generalized variance we define as

$$V(r) = \min_{j=1,\ldots,J} V_j(r).$$

Obviously

$$V(r) = V_j(r) \quad \text{for all } r \in R_j,$$

and this fact is used for the statistical regionalization. It is carried out in two steps:

(1) Typification by classification. At some sample points r_k (k=1,...,K) the selected features are measured. The obtained set of experimental feature vectors

$$K = \{X(r_k), k=1,\ldots,K\}$$

is subdivided by cluster analysis into classes $K_j \subset K$ with the aim that intraclass variances are minimized and interclass variances are maximized. For each class K_j (j=1,...,J*) the vector of expected values

$$m^{(j)*} = (m_1^{(j)*},\ldots,m_n^{(j)*})$$

and the vector of standard deviations (which is used for scaling) can be estimated.

(2) Regionalization by interpolation. For all measure points r_k (k=1,...,K) the j-variances V_j (j=1,...,J*) are estimated according to the estimated expected value vectors $m^{(j)*}$:

$$V_j^*(r_k) = \sum_{i=1}^{n} (X_i(r_k) - m_i^{(j)*})^2 .$$

From the values at the measure points the functions V_j^* are interpolated (e.g. by polynomials) in the whole region R of investigation and the experimental generalized variance

$$V^*(r) = \min_{j=1,\ldots,J^*} V_j^*(r)$$

is defined. Then the classes K_j of measured feature vectors are expanded to subregions R_j of the considered region R by the definition

$$r \in R_j \text{ if } V^*(r) = V_j^*(r).$$

Thus the experimental geological boundary between two or more subregions $R_j, R_k, \ldots, R_l$ is defined by

$$B_{jk\ldots l}^* = \{ r \in R : V_j^*(r) = V_k^*(r) = \ldots = V_l^*(r) = V^*(r) \} .$$

Each subregion may consist of several homogeneous regions with the same expected value vector $m^{(j)*}$.

IMPLEMENTATION OF THE METHODS ON PERSONAL COMPUTERS

The numerical methods described here yield three-dimensional models of a sedimentary structure for different times of its subsidence history and regionalization of a region according to multivariate data. Nevertheless they do not require big computers but can be implemented on personal computers. This is achieved by strong simplifications of the whole process and by the combination of one-dimensional models (e.g. compaction, sum of variances of all features) with interpolation methods. In most practical situations the available data are so rare and inaccurate that the simplifications made in the models are justified.

A principal scheme of information processing is presented in Figure 1. The geologist selects primary data, model parameters (e.g. porosity/depth curves), and an interpolation method. As the result a paleospace model or a regionalization is computed and graphically presented on the monitor. Parameters and conditions of modeling may be changed until the model reflects all the known natural features of the region of investigation. Then the user stops modeling and decides in relation between his conceptional model, the computer result and all his experience about possible search strategies.

For subsidence modeling (including decompaction), for interpolation and graphical presentations, and for classification and regionalization the programs PALEO, ISOPERS, and RECLAS, respectively, were developed by the authors. They are written in Pascal for personal computers.

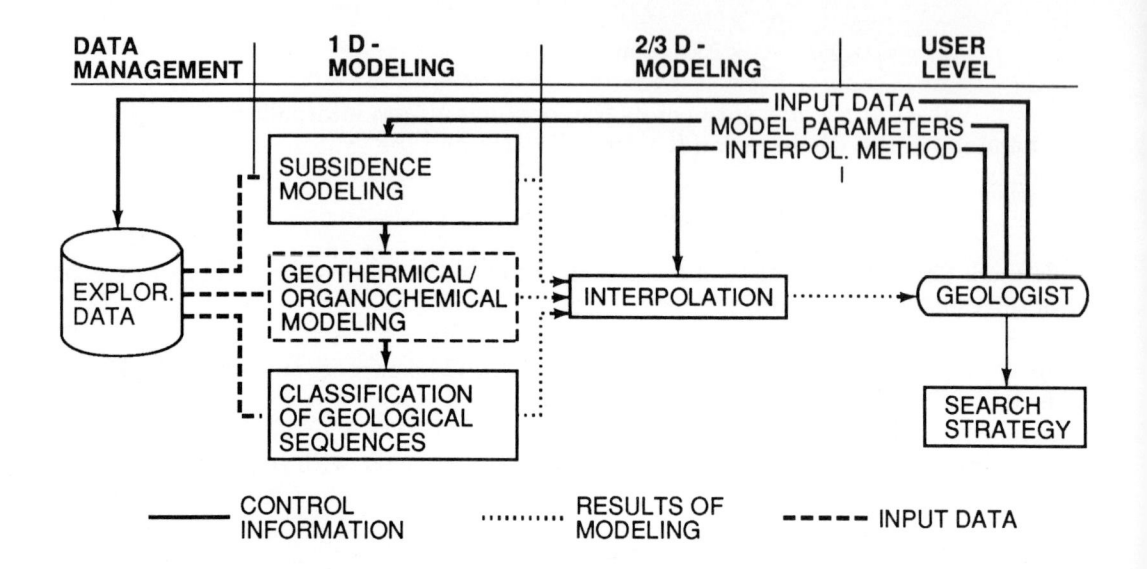

Figure 1. Scheme of information processing for basin analysis.

A CASE STUDY

As a case study we worked with Rotliegend rocks of the SE flank of a structural height in the western part of the GDR. The investigated sequence is formed by Autunian volcanic rocks, Autunian sediments, and Saxonian sediments (Bueste-Sandstein, Hauptsandstein, Elbe-Wechselfolge). The scale of sediment size ranges from conglomerate to shale.

For paleostructural modeling porosity-depth relations for seven lithotypes forming the Rotliegend sediments were determined. Then the structure of the region was modeled for three times: End of Rotliegend time, End of Zechstein time, and Recent time (see Figs. 2-5).

Figure 6 and Figure 7 show experimental difference functions describing the relative vertical movement of Rotliegend subsurface in the time interval from end of Rotliegend time to end of Zechstein time and from end of Zechstein time to Recent time. The fluctuations were estimated on the base of a constant expected value function.

The structure of the sedimentary complex does not change much during Zechstein time (Fig. 6). The vertical movements can be described in general as a tipping of the subsurface to the northern direction around a E-W striking axis.

Figure 7 shows the movement during Mesozoic-Cenozoic time. The picture has changed generally. The main movement in the northern part of the area is a tipping to the E around a N-S striking axis. In the southern part of the area the movement

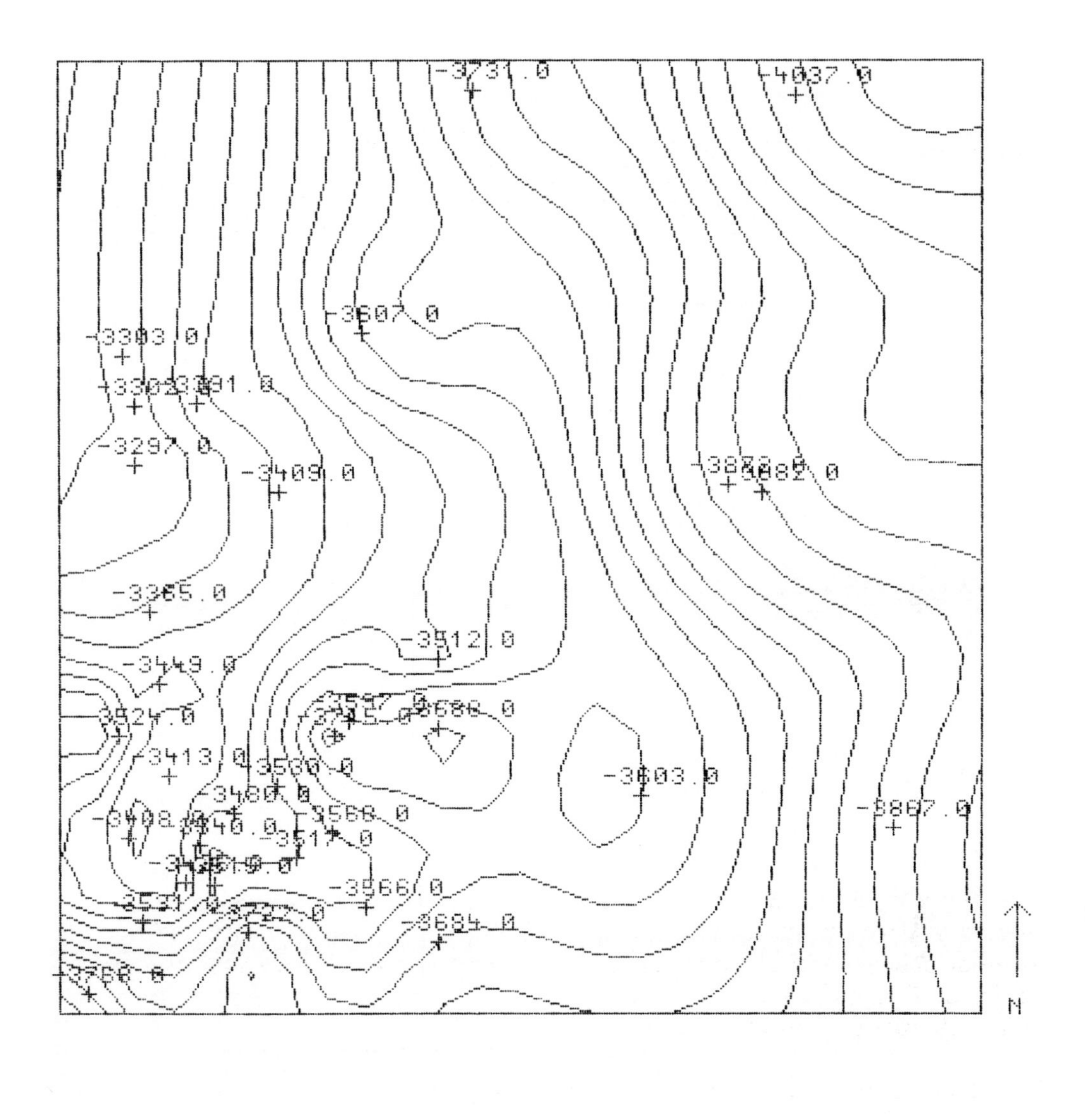

Figure 2. Depth of Rotliegend surface [m], Recent time.

axis turns into even hercynical direction. The isoline with the value 0 marks the line without relative movement and determines a local zone of relative uplifting in the southern part. In the SW direction follow zones of subsidence and uplifting. Structural traps can be identified in regions of uplifting.

For the prognosis modeling by regionalization we used five features:

- depth of Rotliegend subsurface at recent time (ROF),
- thickness of reservoir beds in the Elbe-Folge (h_{eff}),
- porosity of sandstones in the Elbe-Folge (NP),
- water saturation of the reservoir beds (S_w),
- content of free gas in the Elbe-Folge (CH_4).

1 – Autun. volc. rocks, 2 – Autun. sed., 3 – Bueste-Sandstein
4 – Hauptsandstein, 5 – Elbe-Wechselfolge

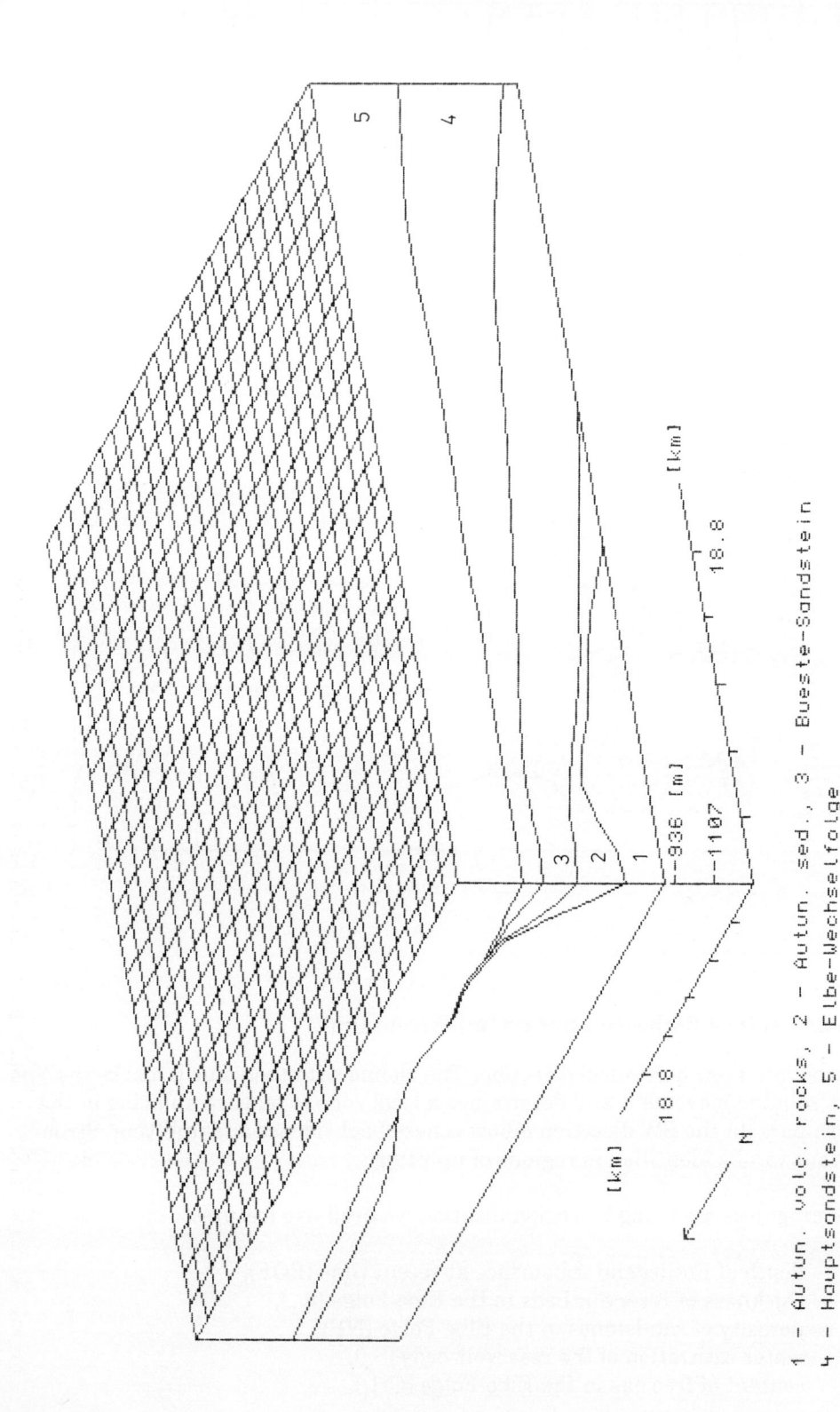

Figure 3. Space model of Rotliegend rock sequence, end of Rotliegend time.

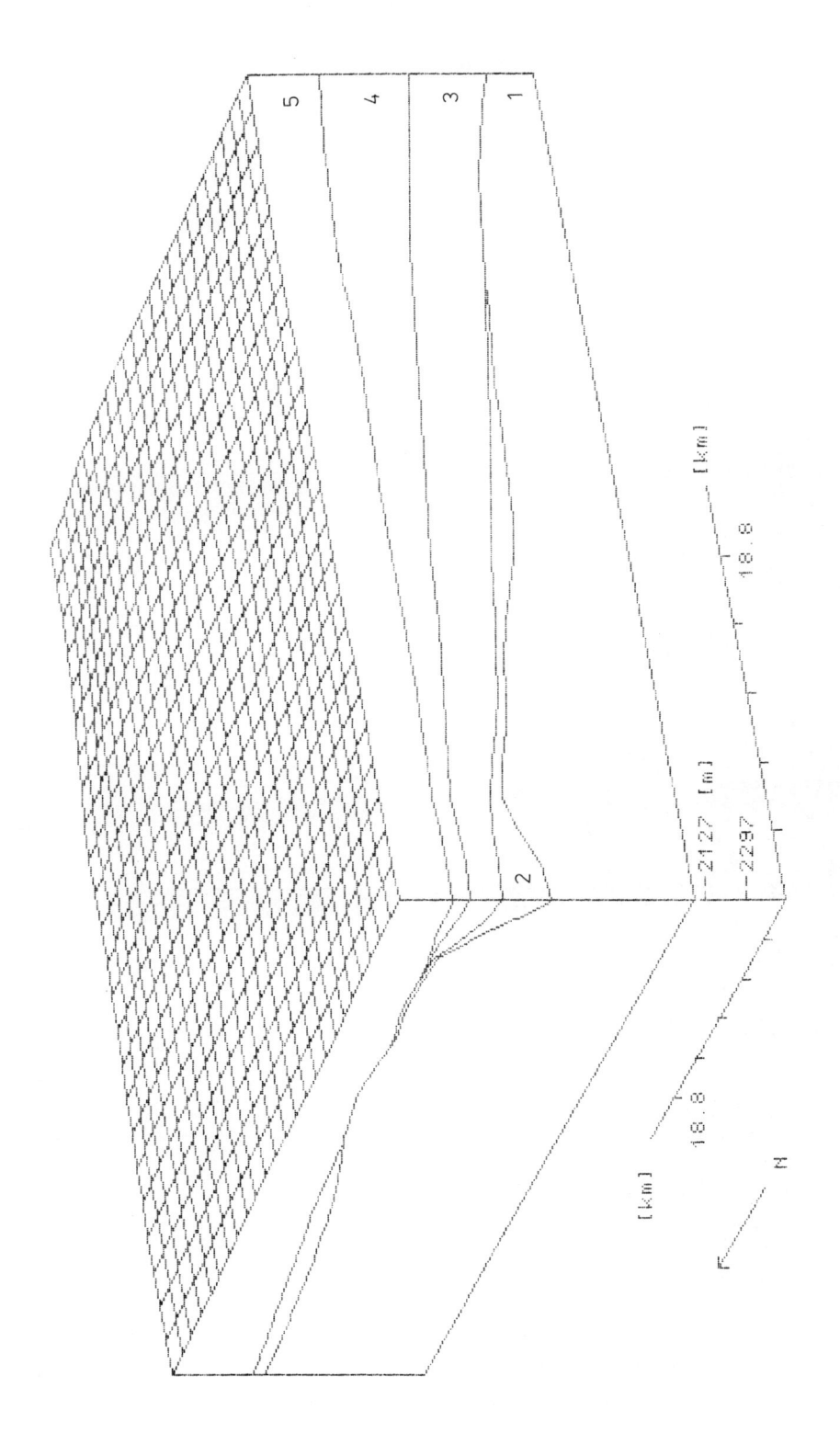

Figure 4. Space model of Rotliegend rock sequence, end of Zechstein time.

1 — Autun. volc. rocks, 2 — Autun. sed., 3 — Bueste-Sandstein
4 — Hauptsandstein, 5 — Elbe-Wechselfolge

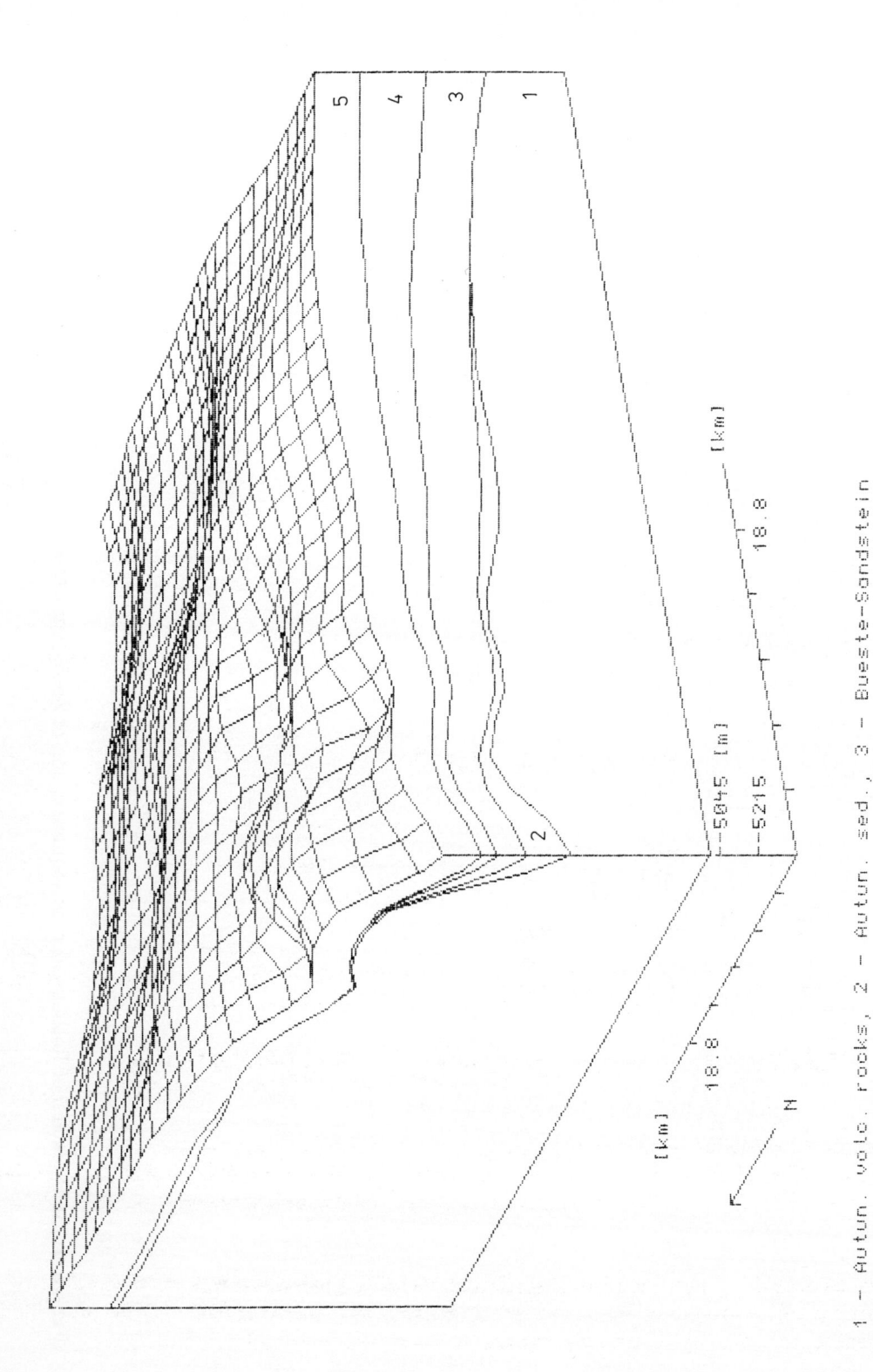

Figure 5. Space model of Rotliegend rock sequence, Recent time.

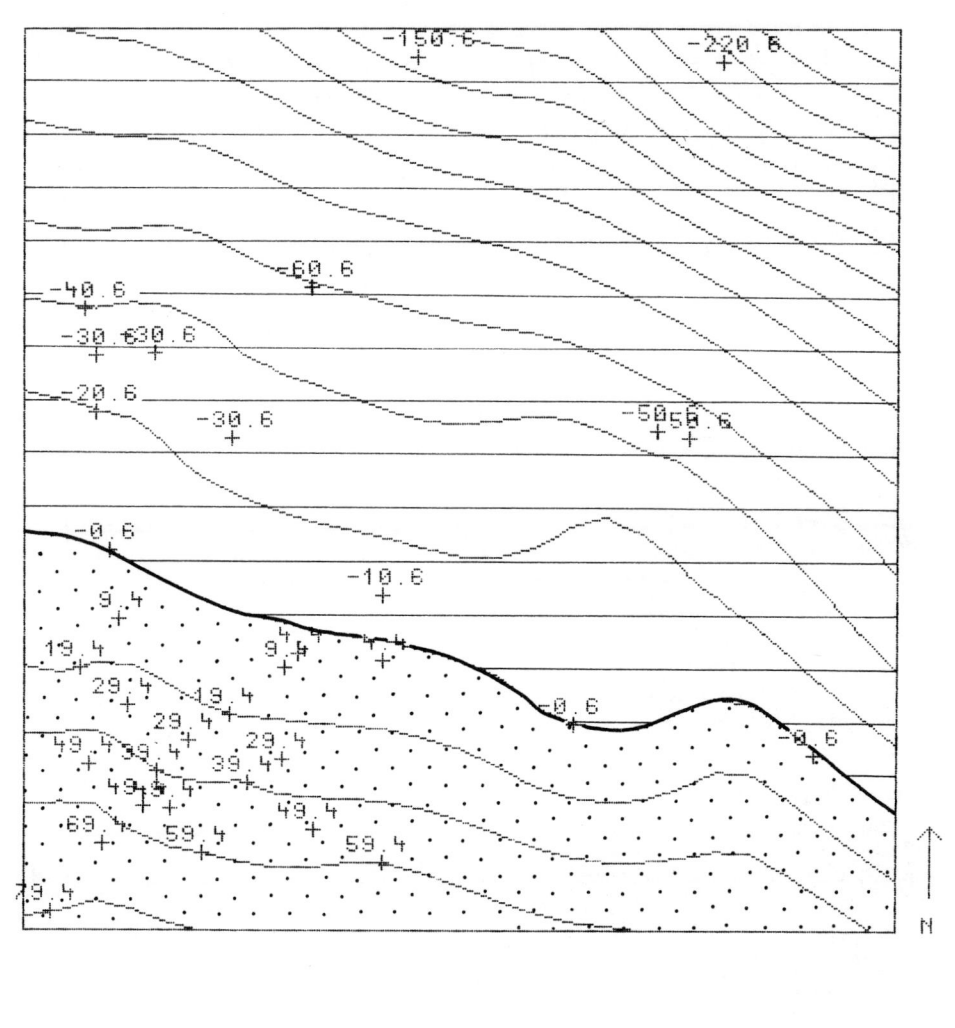

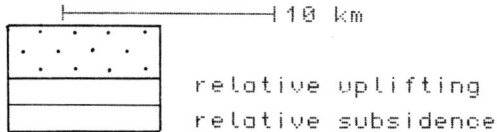

Figure 6. Relative vertical motion of Rotliegend surface during Zechstein time.

We classified 30 drilling profiles on the base of these features by cluster analysis. The dendrogram (Fig. 8) shows three classes (class 1: profiles 6 - 10; class 2: profiles 12 - 3; class 3: profiles 30 - 26). The estimated parameters of these classes are shown in Figure 9.

The classes of drilling profiles show a similar porosity of sandstones. All the gas containing profiles are joined in class 2. In addition class 2 contains the profiles with the lowest water saturation, the maximum thickness of reservoir rocks and the lowest

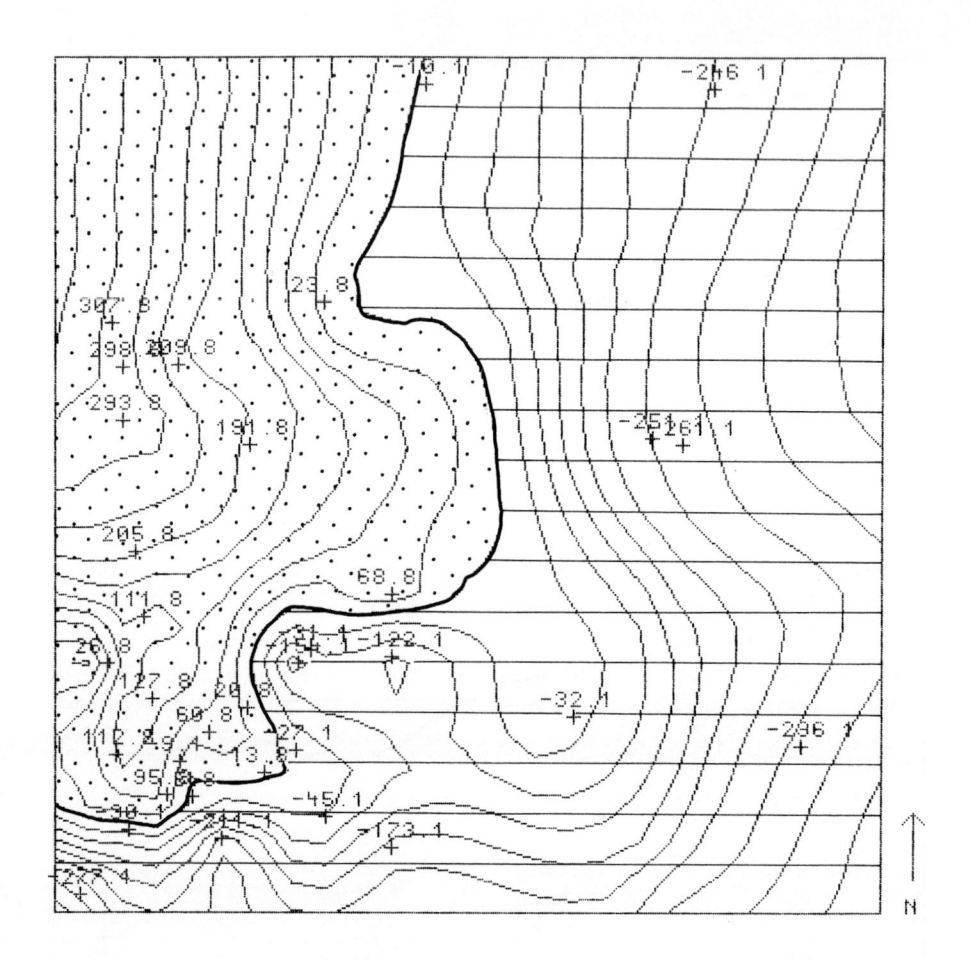

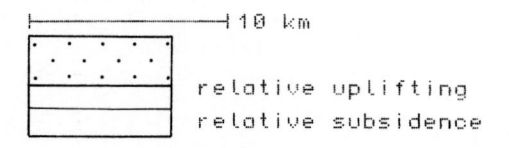

Figure 7. Relative vertical motion of Rotliegend surface interval: Triassic - Recent time.

depth of the Rotliegend subsurface. Therefore class 2 represents the optimal profile type for the search of gas.

On the base of the experimental j-variances and the experimental generalized variance we regionalized the plane. Figure 10 shows the experimental generalized variance; Figure 11 shows the regionalization map. The results can be summarized in the following search strategy:

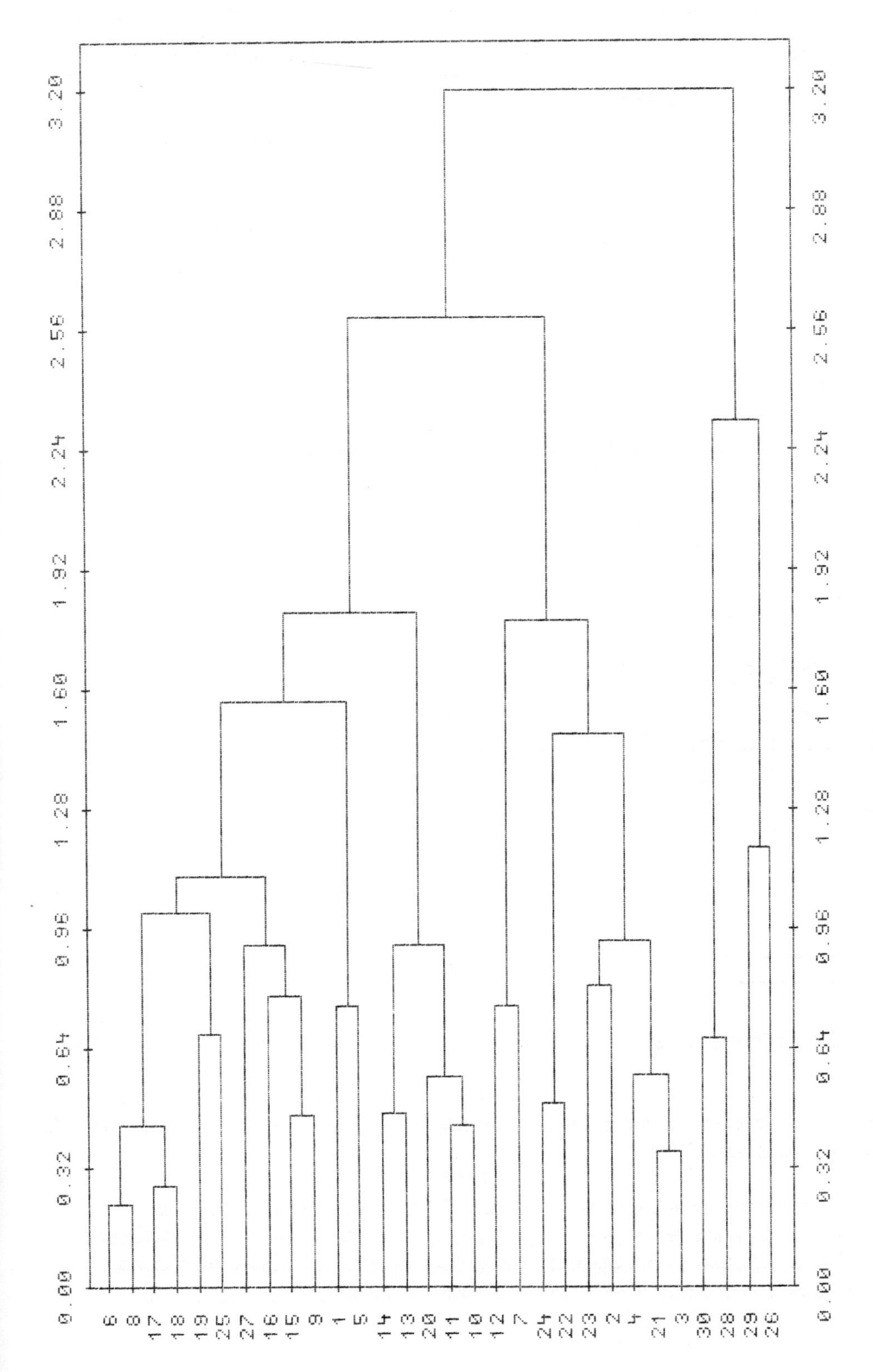

Figure 8. Dendrogram as result of cluster analysis of drilling profiles.

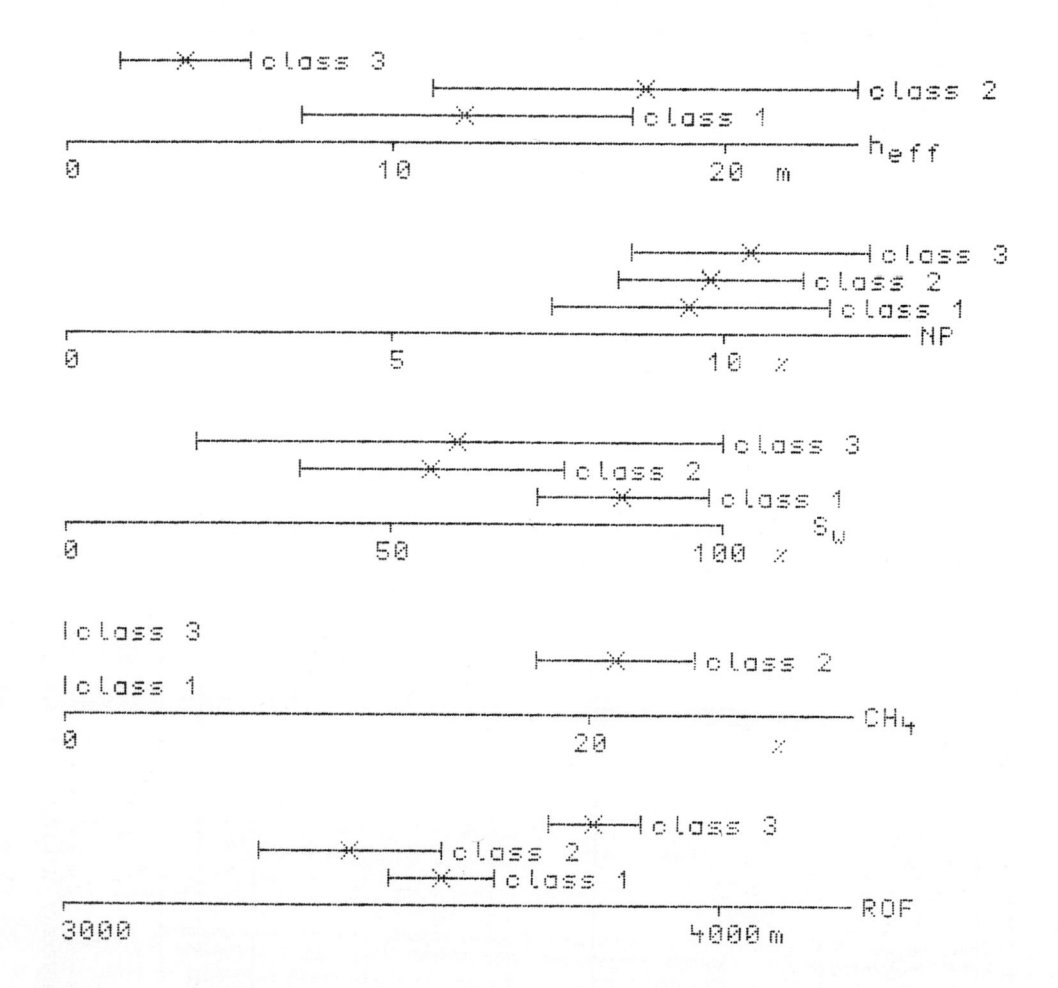

Figure 9. Experimental expection values and standard deviations of drilling profile

(1) Search in the subregion of class 2.
(2) If this search is successful continuing the search in the subregion of class 1, or else stopping the search.

A search in the subregion of class 3 is assessed as not effective because of the unfavorable parameters of this drilling profile class.

SUMMARY

Simple methods for paleostructural modeling and the regionalization of geological complexes in relation to the expection of oil and gas are discussed. The methods combine one-dimensional models of subsidence and compaction, interpolation methods, and the multivariate analysis of prospectivity features. By this manner the application of personal computers is possible.

The methods help geologists to determine optimal search strategies and so to decrease the risk of drilling for oil and gas.

Figure 10. Experimental generalized variance.

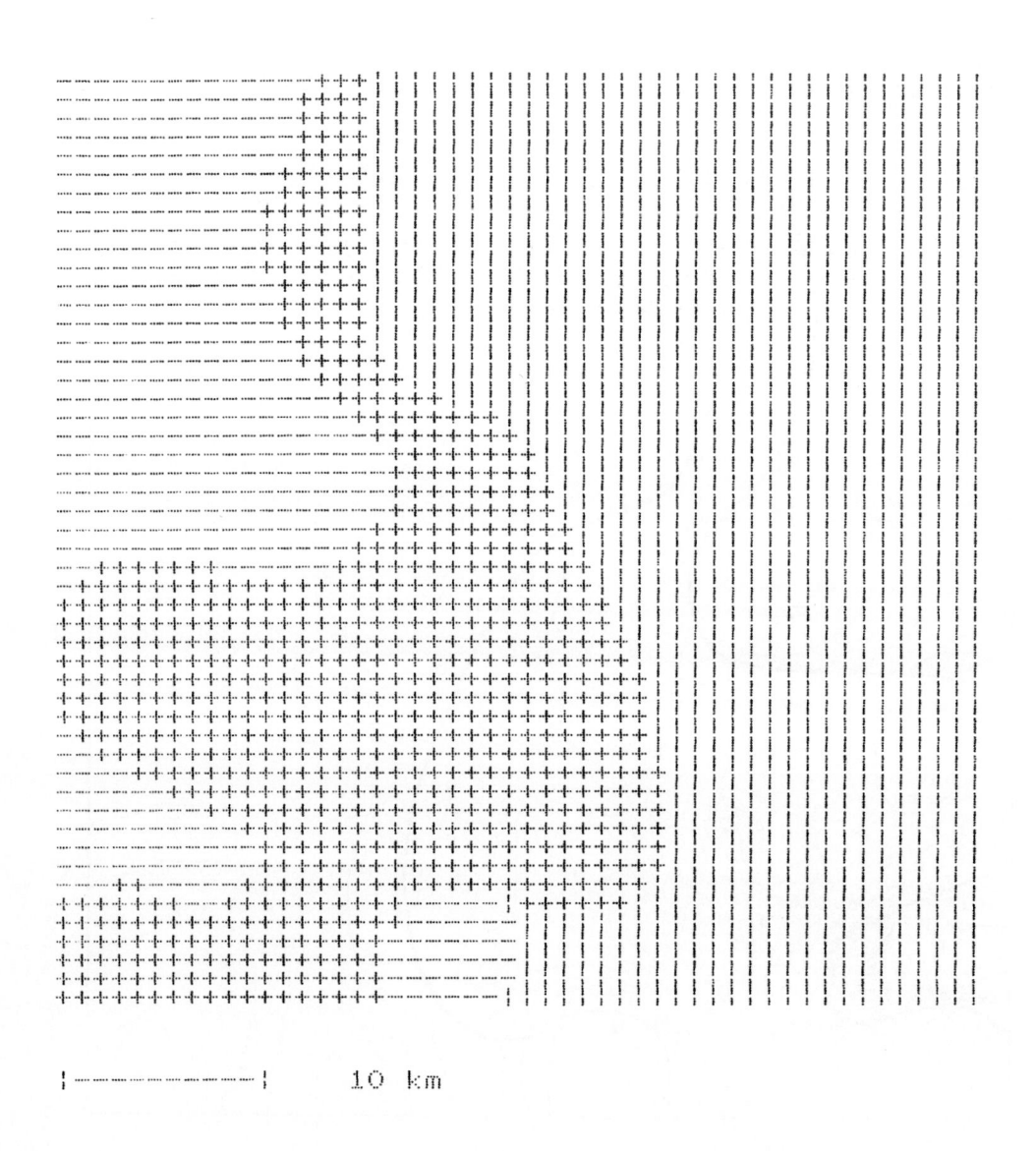

Figure 11. Subdivision of region of investigation into homogeneous parts of drilling profile classes
class 1 +
class 2 -
class 3 :

REFERENCES

Athy, L.F., 1930, Density, porosity and compaction of sedimentary rocks: Am. Assoc. Petroleum Geologists Bull., v. 14, no. 1, p. 1-24.

David, M., 1977, Geostatistical ore reserve estimation: Elsevier, Amsterdam, 356 p.

Perrier, R., and Quiblier, J., 1974, Thickness changes in sedimentary layers during compaction history, methods for quantitative evaluation: Am. Assoc. Petroleum Geologists Bull., v. 58, no. 3, p. 507-520.

Petroleum Prospect Size Estimation by Numerical Methods

T. Jasko
Quartz Scientific Computing Ltd., Watford

ABSTRACT

The size of petroleum prospects usually is evaluated as the product of certain geological parameters
$$Q = Vr * Phi * So...$$
In the simplest situation each of the factors in the product, for example rock volume, porosity, hydrocarbon saturation, is given by a prospect in tonnes or barrels of oil.

If any of the geological factors to be multiplied is uncertain it is no longer given as a single number but as a probabilities variable. In the general situation the formula then becomes a product of functions and the result a stochastic distribution. It can no longer be given by a single number.

The only feasible way to estimate this is through the use of computers. There are various ways to calculate the ultimate answer and their applicability depends on the way the geological parameters are measured. If all parameters are characterized by a handful of numbers then some well-known methods can be applied satisfactorily. For example if the factors have discrete distributions as used in decision tree problems then a full matrix product evaluation may be possible (note that the time required may be excessive). Similarly if all factors have a 2-parameter lognormal distribution then the product distribution also will be lognormal and its parameters can be calculated exactly.

In actual problems, most of these variables are given as continuous distributions belonging to several types of probability distributions. There are good arguments to assume lognormal distribution for factors such as gross rock volume. However, other factors, clearly not lognormal, may have normal or beta distribution or may have empirical distributions given by a curve only; for which the theoretical formula is unknown.

In such situations of practical difficulties probabilistic estimates can be obtained by Monte-Carlo methods. This technique is calculating the target distribution (the answer to the problem) as the distribution of a number of individually computed random samples.

It is a practical method, applicable to a wide range of distributions - even if no formula is known - but the results somewhat depend on the way random numbers are picked for sampling.

INTRODUCTION

Quantitative methods of petroleum prospect evaluation developed in three stages. Matrix methods were used first, then superseded by the Monte-Carlo method, and lately new techniques are emerging that avoid the limitations posed by the matrix methods.

MANUAL CALCULATION - MATRIX METHODS

The size of petroleum prospects usually is evaluated as the product of certain geological parameters

$$Q = Vr * Phi * So * Rfvf...$$

In this formula

> Vr is gross rock volume
> Phi is porosity
> So is oil saturation
> $Rfvf$ is reciprocal of formation volume factor...

Depending on the geological model, other factors can be used, for example instead of gross rock volume, area and thickness can be included as separate factors. Similar formulae are used for gas and condensate prospects. In the simplest situation each of the factors in the product for example rock volume, porosity, hydrocarbon saturation, is given by a precise value and the formula gives a precise answer: the size of the prospect in tonnes or barrels of oil.

If any of the geological factors to be multiplied is uncertain it is no longer given as a single number but as a probabilistic variable. In the general situation the formula then becomes a product of functions and the result a stochastic distribution. It can no longer be given by a single number.

If all parameters are characterized by a handful of numbers, then several methods can be applied satisfactorily. For example if the factors have discrete distributions as used in decision tree problems then a full matrix product evaluation may be possible (note that the time required may be excessive). In typical situations, the discrete values are just a simple approximation to a continuous distribution. Three values may be used to represent minimum, maximum, and most likely cases. These may be termed low, medium, and high values. Apart from nomenclature, the values may indicate different things in statistical terms. The low and high values may be quartiles (P25 and P75) or the complete range (P0 to P100). The middle number may be considered to be the median, the mean, or possibly the modus.

If all the possible values of a factor are grouped in three equiprobable classes, and the three figures are in fact the average values of these three classes then the matrix product of the values will yield a discrete distribution with the correct mean.

On the other hand it was shown that the matrix method distorts variance and it is biased. The bias seems to reduce the tail end of the distribution. This is the part of the curve that shows the (small) chance of locating a large prospect - so it is important for exploration geologists.

Quantitative methods, when first introduced, were based on manual calculation of matrix products. This was not only error prone but also laborious which made it necessary to simplify calculations. With manual calculations it is not possible to generate more than one version of parameters.

MONTE-CARLO METHODS

The first computer programs followed manual methods in attempting to model the prospect by the matrix product of a handful of numbers. These programs, however, were superseded gradually by the more flexible Monte-Carlo methods. This technique is calculating the target distribution (the answer to the problem) as the distribution of a number of individually computed random samples.

It is a practical method, applicable to a wide range of distributions - even if no formula is known - but the results somewhat depend on the way random numbers are picked for sampling. The variants require all variables to be independent, but there is a way to incorporate partial dependencies (see Jasko and Steward, 1990).

Until recently, the Monte-Carlo method generally was considered the best way to estimate the product of more than two continuous distributions. No exact analytical solution of the integral equations was known for these and the Monte-Carlo method avoids the bias present in the matrix method.

The results produced by Monte-Carlo methods are unbiased and converge to the true expectations if the number of random samples grows towards infinity. The convergence is rather slow, in the order of the square root of the number of samples.

In practice any computer implementation can produce a finite number of samples only. Moreover, random number generators do not really produce random numbers, but a predetermined sequence of pseudorandom values. If the numbers were truly random then no two runs of the same program and data could produce the same result. It seems to be a small compromise to give up true randomness for reproducibility.

On the other hand, the use of different algorithms and small sample sizes produces a remarkable scatter of results. Several variants were tested with the same input data. The tests have shown differences ranging from 6 to 20 % depending on the number of samples and the random number generator used. These are typical estimates for the 10 % quantile (or P10). Nearer the expected value the mean and median (P50) have smaller errors. The lower limit of the scatter range relates to tests of 10,000 samples in each run. This number generally is assumed to be sufficient when working with this sort of problem, especially considering the accuracy of the geological input data. Whereas other factors add to the scatter, maybe 5-10 % of this difference is caused by selection of the random number generator alone.

Increasing the number of samples, of course, would decrease the error, but will not eliminate it entirely. To halve the computation error might require four times more memory and up to eight times more computer time. This is because to halve the error one has to compute four times more samples; and the number of the computer program operations increases faster than the number of samples. Under these circumstances it is impractical to expect improvement. Run time and memory required imposes a practical limit on accuracy. Of course if the same program is used with the same 'random' numbers it will (re)produce faithfully the same figures, with the same invisible error. The error may not be visible but the figures will not be any more accurate.

In any situation, it is not enough to standardize on the same random number generator. Using the same algorithm will not guarantee that two programs, perhaps running on two different computer models, will yield the same results. Thus different companies will compute different results from the same geological data. This too, apart from accuracy in the 'absolute' sense, is another difficult issue arising at partnership meetings.

ANALYTICAL APPROXIMATION

Lee and Wang (1983) have started off on a radically different way. Combining lognormal approximations they derived an analytical formula for the product distribution. The principle of the method is based on the fact that if all factors have a 2-parameter lognormal distribution then the product distribution also will be lognormal and its parameters can be calculated exactly. The method is relatively simple to implement on computers.

The Lee and Wang solution has its limitations in the two assumptions that are underlying this model. First, that every input factor has a 2-parameter lognormal distribution. In actual geological problems, the variables may belong to widely different types of probability distributions. There are good arguments to assume lognormal distribution for factors such as gross rock volume. However, other factors may have normal or beta distribution or may have empirical distributions given by a curve only; for which the theoretical formula is unknown.

Some variables, such as oil saturation, have a range of 0 to 100% and are strongly negatively skewed so that the shape of the distribution clearly excludes anything resembling lognormal. In other, less clear-cut situation, statistical tests have indicated a probability distribution type other than lognormal. Jasko and Steward (1990) have tested the distribution of porosity estimates - these were determined to follow the normal law.

Second, it had to be assumed that the factors are independent. This is another generalization that is difficult to satisfy. There are not enough data to deduce generally valid rules on the correlation of the geological variables used in these models. The few situations studied in detail show examples of highly linearly correlated variables, with correlation coefficients of 0.6 and more. To ignore such dependencies would be clearly wrong.

In the original paper, Lee and Wang (1983) considered both of the conditions to be essential for the solution. It is likely that the conditions can be relaxed and a reduced set of assumptions is sufficient. Either the factors have to be lognormal in which situation partial dependencies - or at least certain types of them - can be correctly treated. Or, failing this, they have to be independent and then the lognormal assumption is not necessary. It will be sufficient if skewness and kurtosis of the log transform of the variable is kept within certain bounds. For a lognormally distributed variable, both these values would be zero. It is not known under what circumstances can the formula be applied to the mixed situation where both dependencies and non-lognormal distributions may be present.

THE LATIN-SQUARES APPROXIMATION METHOD

The quest for analytical solutions also can be pursued through other methods. An interesting new technique is based on results of decision theory, and in its form it is related to the latin squares arrangement of experiments. It applies the method of concentrations to compute an accurate estimate of the target distribution curve.

The principle of the method is to represent the target distribution by a set of discrete samples and weights. The samples are taken at predetermined fixed positions. The system of samples and weights is selected to maximize information content. Similiarly to the Monte-Carlo method, this technique can handle almost any distribution. On the other hand, it does not depend on random numbers, and reaches the same level of accuracy much faster.

Rosenblueth and Lind have shown that applying the correct weights and careful selection of representative values removes the inherent bias of equal weight sampling that is present in the matrix method. For some types of univariate and multivariate distributions the exact optimal solutions are known. Applying Lind's method to systems of several variables with unknown distributions may require several hundred pivot points and weights to yield correct estimates for all mixed second moments, that is means, variances, and correlations.

The correct treatment for product distribution is more complex. It requires that the concentrations and pivot points should reproduce skewness (and perhaps kurtosis too). The optimal system of pivot points and weights for this problem is not known yet. It also is difficult to obtain error estimates for suboptimal sets.

Rosenblueth guessed that the number of pivot points needed for product distribution is less then four times than the number of pivot points required for mixed second moments. According to this an optimal set is expected to contain a few thousand points.

I have experimented with various arrangements and determined a system of up to 10,000 points arranged on an N-dimensional grid by the inverse of the cumulative marginal distributions. In situations where all variables are independent all the weights will be equal. If there are dependencies present then the weights will be assigned to correct for correlation between the variables.

It is likely that the system of pivots used by me contains more points for the same accuracy than the, as yet unknown, optimal arrangement. Whereas it may be far from the ultimate solution, it is an improvement on previously used methods.

Why is this new method better than others? The method is certainly superior to purely analytical techniques as the input variables can be of a wide range of families of distributions. Also, they can be correlated.

In comparison to the Monte-Carlo method, both methods allow the same wide range of models; and the results computed by both methods approach the same limits - with precision depending on the number of samples computed. There is good agreement on the general shape of the cumulative curve and the mean, median, and other quantiles for example P10 are all near indeed. It is in the probability density curves that the difference shows up. However many samples are used the curve will be jagged for Monte-Carlo computed results. Even heavy smoothing will not improve it much. The smoothness of the analytically determined results is in striking contrast. A casual look at any test is sufficient to convince that the curve produced by the new algorithm is "nicer" (see Figs. 1, 2).

The figures show two actual examples computed by both methods. The first one (Fig. 1) is an oil prospect with a target distribution nearly normal. Both programs were run with 10,000 samples.

The second example (Fig. 2) is from a gas prospect and here 2000 samples were used in both runs. The Latin-Squares method produces smoother curves with 2000 samples than the Monte-Carlo with 10,000.

The new method requires less memory and it is faster in execution. It can be used equally well on personal computers as on mainframes. An actual implementation runs on IBM PC compatibles as part of the PEREC prospect evaluation program system developed by Quartz Scientific.

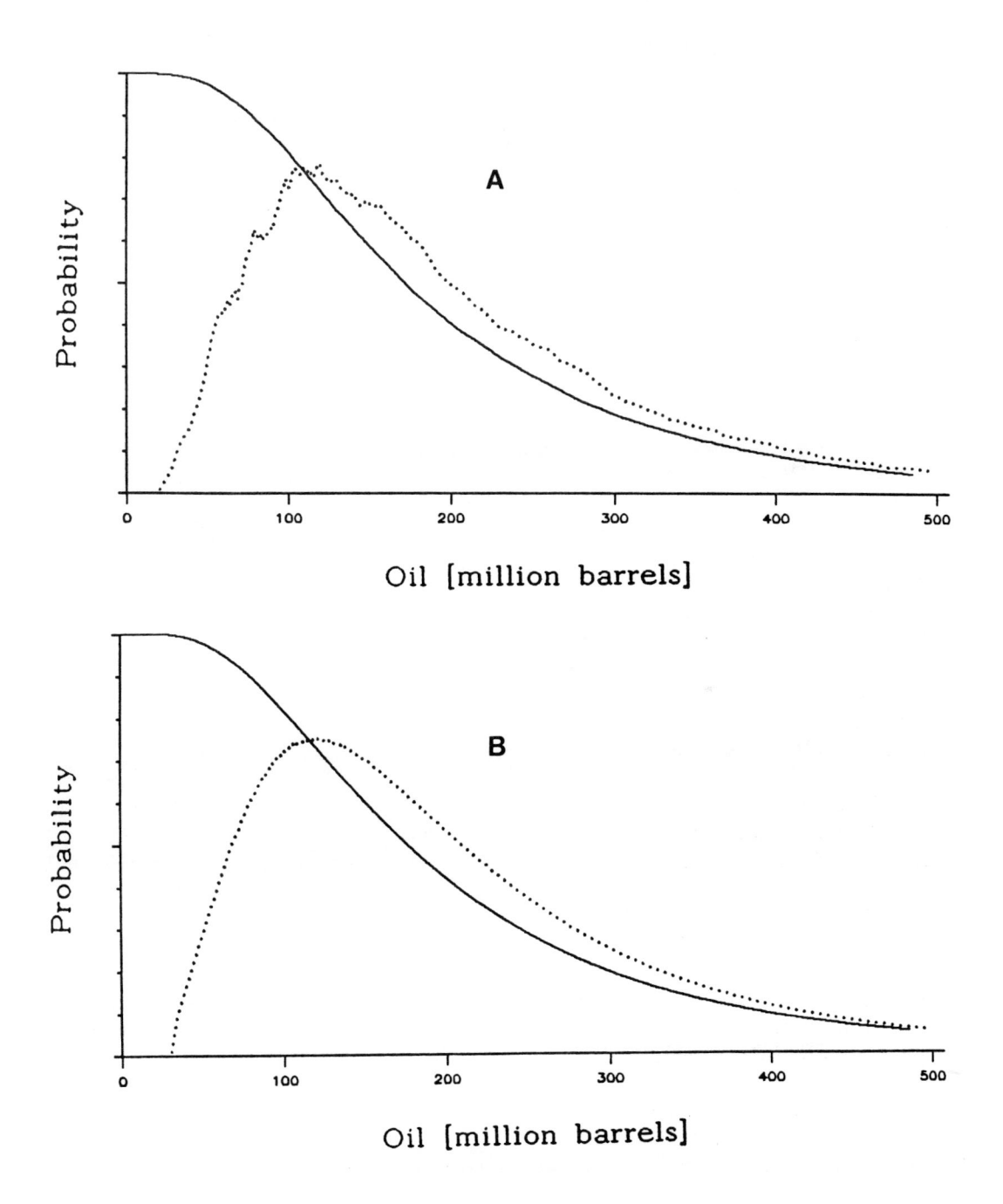

Figure 1. Oil-in-place expectation curves for oil prospect. A, Monte-Carlo method, 10,000 samples; B, Latin-Squares method, 10,000 pivot points.

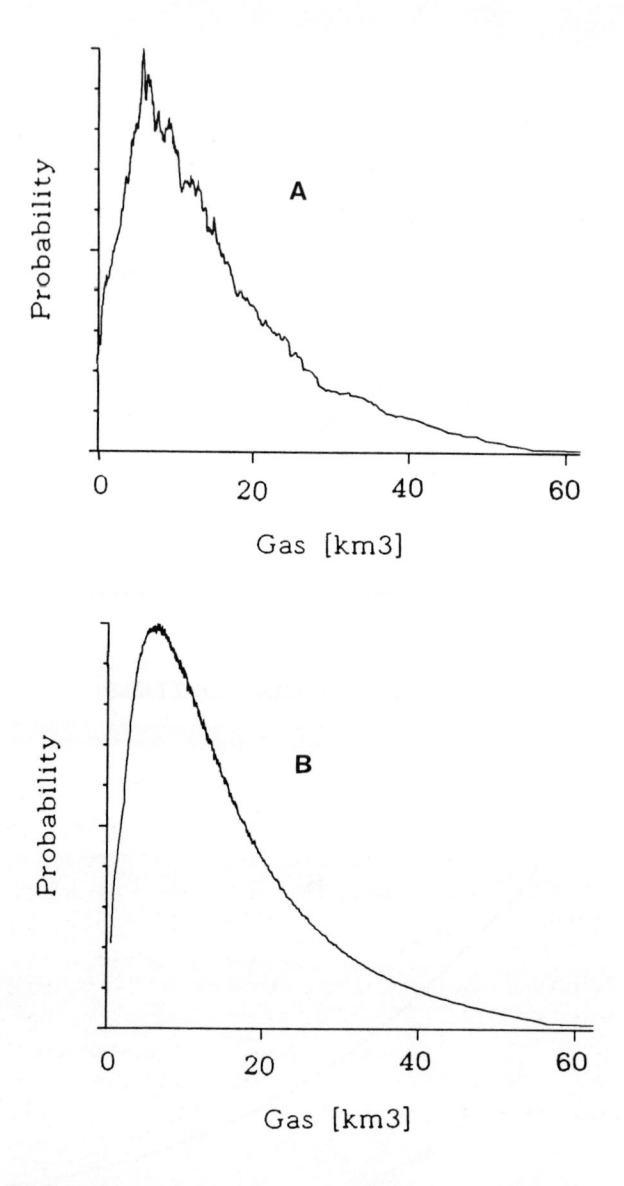

Figure 2. Density distribution curve for gas prospect. A, Monte-Carlo method, 2,000 samples; B, Latin-Squares method, 2,000 pivot points.

REFERENCES

Jasko, T., and Steward, H., 1990, Accounting for statistical dependency in Monte-Carlo prospect evaluation: Computers & Geosciences, in press.

Lee, P.J., and Wang, P.C.C., 1983, Probabilistic formulation of a method for the evaluation of petroleum resources: Jour. Math. Geology, v.15, no.1, p.163-181.

The Use of Expert Systems in the Identification of Siliciclastic Depositional Systems for Hydrocarbon Reservoir Assessment

P.G. Sutterlin
Wichita State University

G.S. Visher
Geological Services & Ventures Inc., Tulsa

ABSTRACT

The use of expert systems in geology, although not yet widespread, has great potential because these systems can be adapted in situations where data are incomplete, ambiguous, and missing. This may be the situation in siliciclastic deposits which are the sites of hydrocarbon accumulations. Identification, on the basis of available data, of the environments in which these deposits were formed can aid in inferring the presence of geologic features which are not measurable or observable directly, but have a significant influence on recovery efficiency. The use of two expert systems "shells" demonstrates not only the utility of an expert system for siliciclastic depositional environment identification, but also some of the limitations of the "shells". However, additional limitations are the result of incomplete understanding of the structure of geological knowledge.

INTRODUCTION

The dramatic fall in oil prices internationally which occurred in the Spring of 1986 has led to a situation in North America in which the search for new petroleum resources, the initiation of enhanced oil recovery (EOR) projects, and (in many instances) production from "stripper" wells has become uneconomic. As a result, replacement of recoverable crude oil reserves has not kept pace with their rate of depletion. Nevertheless, North Americans continue to be, per capita, the world's leading consumers of petroleum. Thus, as production potential has decreased, reliance on offshore petroleum supplies has increased. There is a growing awareness of the possible negative consequences of a continuation of this trend.

Part of the problem is because of the inhomogeniety of virtually all petroleum reservoirs. Conventional field development practices usually result in drilling patterns which are able to extract effectively only a portion of the moveable oil in most reservoirs. This was reemphasized recently in *Science* by Fisher (1987). In the paper, Fisher suggested that recovery of a significant amount of mobile oil in existing reservoirs by "infill" drilling might be an alternative. In many situations, infill drilling could be a less costly alternative to either exploration for new (and likely relatively small) fields, or to EOR projects. The primary impediment, however, in assessing the effectiveness of an infill drilling program in most fields is a basic lack of understanding of the nature of the reservoir's inhomogeneities, especially those which are essentially dictated by geological factors. The discussion which follows outlines an aspect of this problem in which expert systems can play a significant role.

THE BASIS FOR THE APPLICATION OF EXPERT SYSTEMS TECHNOLOGY

The majority of sedimentary rocks, including those which make up hydrocarbon reservoirs, are not internally homogeneous. Basic textural and fabric variations, resulting from different depositional and diagenetic histories, impart various degrees of petrologic heterogeneity which affect fluids flow. In reservoirs with abundant heterogenieties, much mobile oil is not drained by primary and secondary recovery programs (Finney and Tyler, 1986). Describing, defining, and quantifying (where possible) the spatial distribution of heterogeneities leads to an understanding of reservoir properties. On this basis, better models, could be devised with which to design programs to optimize drainage and recoverability.

Variation in geologic elements such as mineralogy, grain-size distribution, pore geometry and distribution, lithologic continuity, facies variation, fractures and faults, and rock-fluid interactions contribute to reservoir heterogeneity. All have parameters which are scale-dependent. Scales of heterogeneity have been defined by Krause and others (1987) as:

Megascale - field-wide heterogeneities which have to do with the overall reservoir geometry, and which usually are controlled by tectonic or paleotopographic features.

Macroscale - heterogeneities on an interwell scale, which dictate the amount and distribution of mobile oil in a reservoir, and and which are controlled by faults and other major field-wide permeability barriers which restrict fluid flow.

Mesoscale - heterogeneities that affect strata adjacent to the borehole, which occur as either fluid conduits, or as fluid baffles within genetic facies. Sedimentary structures such as cross-bedding and bioturbation, stylolites, and fractures, which are a reflection of the depositional and diagenetic environments, are the most abundant mesoscale features which can affect recovery, residual mobileoil saturation, and sweep efficiency. Mesoscale heterogeneities are the most difficult to quantify.

Microscale - heterogeneities at the microscopic level are probably the geological features which have the greatest influence in hydrocarbon reservoir behavior because they establish fluid transmissibility, and hence they directly dictate recovery efficiency, distribution of residual mobile oil,and rock-fluid interactions. Important at this scale are rock texture and fabric, the nature and distribution of pore networks, and the nature of the fluids inherent in the reservoir, and those introduced during drilling, completion, stimulation, and enhanced recovery.

Any program of infill drilling must be based on as accurate a model of the reservoir as possible. However, the data that are obtained from any tests (including core analyses), from geophysical well-logs, and from macroscopic and microscopic examination of well cores and cuttings represent only an infinitesimal fraction of the entire volume of any reservoir. Many features pertaining to stratigraphic and lithologic variation either are unobservable in the subsurface, or are describable only in qualitative, or (at best) in semiquantitative terms. Moreover, it must be acknowledged that even the quantitative values obtained do not always reflect reservoir heterogeneities, particularly those at the meso-and microscales, which are the most difficult to assess (Sutterlin, Linehan, and Sondergard, 1987). Therefore, a complete, detailed, and fully quantitative description of an entire reservoir to use in simulating (presumably using a computer) and predicting the behavior of the reservoir as yet is not achievable. It is in this context that expert systems can be utilized to advantage. This paper describes an application of expert systems concepts to assist the geoscientist in identifying siliciclastic depositional systems, and thereby provide a better basis from which to infer the nature and extent of mesoscale features in siliciclastic hydrocarbon reservoirs.

Mesoscale features reflect the depositional and diagenetic environments in which a sedimentary deposit is formed. The example nakes use of two expert system-building tools or "shells" to develop an expert system designed to aid in the identification of siliciclastic depositional systems. By studying present-day environments in which siliciclastic sedimentary deposits are being formed, and by documenting mesoscale features characteristically developed in the different depositional settings, criteria have been developed (Visher, 1984) which result in the recognition of 15 major siliciclastic depositional systems (Table 1), some of which can be further divided into subsystems (Table 2). The objective is to be able to, by analogy, recognize these same depositional systems in fossil siliciclastic deposits, and thereby infer the nature and extent of mesocale features and predict their influence on fluid flow and recovery efficiency.

Table 1. List of siliciclastic depositional systems included in the expert system.

Valley Fill Fluvial System
Low Sinuosity Fluvial System
Alluvial Fan System
Wave Dominated Shoreface Siliciclastic System
Tide Dominated Shoreface Siliciclastic System
Tidal Siliciclastic Shelf System
Siliciclastic Storm Dominated Shelf System
Coastal Plain Alluvial Valley System
Riverine Dominated Delta System
Wave Dominated Strandplain Delta System
Tide Dominated Delta System
Fan Delta System
Submarine Fan System
Siliciclastic Trench System
Siliciclastic Lacustrine System

Table 2. The subsystems of riverine dominated delta siliciclastic depositional system.

Distributary Valley Subsystem
Bay Fill Subsystem
Shoreface Subsystem
Crevasse Splay Subsystem

The features which characterize siliciclastic depositional systems, and which can be described and measured, have been assigned to six catagories. These are:

Textural patterns - features having to do with the vertical and lateral variations in particle size within stratigraphic sequences, intervals, and individual beds of the depositional system.

Mineralogy - the mineralogical composition of the particles and cements which constitute the deposits.

Geometry - the three-dimensional configuration, and the areal extent, of sequences, intervals, and beds of the deposits.

Sedimentary structures - features associated with the stratification of the deposits, formed during or shortly after the deposits were emplaced.

Geophysical log responses - The characteristic "signatures" of the traces of a variety of logs which are considered diagnostic of particular siliciclastic deposits.

Seismic responses - The continuity and strength of seismic waves in response to specific siliciclastic deposits.

Each of the recognized depositional systems has a set of features which, although not mutually exclusive, serve in sum total to distinguish one system from another. For example, the mineralogical features of a Storm Dominated Siliciclastic Shelf System (SDSSS) are that:

- more then 90% of the particles of the deposit consist of quartz

- glauconite or phosphate grains occur throughout the deposit

- there are rare (less than 1%) feldspar grains present

- micaceous, carbonaceous, clay mineral, and rock fragment

- particles constitute up to 5% of the deposit's particles

The mineralogical features of a Tidal Siliciclastic Shelf System (TSSS) are that:

- more than 90% of the particles of the deposit consist of quartz

- glauconite grains occur throughout the deposit

- there is less than 5% micaceous or carbonaceous material

- there are illitic clay mineral coatings on the quartz particles

- the cementing material is silica or iron oxide

- there are thin-bedded and irregularly bedded limestone intervals throughout the deposit

- the topmost bedding unit consists of calcite (limestone), or limestone and anhydrite

Although some of these features are the same or similar (e.g. the proportion of quartz, and the presence of glauconite), there are enough mineralogical dissimilarities that one might be tempted to consider distinguishing between these two systems based on the mineralogical features alone. However, because there are features in the five other catagories which also characterize each system, identification of any one of the systems based on features in only one catagory is marginally reliable at best, and would carry with it a low level of confidence. Confidence is increased as data about features from all six catagories become incorporated in an analysis.

Unfortunately, it is not always possible that data in all catagories are available in a given situation. Nevertheless, it may be necessary for the reservoir geologist or engineer to make a recommendation based on whatever data are at hand, regardless of how sparse, incomplete, or ambiguous they may be. It is the recognition of this fact that suggested the application of expert systems technology as a tool to guide the geologist– particularly one whose experience with siliiclastic depositional systems is limited– in the identification of these systems, and to provide some measure of reliability and confidence based not only on the available data, but on the inferential capabilities of the expert system which takes into account the sparsity, incompleteness, and ambiguity in the data.

THE EXPERT SYSTEM

Two expert systems "tools" or "shells" have been used. "The Deciding Factor", is a software system marketed by "Power Up", a registered Trademark of Channelmark Corporation, 2929 Campus Drive, San Mateo, California (Campbell and Fitzgerrell, 1985). The Deciding Factor, written in the LISP programming language, was developed by Todd Glover and Alan Campbell, the latter having been associated with development of the well known PROSPECTOR (Gaschnig, 1981) Expert System. The cost is $95.00 U.S.

The Deciding Factor is rule-based, and consists of a Knowledge Base structured in the form of an inverted tree (illustrated by the example of a Tide Dominated Delta System; Fig. 1), and a backward-chaining inference engine. The system is in two

IDENTIFICATION OF SILICICLASTIC DEPOSITIONAL SYSTEMS 353

parts. The EDITOR portion guides the user in constructing the knowledge base by generating graphic models of the "knowledge structure" (Fig. 2). Subsequently, graphic input templates are generated (e.g. Fig. 3) into which are inserted first the facts, and then the rules in the form of a "Weight" and a "Relative Importance" (which have a default value of plus 1.00 and 50% respectively) assigned to each fact. The CONSULTANT portion is an inference engine which conducts a question-and-answer session with the user (Fig. 4), the purpose of which is to evaluate the facts and rules of the knowledge base. The degree of certainty which a user has in the

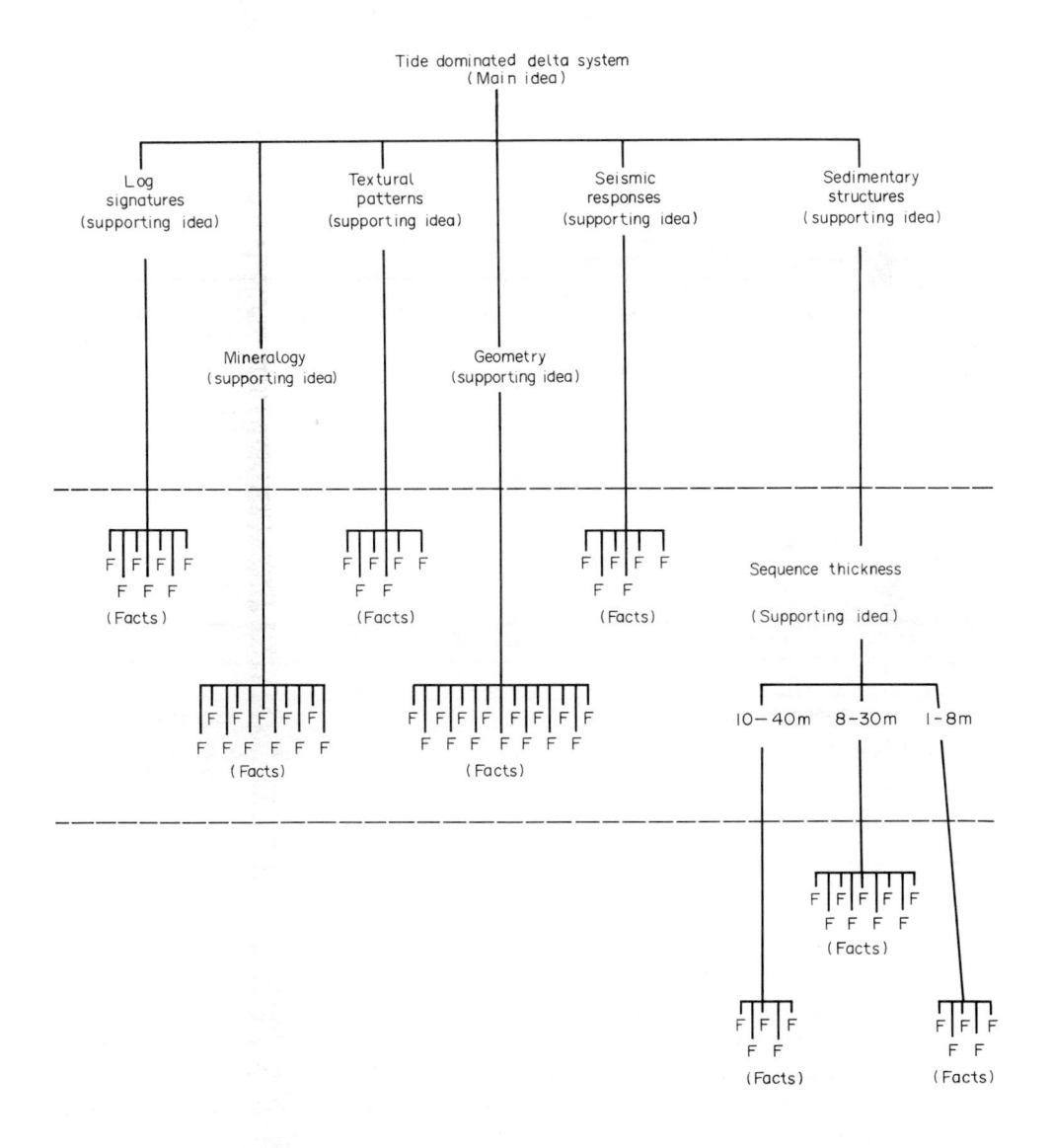

Figure 1. Representation of the structure of The Deciding Factor knowledge base for tide dominated delta system.

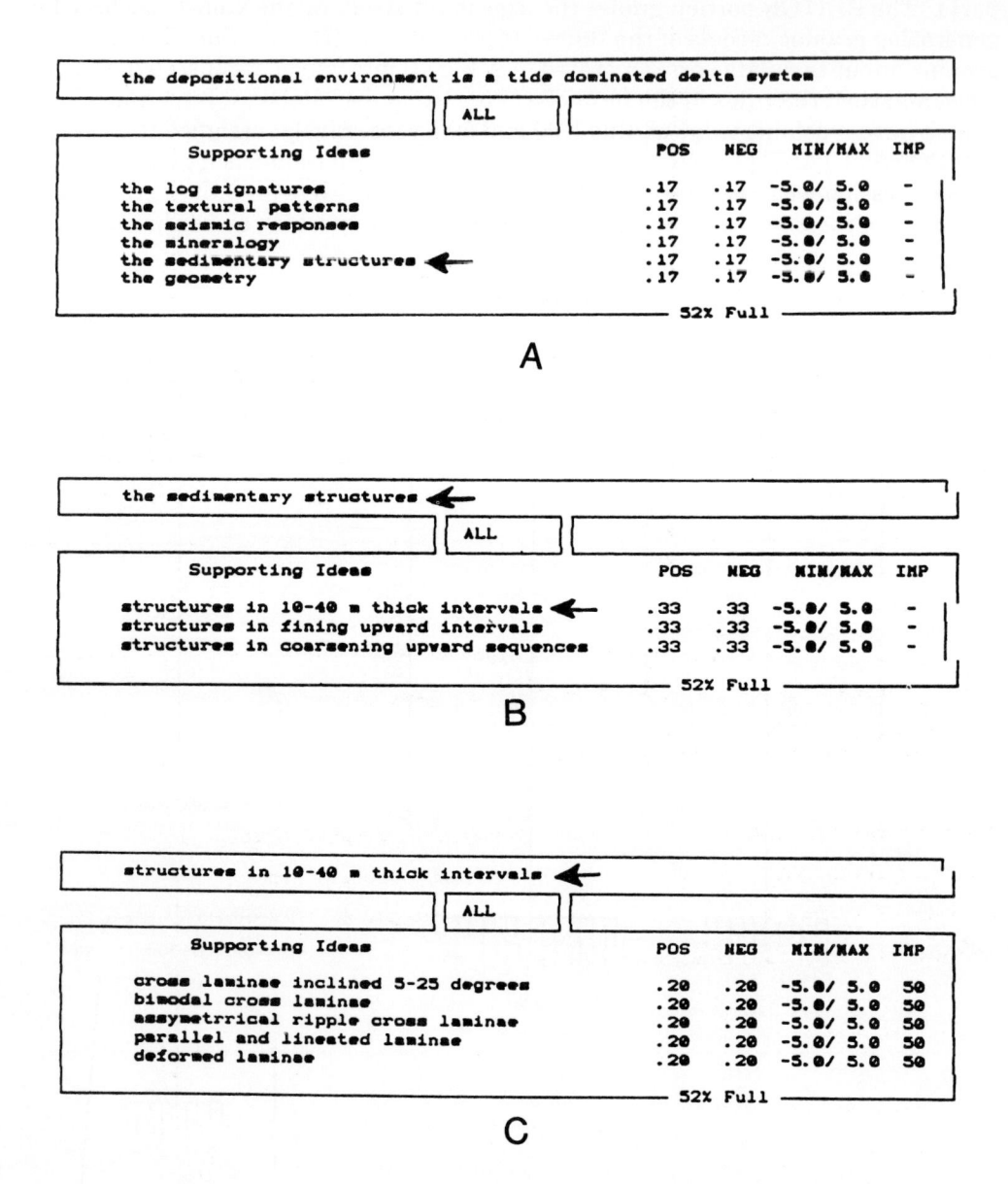

```
┌─────────────────────────────────────────────────────────────────┐
│  the depositional environment is a tide dominated delta system │
│                         ┌─────┐                                 │
│                         │ ALL │                                 │
│    Supporting Ideas              POS    NEG    MIN/MAX    IMP  │
│    the log signatures            .17    .17   -5.0/ 5.0    -   │
│    the textural patterns         .17    .17   -5.0/ 5.0    -   │
│    the seismic responses         .17    .17   -5.0/ 5.0    -   │
│    the mineralogy                .17    .17   -5.0/ 5.0    -   │
│    the sedimentary structures ←  .17    .17   -5.0/ 5.0    -   │
│    the geometry                  .17    .17   -5.0/ 5.0    -   │
│                                          ──── 52% Full ────    │
└─────────────────────────────────────────────────────────────────┘
                                  A
```

```
┌─────────────────────────────────────────────────────────────────┐
│  the sedimentary structures  ←                                  │
│                         ┌─────┐                                 │
│                         │ ALL │                                 │
│    Supporting Ideas                    POS    NEG   MIN/MAX  IMP│
│    structures in 10-40 m thick intervals ← .33 .33 -5.0/5.0  -  │
│    structures in fining upward intervals   .33 .33 -5.0/5.0  -  │
│    structures in coarsening upward sequences .33 .33 -5.0/5.0 - │
│                                          ──── 52% Full ────    │
└─────────────────────────────────────────────────────────────────┘
                                  B
```

```
┌─────────────────────────────────────────────────────────────────┐
│  structures in 10-40 m thick intervals ←                        │
│                         ┌─────┐                                 │
│                         │ ALL │                                 │
│    Supporting Ideas                  POS    NEG    MIN/MAX  IMP │
│    cross laminae inclined 5-25 degrees .20  .20  -5.0/ 5.0   50 │
│    bimodal cross laminae              .20  .20  -5.0/ 5.0   50 │
│    assymetrrical ripple cross laminae .20  .20  -5.0/ 5.0   50 │
│    parallel and lineated laminae      .20  .20  -5.0/ 5.0   50 │
│    deformed laminae                   .20  .20  -5.0/ 5.0   50 │
│                                          ──── 52% Full ────    │
└─────────────────────────────────────────────────────────────────┘
                                  C
```

Figure 2. On-screen representation of structure of deciding factor knowledge base showing (A) main idea and 1st-level supporting ideas, (B) 2nd-level supporting ideas corresponding to sedimentary structures, and (C) facts corresponding to sedimentary structures in 10-40 m thick intervals.

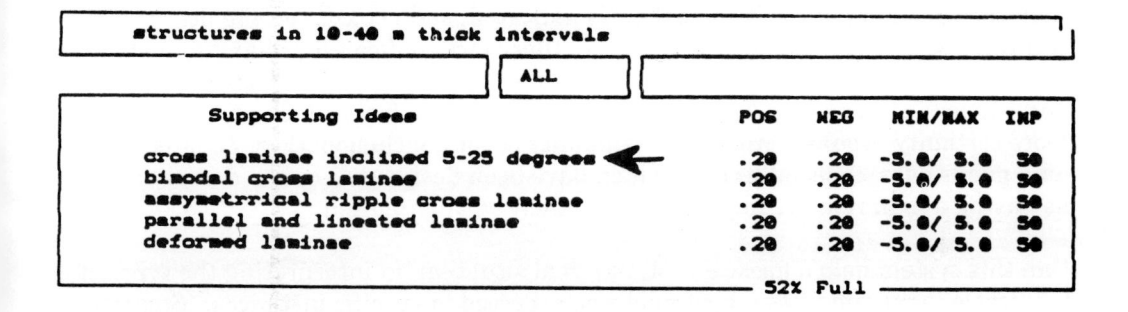

IDENTIFICATION OF SILICICLASTIC DEPOSITIONAL SYSTEMS 355

```
┌─────────────────────────────────────────────────────────────────┐
│  structures in 10-40 m thick intervals                          │
│                    ┌─────┐                                       │
│                    │ ALL │                                       │
│                    └─────┘                                       │
│       Supporting Ideas              POS    NEG   MIN/MAX   IMP  │
│                                                                  │
│    cross laminae inclined 5-25 degrees ◄─  .20  .20  -5.0/ 5.0  50
│    bimodal cross laminae            .20    .20   -5.0/ 5.0   50 │
│    assymetrrical ripple cross laminae .20  .20   -5.0/ 5.0   50 │
│    parallel and lineated laminae    .20    .20   -5.0/ 5.0   50 │
│    deformed laminae                 .20    .20   -5.0/ 5.0   50 │
│                                              ─── 52% Full ───   │
└─────────────────────────────────────────────────────────────────┘

Hypothesis: structures in 10-40 m thick intervals
Logical relationship: ALL
----------------------------------------------------------------

Factor: cross laminae inclined 5-25 degrees  ◄──

The Consultant will ask:
Are there cross-laminae inclined from 5 to 25 degree?


Positive weight: 1.00            Negative weight: 1.00

Minimum answer: -5.0             Maximum answer: 5.0

Hypothesis (Y/N): N              Importance: 50

Optional introductory statement:
```

Figure 3. On-screen template for facts and rules entry to deciding factor knowledge base.

data, in response to each question, is taken into account in the form of a sliding scale ranging from -5 (absolute certainty that the feature is not present) through 0 (complete uncertainty i.e. "don't know") to + 5 (absolute certainty that the feature is present). The CONSULTANT then generates a report (Fig. 5) which indicates, by a value ranging from 0.0 to 5.0, the degree to which the answers support the model contained in the knowledge base. Although this is not a statistical measure in the true sense, it can be regarded as a relative probability because it is based on the weights assigned by the domain expert to the facts in the knowledge base. It addition, the reliability of the "conclusion" is indicated, and is based on the degree of certainty with which the user responds to those facts indicated as having the highest relative importance. The CONSULTANT then summarizes the responses to the key and supporting facts (e.g. Fig. 6), and finally lists those facts which, if known with more certainty, would increase the reliability of the conclusion (Fig. 7). Missing and ambiguous data, especially those which have been designated by the domain expert as having a high relative importance, are thereby highlighted.

Can this system help a geoscientist, in a real situation, in interpreting the types of mesoscale hetergenieities which might be expected in specific instances? Suppose that available data about a given reservoir is confined to that derived from geophysi-

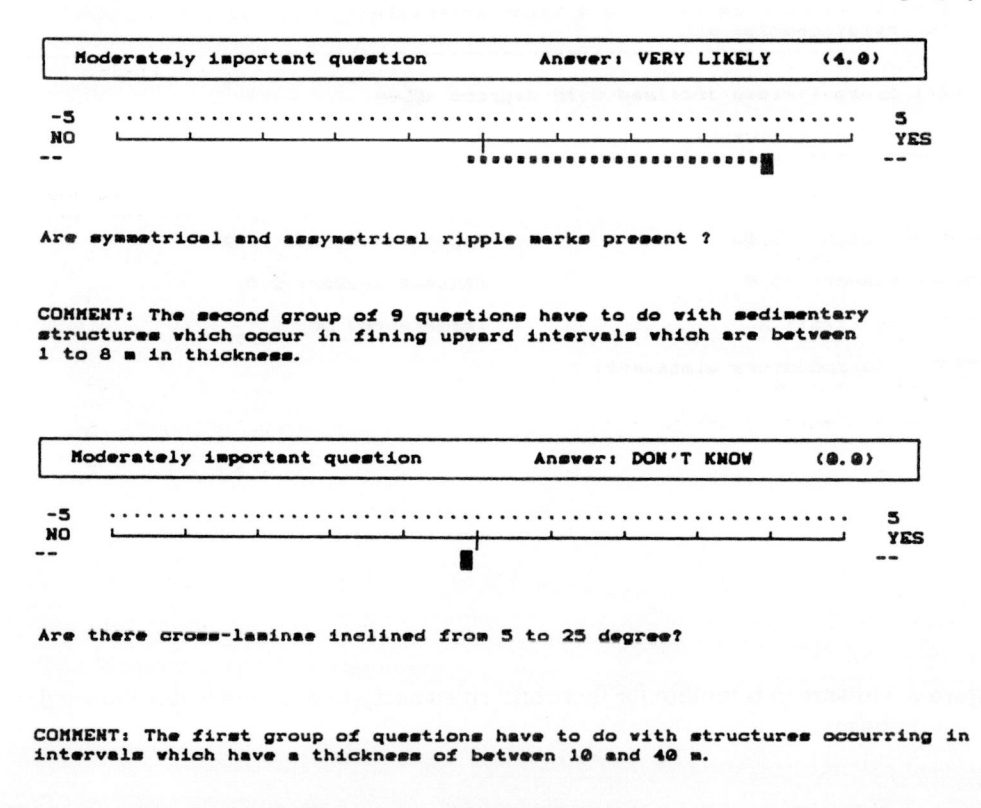

Figure 4. On-screen display produced by deciding factor consultant showing ranges of responses to facts and rules in knowledge base.

```
                        FINAL CONCLUSION
                        ----------------

At this point, enough information has been gathered to come to a final
conclusion.

Your information has led to a slightly positive (1.6) evaluation of whether the
depositional environment is a tide dominated delta system.

The conclusion is considerably reliable (85%) because you tended to answer the
more important questions with substantial certainty.


                            KEY IDEAS
                            ---------

    There is a slightly positive (1.6) evaluation that the depositional
    environment is a tide dominated delta system.

    This is based on all of the following.

         Idea                                 Response      Influence
         ----                                 --------      ---------

    The log signatures                          1.0            .2

    The textural patterns                        .5            .1

    The seismic responses                       1.7            .3

    The mineralogy                               .7            .1

    The sedimentary structures                  3.1            .5

    The geometry                                2.8            .5
```

Figure 5. Deciding factor on-screen evaluation.

SUMMARY

There is a slightly positive (1.6) evaluation that the depositional environment is a tide dominated delta system.

These are the ideas that led to this conclusion.

Idea	Response
Logdata	5.0
Clay content from SP and GR curves in uniform texture beds	2.0
Clay content from SP and GR in coarsening upward intervals	4.0
SP and GR responses in fining upward intervals	-4.0
Dipmeter patterns	0.0
Dipmeter orientation and range	5.0
Porosity distribution	-5.0
Irregular SP and GR curve responses show little textural pattern	5.0
Fining upward top 1-8 m thick	3.0
Coarsening upward 1-5 m units within 8-30 m intervals	-5.0
Vertical change in 10-40 m intervals	0.0
Basal 1-5 m with grains gt 500 um	5.0
Maximum grain size in coarsening intervals is 300 um	-5.0
Coarsening upward intervals, 90% of grains 100-300 um	5.0
Impedence reflections	5.0
High and low velocity wavelet responses	5.0
Reflectors over 500 km2	5.0
Greater than 5000 km2	5.0
Low velocity bedding units	-5.0
Scour patterns	-5.0

Figure 6. Partial listing of deciding factor on-screen summary of user responses.

```
                    INCREASING RELIABILITY
                    ----------------------

Stronger responses to these ideas could increase reliability from 85%
to 98%
                                            Your        Potential
                                            Current     Increase in
            Idea                            Response    Reliability
            ----                            --------    -----------

Dipmeter patterns                             0.0           2%

Vertical change in 10-40 m intervals          0.0           1%

Less than 90% quartz grains                   0.0           1%

Clay and carbonaceous                         0.0           1%

K-feldspar grains                             0.0           1%

Shape                                         0.0           1%

100                                           0.0           1%

Clay content from SP and GR curves in         2.0           1%
uniform texture beds

Fining upward top 1-8 m thick                 3.0           1%
```

Figure 7. Deciding factor on-screen listing of facts in knowledge base considered of high enough importance to influence reliability of conclusion.

cal well logs, seismic responses, and well cuttings, which yield a certain amount of mineralogical and textural pattern data. If these data produce a positive evaluation, of say 3.5, that the depositional environment can be interpreted as, for example, a Tide Dominated Delta System, with a reliability of 75%, the user then might be inclined to accept the evaluation. Further by analogy, and using the model of this depositional system incorporated in the knowledge base, the user then might infer features of the geometry and sedimentary structures in the reservoir - features which are a reflection the depositional and diagenetic environment, and which exert a major control on mesoscale heterogeneities. In a Tide Dominated Delta System, these would be, in respect to sedimentary structures, as follows:

- cross-beds inclined from 5-25 degrees

- bimodal cross-beds

- beds more than 1 m thick may be deformed

- trough cross-beds in middle portion of sequence

- flaser bedding

- upper and lower bedding units bioturbated

It is easy to see that, at the mesoscale, these features produce heterogeneities which have the potential to impede fluid flow, especially if their scale is less than the distance between wells. Mobile oil between wells then could be beyond the sweep of individual boreholes, and thus might be essentially unrecoverable by any methods other than infill drilling.

Because each silicilastic depositional system constitutes a separate knowledge base, the The Deciding Factor can be used to generate an expert system to evaluate the facts and rules of only one depositional system at a time. This presupposes that the user has some prior notion as to which of the 15 depositional systems is indicated by the data at hand. The user is able only to ascertain, based on expert knowledge, the extent to which the initial supposition is supported in the face of incoplete, missing, and ambiguous data. The Deciding Factor-based expert systems serve to either reinforce or cast doubt on the initial perception, and indicate which additional data might improve the reliability of the conclusion. In many instances, this limits the use of the program to geoscientists who already are more than casually familiar with siliciclastic depositional systems.

In order to overcome this drawback, another expert systems building tool is employed as a "front end" to the expert system. 1st CLASS is an expert systems shell for the IBM PC and compatible micro-computers (Thomas, 1987). Developed by William

Hapgood of Programs in Motion Inc. 10 Sycamore Road, Weyland, Massacheusetts, is written in Pascal and Macro Assembler, and is available for a cost of about U.S. $500. The system is both an example-based and a rule-based tool. The EDITOR part of 1st CLASS uses a spreadsheet approach which generates knowledge bases by building a graphic "decision tree" from either a set of "definitions" (facts) and "examples" (rules), or directly from a rule . The path taken by the inference engine can then be stipulated to proceed by simple matching, from left to right, or on the basis of either rule optimization or rule customization. Both forward and backward chaining can be used to link knowledge bases.

In the 1st CLASS context, each individual file constitutes a knowledge base. The Definitions spreadsheet (Fig. 8) contains, in the right-hand column, a list of possible "results" or "outcomes", of which there can be up to 31. The columns to the left contain the values for the "factors". There can be up to 32 factors, and up to 32 values for each factor. In the example, the results are the siliciclastic depositional systems and subsystems, the factors correspond to the catogories of features, and the values correspond to the features themselves.

Each line of the Examples spreadsheet, which is in effect an example or "rule" in itself, contains in the right-hand column a possible result, and in the columns to the left, values for each factor which lead to that result. It is immediately evident that, in the situation with which we are concerned, the number of possible examples is far greater than the number of depositional systems and subsystems. This is, of course, because of the number of different values which each factor can assume. Each combination of values, produces an example, with its attendant result. In the instance of the Storm Dominated Silici-clastic Depositional System mentioned previously, the values for the mineralogy factor (criterion), of which there are 4, will by themselves generate 21 examples, and does not even take into account examples in which data values are missing. The sedimentary structures factor has 11 possible values. It quickly becomes apparent that, to accomodate all possible combinations of values (and missing values) which could produce a valid result, the maximum allowable number of 255 examples would be easily exceeded.

To avoid this constraint without risking a significant loss of information, it was decided to limit the number of values in each factor to those which have the highest relative importance in The Deciding Factor systems and subsystems knowledge bases. For the mineralogy criterion, the following values were defined:

- the mineralogy of the system is not diagnostic

- quartz particles constitute more then 95% of the deposit

- quartz particles constitute less than 95% of the deposit

- the deposit contains rock fragment particles

- the deposit contains carbonate particles

- the diagnostic mineralogy is other than any of the ones listed

In this manner, and "factoring" the other criteria to isolate those values considered most important, many of the systems and subsystems have criteria in common. This serves to limit the number of possible combinations which could produce a rational result, thereby reducing the number of examples required.

The resultant 1ST CLASS expert system ADVISOR then is used to effect, on the basis of the collectively most significant attributes, a preliminary "screening" to indicate the most probable system or systems which best match the values as they are selected, one at a time (Fig. 9). Moreover, unlike The Deciding Factor, more than one conclusion is possible, and the conclusion can change as more data become available, or new data and rules are added to the system.

```
new_Factor,   new_Value,   edit_Text,   Change,   Activate,   Move,   Delete
        Files       Definitions  Examples    Methods     Rule    Advisor
[F1=Help]           4  Factors in ENVIRNMT               [F9=Files] [F10=Examples]
            (inactive)
        MEMO            #TEXTURE    #MINERAL    #GEOM       ENVIRONMENT
                        TP0         M0          G0          TDSSS
                        TP1         M1          G1          TSSS
                        TP2         M2          G2          CPAVS
                        TP3         M3          G3          RDDS
                        TP4         M4          G4          TDDS
                                                G5          STS
                                                G6          SFS
                                                G7          WDSSS
                                                            WDSDS
                                                            SSDSS
                        A
```

```
Complete the definitions, then
press F10 to give some examples.

For more help, press F1.
```

```
new_Example,  Replicate,   Change,   Activate,   Move,   Delete
        Files       Definitions  Examples    Methods     Rule    Advisor
[F1=Help]           14 Examples in ENVIRNMT              [F9=Definitions] [F10=Methods]
            (inactive)
        MEMO            #TEXTURE    #MINERAL    #GEOM       ENVIRONMENT  Weight
         1:             TP0         M2          G5          TSSS         [1.00]
         2:             TP0         M3          G0          SFS          [1.00]
         3:             TP1         M3          G2          STS          [1.00]
         4:             TP1         M4          G4          TDSSS        [1.00]
         5:             TP2         M0          G6          RDDS         [1.00]
         6:             TP2         M0          G4          RDDS         [1.00]
         7:             TP2         M3          G0          SFS          [1.00]
         8:             TP4         M0          G4          RDDS         [1.00]
         9:             TP4         M0          G6          RDDS         [1.00]
        10:             TP3         M3          G0          CPAVS        [1.00]
        11:             TP4         M4          G5          TDDS         [1.00]
        12:             TP2         M2          G1          WDSSS        [1.00]
        13:             TP2         M2          G4          WDSSS        [1.00]
        14:             TP2         M2          G6          SSDSS        [1.00]
                        B
```

Figure 8. 1st-CLASS On-screen display of (A) definitions spreadsheet, and (B) examples spreadsheet.

IDENTIFICATION OF SILICICLASTIC DEPOSITIONAL SYSTEMS 363

```
[F1=Help]          1st-CLASS Advisor for MINERAL         [F9=Rule] [Esc=Stop]

There are five major mineralogical indicators which are diagnostic of
certain clastic sedimentary environments.  Please select the one that is
most appropriate.

If no data regarding the mineralogy of the sand bodies are available,
please ENTER A QUESTION MARK (?).  The program will then continue.

 ┌────────────────────────────────────────────────────────────────────┐
 │ The mineralogy of the sandstone units is not diagnostic.           │
 │ The sandstone units contain less than 90% quartz grains.           │
 │ The sandstone units contain more than 90% quartz grains.           │
 │ The sandstone units contain clay and rock fragments.               │
 │ The sandstone units contain carbonate fragments.                   │
 └────────────────────────────────────────────────────────────────────┘
```

A

```
[F1=Help]          1st-CLASS Advisor for TEXTURE         [F9=Rule] [Esc=Stop]

There are five major textural patterns in the sandstone units which
describes variation in grain size which is diagnostic of certain clastic
depositional environmnets.  Please select the one which is most appropriate.

If no data regarding textural patterns are available, please ENTER A
QUESTION MARK (?).  The program will then continue.

 ┌────────────────────────────────────────────────────────────────────┐
 │ The variation in grain size is not diagnostic.                     │
 │ The sandstone units show a fining upward texture.                  │
 │ The sandstone units show a coarsening upward texture.              │
 │ The sandstone units show both a fining and coarsening upward texture. │
 │ The sandstone units show little textural change.                   │
 └────────────────────────────────────────────────────────────────────┘
```

B

```
[F1=Help]          1st-CLASS Advisor for ENVIRNMT        [F9=Rule] [Esc=Stop]

 ┌────────────────────────────────────────────────────────────────────┐
 │ The depositional environment is probably a Tide Dominated Delta System. │
 └────────────────────────────────────────────────────────────────────┘
```

C

Figure 9. 1st-CLASS Advisor on-screen display of (A) mineralogy criteria (B) texture criteria, and (C) siliciclastic depositional system selected by 1st-CLASS.

CONCLUSIONS

Expert systems in the geosciences have great potential (Campbell and others, 1982). Present commercially available systems-building software, however, has limitations. However, this may not be entirely the fault of the software itself. Rather, it may be because of a lack of understanding of geoscience knowledge and its structure. Not too long ago, analagous limitations of commercial database-management systems were evident. It was not until geoscientists began to understand the structures of geoscience data, coupled with the advent of the relational database concept, that many and varied data structures could be accomodated readily without the risk of information loss. In the same vein, efforts must be made to better understand the nature of geoscience knowledge stuctures. Real progress will require that geoscientists, and especially domain experts, not only examine their data even more closely, but also attempt to articulate as clearly as possible the "rules" they use to arrive at a conclusion - a geological, not a computer science, pursuit.

REFERENCES

Campbell, A.N., Hollister, V.F., Duda, R.O., and Hart, P.E., 1982, Recognition of a hidden mineral deposit by an artificial intelligence program: Science, v. 217, no. 4563, p. 927-929.

Campbell, A.N., and Fitzgerrell, W., 1985, The deciding factor user's manual: Channelmark Corporation, San Mateo, California, 53 p.

Fisher, W.L., 1987, Can the U.S. oil and gas resource base support sustained production?: Science, v. 236, no. 4809, p. 1631-1636.

Finley, R.J., and Tyler, N., 1986, Geological characterization of sandstone reservoirs in Reservoir characterization: Academic Press Inc., Orlando, Florida, p. 1-37.

Gaschnig, J., 1981, PROSPECTOR: an expert system for mineral exploration, in Machine intelligence: Infotech State of the Art Report 9, no. 3, p.

Krause, F.F., Collins, H.N., Nelson, D.A., Machemer, S.D., and French, P.R., 1987, Multiscale anatomy of a reservoir: geological characterization of Pembina-Cardium Pool, west-central Alberta, Canada: Am. Assoc. Petroleum Geologists Bull., v. 71, no. 10, p. 1233-1260.

Sutterlin, P.G., Linehan, J.M., and Sondergard, M.A., 1987, An expert-systems approach to estimating recovery efficiency in hydrocarbon reservoirs: The Compass, v. 65, no. 1, p. 18-27.

Thomas, W., 1987, 1st CLASS Instruction Manual: Programs in Motion Inc.,Wayland, Massachusetts.

Visher, G.S., 1984, Exploration stratigraphy: Pennwell Company, Tulsa, Oklahoma, 331 p.

INVENTORIES

Geological Comparison of Brazil and China by State

J. C. Griffiths,
Pennsylvania State University
H. Hu, and
China University of Geology
H. C. Chou
Pennsylvania State University

ABSTRACT

The geological characteristics of different regions have been analyzed quantitatively using a standard nomenclature for their contained rock types ($s_g <= 65$) and their mineral resources ($s_m <= 75$). On the basis that tectonic processes determine both rock types and mineral resources in different regions, they may be compared, using well-developed areas, such as the U.S.A. and Europe, with those which are relatively less well developed.

On this basis, diversity (s_g - 1) may be used as a criterion for comparison; for example, the diversity of the rock types in Europe (s-1 = 52) and U.S.A. (s-1 = 51) are compared with Brazil (s -1 = 42), China (s-1= 41), and the Middle East (s-1 = 35). In addition, rock types, common to pairs of regions, range from s-1 = 26-46. Brazil and China each has 41 rock types in common with Europe and 37 and 39 respectively in common with the U.S.A. They also mutually possess (s-1 = 34) rock types, or about 80 percent, in common.

Diversity is related approximately linearly to $\log_{10}$ area in km.² through the range of 3.5 to 6.0 (3000 to 1 million km.²) but in the larger areas, to which these regions belong, diversity seems to be independent of area.

Selecting five states of high geological diversity, three from Brazil and two from China, shows about 70 percent of the rock types are common among the three states of Brazil and some 60 percent are common to the two Chinese provinces; 60 percent are common to the three Brazilian states and Hebei, China, whereas only 45 percent are common to Bahia, Minaes Gerais, and Zhejang. Some 60 percent are common to Matto Grosso, Brazil, and Zhejang, China.

It has been determined that the number of rock types ($X = S_g$ -1) is related linearly to the number of mineral resources ($Y = s_m$ -1) in well-developed regions, according to the equation, $Y = 11.28 + 1.85X$ ($r^2 = 80\%$). Unfortunately, the records of mineral-resource production are not always complete, particularly in the sectors of construction materials (e.g. cement in the states of Brazil), some fuels and various nonmetals in the states of both countries. Nevertheless, applying this relationship to the states of Brazil and China shows a wide scatter of points in both examples; the departures of observed from expected values yield similar distributions. Apparently, both countries are equally well developed; nearly all the states are under developed relative to that expected from the linear relationship. Six Chinese provinces fall within the lower one-half of the 95 percent confidence intervals and three (Kwantung, Kuangsi, and Sechuan) closely approach their expected values. Only three states (Minaes Gerais, Rio de Janeiro, and Espirito Santo) from Brazil fall within the confidence limits.

It may be concluded from this analysis that Brazil and China are geologically similar and possess similar resources; the Chinese provinces, on the whole, are somewhat better developed than the states of Brazil, possibly because of China's longer recorded history or to the lack of adequate published records of production of the less glamorous commodities.

INTRODUCTION

We have described elsewhere one procedure which may be used to perform first approximation estimates of the value and amount of the mineral resources of a region based on its past production records (Griffiths, 1978; Gill and Griffiths, 1984); the black box-model (Fig. 1, ibid) used in that series of investigations contained an input from "geology" and, together with the assumption that tectonic events determine both the geology (rock types) and the mineral resources in a region, this forms an adequate basic philosophy for estimating the mineral resources from the geology, or its equivalent, the rock types present in a region. To operationalize this hypothesis it is necessary to possess a knowledge of the rock types in a region in a standardized form so that different regions may be compared (Griffiths, 1983). Then, selecting well-developed regions, such as the states of the U.S.A. as a basis and areas of similar geology in the region to be analyzed, it should be possible to determine the mineral resource potential of less well-developed regions of interest.

In our earlier studies we determined a simple empirical relationship, a linear regression, between geological (X) and mineral resource (Y) diversity which led to an estimate of the number of commodities in a standard set of 75 which were to be expected for a given geological diversity, that is for a given number of rock types. The regression is $Y = 1.85X + 11.28$ (Griffiths, Watson, and Menzie, 1980).

To use this approach it becomes necessary, therefore, to compile a global databank of geological diversity by which geological similarity among different regions may be established. This report and the one presented by A.N. Pilant, J.C. Griffiths, and C.M. Smiths (this volume) are two examples which summarize the analysis and application of this type of information.

METHOD OF ANALYSIS

The procedure is presented in a flowdiagram in Figure 1; a region is selected and a suitable geological map is located. The first step is to print the map explanation; each explanation symbol is given a number, may be more than one if there are several lithologies (rock types) under one symbol. Then the lithological description in the map explanation, amplified by relevant literature where necessary, is used to assign each number to a standardized set of 65 rock types; these are known as time-petrographic index numbers. The number of symbols in the explanation (= S) yields one measure of the amount of information in the map.

The map symbols are transduced to time-petrographic numbers. Igneous rocks are assigned on the basis of the Rosenbusch-Zirkel classification (Hurlbut, 1966). From preliminary testing only one class of metamorphic rocks, namely regional metamorphic is used. There are four detrital sediments and assignments follow the Krynine classification with some modifications (Griffiths, 1978, p.452). The chemical sediments are subdivided into carbonates and evaporites; the latter include sedimentary ore deposits such as bauxites, iron ores, and diatomites and radiolarites.

Another important aspect of the map is its scale; in our earlier work we attempted to standardize the sampling unit at 12.7 km. "ground" scale; this leads to a 1/4 inch grid at a 1:2 million map scale. Recently we have attempted to use a 6.35 km. "ground" scale which is equivalent to a 1/4 inch grid at a scale of 1:1 million. Needless to say, we have had to compromise when a map of suitable scale is not available. After selecting a suitable grid the map is point-counted and the collected data are submitted to a computer program (COUNT written by C. M. Smith). The output is printed in three forms:

(1) A list of the symbol numbers in grid position in terms of rows and columns. The map is oriented by giving a start position.
(2) A digitized map of the symbol numbers arranged spatially in rows and columns.
(3) A frequency distribution of number of points which fall on each symbol number.

Step 3 yields another measure, the number of symbol numbers present in the point-count (S'); because S is nearly always greater than S' the difference is a measure of how many lithologies have been missed because of sampling or other errors. The point-count, as frequencies from step 3, are inserted into the explanation using a Watfile program (Wilkinson, 1986) and this may be manipulated on the computer to yield frequencies of points on rock types. These data are entered into the 'time petrographic' (TPI) table of 65 standardized rock-types; this TPI table, in turn, yields a measure (s_g) of geological diversity of a region. By convention diversity $s_g = 0$ when there is only one rocktype in a region or, more generally, geological diversity = s_g-1; this definition of diversity also reflects the geometric feature that any region may be represented by a simplex which possesses s -1 dimensions with s apices (= rock types).

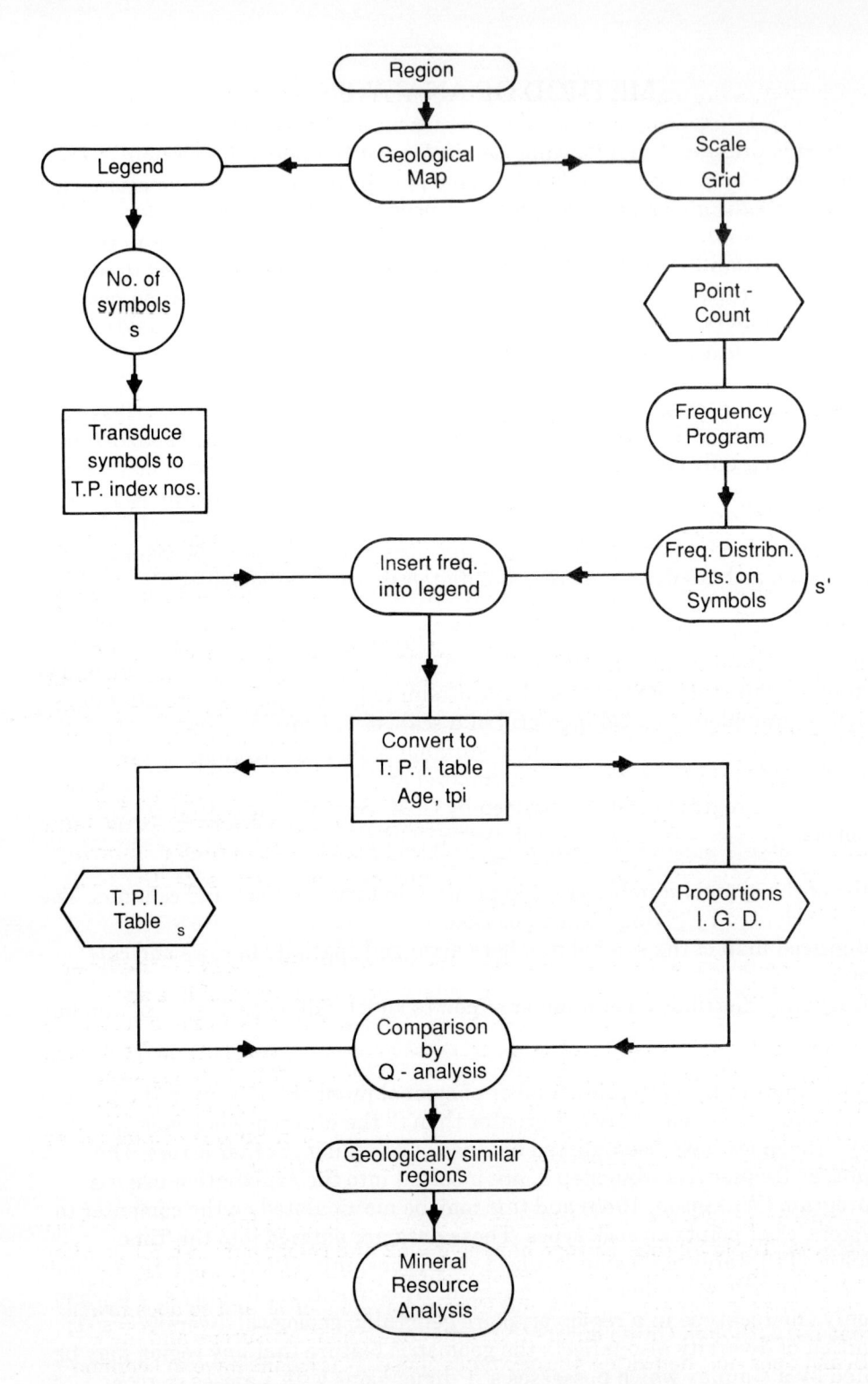

Figure 1. Flow diagram summarizing analytical procedure.

In the subsequent analysis only the presence or absence of rock types (and, later, equivalently, mineral resources) is used as the fundamental variable. Classification of these zero/one data is accomplished through the use of Atkin's Q-analysis (1974; Gould, 1981). A computer program written by Mr. S. L. Mann, Department of Geography, the Pennsylvania State University, and used through the courtesy of Professor P.R.Gould, Professor of Geography, facilitates this analysis. An example and explanation of Q-analysis in a similar context is given in Griffiths (1983).

DESCRIPTION OF THE RESULTS

An illustrative example using the five large regions of Brazil, China, U.S.A., Europe and the Middle East

In general, as would be expected, the larger the region, the more rock types it contains; these five regions are of continental to subcontinental size and so contain many rock types. When the incident matrix of Table 1 is submitted to Q-analysis it gives rise to the connectivity matrix $K_c(R:\lambda_{ij})$ of Table 2 and, using Atkin's notation, top q = s_g-1 and lies along the diagonal; the off-diagonal terms are termed bottom q and measure the number of rock types minus one which are common to each pair of regions (R). For example, Brazil has a diversity (s_g-1 = top $\hat{q}$) of 42 or, it possesses 43 different rock types; China possesses a diversity of 41 whereas bottom $\check{q}$ for this pair of regions q = 34 or, equivalently, they have 35 rock types in common. That is some 80 percent of the rock types are common to both regions. Brazil is least like the Middle East in that their mutual bottom $\check{q}$ = 26 and the Middle East has a top $\hat{q}$ of 35; they have about 75 percent of their rock types in common. Obviously, these large regions are similar, suggesting perhaps, that all large regions tend to contain many of the same rock types,that is any region this large tends to be representative (= a random sample) of the Earth's surface crust.

An alternative geometrical expression of these relationships is by way of simplicies representing the five regions as in Figure 2; each simplex is formed of five ages (rows) and 13 rock types (columns). Filled circles indicate the presence of a rock type and, for example, Brazil has top $\hat{q}$ of 42 or possesses 43 rock types; bottom $\check{q}$ between Brazil and China = 34 or they have 35 rock types in common. Filled circles with extensions pointing to each region indicate which rock types are common to both. There are 52 rock types in the U.S.A. (top $\hat{q}$ = 51) and q = 52 in Europe whereas they both possess bottom $\check{q}$ = 46 or they have 47 rock types in common.

The successive steps of the Q-analysis of these 5 regions may be followed in the table of equivalences (Table 3); thus Europe has top $\hat{q}$ = 52 and comes out first, followed by U.S.A. at q = 51; they join in the same component at q =46 which implies that they have 47 rock types in common. China and Europe have a q of 41 and so does Brazil and Europe but although China enters the same component as Europe and the U.S.A., Brazil does not; hence the 42 rock types these two regions have in common with Europe are not all the same rock types.

Table 1. Zero/one data matrix of rock types in five large regions.

| Name | Cat.No. | Rocktypes |||||| s |
		501-513	401-413	301-313	201-213	101-113	
Braz	310600000	1101110101010	1100111111100	1100111111000	1100111111101	1000101111101	43
Chin	510300000	1100111011100	1101111111100	1101111111101	1100101111101	0000001010101	42
USA	211500000	1111111111100	1111111111111	1111111111111	1101101111101	0001001011001	52
EURO	620300000	1101111111100	1101111111101	1111111111101	1101101111111	1100101111101	53
MEST	540300000	1101111111001	1101110011111	1101101100001	0101101111101	0000000000000	36

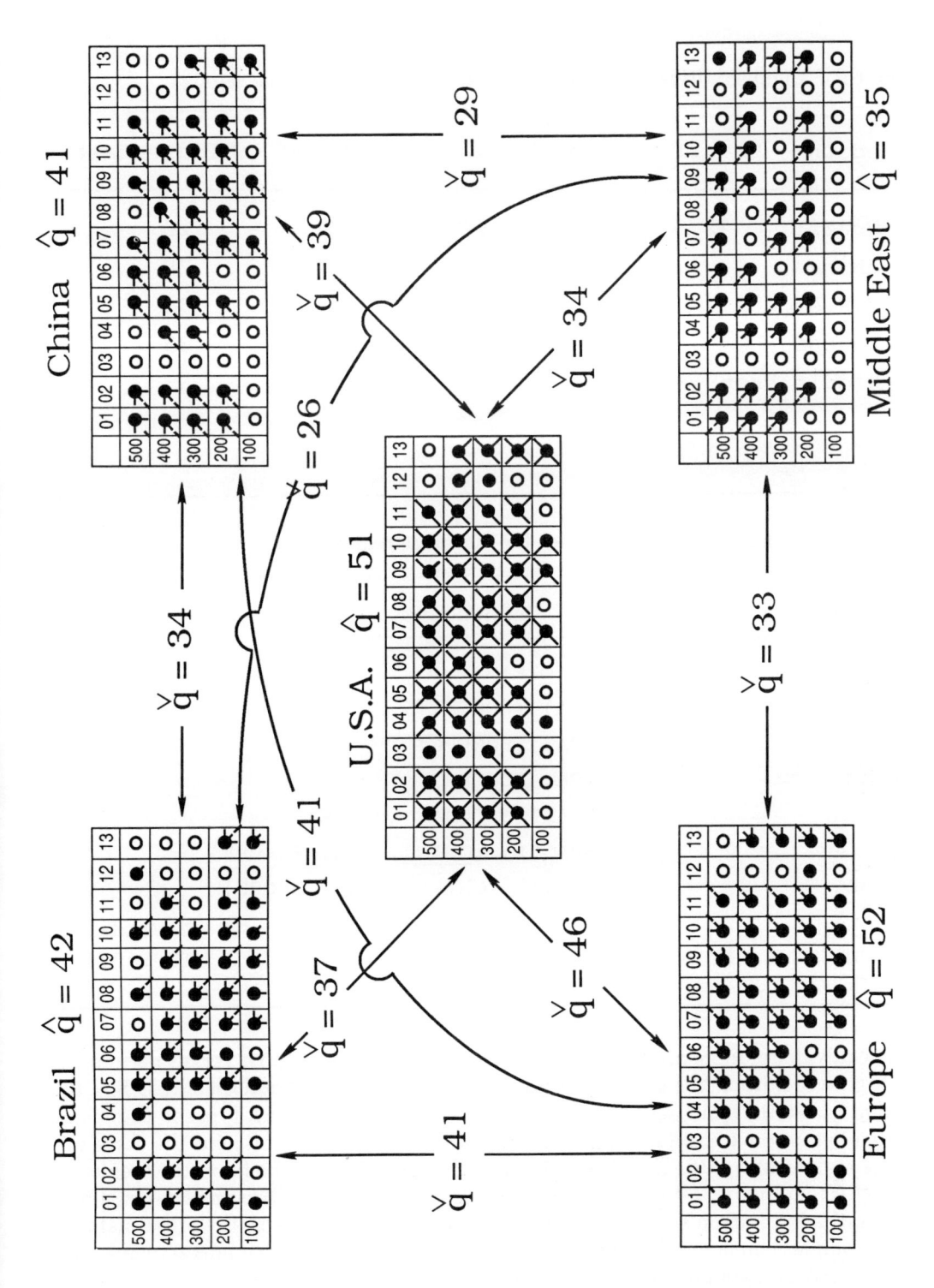

Figure 2. Comparison of simplicies for five large region: Brazil, China, U.S.A., Europe, and Middle East.

Table 2. Connectivities ($\hat{q}$ and $\check{q}$) of diversity (s-1) for five large regions

Region	Brazil	China	U.S.A.	Europe	Middle East
Brazil	42	34	37	41	26
China		41	39	41	29
U.S.A.			51	46	34
Europe				52	33
Middle East					35

Table 3. Q-analysis: equivalencies for five large regions.

$\hat{q}$ value	Q value	Components.
52	1	(EUROPE)
51	2	(USA) (EUROPE)
50-47	2	(USA) (EUROPE)
46-43	1	(USA, EUROPE)
42	2	(BRAZIL) (USA, EUROPE)
41	2	(BRAZIL) (CHINA, USA, EUROPE)
40-36	1	(BRAZIL, CHINA, USA, EUROPE)
35	2	(" " " ") (MID-EAST)
34- 0	1	(" " " " , MID-EAST)

The equivalence table also may be represented as a graph of top $\hat{q}$ versus bottom $\check{q}$ as in Figure 3; Europe has the largest number of rock types (52), followed by U.S.A. (51) and at q = 46 they join along the diagonal (i.e., they have 47 rock types in common). China lies on the diagonal at q = 41. Brazil possesses a q = 42 and joins the three previous regions at q = 40. The Middle East, with a q = 35, joins the four other regions at q = 34. From Figure 2 it may be seen that whereas the Middle East has a q = 34 or 35 rock types in common with the U.S.A., it has less in common with the other members of the single component; it is necessary always to examine the original zero/one data as a matrix or a simplex to clarify the exact relationships between the several regions.

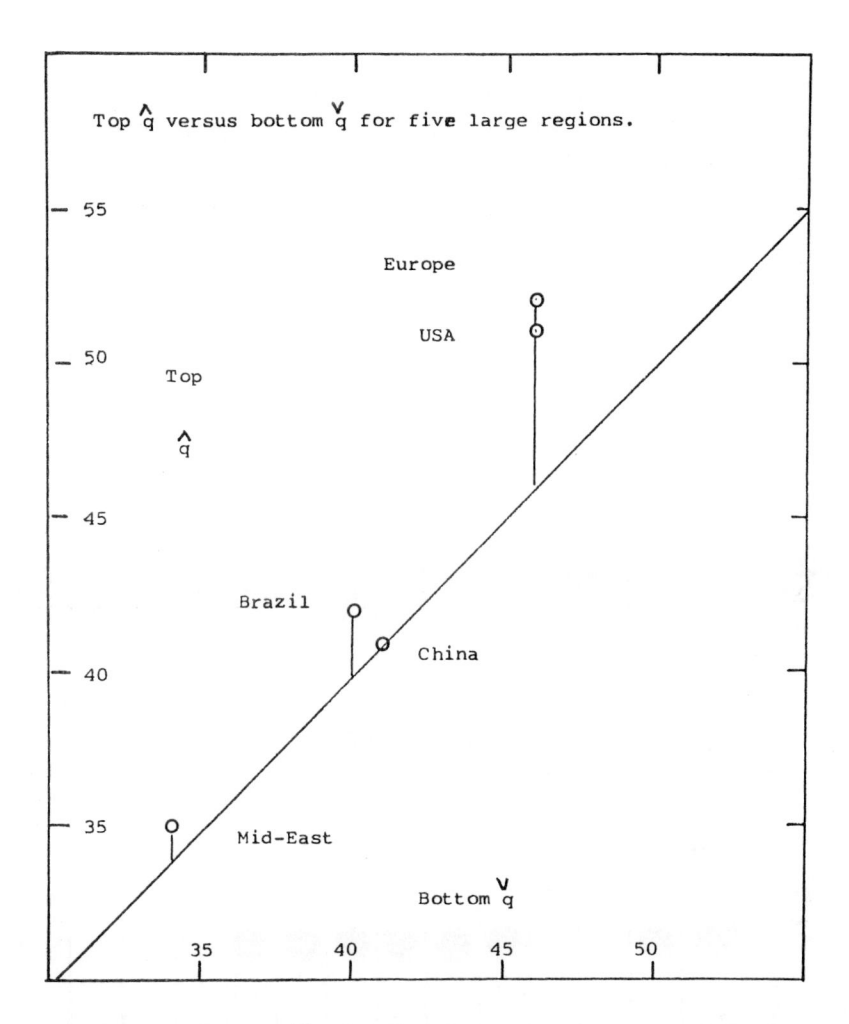

Figure 3. Top $\hat{q}$ versus bottom $\check{q}$ (= eccentricity) for five large regions; Brazil, China, U.S.A., Europe, and Middle East.

Comparison of Brazil and China by state

Brazil has 27 political units, 23 states, and three territories and there are 26 in China, 21 provinces and 5 autonomous regions. The incident zero/one matrix therefore consists of 53 regions by 65 rock types (Table 4). Q-analysis displays their relationships as connectivity matrix $K_c(R:\text{lambda}_{ij})$ in Table 5 and as a table of equivalencies in Table 6. The state of Bahia, Brazil comes first with 27 rock types (q = 26) and several other states from Brazil follow until q = 23 when Zhejang, China occurs in a separate component. The first component to contain two states occurs at q = 20 where Bahia and Minas Gerais possess 21 rock types in common. At q = 18 Hubei enters this component and when q =16, ten states, seven from Brazil and three from China occur in a single component. These arrangements are displayed graphically in the eccentricity plot of top $\hat{q}$ versus bottom $\check{q}$ in Figure 4. The last state to occur from China is Fukien at q = 12 but there are ten Brazilian states with q less than 12; it seems clear that the range in geological diversity for Brazil is much larger than that for China.

Table 4. Time-petrographic tables for (A) Brazil and (B) China showing s = presence and number of rock types.

TRANSDUCER FOR CONVERTING GEOLOGICAL MAP UNITS TO TIME PETROGRAPHIC UNITS

LOCALITY BRAZIL   No. of rock-types = s = 43 ;   27 States

ROCK-TYPES = (TIME) PETROGRAPHIC UNITS

TIME INTERVALS	NO.	SEDIMENTARY						IGNEOUS						META.		MARGINAL TOTAL	PROBABILITY (pi)
		QUARTZITE	ARKOSE	GRAYWACKE LOW RANK	GRAYWACKE HIGH RANK	CARBONATE	EVAPORITE	ACID INTRUSIVE	ACID EXTRUSIVE	MAFIC INTRUSIVE	MAFIC EXTRUSIVE	ULTRAMAFIC INTRUSIVE	ULTRAMAFIC EXTRUSIVE	REGIONAL	OTHERS		
		01	02	03	04	05	06	07	08	09	10	11	12	13	14		
CENOZOIC	5	●	●		●	●	●		●		●		●				
MESOZOIC	4	●	●			●	●	●	●	●	●	●					
PALEOZOIC	3	●	●			●	●	●	●	●	●						
PROTEROZOIC	2	●	●			●	●	●	●	●	●	●		●			
ARCHAEOZOIC	1	●				●		●	●	●	●	●		●			
MARGINAL TOTAL																	
PROBABILITY (pi)																	1.0000

A

Table 4. Continued.

TRANSDUCER FOR CONVERTING GEOLOGICAL MAP UNITS TO TIME PETROGRAPHIC UNITS
LOCALITY C H I N A   No. of rock-types = s = 42 ; 26 provinces .

ROCK-TYPES = (TIME) PETROGRAPHIC UNITS

TIME INTERVALS	NO.	SEDIMENTARY						IGNEOUS						META.		MARGINAL TOTAL	PROBABILITY (pi)
		QUARTZITE	ARKOSE	GRAYWACKE LOW RANK	GRAYWACKE HIGH RANK	CARBONATE	EVAPORITE	ACID INTRUSIVE	ACID EXTRUSIVE	MAFIC INTRUSIVE	MAFIC EXTRUSIVE	ULTRAMAFIC INTRUSIVE	ULTRAMAFIC EXTRUSIVE	REGIONAL	OTHERS		
		01	02	03	04	05	06	07	08	09	10	11	12	13	14		
CENOZOIC	5	●	●			●	●	●		●	●	●					
MESOZOIC	4	●	●		●	●	●	●	●	●	●	●					
PALEOZOIC	3	●	●		●	●	●	●	●	●	●	●		●			
PROTEROZOIC	2	●	●			●		●	●	●	●	●		●			
ARCHAEOZOIC	1							●		●		●		●			
MARGINAL TOTAL																	
PROBABILITY (pi)																	1.0000

B

Table 5. q-connectivities among provinces, KC (R: λ_{ij}): Brazil and China.

Province: No. Name	1	2	3	4	5	6	7	8	9	10	11	12	13	14	15	16	17
1. Acre	4	1	1	3	4	2	1	3	0	3	1	4	4	4	2	3	2
2. Alag.		8	3	5	8	8	3	2	-1	6	3	8	8	7	6	7	6
3. Amap.			7	7	7	5	1	4	-1	7	2	7	5	7	7	7	4
4. Amaz.				17	12	10	4	5	0	14	4	15	12	12	11	9	7
5. Bahi.					26	16	5	8	1	18	5	19	15	20	13	13	9
6. Cear.						20	4	4	-1	15	7	17	13	14	12	10	9
7. D.Fd.							5	2	0	5	1	5	5	5	2	3	1
8. EspSt								8	1	6	0	6	5	8	4	6	3
9. F.d.N.									3	0	-1	0	0	2	-1	0	-1
10. Goias										23	6	18	14	17	12	11	10
11. Maran											7	6	4	5	6	4	4
12. MtGr.												25	19	17	13	12	12
13. MGdS.													20	14	10	11	10
14. MnGr.														24	11	14	8
15. Para															16	9	8
16. Parai																15	6
17. Parn.																	13
18. Pern.																	
19. Piaui																	
20. RdJan																	
21. RGDN.																	
22. RGDS.																	
23. Rondo																	
24. Rora.																	
25. StCat																	
26. SaPl.																	
27. Serg.																	
28. Anhu.																	
29. Fuki.																	
30. Gans.																	
31. Kwant.																	
32. Guanx.																	
33. Guizo.																	
34. Hebei																	
35. Heilj.																	
36. Honan																	
37. Hubei																	
38. Hunna.																	
39. Kiang.																	
40. Jiang.																	
41. Jilin																	
42. Liaon.																	
43. NMngl.																	
44. Ningx.																	
45. Qingh.																	
46. Shaanx.																	
47. Shant.																	
48. Shansi																	
49. Szech.																	
50. Xingj.																	
51. Yunnan																	
52. Zheji.																	
53. Xizang																	

A

Table 5. Continued.

Province: No. Name	18	19	20	21	22	23	24	25	26	27	28	29	30	31	32	33	34	35
1.Acre	3	4	3	2	2	2	2	1	4	2	3	1	2	1	1	1	3	3
2.Alag.	6	8	2	7	5	5	4	4	6	6	6	4	6	4	6	6	6	6
3.Amap.	6	6	3	3	5	6	5	3	5	5	5	2	3	1	3	1	5	5
4.Amaz.	8	12	6	6	8	9	10	6	11	6	9	5	9	5	9	7	11	9
5.Bahi.	13	18	6	9	11	11	8	6	14	12	12	5	10	6	8	9	15	10
6.Cear.	10	16	4	9	12	9	8	7	11	10	13	8	10	7	9	10	14	11
7.D.Fd.	3	5	3	3	2	2	2	1	4	2	3	2	2	1	3	3	4	3
8.EspSt	6	7	5	4	3	4	4	2	5	4	5	1	3	0	1	1	5	4
9.F.d.N.	-1	0	0	-1	-1	-1	-1	-1	1	-1	0	-1	-1	0	-1	-1	0	0
10.Goias	10	16	6	8	11	10	9	10	15	10	14	9	11	8	11	8	15	12
11.Maran	3	6	0	2	5	4	4	4	6	3	6	4	3	4	4	3	5	6
12.MtGr.	11	19	7	9	12	11	10	10	15	8	13	9	14	9	11	12	15	14
13.MGdS.	10	14	7	9	11	9	8	8	13	7	11	8	10	8	10	10	12	12
14.MnGr.	12	17	7	9	10	11	9	6	15	12	14	6	9	6	8	7	15	12
15.Para	7	11	3	6	11	10	10	6	9	7	10	5	7	4	7	6	11	10
16.Parai	12	12	4	8	8	8	7	4	10	9	12	5	7	6	8	6	12	11
17.Parn.	6	10	4	7	7	6	6	10	11	6	8	6	9	6	7	5	8	9
18.Pern.	14	13	5	7	6	6	6	4	9	9	10	5	8	4	6	6	11	9
19.Piaui	17	21	7	10	9	9	9	7	14	9	14	8	12	7	9	10	15	12
20.RdJan			7	4	3	3	4	3	7	3	5	2	4	1	2	2	5	5
21.RGDN.				10	5	5	6	4	7	7	8	4	6	4	6	5	8	7
22.RGDS.					15	10	7	7	8	6	10	7	5	6	8	6	10	10
23.Rondo						12	8	5	8	6	7	4	4	3	5	3	7	7
24.Rora.							12	5	8	5	9	5	6	4	6	3	9	10
25.StCat								11	9	4	7	7	8	6	7	4	7	9
26.SaPl.									18	8	12	6	9	8	8	7	12	12
27.Serg.										12	9	4	5	3	5	4	8	7
28.Anhu.											19	10	9	10	10	7	15	15
29.Fuki.												12	8	9	10	6	9	10
30.Gans.													17	8	11	10	13	11
31.Kwant.														14	10	6	9	9
32.Guanx.															15	9	12	11
33.Guizo.																15	10	8
34.Hebei																	22	15
35.Heilj.																		18
36.Honan																		
37.Hubei																		
38.Hunna.																		
39.Kiang.																		
40.Jiang.																		
41.Jilin																		
42.Liaon.																		
43.NMngl.																		
44.Ningx.																		
45.Qingh.																		
46.Shaanx.																		
47.Shant.																		
48.Shansi																		
49.Szech.																		
50.Xingj.																		
51.Yunnan																		
52.Zhej.																		
53.Xizang																		

B

Table 5. Continued.

Province: No. Name	36	37	38	39	40	41	42	43	44	45	46	47	48	49	50	51	52	53
1.Acre	3	2	2	1	3	3	2	4	4	3	3	3	3	2	3	3	3	3
2.Alag.	8	5	6	7	6	6	6	6	7	7	7	6	6	7	6	4	6	4
3.Amap.	6	6	2	3	2	4	5	4	4	4	5	5	5	4	3	3	4	5
4.Amaz	9	10	6	7	5	8	9	8	9	10	10	11	9	10	8	7	11	7
5.Bahi.	13	15	9	8	10	10	10	11	12	11	13	11	13	10	10	8	11	8
6.Cear.	13	14	10	9	10	12	11	11	11	10	11	10	11	9	9	8	9	8
7.D.Fd.	3	3	3	3	2	3	3	3	4	4	3	3	2	3	4	1	3	1
8.EspSt	5	6	2	2	4	4	4	4	4	4	5	4	4	4	4	3	5	4
9.F.d.N.	0	-1	-1	-1	0	0	0	1	1	1	-1	0	0	-1	0	0	1	0
10.Goias	12	14	10	10	9	12	12	11	10	13	12	12	10	11	9	10	13	9
11.Maran	6	4	5	4	4	4	4	5	5	5	3	5	5	4	4	4	3	5
12.MtGr.	15	18	12	10	11	14	13	14	14	14	15	13	14	13	12	11	15	10
13.MGdS.	12	13	10	10	10	13	12	12	12	13	12	11	11	10	11	9	12	9
14.MnGr.	13	14	9	8	11	11	11	12	11	12	14	12	13	10	8	9	11	10
15.Para	12	10	6	6	7	8	10	9	9	10	9	10	10	7	8	6	7	8
16.Parai	12	10	7	8	9	9	9	9	10	9	9	8	11	9	7	7	9	9
17.Parn.	9	10	7	8	8	9	9	9	7	9	11	9	7	8	7	10	9	8
18.Pern.	10	11	7	7	7	9	9	8	8	6	11	7	9	8	7	6	8	8
19.Piaui	14	15	12	9	12	13	12	13	12	12	14	12	13	12	10	10	12	10
20.RdJan.	4	6	3	3	4	6	5	5	4	5	6	5	4	4	5	4	5	5
21.RGDN.	8	7	6	7	9	7	7	7	7	7	8	7	7	7	5	5	6	5
22.RGDS.	9	9	7	8	7	9	10	7	7	10	7	8	9	6	9	6	7	8
23.Rondo	8	8	5	5	5	6	7	5	5	7	6	6	8	6	5	5	5	6
24.Rora.	10	10	6	6	7	7	9	8	7	8	8	9	9	7	6	6	7	8
25.StCat	7	9	7	8	5	8	8	7	5	9	9	7	5	7	7	8	9	7
26.SaPl.	12	12	9	9	10	11	10	12	11	12	11	11	10	10	9	12	12	11
27.Serg.	8	8	5	6	6	6	6	7	7	5	7	6	6	6	4	5	6	6
28.Anhu.	13	12	11	10	12	14	12	12	11	12	10	12	13	11	9	12	12	13
29.Fuki.	8	8	10	9	6	11	10	9	7	10	8	8	7	8	7	9	9	8
30.Gans.	12	13	10	10	9	12	12	11	10	12	16	9	10	12	11	9	14	7
31.Kwant.	8	7	9	9	7	9	8	8	8	9	7	9	8	8	7	11	11	7
32.Guanx.	10	9	11	12	7	10	10	10	10	12	11	10	8	11	9	9	12	7
33.Guizo.	10	8	9	7	7	9	9	9	11	10	10	7	6	9	10	7	11	4
34.Hebei	15	15	10	11	11	14	14	14	13	14	13	12	13	11	12	10	13	11
35.Heilj.	15	14	11	11	11	14	14	14	13	14	13	12	13	10	12	10	12	13
36.Honan	19	15	11	10	12	13	15	16	15	15	15	11	13	11	12	10	13	11
37.Hubei		22	10	9	10	12	14	14	11	12	16	10	12	10	11	10	13	10
38.Hunna.			14	10	9	11	9	11	10	11	11	8	8	11	8	10	11	8
39.Kiang.				14	8	10	10	9	9	11	11	9	9	11	9	8	12	8
40.Jiang.					15	11	11	11	10	12	10	8	11	9	8	9	10	9
41.Jilin						17	15	13	11	13	13	11	12	10	11	11	12	12
42.Liaon.							19	14	11	15	15	11	12	9	13	9	13	11
43.NMngl.								18	15	15	14	11	11	9	11	11	13	11
44.Ningx.									17	13	12	11	11	10	11	9	13	9
45.Qingh.										22	14	11	11	11	15	12	16	10
46.Shaanx.											20	10	11	12	13	10	14	9
47.Shant.												16	11	9	11	9	11	10
48.Shansi													17	10	9	8	10	11
49.Szech.														15	8	10	14	8
50.Xingj.															19	8	12	8
51.Yunnan																19	14	10
52.Zheji.																	23	9
53.Xizang																		13

C

Table 6. Q-analysis (equivalencies) of 27 states in Brazil and 26 provinces of China.

q - value	Q - value	Components [a]
26	1	(Bahia)
25	2	(Bahia) (MtGr)
24	3	(Bahia) (MtGr) (MnGr)
23	4	(Bahia) (Goias)(MtGr) (MnGr) (Zhej)
22	8	(") (")(") (") ((Hebei) (Hubei) (Qing) (Zhej)
21	9	(") (")(") (") (Piaui)(Hebei) (Hubei) (Qing) (Zhej)
20	11	(Bahia,MnGr) (Ceara)(Goias)(MtGr) (MGdS)(Piaui)(Hebei) (Hubei) (Qing) (Shaanxi) (Zhej)
19	13	(Bahia,MtGr,MGdS,MnGr,Piaui)(Ceara)(Goias)(Anhui)(Hebei)(Honan)(Hubei) (Liaon)(Qing)(Shaanxi)(Xing)(Yunn)(Zhej)
18	14	(Bahia,.....Goias,..Hubei)(Ceara)(SaP1)(Anhui)(Hebei)(Heil)(Honan) (Liaon)(NeiMng)(Qing)(Shaanxi)(Xing)(Yunn)(Zhej)
17	18	(Amaz)(Bahia,..Ceara,..Hubei)(SaP1)(Anhui)(Gans)(Hebei)(Heil)(Honan) (Jilin)(Liaon)(NeiMng)(Ningx)(Qing)(Shaanxi)(Shansi)(Xing)(Yunn)(Zhej)
16	16	(Amaz)(Bahia,..Ceara,..Qing,..Shaanxi,..Zhej)(Para)(SaP1)(Anhui) (Jilin)(Hebei)(Heil)(Honan,NeiMNg)(Liaon)(Ningx)(Qing,Zhej)(Shant) (Shansi)(Xing)(Yunn)
15	10	(Amaz,Bahia,..MtGr,MGdS,MnGr,Piaui,SaP1,..Zhej)(Para)(Paibo)(RGdS) (Guangx)(Guiz)(Jiang)(Shant)(Szech)(Yunn)
14	11	(Amaz,.........,Zhej)(Para)(Pernam)(RGdS)(Kwant)(Guangx)(Guiz) (Hunnan)(Kiangsi)(Jiang)(Shant)
13	9	(Amaz,....Xizang)(Parana)(RGdS)(Kwant)(Guangx)(Guiz)(Hunnan)(Kiangsi) (Jiang)
12	5	(Amaz,...Serg,..Xizang)(Rondo)(Rorai)(Fukien)(Kwant)
11	3	(Amaz,....Xizang)(Rorai)(StCat)
10	1	(Amaz,..RGdN,...Xizang)
9	1	(Amaz,....Xizang)
8	1	(Alaga,Amaz,...EspSt,...Zhej)
7	1	(Alaga,Amap,...Mara,...RGdJ,..Xizang)
6	1	(")
5	1	(Alaga,...DFed,....Xizang)
4	1	(Acre,Alaga,.......Xizang)
3	2	(Acre,.....Xizang)(FdN)
2	1	(Acre,....Xizang)
1	1	(")
0	1	(")

a. Underlining marks the first entry of a state into a simplex.

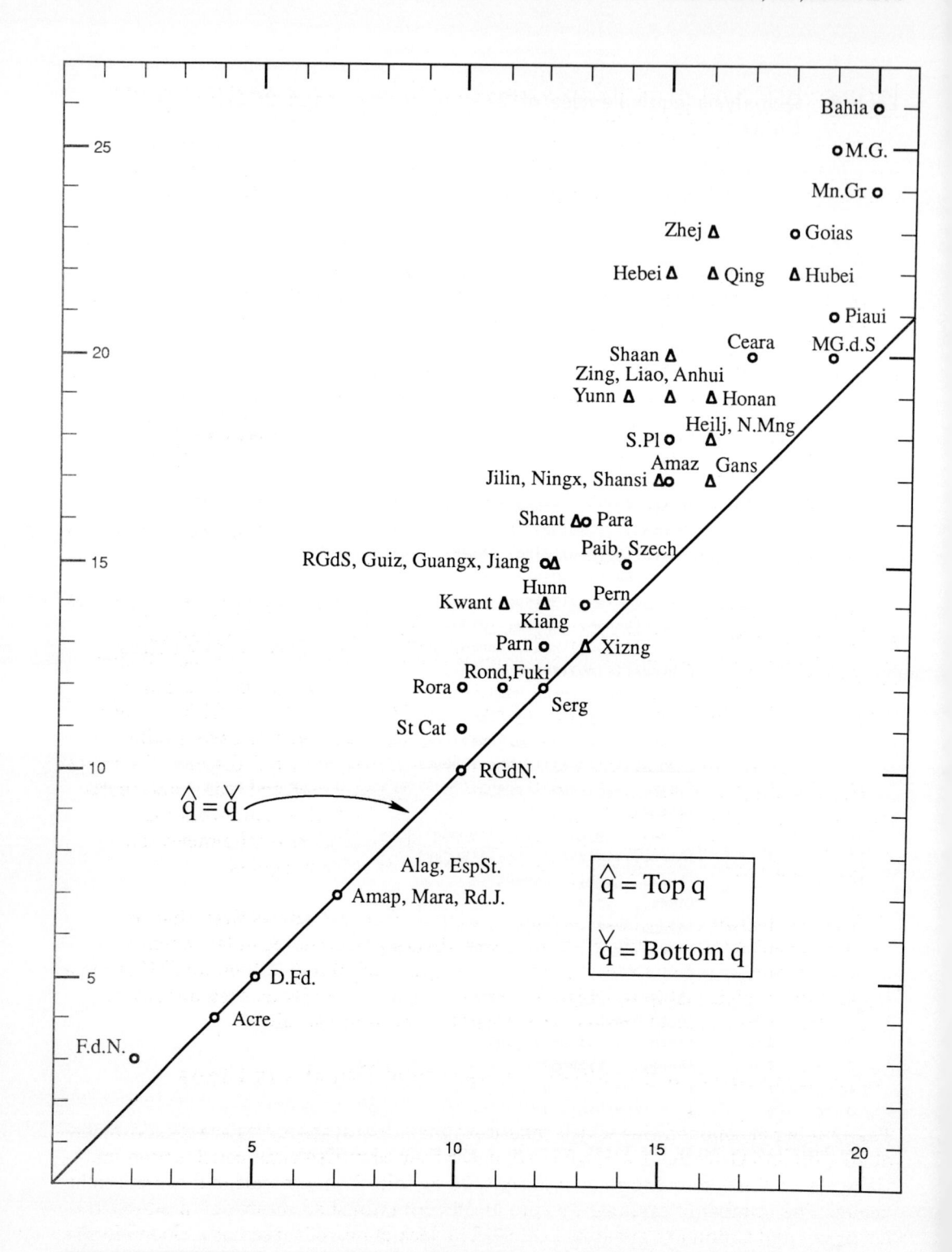

Figure 4. Eccentricity plot of top $\hat{q}$ versus bottom $\check{q}$ for states of Brazil and China.

In order to determine the exact similarities in rock types it is necessary to examine the original zero/one matrix and this may be illustrated by the simplicies for each region; an example of this comparison is given in Figure 5 where three states from Brazil are compared with two from China, all taken from the ten states in a single component at q =16. Each circle represents a rock type and the extension spikes pointing at one or another simplex indicate which rock type is common to a pair of regions (=simplicies). For example, those circles with four spikes are common to all five simplicies (regions); there are seven rock types common to all five regions, Cenozoic arkose (502) and carbonate (505), Mesozoic arkose (402), Paleozoic arkose (302),and Proterozoic carbonate (205), mafic intrusive (209) and regional metamorphic (213). In other words about one-quarter of the rock types are common to all regions. Some 12-16 rock types are common to at least three regions or, in total, some 50-60 percent are common. At the other extreme only one or two rock types are unique to one region (i.e. their circles possess no spikes) except for Zhejang which possesses four rock types not present in any of the other four regions. One may conclude that, on the whole, the rock types in these states are more similar than different and this holds true for most of the high diversity regions in Brazil and China.

Relationship between area and geological diversity

It seems clear that geological diversity is a fundamental characteristic of these regions and a comparison of diversity among the five large regions expressed in terms of their contained states and or countries is exhibited in the box plot (Tukey, 1977) of Figure 6. U.S.A., Europe, and Brazil show similar ranges in diversity whereas the Middle East shows a lesser range and China much less; China is more homogeneous in diversity than the other large regions. Europe is clearly similar to Brazil and both differ from the U.S.A. A similar box plot, Figure 7, displays the range in area (log km.^{2}) of these large regions in terms of their contained states and countries and, again, Europe and Brazil are similar with the U.S.A. somewhat more homogeneous and the Middle East less in terms of their central tendencies. The states of China are more homogeneous than any of the other regions.

A plot of geological diversity (s) versus log area in Figure 8, shows first, that in regions of continental to subcontinental size, diversity fluctuates independently of area. However, in regions ranging from 3.5 to 6 (i.e., 3000 to 1 million km.^{2}) there is a strong linear relationship in which diversity increases steadily as area increases. This relationship seems to hold over most of the countries studied.

Mineral resource and geological diversity of Brazil and China

The number of commodities which constitute the mineral resources are standardized at 75 (Gill and Griffiths, 1983, table 10, p.67); they comprise eight construction materials, eleven fuels, four precious materials, 26 metals, 26 nonmetals and one miscellaneous. The number of commodities produced with published records of production for Brazil and China are given in Table 7. The lack of any of those not included in the list is mainly because of inadequate published records.

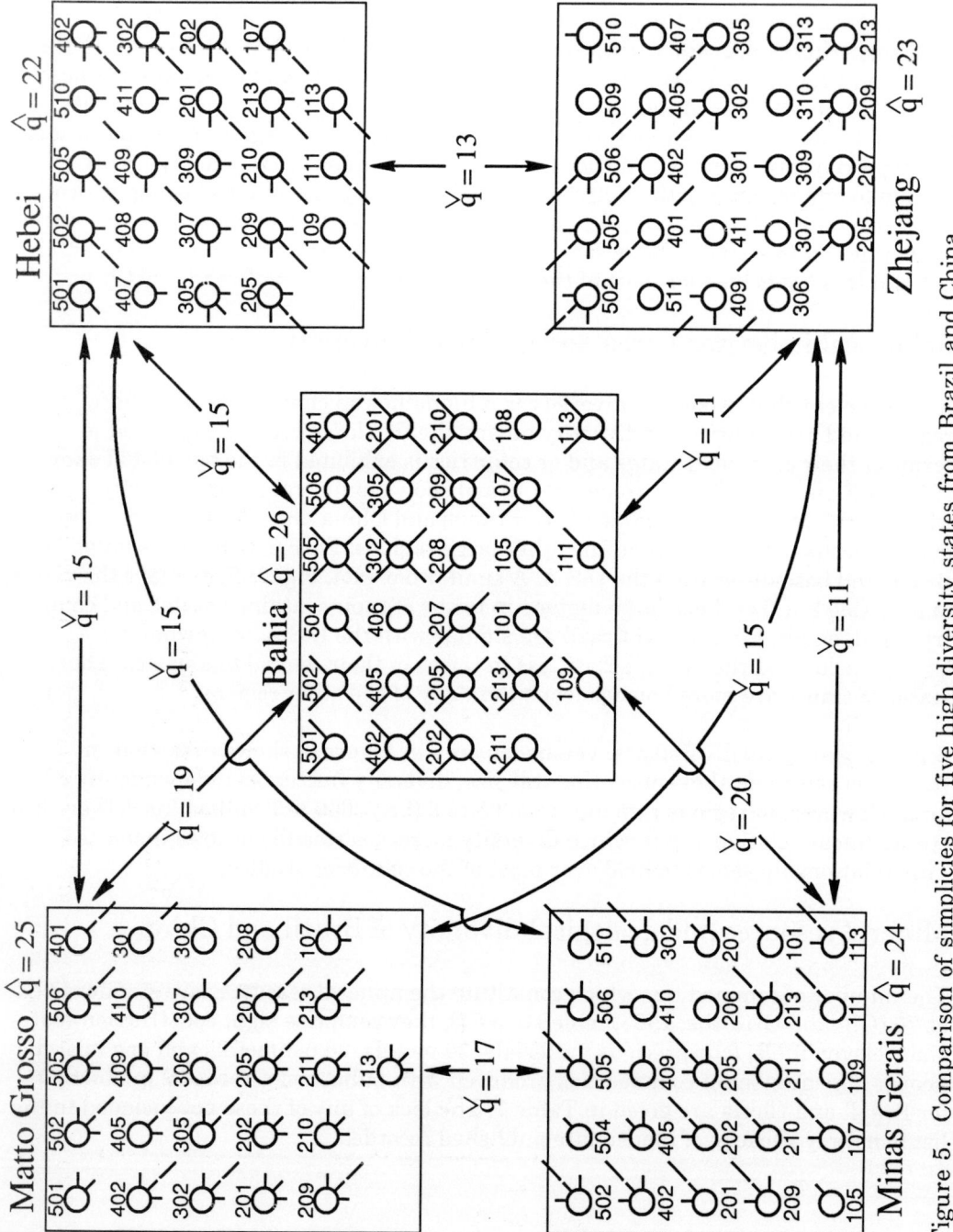

Figure 5. Comparison of simplicies for five high diversity states from Brazil and China.

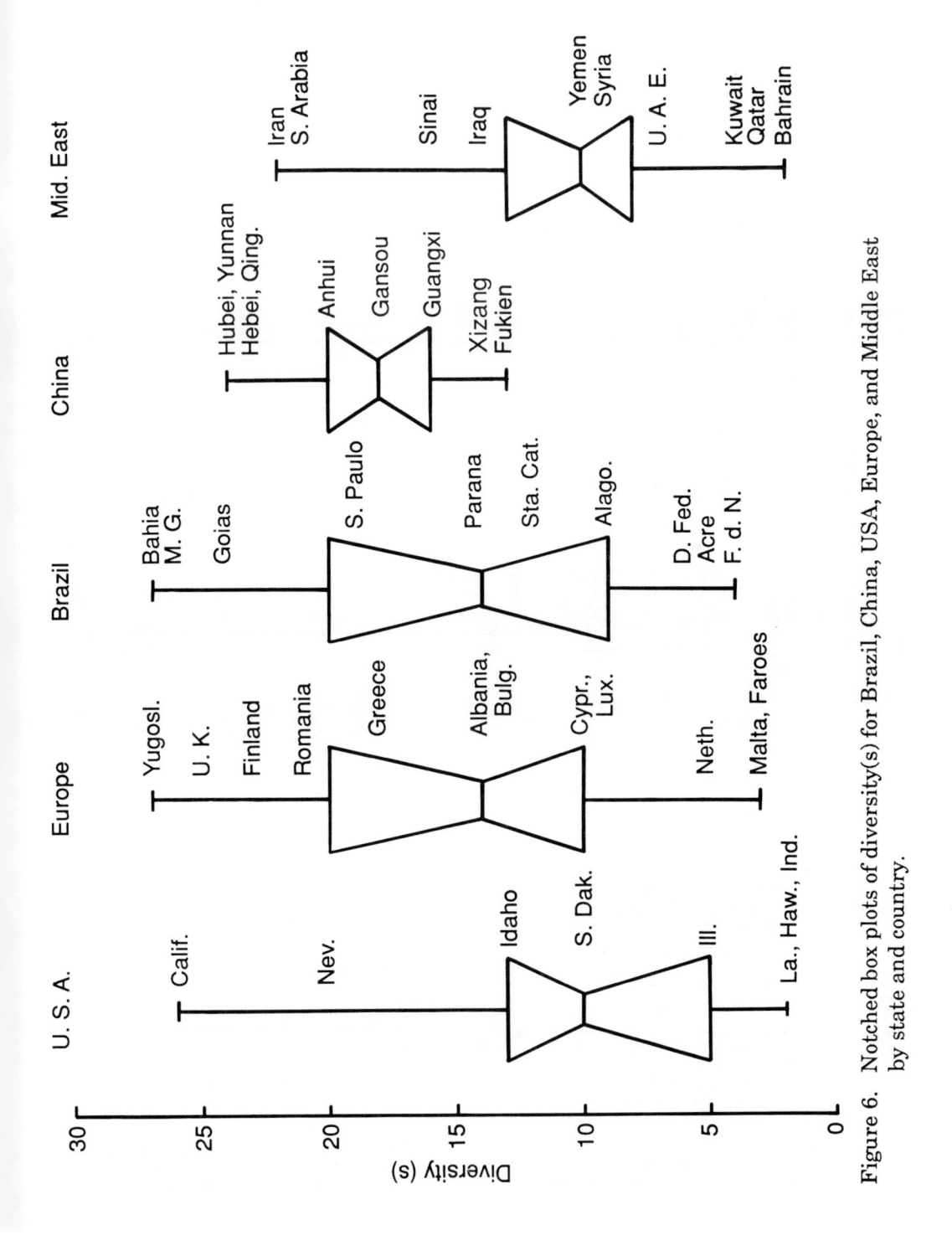

Figure 6. Notched box plots of diversity(s) for Brazil, China, USA, Europe, and Middle East by state and country.

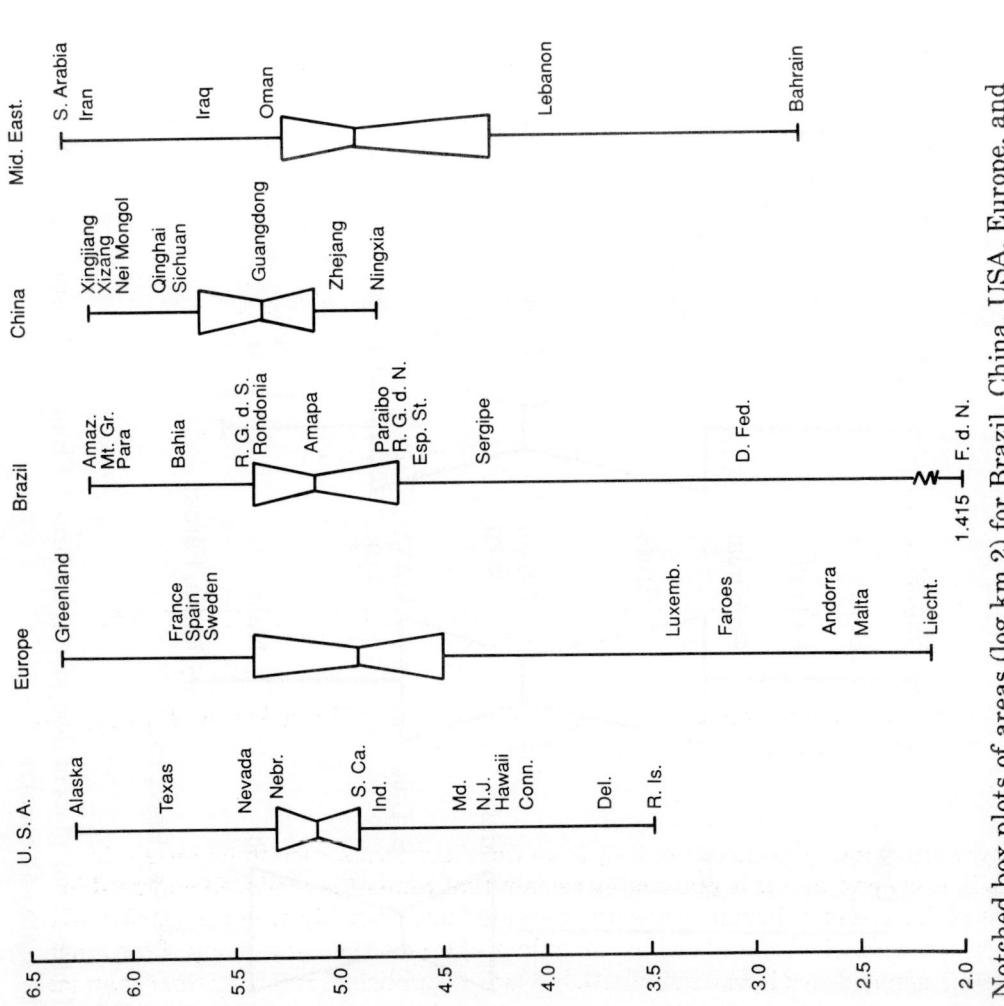

Figure 7. Notched box plots of areas (log km.2) for Brazil, China, USA, Europe, and Middle East by state and country.

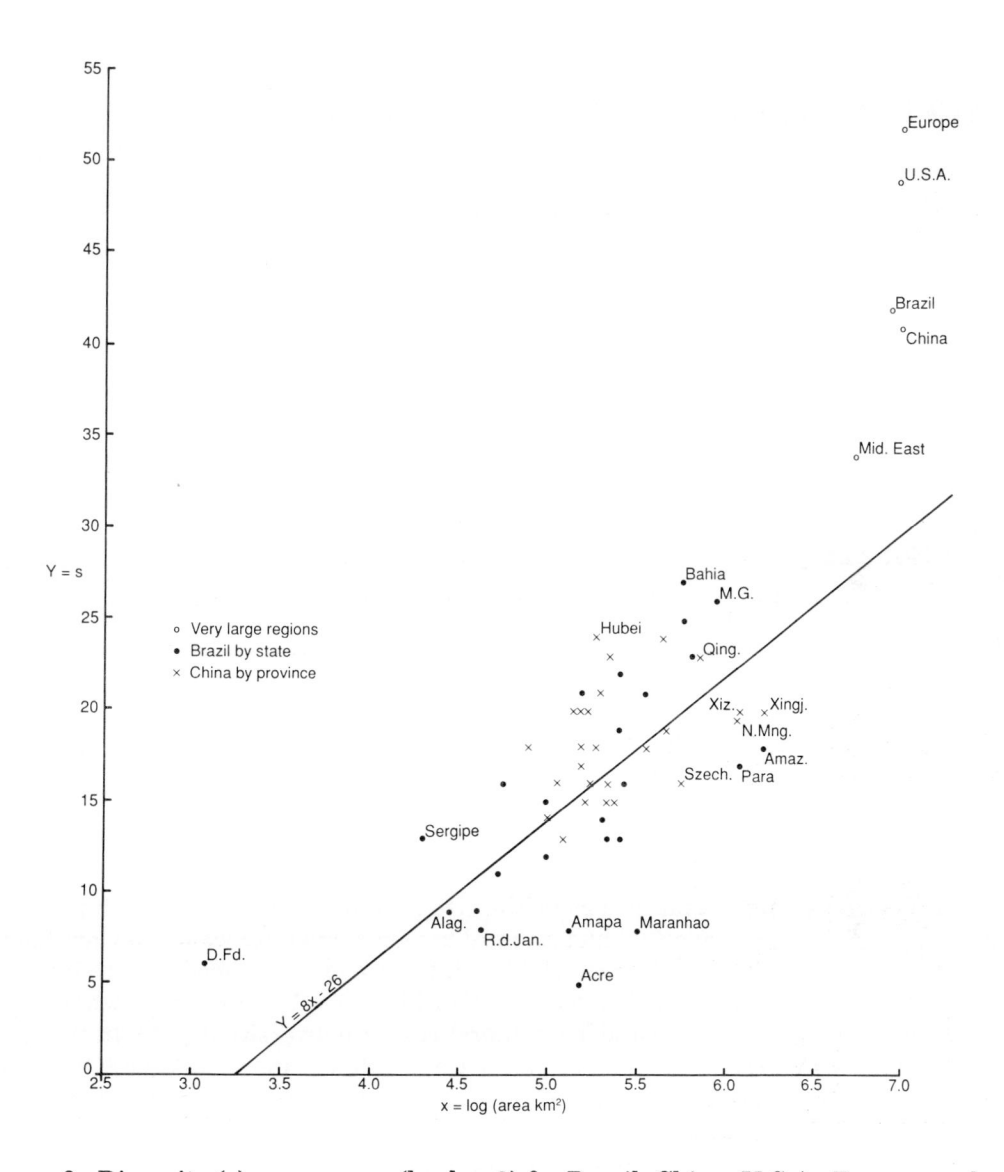

Figure 8. Diversity (s) versus area (log km.2) for Brazil, China, U.S.A., Europe, and the Middle East; Brazil and China by state.

For example, the missing commodity from the sector, construction materials, in Brazil, is cement and it is reasonably certain that cement is produced and used in most of the states in Brazil. Similarly, missing fuels probably are to be attributed to absence of published records rather than lack of production; how many of the other missing commodities is to be attributed to lack of published records rather than lack of production, is not known. Of course, it is clear that the number of commodities produced by such large regions is likely to be a large proportion of the total 75 commodities.

Table 7. Mineral resources produced in Brazil and China.

Commodities	Brazil	China	Total
Construction materials	7	8	8
Fuels	7	9	11
Precious materials	4	4	4
Metals	24	24	26
Non-metals	16	22	26
Total	58	67	75

It has been determined that in regions the size of states there is a simple linear relationship between geological and mineral-resource diversity when they are both expressed in the standardized form of 65 rock types and 75 commodities (Griffiths, Watson, and Menzie, 1980); this linear relationship is $Y = 11.28 + 1.85X$, where X = geological diversity ($q = s_g - 1$) and Y = mineral-resource diversity ($s_m - 1$). In the original article the 95 percent confidence limits are inserted in the regression (Griffiths, Watson, and Menzie, 1980, fig.1, p. 334). We then may calculate the expected value of this number of mineral resources from the observed geological diversity and so estimate how many are likely to be missing.

The geological diversity of the 53 states from Brazil and China is inserted in Figure 9 and the number of commodities expected are included as a regression line together with the previously determined 95 percent confidence limits; the number of commodities produced in each state is indicated by the filled circles for Brazil and triangles for China. No state exceeds its expected value in Figure 9; Kwantung achieves its expected value and three Brazilian states and ten Chinese provinces fall on or within the 95 percent confidence limits of Figure 9. There are some obvious discrepancies; for example, Fernando do Noronha has no record of production for any commodity but because it has a geological diversity ($s_g - 1$) = 4 then it would be expected to produce some 18 commodities. It seems likely that lack of published records accounts for the larger part of this discrepancy.

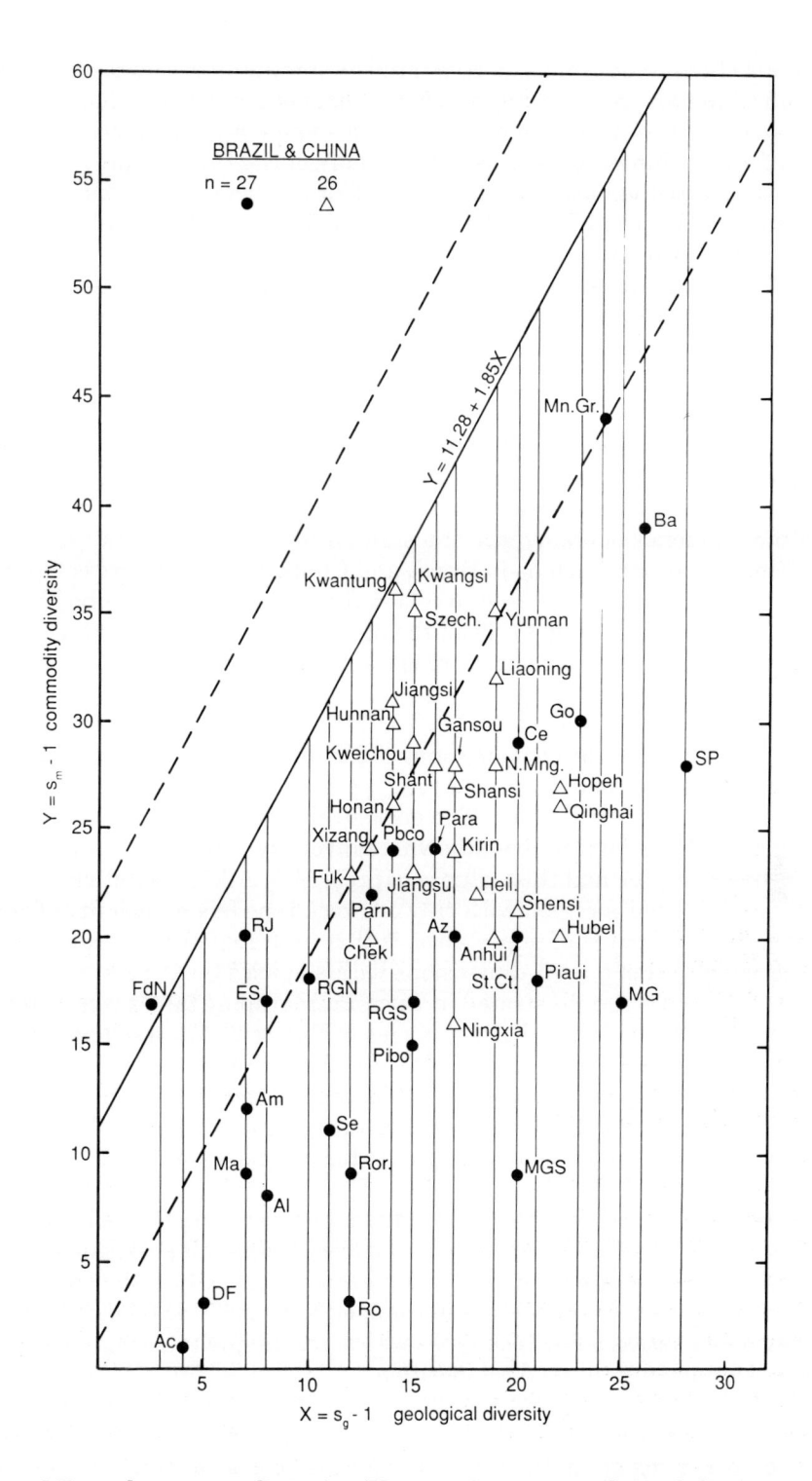

Figure 9. Mineral-resource diversity (Y = sm-1) versus geological diversity (X = sg-1) for states Brazil and China.

The deviations of observed from expected number of mineral resources for each state in Brazil and China are given in Figure 10; the median for the Brazilian states is 19, that is 50 percent of the states are better than this value and 50 percent are worse. The median for the Chinese provinces is 15 so more of the Chinese provinces are closer to their expected values than the states of Brazil. The 'central tendencies',that is the bounds of the boxes in Figure 10 are similar. There are 6 states beyond the notched boxes in Figure 10 for both Brazil and China. It may be concluded that on the whole, Chinese provinces are somewhat better developed than their Brazilian counterparts; this may be because of a longer history of development in China, or a poorer published record of commodities produced in Brazil.

SUMMARY AND CONCLUSIONS

We now have compiled a databank for most of the countries in the world which consists of the frequency of occurrence of a standardized set of 65 rock types in each region; in some situations, such as in Brazil and China, these data are compiled by state or province. The procedure used to collect this databank is described briefly and summarized in a flowsheet (Fig. 1). Then the data, which are presented as the presence-absence of rock types in a region, are classified by using Atkin's Q-analysis (1974).

An illustrative example of the comparison is presented using five large continental sized regions in terms of their contained rock types; the regions are Brazil, China, U.S.A., Europe, and the Middle East. The geological characteristics of each region are summarized as their geological diversity where diversity is defined as number of rock types minus one or (s_g -1), and this value is expressed in Atkin's notation as top q = s_g -1. It then is clear that a region with only one rock type has a diversity of zero; what is not so evident, but is of importance, is that each region can be displayed geometrically as a simplex of q = s-1 dimensions with s apicies (i.e.,s rock types). The standardized rock types are illustrated in Time-Petrographic tables (see Table 4); these five regions are represented as simplicies in Figure 2. Atkin's Q-analysis may be followed from the initial data matrix (Table 1) through the connectivity matrix of Table 2 and, by following the equivalence table, the step by step comparison of each region is exhibited in Table 3. The equivalence table is presented graphically in Figure 3 which is termed an eccentricity plot of top q versus bottom q.

The same analytical procedure then is presented for Brazil and China by state or province; five states, three from Brazil and two from China, all possessing the same high diversities of q = 16, are compared as simplicies (Fig. 5). It is shown that these states possess some 12 to 16 rock types in common or some 50 to 60 percent. Additionally, they only possess a few rock types which are unique to a single state. Therefore, they are similar in geological composition.

An interesting spin-off from these studies is the relationship between geological diversity and area of region; in the five large regions geological diversity is independent of area whereas in regions the size of states diversity increases directly with increase in area.

GEOLOGICAL COMPARISON OF BRAZIL AND CHINA BY STATE

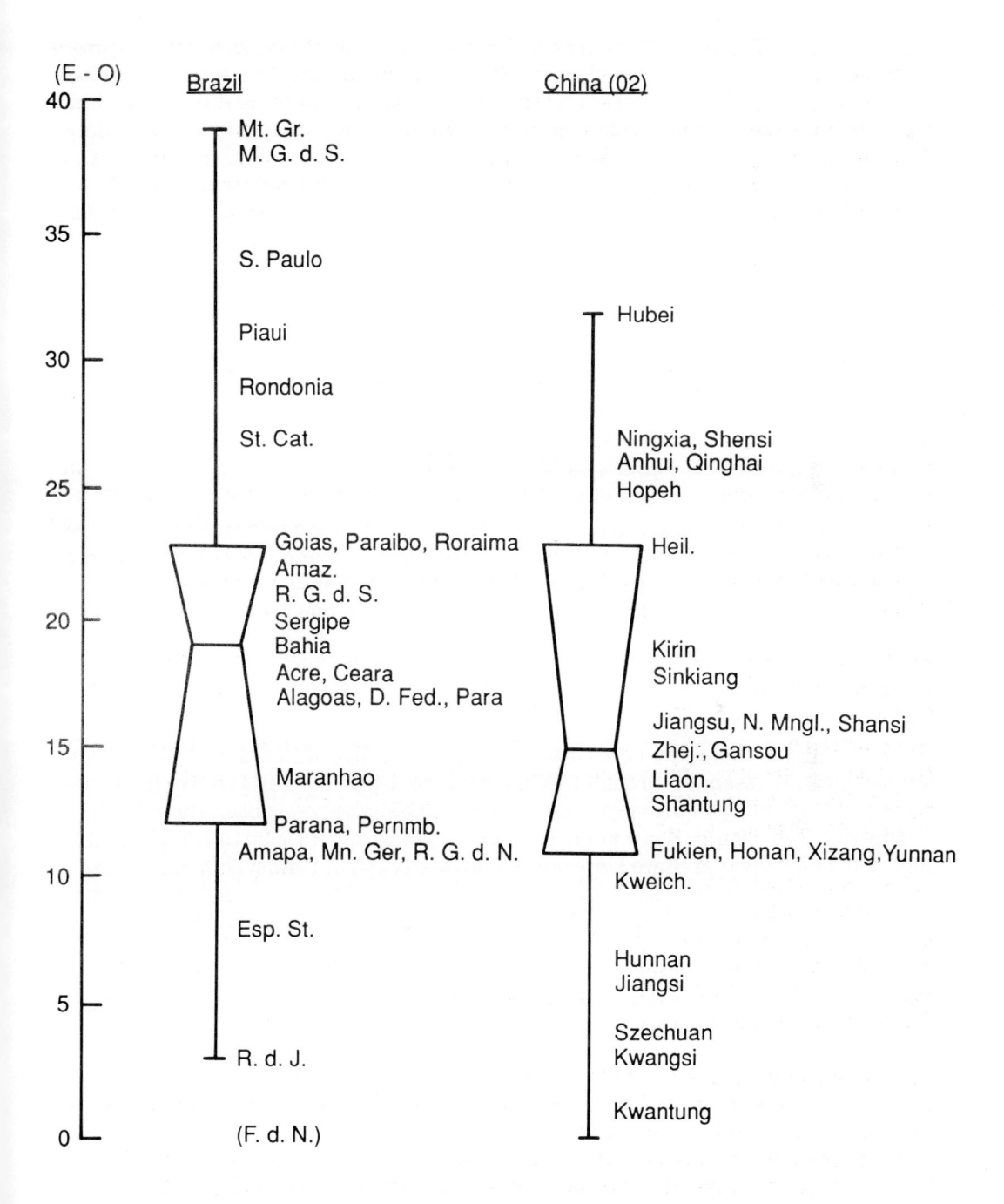

(E - O)

Brazil	China (02)
40 — Mt. Gr. / M. G. d. S.	
35	
S. Paulo	
	Hubei
Piaui	
30	
Rondonia	
St. Cat.	Ningxia, Shensi / Anhui, Qinghai
25	Hopeh
Goias, Paraibo, Roraima / Amaz. / R. G. d. S.	Heil.
20 — Sergipe / Bahia	Kirin
Acre, Ceara / Alagoas, D. Fed., Para	Sinkiang
	Jiangsu, N. Mngl., Shansi / Zhej., Gansou
15 — Maranhao	Liaon. / Shantung
Parana, Pernmb. / Amapa, Mn. Ger, R. G. d. N.	Fukien, Honan, Xizang, Yunnan
10	Kweich.
Esp. St.	Hunnan / Jiangsi
5	Szechuan / Kwangsi
R. d. J.	Kwantung
0 (F. d. N.)	

Figure 10. Comparison of deviations of expected (E) and observed (O) mineral resources for Brazil and China by state.

It then is shown that somewhat similar relationships hold for the diversity of mineral resources (s_m-1) or the number of commodities from a standardized set of 75 produced in a region; by way of a regression relation, previously established, it is shown that only a few states from both Brazil and China achieve their expected value calculated from the regression equation (Fig. 9); by plotting the deviation of expected minus observed commodities (Fig. 10) it is shown that whereas both countries are at a somewhat similar stage of development, China seems to produce more commodities than Brazil.

The reasons why are not clear, however China has a longer historical record of production than Brazil. Then, frequently, the less glamorous commodities (usually construction materials and some nonmetals) do not have as readily accessible published records as those that are internationally traded.

POSTSCRIPT

There may seem to be an inconsistency in the use of Atkin's q notation; we have used his top q = q and bottom q = q; except where they occur in, and refer to, equivalence tables and eccentricity graphs. In these situations top q for one region equals bottom q for another, so they are left plain q. The text usually indicates when this occurs.

ACKNOWLEDGMENTS

We would like to thank the U.S. Geological Survey, Office of Resource Analysis, for support for this research through USGS Grant No. 14-08-0001-01115. We have indicated that the computer program for Atkin's Q-analysis was used through the courtesy of P. R. Gould, Evan Pugh Professor of Geography, Department of Geography, The Pennsylvania State University, University Park, Pennsylvania. We also would like to thank Messrs. H. Robert Ensminger and E. Chin, specialists on Brazil and China respectively, from the Division of International Minerals, U. S. Bureau of Mines, for guidance and advice on the sources for production records. We would like to thank Professor J. Sturdevant, Universidade Federal Fluminense, Rio de Janeiro, Brazil, for arranging for and overseeing the point-count of several sheets of Brazil which were only available as Open File reports in Brazil.

REFERENCES

Atkin, R.H., 1974, Mathematical structure in human affairs: Heinemann Educational Books Ltd., London, 212p.

Buerlen, K., 1970, Geologie von Brasilien: Gebunder Borntraeger, Berlin, 427p.

Clark, A.L., Dorian, J.P., and Fan Pow-foong, 1987, An estimate of the mineral resources of China, Resource Policy, v. 13, p.68-84.

Gill, D., and Griffiths, J.C.,1984, Areal value assessment of the mineral resource endowment of Israel: Jour. Math. Geology, v.16, no. 1, p. 37-89.

Gould, P.R., 1981, A structural language of relations, *in* Craig, R.G., and Labovitz, M.L., eds., Future trends in geomathematics: Pion Ltd., London, p.281-312.

Griffiths, J.C., 1978, Mineral resource assessment using the unit regional value concept: Jour. Math. Geology, v.10, no. 5, p. 441-472.

Griffiths, J.C., 1983, Geological similarity by Q-analysis: Jour.Math. Geology, v. 15, no. 1, p.85-108.

Griffiths, J.C., Watson, A.T., and Menzie, W.D., 1980, Relationship between mineral resource and geological diversity, *in* Miall, A.D., ed., Facts and principles of world petroleum occurrence: Can. Soc., Petrol. Geol., Mem. 5, p.329-341.

Hurlbut, C.S., 1966, Dana's manual of mineralogy: John Wiley & Sons, Inc., New York, 609p.

Tukey, J.W., 1977, Exploratory data analysis: Addison-Wesley Publ.Co., Reading, Massachusettes, 506p.

U.S. Bureau of Mines, 1963-84, Mineral Yearbooks, Vol.IV, Area Reports Intl., Dept. Interior, Washington, D.C.

Wilkinson, T., 1986, WATFILE/PLUS data manipulation system: Watcom Publ. Ltd., Waterloo, Ontario, Canada, 353p.

Application of Q-Analysis to the GLOBAL Databank: A Geological Comparison of the U.S.S.R. and the U.S.A.

D. N. Pilant, J. C. Griffiths, and
Pennsylvania State University
C. M. Smith, Jr.
Pennsylvania State University

ABSTRACT

The geology of regions of the U.S.S.R. and U.S.A. are characterized in terms of the presence and absence of 65 standardized rock types obtained from point-counts of geologic maps. Regions are compared and geological similarity estimated in terms of the number of shared rock types using Atkin's Q-analysis. Assuming that a region's geology and mineral resource endowment are related through their common geotectonic evolution, comparison of geologically similar regions may provide guidelines relevant to exploration for undiscovered mineral resources. The point-counts residing in the GLOBAL databank constitute an inventory of the rock types present at the Earth's surface, providing data necessary for comprehensive examination of distributions and associations of rock types on a global scale. In this paper we demonstrate an application of the GLOBAL databank in a geological comparison of regions of the U.S.S.R. and U.S.A.. Tables and maps are presented showing U.S.S.R. regions in terms of their geological diversities and in terms of their geological similarities with states of the U.S.A.

INTRODUCTION

The GLOBAL computer databank consists of point-counts of geologic maps for nearly every subareal region on Earth. It has been compiled over the last several years by Griffiths and his students at the Pennsylvania State University in connection with studies utilizing the unit regional value approach to mineral resource assessment (Griffiths, Watson, and Menzie, 1980; Griffiths, 1978; Griffiths, 1983). The GLOBAL databank represents a systematic digital compilation of certain lithological aspects of the geological information residing in the worlds inventory of geologic maps. The underlying objective is to quantify the concept of geological similarity in terms of standard rock types so that it may be applied at a gross level in mineral-resource appraisal.

The fundamental data are the binary presence or absence at grid points of 65 possible standard rock types. These '0/1' data are not well suited to analysis with traditional parametric statistical methods. For this reason, an algebraic procedure known as Q-analysis has been adopted from Atkins (1974; see also Griffiths, Hu, and Chou, this volume; and Griffiths, 1983). Q-analysis provides a method of examining the connectivities among regions and rock types and permits these (complex) relationships to be expressed algebraically and geometrically.

The objective here is to give an overview of the features of the GLOBAL databank, and to present some preliminary results of a geological comparison of regions of the U.S.S.R. and U.S.A.

DATA ACQUISITION AND TRANSFORMATIONS

Figure 1 outlines the basic components of data acquisition and analysis. The databank resides in the university's IBM 3090 mainframe computer.

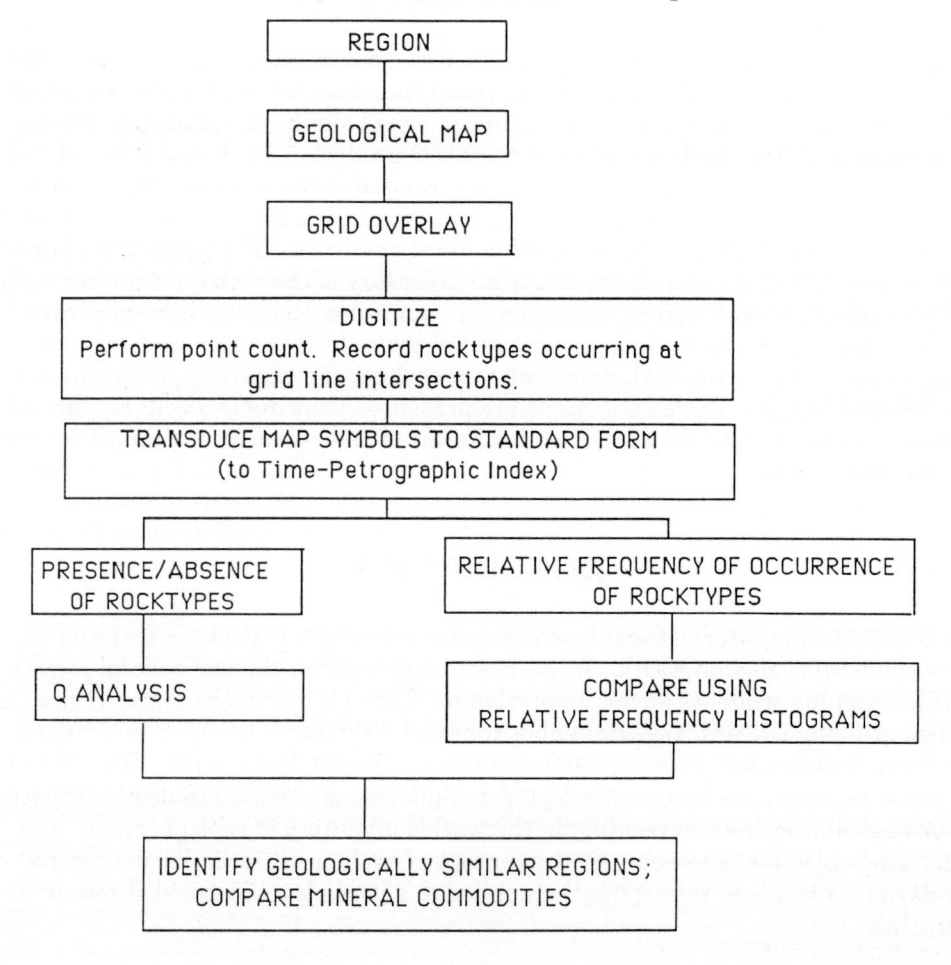

Figure 1. Flow diagram showing basic components of data acquisition and analysis.

Geologic maps depicting bedrock lithology are obtained from the regions of interest. They range in scale from 1:100,000 to 1:2,500,000; scales of 1:250,000 to 1:1,000,000 are most widely used. A digital point-count is performed using a mylar grid laid over the geologic map; the coordinates and map symbols from the legend occurring at each grid line intersection are recorded in a computer data file. The grid spacing of 0.83 cm (0.25 in) or 1.27 cm (0.5 in) yields ground-sampling intervals ranging from 6 to 15 km depending upon map scale. The map units from the map explanation and point-count then are transduced to standard form (time-petrographic index, TPI), based on map explanation and literature descriptions, following the criteria outlined next. The transducer for converting map units to TPI rock types (Table 1) may be represented as a matrix consisting of 5 time intervals along the time dimension and 13 rock types (or sequences) along the petrographic dimension, generating a space of 65 possible rock types. (A fourteenth column, 'OTHER', is designated for special applications or extremely unusual rock types but it has not been necessary to utilize it in these studies.) A summary table such as Table 1 (Kazakhstan) is prepared for each region, with numbers in the cells indicating presence or absence or, if desired, relative frequency of occurrence of each rock type in the region of interest.

The igneous rocks are classified using the Rosenbusch-Zirkel classification (Hurlbut, 1966). The regional metamorphic class includes contact metamorphic rocks; contact metamorphic rocks rarely occur as individual map units in the range of scales used here. The chemical sedimentary rocks are represented by the carbonate and evaporite classes. The detrital sedimentary rock types (rock sequences) follow the classification proposed by Krynine (1948, 1951) and later modified by Griffiths (1968, 1978). The detrital rock types are transduced following two main criteria: tectonic background and composition. Each map unit must be considered in the context of its host-rock sequence to ensure accurate classification. More detailed explication of transduction and the time-petrographic index is presented in Griffiths (1968, 1978), Pilant (1988), and Clark, Dorian, and Fan (1987).

Several Soviet geologic maps were point-counted to compile the U.S.S.R. data set (Pilant, 1988). In general, the Russian republic was divided into regions by west-center-east subdivision of each of 16 sheets comprising the Geologic Map of the U.S.S.R. (Churinov, 1968). The other republics were subdivided similarly, or in accordance with the portion of each republic occurring on a given sheet of their respective maps.

The mechanics of the Q-analysis algorithm are outlined in Figure 2 and the theory described in detail by Atkin (1974) and Griffiths (1983). The input matrices consist of A, A^T (A transpose), and U. Matrix A consists of n regions (rows) by 65 TPI rock types (columns) and is filled by zeroes and ones indicating presence and absence of rock types. This matrix is generated from files such as Table 1. The analysis also may incorporate relative frequencies of occurrence by using slicing parameters. For instance, thresholds may be defined where if the relative frequency of a rock type (see Table 1) is greater than or equal to a some value, it is entered as 1, and if less than that value, it is entered as 0. U is a n by n identity matrix subtracted from matrix products to yield results in terms of diversity. The output matrices of the operations in Figure 2 yield two variables: top $\hat{q}$ (diagonal elements) and bottom $\check{q}$

Table 1. Time-petrograpic index with data for Kazakhstan.

ROCK TYPES = (TIME) PETROGRAPHIC UNITS

TIME INTERVALS	NO.	SEDIMENTARY						IGNEOUS						META.		MARGINAL TOTAL	PROBABILITY (pi)
		QUARTZITE 01	ARKOSE 02	GRAYWACKE LOW RANK 03	GRAYWACKE HIGH RANK 04	CARBONATE 05	EVAPORITE 06	ACID INTRUSIVE 07	ACID EXTRUSIVE 08	MAFIC INTRUSIVE 09	MAFIC EXTRUSIVE 10	ULTRAMAFIC INTRUSIVE 11	ULTRAMAFIC EXTRUSIVE 12	REGIONAL 13	OTHERS 14		
CENOZOIC	5	.045	.442			.072		.0002	.0001								.559
MESOZOIC	4	.009	.041			.047		.003	.0001	.001	.0001						.101
PALEOZOIC	3		.094		.056	.029	.002	.056	.055	.004	.018	.002		.006			.316
PROTEROZOIC	2		.002		.010	.00004		.002	.001		.004	.00007		.002			.022
ARCHAEOZOIC	1																.002
MARGINAL TOTAL		.054	.579		.066	.148	.002	.061	.056	.005	.018	.002		.008		28,721 pts.	
PROBABILITY (pi)																	1.0000

(off-diagonal elements). Top $\hat{q}$ (diversity) equals the number of rock types minus one present in a region (or, in the second operation, the number of regions minus one hosting a rock type). Bottom $\check{q}$ (connectivity) equals the number of rock types in common minus one between two regions (or, alternatively, the number of regions minus one hosting two given rock types). Bottom $\check{q}$ is used to estimate similarity between regions. Figures 3-6 and Tables 2-5 were generated from synthesis of the Q-algorithm output matrices.

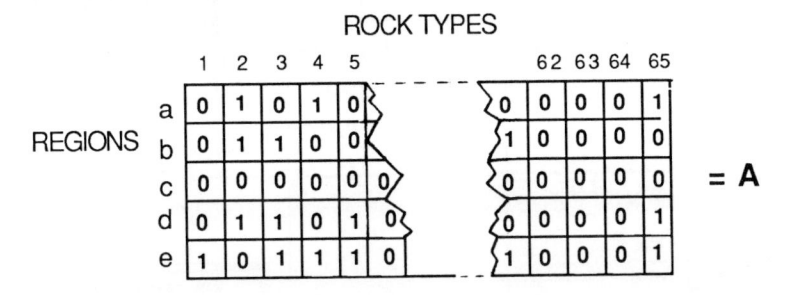

ROCK TYPES

	1	2	3	4	5		62	63	64	65	
a	0	1	0	1	0		0	0	0	1	
b	0	1	1	0	0		1	0	0	0	
c	0	0	0	0	0		0	0	0	0	= A
d	0	1	1	0	1		0	0	0	1	
e	1	0	1	1	0		1	0	0	1	

REGIONS (rows a–e)

Data matrix = A = n regions (rows) by 65 rocktypes (columns)

Unit matrix = U = n rows by n columns (filled with ones)

Q analysis algorithm

1. Connectivity pattern among regions:

$$(A \times A^T) - U = K_c(C:\lambda_{ij})$$

nx65   65xn   nxn   nxn

2. Connectivity pattern among rocktypes:

$$(A^T \times A) - U = K_R^{-1}(C:\lambda_{ij}^{-1})$$

65xn   nx65   65x65   65x65

(where n = number of regions and c = number of rocktypes)

Figure 2. Outline of Q-analysis algorithm.

DISCUSSION

The maps in Figures 3 and 4 display the geological diversities of selected regions of the U.S.S.R.. Geological diversity (top $\hat{q}$) is defined as the number of rock types minus one present in a region. Using mineral production statistics from the U.S.A., New Zealand, Zimbabwe, and South Africa, Griffiths, Watson, and Menzie (1980) formulated the following expression relating mineral commodity diversity to geological diversity:

mineral commodity diversity = 11.28 + 1.85*(geological diversity) r^2 = 80%.

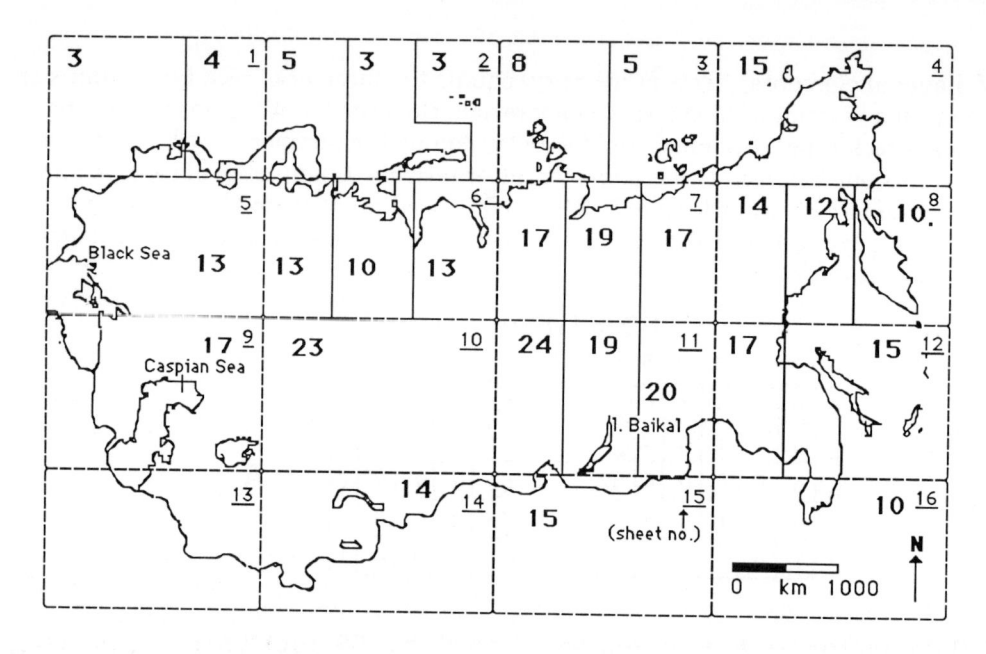

Figure 3. Map showing U.S.S.R. territory and geological diversities (top $\hat{q}$) of regions comprising Russian republic.

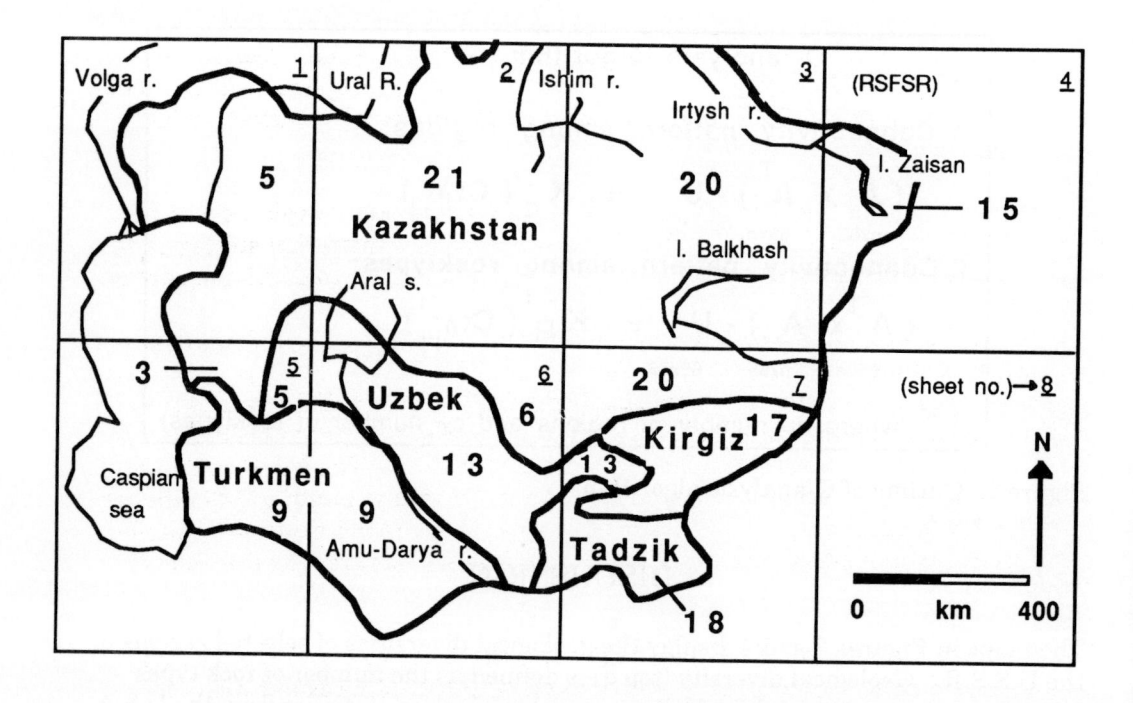

Figure 4. Map showing geological diversities (top $\hat{q}$) of regions comprising republics of Soviet Central Asia.

Mineral commodities are expressed in terms of a standardized list of 75 commodities comprising the following groups: construction materials, fuels, precious materials, metals, and nonmetals (see Griffiths, 1978). As the geological diversity of a region increases, so does the expected mineral commodity diversity. One objective of near-term study is to delimit the global range of geological diversities of large regions.

Tables 2 and 3 and Figures 5 and 6 depict regions of the U.S.S.R. in terms of geologically similar states of the U.S.A.. Similarity of regions may be recognized intuitively through subjective comparison of selected attributes such as landforms, lithology, or tectonic background, but the objectification of this subjective process is a complex problem. The similarity criterion used in this analysis is simple: selection of maximum bottom $\check{q}$ (connectivity) between regions while seeking equitable top $\hat{q}$ (diversity) between subject (U.S.S.R.) and similar (U.S.A.) regions. There is a tendency for regions with high geological diversities (large top $\hat{q}$) to dominate as geologically similar regions (e.g., California, Nevada, Washington, etc.). This effect can be accommodated by seeking equitable values of topq of subject and similar regions. For instance, in Table 2 Kaz2 (the region on sheet two of the Kazakhstan map) hosts 21 rock types (top $\hat{q}$ = 20) and shares 13 in common with California (top $\hat{q}$ = 25). North Carolina (top $\hat{q}$ =14) shares 11 rock types with Kaz2. California hosts (26-13)=13 rock types not present in Kaz2; North Carolina hosts (15-11)=4 rock types not present in Kaz2. With fewer extraneous rock types (4 vrs. 13), North Carolina may provide a more suitable model for a commodity comparison with Kaz2.

An interesting match occurs in Table 3. Russia 9 hosts all 8 of the 8 rock types present in Arkansas. In theory, then, the mineral endowment of Arkansas could be expected to have a relatively high correlation with the appropriate subset of the mineral endowment of Russia 9.

Most of the regions of the U.S.S.R. and U.S.A. are relatively well explored and well developed. They provide potential models for formulating exploration strategies when matched with geologically similar, underexplored regions in developing areas.

Equivalence Tables 4 and 5 examine communality and uniqueness. Table 4 displays connectivities among regions in terms of rock types, and Table 5 displays connectivities among rock types in terms of regions. Referring to Table 4, California is the most geologically diverse region, entering the table at q=25 (26 rock types). It occupies a unique position until it is joined by Russia 11 at q=23. The fact that California and Russia 11 exist in separate components (parenthetic units) until q=13 (data not shown) indicates that despite their high diversities, they are not necessarily similar. In this pair, the number of rock types not shared is almost as great as the number shared.

California and Nevada (top $\hat{q}$=18) enter the same component at q=17 (18 rock types in common). This is a quantitative confirmation of the intuitive expectation that adjacent regions tend to be geologically similar. Russia 11W and adjacent Russia 10 exhibit a similar structure in that they join the same component when Russia 10 enters the table at q=16. As we progress down the table, each q-level has

Table 2. Regions of Russian republic and geologically similar states of U.S.A. (decreasing similarity left to right). (7) = top $\hat{q}$ = number of rock types minus one present in region. 4 = bottom $\hat{q}$ = number of rock types minus one in common between two regions.

Russia 1	(4)	Massachusetts (10) 3; Illinois, Iowa (4) 1; Alabama (12) 3; New Jersey, New Mexico (10) 2
Russia 2	(8)	New Jersey (10) 4; Idaho, Alabama (12) 4; Montana (18) 5; Nevada, Washington (19) 5
Russia 3	(10)	New York (13) 6; Arkansas (7) 4; Oklahoma, New Jersey (10) 5; Virginia (13) 5
Russia 4	(14)	Nevada (19) 8; New Jersey (10) 6; Montana, Utah (18) 7; Colorado (11) 5
Russia 5	(12)	Alabama (12) 7; New Jersey, New Mexico (10) 6; Idaho (12) 5; Montana, Utah (18) 7
Russia 6	(20)	Utah (18) 9; New Jersey (10) 8; Alabama (12) 8; Alaska (18) 8; New York (13) 7;
Russia 7	(28)	California (25) 13; N. Carolina (14) 11; Montana (18) 11; Alabama (12) 10
Russia 8	(19)	Alaska (17) 13; Nevada, Washington (19) 13; California (25) 14; Alabama, Idaho (12) 7
Russia 9	(16)	Arkansas (7) 7; Alabama (12) 9; New Jersey (10) 8; Virgina (13) 7
Russia 10	(22)	Utah (18) 11; California (25) 11; Alabama (12) 10; Georgia (11) 9
Russia 11	(29)	Alaska (17) 12; California (25) 16; Massachusetts (10) 9; Colorado, Georgia (11) 9
Russia 12	(21)	Washington (19) 13; California (25) 14; Utah (18) 11; Nevada (19) 12; Alaska (17) 10
Russia 14	(13)	Washington (19) 10; Maine (11) 7; Vermont (12) 7; California (25) 8
Russia 15	(14)	Washington (19) 11; California (25) 10; Alabama (12) 8; Massachusetts (10) 7
Russia 16	(9)	Washington (19) 7; Nevada (19) 6; Oregon (10) 4

Table 3. U.S.S.R. republics and geologically similar states of U.S.A. (see Table 2 for explanation).

Armenia	(7)	Idaho (12) 4; Texas (9) 3
Azerbaijan	(17)	Nevada (19) 13; Washington (19) 11; Montana (18) 10
Byelorussia	(0)	Louisiana (0) 1
Estonia	(1)	Louisiana (0) 1
Georgia	(10)	New Jersey, New Mexico (10) 4; Washington (19) 6
Kazakhstan	(29)	California (25) 17; Nevada (19) 14; Washington (19) 11
Kaz 1	(4)	Mississippi (5) 3; N. Dakota (6) 3
Kaz 2	(20)	California (25) 12; N. Carolina (14) 10; Utah (18) 9
Kaz 3	(19)	California (25) 12; Arizona (18) 10; Maine (11) 8
Kaz 4	(14)	Vermont (12) 9; North Carolina (14) 8; Maine (11) 7
Kaz 5	(2)	Mississippi (5) 2; North Dakota (6) 2; Delaware, Florida (2) 1
Kaz 6	(6)	Massachusetts (10) 4; Colorado (11) 4; New Hampshire (5) 3; North Dakota (6) 3
Kaz 7	(19)	California (25) 13; Montana (18) 9; Nevada (10) 8; Vermont (12) 9
Kirgiz	(16)	Vermont (12) 8; New York (13) 8; Colorado (11) 7; Alaska (17) 5
Latvia	(4)	Wisconsin (5) 1; New York (13) 3; Oklahoma (10) 2; Nebraska (4) 0
Lithuania	(3)	Wisconsin (5) 1; Tennessee (8) 1; Michigan (10) 2; Minnesota (12) 2;
Moldavia	(5)	Kentucky (4) 2; Georgia (11) 3; Alabama (12) 3; Mississippi (5) 1
Tadzik	(18)	Alaska (17) 10; Nevada,Washington (19) 10; Vermont (12) 10
Turkmen	(8)	N. Dakota (6) 5; Texas (9) 5; Vermont (18) 7
Ukraine	(13)	Arkansas (7) 6; Utah (18) 8; New Jersey (10) 6
Uzbek	(17)	Georgia (11) 8; California (25) 11; Alabama, Vermont (12) 7

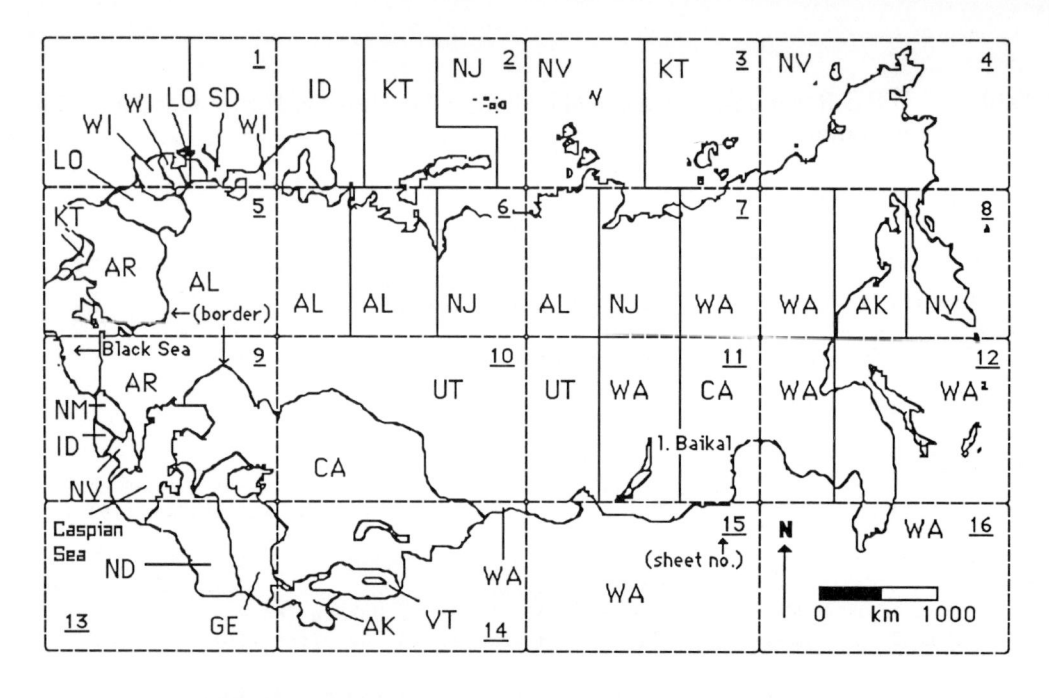

Figure 5. Map showing U.S.S.R. regions in terms of geologically similar states of U.S.A.

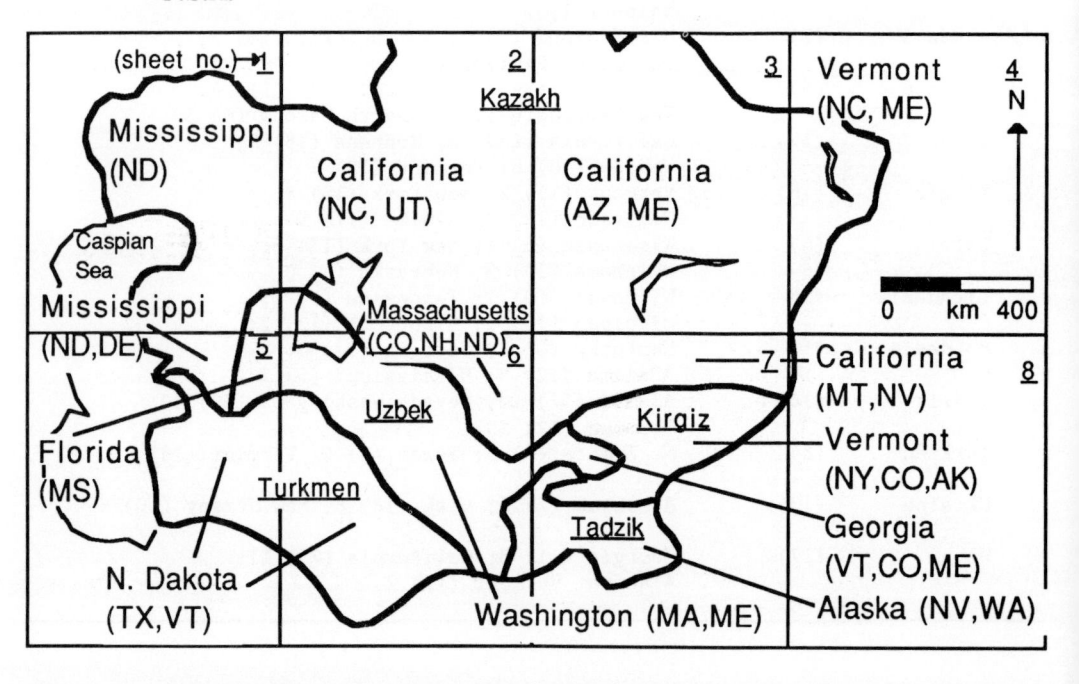

Figure 6. Map showing regions of Soviet Central Asia in terms of geologically similar states of U.S.A. Most similar regions listed first, followed by other similar regions in parentheses.

fewer components as more regions combine into the same components. This is an expression of the fact that the regions seems more similar with relaxation of similarity criteria (i.e., with decreasing q at lower levels). Combination of regions into components constitutes a type of clustering. At q=18 there are 14 components defined by 14 regions; each region is represented as a unique component at that level. At the level where q=14 there are 27 regions but only 13 components. Individual regions have combined ('clustered') with geologically similar regions. Regions (such as Kirgiz) that persist alone in their components from higher to lower q-levels are more unique; they enter components with other regions only at lower q-levels (relaxed similarity criteria).

Table 4. Equivalence table showing relationships among more geologically diverse regions of U.S.S.R. and U.S.A. in terms of rock types. q represents top $\hat{q}$ when region enters table and bottom $\check{q}$ when referring to connectivity between Qq = number of components at level of q. Underlining indicates first entry of region into table.

```
Value of
---------
q     Qq              Components

25-24  1   (CA)
23     2   (CA)(Russia11w)
22-21  3   (CA)(Rus11w)(Rus10)
20     4   (CA)(Rus11w)(Rus10)(Kazakh2)
19     9   (CA)(NV)(WA)(Rus11w)(Rus10)(Rus11e)(Kaz2)(Kaz3)(Kaz7)
18     14  (AZ)(MT)(UT)(CA)(NV)(WA)(Rus7c)(Rus11w)(Rus11c)(Rus11e)(Rus10)
           (Kaz2)(Kaz3)(Kaz7)
17     16  (CA,NV)(AK)(AZ)(MT)(UT)(WA)(Rus7c)(Rus11w)(Rus11c)(Rus11e)
           (Rus10)(Kaz2)(Kaz3)(Kaz7)(Azerbaijan77)(Tadzik)
16     19  (CA,NV)(AK)(AZ)(MT)(UT)(WA)(Rus10,Rus11w)(Rus7c)(Rus11c)
           (Rus7w)(Rus7e)(Rus11e)(Rus9)(Rus12w)
           (Azer77)(Kirgiz)(Kaz2)(Kaz3,Kaz7)(Tadzik)
15     15  (AZ,CA,NV,WA)(AK)(MT)(UT)(Rus10,Rus11w,Rus11e)(Rus7c)(Rus11c)
           (Rus7w)(Rus7e)(Rus9)(Rus12w)
           (Azer77)(Kirgiz)(Kaz2,Kaz3,Kaz7)(Tadzik)
14     13  (Azer77,AZ,CA,NV,WA)(AK)(MT)(NC)(UT)
           (Rus4)(Rus7w,Rus7c,Rus7e,Rus9,Rus10,Rus11w,Rus11c,Rus11e,Rus12w)
           (Rus12e)(Rus15)(Kirgiz)(Kaz4)(Kaz2,Kaz3,Kaz7)(Tadzik)
```

Table 5. Equivalence table showing relationships among more frequently occurring rock types in U.S.S.R. and U.S.A. in terms of number of regions in common between pairs of rock types. q represents top q̂ when rock type enters table and bottom q̌ when referring to connectivity between rock types within component. Qq = number of components at level of q. Underlining indicates first entry of rock type into table.

Value of q	Qq	Components
73-55	2	(<u>Cen.Arkose</u>)(<u>Pal.Carb</u>)
54-53	1	(Cen.Ark,Pal.Carb)
52-49	2	(Cen.Ark,Pal.Carb)(<u>Mes.Ark</u>)
48	3	(Cen.Ark,Pal.Carb)(<u>Mes.Ark</u>)(Pal.Ark)
47-45	4	(Cen.Ark,Pal.Carb)(Mes.Ark)(<u>Pal.Ark</u>)(<u>Pro.Reg.Meta.</u>)
44-43	2	(Cen.Ark,Mes.Ark,Pal.Ark,Pal.Carb)(<u>Pro.RM</u>)
42	3	(Cen.Ark,Mes.Ark,Pal.Ark,Pal.Carb)(<u>Pal.Quartzite</u>)(Pro.RM)
41-40	2	(Cen.Ark,Mes.Ark,Pal.Ark,Pal.Carb,Pro.RM)(Pal.Qzt)
39	1	(Cen.Ark...Pal.Qzt...Pro.RM)
38-37	2	(Cen.Ark...Pro.RM)(<u>Mes.Carb</u>)
36	3	(Cen.Ark...Pro.RM)(<u>Mes.Carb</u>)(Pal.LRG)
35	4	(Cen.Ark...Pro.RM)(Mes.Carb)(<u>Pal.LRG</u>)(Pal.HRG)
34	5	(Cen.Ark...Pro.RM)(Cen.LRG)(Mes.Carb)(<u>Pal.LRG</u>)(Pal.HRG)
33	4	(Cen.Ark...Pal.LRG...Pro.RM)(Cen.LRG)(Mes.Carb)(Pal.HRG)
32-31	4	(Cen.Ark...Mes.Carb...Pro.RM)(Cen.LRG)(Pal.HRG)(Pal.RM)
30-29	4	(Cen.Ark...Pal.HRG...Pro.RM)(Cen.LRG)(Pal.RM)(<u>Pro.Acid In</u>)
28	7	(Cen.Ark...Pro.RM)(Cen.LRG)(<u>Mes.Mafic Ex</u>)(<u>Pal.Acid In</u>)(Pal.RM)(Pro.Acid Int)(<u>Arc.RM</u>)
27	6	(Cen.Ark...Pro.Ac.In...Pro.RM)(Cen.LRG)(Mes.Mafic Ex)(Pal.Ac.In)(Pal.RM)(Arc.RM)
26	7	(Cen.Ark...Cen.LRG...Pro.RM)(<u>Cen.Maf.Ex</u>)(<u>Mes.Ac.In</u>)(Mes.Mafic Ex)(Pal.Ac.In)(Pal.RM)(Arc.RM)

Another feature of Table 4 is the geographic clustering into components of the western U.S.A. states (AZ,...,WA), regions in the Russian republic (Russia 7W,...,Russia 12 W) and regions in Kazakhstan (Kaz2,...,Kaz7). This tendency also is reflected in Figures 5 and 6. In the U.S.A. the western states, by virtue of their Phanerozoic tectonic histories, form a set of regions distinct from the states of the U.S. continental interior. A similar pattern occurs in the U.S.S.R. where regions of recent orogenic development (i.e., the Uralo-Mongolian orogenic zone) contrast with regions located within the more quiescent platforms and shields. The western states dominate the similar regions maps in Figures 5 and 6.

Equivalence Table 5 parallels Table 4 but with the interchange of rock types and regions. This table provides a method of displaying connectivities among rock types in terms of regions. Cenozoic arkose and Paleozoic carbonate are the most abundant rock types by a significant margin, occurring in 73 of the 89 regions. At q=54 they enter the same component; 55 regions host both rock types. Overall, the dominant rock types are Phanerozoic arkoses, carbonates, and quartzites. This reflects the predominance in the data set of geotectonic environments consisting of mature continental crust. Preliminary analysis using the databank indicates that these rock types dominate on a global level. Igneous rocks come in at q=30; approximately one third of the regions host Proterozoic acid intrusives. However, igneous rock types are present in high-rank graywacke sequences (entered at q=35), so, by implication, igneous rocks occur in at least 36 regions. Cenozoic arkose, Mesozoic arkose, Paleozoic arkose, and Paleozoic carbonate form a strong association (component) occurring together in approximately 50% of the study regions. This is predictable in that the detrital sequences tend to persist through long intervals of time, particularly the arkose sequence, and carbonates usually are associated with arkoses. Persistence of separate components to lower q-levels in Table 5 suggests that those regions associated with the rock types of a given component may be dissimilar from regions associated with other components at the same q-level. The 17 most abundant rock types in the data set are listed in descending order in Table 5, recognized by following the underlined entries from top to bottom.

SUMMARY

The GLOBAL databank and the approach demonstrated in this paper provide:

- a comprehensive inventory of the rock types existing at the surface of the Earth;
- a way of making objective comparisons of regions in terms of geological factors;
- a framework in which to make mineral commodity comparisons based upon geological similarity;
- a method of quantifying relationships and associations among rock types.

ACKNOWLEDGMENTS

Research involving the GLOBAL databank is funded by United States Geological Survey Grant No. 14-08-0001-G1115.

REFERENCES

Atkin, R.H., 1974, Mathematical structure in human affairs: Heinemann Educational Books, Ltd., London, 212 p.

Churinov, M.B., ed., 1968, The engineering and geologic map of the U.S.S.R: Moscow, V.S.E.G.E.I., 16 sheets, 1:2,500,000.

Clark, A.L., Dorian, J.P., and Fan., P., 1987, An estimate of the mineral-resources of China: Resources Policy, v. 13, p.68-84.

Griffiths, J.C., 1968, Geological data for classification: Western Miner, v. 41, p.37-42.

Griffiths, J.C., 1978, Mineral resource assessment using the unit regional value concept: Jour. Math. Geology, v. 10, no. 5, p.441-472.

Griffiths, J.C., 1983, Geological similarity by Q analysis: Jour. Math. Geology, v. 15, no. 1, p.85-108.

Griffiths, J.C., Watson, A.T., and Menzie, W.D., 1980, Relationship between mineral resource and geological diversity, in Miall, A.D., ed., Facts and principles of world petroleum occurrence, Can. Soc. Petrol. Geol.Mem. 5, p. 329-341.

Hurlbut, C. S., 1966, Dana's manual of mineralogy: John Wiley & Sons, Inc., New York, 609p.

Krynine, P.D., 1948, The megascopic study and field classification of sedimentary rocks: Jour. Geology, v. 56, no. 2, p.130-165.

Krynine, P.D., 1951, A critique of geotectonic elements: Am. Geophys. Union, Trans. v. 32, no. 5, p.743-748.

Pilant, D.N., 1988, Regional geology of the U.S.S.R. and comparisons with other regions in the GLOBAL databank: unpubl. masters thesis, Pennsylvania State Univ., in progress.

Explorational Databases at the Geological Survey of Finland

B. Saltikoff and T. Tarvainen
Geological Survey of Finland, Espoo

ABSTRACT

The Exploration Department of the Geological Survey of Finland develops and maintains explorational databases both for internal use and as nation-wide reference databanks. Three national databanks on ecomonic materials in Finland are in operation, namely the Ore Deposit Databank, the Industrial Mineral Databank, and the Mineral Indication Databank. Some databases on explorational GSF materials, as the drilling material database and the thin-section database, are under construction. Close cooperation is kept up with other reference databanks at GSF, such as the Report Reference Database, the Project Register, and the Finnish Geological Bibliographical Database. The three databanks on economic minerals are created in cooperation with all exploration and geological organizations in Finland. Data are open for public use. Therefore the databanks are realized as relational databases with multiple access, and easy-to-use menu driven data retrieval programs are provided for ordinary users. These databanks are used regularly for reference in practical exploration by GSF and mining companies, as well as in regional planning. They also are used in geological research and mineral-assessment applications.

Job-defined databases are under construction. These will be small, flexible databases for individual exploration projects. They have a common part of geographic, etc., fields plus special fields designed according to needs of each project. Background data are extracted from reference databanks, and field data are added in course of the project. All databases by the Exploration Department are implemented using the Digital Rdb/YMS Relational Database Management System and standardized co-ordinate and abbreviation systems. Our explorational database system is being built step by step, with an integrated database system as the ultimate goal.

BACKGROUND: HISTORY AND BASIC PRINCIPLES

Construction of databases within the Exploration Department of GSF has started with nation-wide databanks on mineral data.

The Geological Survey of Finland is a major geological research and exploration center in the country. Because of its close cooperation with universities and, on the other side, mining companies and other exploration organizations, GSF traditionally has been regarded as a general information center for geological activities. Thus it was more than natural that national explorational databanks were concentrated within GSF.

The Ore Deposit Databank was initiated in 1973 by a special project for study of ore geology in Northern Finland (Gaál and others, 1977a, 1977b). Its goal was set as to provide detailed data on ore deposits and their surroundings for strategic planning of

```
GENERAL DATA
        Name, location, ore type, right owner

CHEMICAL CHARACTERISTICS
        For each element assayed:
        arithmetic and geometric mean, deviations, distribution
        law, no. of assays

ORE GEOLOGY
        Ore minerals, gangue minerals
        Host rocks, wall rocks, alterations, textures

DIMENSIONS
        Geometry: size, orientation
        Reserves: tonnage, cut-off

PETROPHYSICAL PROPERTIES

REFERENCES
        Investigations carried out
        Literature references

CASE HISTORY
        Discovery and mining history
        Indicative anomalies

GEOLOGY OF SURROUNDINGS
        2x2 km cell: rock types, age, metamorphism, orientations,
        lineaments
        10x10 km cell: rock types, rocks related to mineralization,
        tectonic structures, lineaments

GEOPHYSICS OF SURROUNDINGS
        For 2x2 km, 10x10 km and 20x20 km cells:
        aeromagnetic field statistics (average, deviation, no. of
        peaks, range, gradient, trends)
        gravimetric field statistics
```

Figure 1. Data structure of Ore Deposit Databank (after Gaál, Autio, and Lehtonen, 1977b).

prospecting and for ore geological studies, particularly for regional mineral-resource assessment. The data collection was designed to be as comprehensive as possible, with a scope to serve many various fields of interest (Fig. 1). As the number of objects of study was not too high, some derivative data were included (such as statistical summaries of assay data) which require initial calculation prior to inputting.

The scheme for the Ore Deposit Databank was applicable with minor changes to deposits of industrial minerals. The Industrial Mineral Databank was built up by a project for assessment of industrial minerals and rocks in Finland in 1977 to 1979 (Söderholm, Sotka, and Eloranta, 1980).

A different approach was followed in creation of the Mineral Indication Databank. It goes back to the traditional mode of prospecting in Finland, which includes: intensive cooperation with amateurs for initial clues; tracing groups of glacial ore-bearing boulders for locating of probable source areas; and detailed investigations on these. Such prospecting always leaves behind a number of unconnected, poorly investigated mineral indications, which need to be registered for possible use in future and for avoiding of duplicate studies. On initiative by all major exploration organizations in Finland, a national databank on mineral indications was set up in 1978 and realized at GSF (Saltikoff, 1979; Saltikoff, 1984).

The volume of the databanks is as follows: Ore Deposit Databank ca. 300 items (deposits), on average 900 data fields in each; Industrial Mineral Databank - ca. 500 items, 500 data fields in each; Mineral Indication Databank - ca. 9500 items (indications), 40 data fields in each.

These databanks provided the basis for a territorial information system on geological data useful in exploration. More such data are contained in geological maps, published and unpublished reports and geological collections. Computerizing of this information is in progress. In fact, a database on the geological bibliography for Finland and one on unpublished (but publicly available) reports by GSF are built up at the Information Bureau of GSF (Tiainen and Aumo, 1988). Within the Exploration Department, under construction is a databank on drilling results and a databank on thin section data, both of which are recalled for exploration.

GEOLOGICAL CONTENT AND DATA DESCRIPTION LANGUAGE

Always when constructing a geological databank a problem arises in computerizing of nonnumerical, descriptive geological concepts. It requires standardization of definitions and abbreviations used and familiarizing of the users with the code system.

The nomenclature and abbreviations system used in the GSF databases is determined by the requirements of geological public. All terms and codes are self-explained. Verbal codes are preferred to numerical ones.

Mineralogical descriptors (characteristic mineral names) are given as mnemonic abbreviations. Because of the limited number of various mineral species and the large number of descriptors in a list (4 to 6 per rock), an authorized list of short and pithy abbreviations was developed for about 1000 most abundant minerals. This was implemented within a glossary of mineral names (Saltikoff, 1976).

On the contrary, petrological description (host and wall rock specification) is solved on a blank sheet principle. This is because the rock nomenclature used in geology is not stabilized, not least because of a large number of geological schools and various interpretations of the genesis of metamorphic rocks. Therefore, no lists of 'legal' rock names were introduced, but rather a grammar for uniform spelling of any formally valid rock name (Fig. 2). Deciphering of rock names was left to the user of the databanks.

```
Abbreviation rules for rock names
        (Adapted from Finnish)

Space reserved for rock name:            [              ]

(1) Short name - unchanged               [GRANITE       ]

(2) Long name - truncated                [GABBRO-PERIDOTIT]

(3) Mineral name attirbutes - use
        abbreviations                    [GARN-BIOT GNEISS]

(4) Adjective attributes - truncated to
        5 letters                        [PORPH GRANODIORI]
```

Figure 2. Abbreviation grammar used in explorational databanks.

Experience on the Mineral Indication Databank however showed a need in formalized rock classification and grouping. The problem was solved by introduction of a numerical multidimensional rock index, which describes the rock in terms of genesis, chemical composition, metamorphic grade and superimposed alterations (Fig. 3). It allows retrieval of objects with related host rock types.

A link from point-like deposits to regional geological features is facilitated in the Ore Deposit Databank by the concept of surrounding cell of area characterized by percentages of various rock groups, structural geological characteristics (orientations of foliation and lineation, lineaments, etc.) and statistical parameters of geophysical fields (see Fig. 1).

The philosophy on the informative nature of economic parameters of mineralizations can be summed up as follows. Firstly, we know that major mineralizations are well-studied and provided with a lot of numerical investigation data, whereas mineral

```
Indexes describing a rock type:

1. - GENESIS:              1 = sedimentary rocks
                           2 = volcanogenic rocks
                           3 = dyke and vein rocks
                           4 = plutonic rocks
                           5 = metasomatic rocks

2. - COMPOSITION:          1 = felsic
                           2 = intermediate
                           3 = mafic
                           4 = ultramafic
                           6 = carbonate rocks
                           7 = heavy metal rocks

3., 4. - DETAILS, varying from group to group

5. - METAMORPHIC GRADE:    1 = unmetamorphosed
                           2 = lower greenschist facies
                           3 = greenschist facies in general
                           4 = upper greenschist facies
                           5 = lower amphibolite facies
                           6 = amphibolite facies in general
                           7 = upper amphibolite facies
                           8 = granulite facies

6. - SUPERIMPOSED ALTERATIONS:
                           1 = silification
                           2 = albitisation
                           3 = skarn formation
                           4 = tourmalinization
                           5 = scapolitization
                           9 = mechanical alterations.

For example, MICA GNEISS (*intermediate* rock of *sedimentary* origin,
metamorphosed under *amphibolite facies*) is expressed as
        (1.=) 1  (2.=) 2  (3.=) 1  (4.=) -  (5.=) 6  (6.=) -

MICA SCHIST (the same but metamorphosed under *upper greenschist
facies*) is expressed as
        (1.=) 1  (2.=) 2  (3.=) 1  (4.=) -  (5.=) 4  (6.=) -

INTERMEDIATE METAVOLCANIC SCHIST may be expressed as
        (1.=) 2  (2.=) 2  (3.=) -  (4.=) -  (5.=) 4  (6.=) -
```

Figure 3. Rock index systems.

indications are pieces of preliminary information giving only hints on grade and size. Secondly, the quality of ore mineralizations is described fully by grade, whereas 'goodness' of an industrial mineral deposit is mainly determined by end use and technology applied and cannot be expressed as one number. In accordance with this knowledge, the Ore Deposit Databank was designed to contain data on tonnage and size of the mineralization plus statistics on assay results; the Industrial Mineral Databank provides a list on suggested end use ways and a summary on assays and physical test results; and the Mineral Indication Databank simply lists the commodities met plus a rough estimate of the grade of the discovery.

TECHNICAL SOLUTION OF THE DATABASES

In the beginning of the 1980's most of the data in the computers at the GSF were stored in sequential or indexed files. The Ore Deposit Databank was a big card image file and the data could be retrieved by few specialists only. The retrieval program was written in FORTRAN and it needed a card selection control file.

The Industrial Mineral Database and the Mineral Indication Database were stored originally in a Hewlett-Packard computer using HP Image/3000 Database Management System, a hierarchial database with some CODASYL features (Eloranta, 1979). An interactive query languag (Query/3000) was used to retrive the data.

In a hierarchial data structure the path to an information item is started from the root. Every parent item can have several children, but every child has only one parent. In CODASYL-type databases other paths can be defined as well.

In this type of DBMS data retrieval is fast if the standard path can be used. When a parent data record is located, its children are retrieved easily using physical pointers. On the other hand, every change in the data structure is time consuming, so that the structure of the database must be well defined and stable.

Usually geological data can be regarded as hierarchial ones, but from one point of view at a time only. One explorational area comprises many outcrops and boulders; many samples are taken from a single outcrop; several assays and thin sections may be made from one sample. It would be easy to make a hierarchial data structure in this way. However, if we later need a way to retrieve and verify the information grouped in a different way, for exampleon a specific ore type, this might be laborious.

In 1986 the Industrial Mineral Database was converted from Hewlett-Packard to a Digital VAX computer. The hierarchial DBMS was incompatible with VAX, and the new version of the database was made using the Digital Rdb/VMS relational database-management system (Tarvainen, 1986).

In a relational database the data are stored in tables (relations). Each table is treated separately, there are no physical links (address information) from one relation to the others. Whenever needed, the information from two or more relations can be linked with a common field (foreign key).

In the Industrial Mineral Database we used three relations: (1) industrial mineral deposits; (2) assay information; and(3) physical properties of the minerals (Fig. 4). Information on each deposit comprises one line (record) in the first relation and 0 - n lines in other relations. The ADP-keynumber of the deposit is used as the linking key between relations.

The Ore Deposit Database was organized in the same way in 1987. The information was stored in a more normalized form; there are separate relations for rock types, main minerals, structural features, explorers, etc. All together the database comprises 53 different relations (Tarvainen, 1987a).

"DEPOSIT" -relation

ID NO.	NAME	DISTRICT	MAP SHEET	X
01001	HIRVIVAARA	ENO	4313 10 D	6987.930
01002	HALLAVAARA	SUOMUSSALMI	3533 05 D	

"ASSAY DATA" -relation

ID NO.	MATERIAL ASS.	OXIDE	MEAN	DEV	N
01001	KYAN	SIO2	42.0	3.76	18
01001	KYAN	AL2O3	54.8	8.30	18
01001	KYAN	FE2O3	0.6	1.21	18
01001	KYAN QUARTZITE	SIO2	74.0	11.25	63
01002	KYAN	SIO2	41.7		
01002	KYAN				

"PHYS PROP" -relation

ID NO.	MINERAL	COLOR	SHAPE	SIZE MM
01001	KYAN	GREYBLUE	NEEDL	50.00
01001	PYPH	PINKISH	FLAKE	-
01002	KYAN			

Figure 4. Relations in Industrial Mineral Database (first fields of each relation shown only).

Data from these databases are retrieved using VAX/VMS Datatrieve system. This is a fourth generation general purpose query language. The user defines the search strategy and the format of the output using simple logical expressions (Fig. 5). Our experience shows that writing programs with this type of 4GL query language is easy; however, retrieving a big amount of data with them requires a lot of CPU-time.

With the experience of the two explorational databases the Mineral Indication Database was moved from HP to VAX in late 1987 (Tarvainen, 1987b). Information was stored in a relational database with six relations. The query program was written in precompiled Pascal with RDML query statements translated into Pascal. The program uses predefined forms to ask the user for search strategy (Fig. 6). Datatrieve search can be used in special situations as well.

Technical solutions used here proved well-suited for management of geological data. Relational database-management system, predefined forms and the Pascal precompiler have become standard tools in ADP practices at GSF. FORTRAN and RDML-

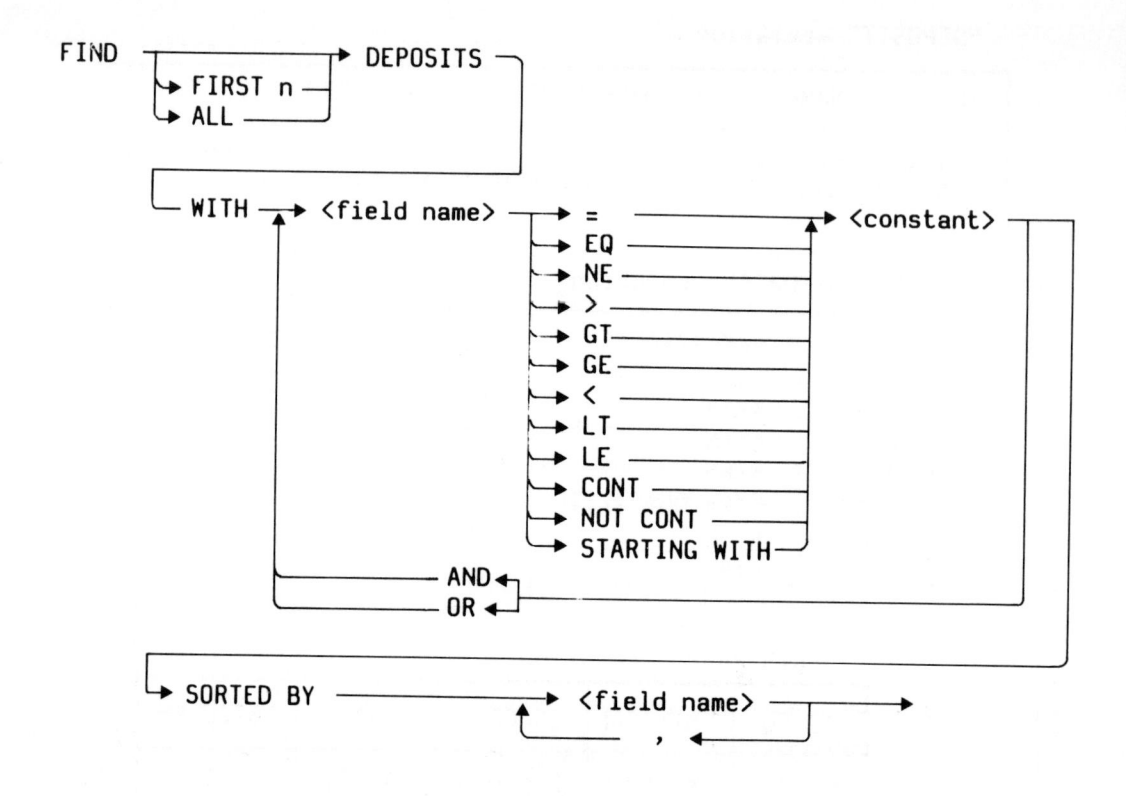

Figure 5. Format of Datatrieve query statement.

precompiled FORTRAN are used in some programs. Query language procedures and the Finnish SETTI-1-2 database-management program are used in smaller applications. Ordinary files are used for example for picture data.

UTILIZATION OF THE DATABANKS

The explorational databases are used for reference in exploration, geological mapping, geochemical investigations, and even by regional planning authorities. They also are used in research.

Sophisticated, geomathematical use of explorational databases is demonstrated elsewhere (e.g. Tontti, Koistinen, and Lehtonen, 1979; Tontti, Koistinen, and Seppänen, 1981). Thus, here we concentrate on "every man's" ways of use.

Normally a geologist selects a set of targets of interest, verifies the relevance of selection by studying through the descriptions (lists) concerned and plots the targets on a map.

EXPLORATIONAL DATA BASES AT THE GEOLGICAL SURVEY OF FINLAND 417

```
┌─────────────────────────────────────────────────────────────┐
│                                                             │
│            Mineral Indications Data Base                    │
│                                                             │
│   Retrieve INDICATIONS (outcrops □ /boulders □) containing  │
│   metal     "[    ]"                                        │
│           (Specify details if desired:)         PASS BY     │
│           1) as main metal □    (<- cross X)    FIELDS      │
│           2) ahead of metal "[    ]"            YOU DON'T   │
│           3) accompanied by metal "[    ]"      NEED!       │
│           4) accompanied by mineral "[    ]"                │
│   or mineral  "[    ]"                                      │
│           (Specify details if desired:)                     │
│           1) as main mineral □    (<- cross X)              │
│           2) ahead of mineral "[    ]"                      │
│           3) accompanied by mineral "[    ]"                │
│                                                             │
│   in LOCATION specified as                                  │
│           map sheet    [    ][  ] to [    ][  ]             │
│               or       [    ][  ] to [    ][  ]             │
│           or district is "[          ]"                     │
│                                                             │
│   with ID NUMBERS between [    0] and [999999].             │
│                                                             │
│   Wish to specify the rock type (Y/N)?   [N]                │
│                                                             │
└─────────────────────────────────────────────────────────────┘
```

Figure 6. Query menu for Mineral Indication Database.

Our system provides a range of list formats from short notes (1 line per object) to comprehensive data packages (up to 11 pages on line printer for an ore deposit description). Map plotting procedures are linked to the general system via intermediate data files, which can be subject to "manual" editing. This is because formulating of a precise request within a geological data system may be a desperate job; it is easier to make a broad request and then to delete the nondesired items using geological intuition.

All the query programs and the map plotting program are started from one main menu (Fig. 7). Available are individual query programs for each database and an easy-to-use general territorial query. The latter displays a collection of information from the three databases described plus from the Report Database, for a user-defined locality unit. The output file can be printed and used as input for the map plotting program.

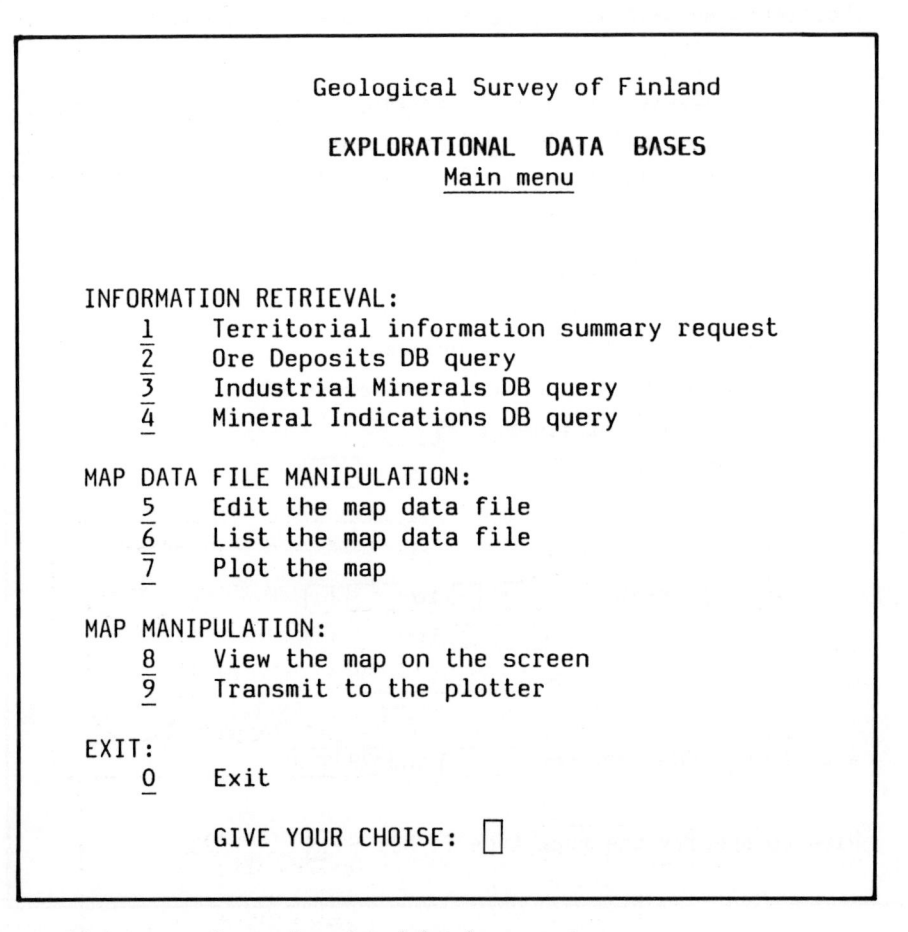

```
                    Geological Survey of Finland

                    EXPLORATIONAL   DATA   BASES
                              Main menu



    INFORMATION RETRIEVAL:
        1      Territorial information summary request
        2      Ore Deposits DB query
        3      Industrial Minerals DB query
        4      Mineral Indications DB query

    MAP DATA FILE MANIPULATION:
        5      Edit the map data file
        6      List the map data file
        7      Plot the map

    MAP MANIPULATION:
        8      View the map on the screen
        9      Transmit to the plotter

    EXIT:
        0      Exit

               GIVE YOUR CHOISE:  []
```

Figure 7. Main menu for explorational database system.

JOB-DEFINED DATABASES

Full benefit of national mineral databases and other computerized information will be obtained during introduction of job-defined databases. These are small, flexible databases for managing of data within individual exploration projects. All of them have a common general scheme, but each project can supplement its database with special files and relations. Design of job-defined databases is going on using experience gained during work on national databases. The first job-defined database is at pilot stage.

Job-defined databases serve also as a link to the general Project Register, a system for administrative and financial management of research projects, which has been developed at GSF.

EXPLORATIONAL DATA BASES AT THE GEOLGICAL SURVEY OF FINLAND 419

Outlines of use of job-defined databases are presented in Figure 8. A new project is included first in the Project Register. A job-defined database is opened for the project. Then the project manager designs the new database according to the project needs and copies here the relevant data from the large informational databanks. Although the project is running, more information from the field and laboratories is collected and stored. Pretailored output routines are used to produce progress reports, maps, etc., to provide array data for statistical calculations and to feed back the Project Register with administrative and financial information. At the end of the project the database is closed after preparing of the final report and returning of relevant new data to the informational databanks.

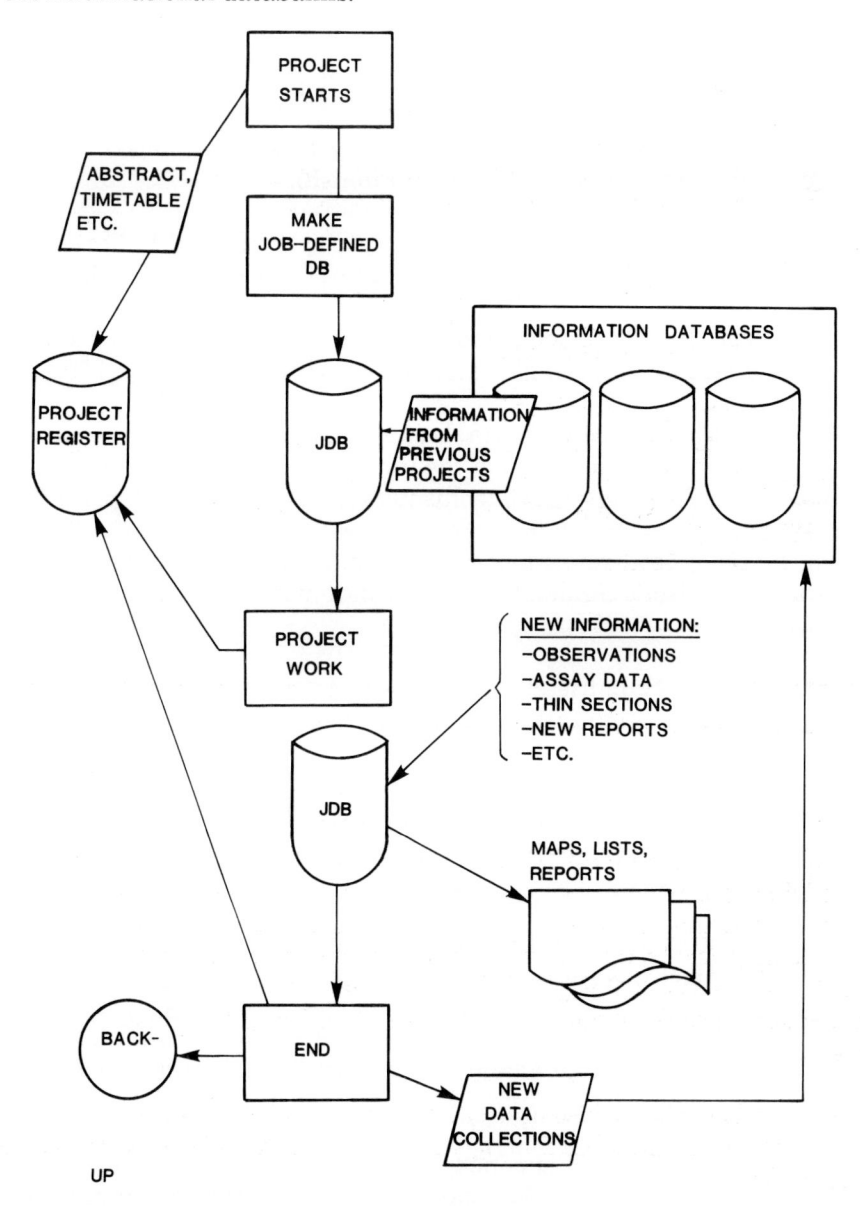

Figure 8. General flowchart for job-defined database.

REFERENCES

Eloranta,E., 1979, Erään tiedonhallintajärjestelmäpohjaisen geologisen tietorekisterin suunnittelu (Designing of a DBMS-based geological data bank): unpubl. Helsinki University of Techonology, 76 p. (in Finnish).

Gaal, G., Autio, H., Lehtonen, M., Mäkinen, A., Oksama, M., Saltikoff, B., Tontti, M., Vuorela, P., 1977a, Pohjois-Suomen malmitiedostoprojekti. Summary: ore data file project for northern Finland: Geol. Survey Finland, Rept. of Invest. No. 26, 43 p.

Gaal, G., Tontti, M., Autio, H., Lehtonen, M., 1977b, Establishing a national ore data file in Finland: Jour. Math. Geology, v. 9, no. 3, p. 319-325.

Saltikoff, B.,1976, Mineraalinimisanasto (glossary of mineral names): Geol. Survey Finland, Rept. of Invest. No. 11, 35 p. (in Finnish).

Saltikoff, B., 1979, Malmiaihetiedoston käyttäjän opas Mineral Indication Data Bank, User's Guide: Geologinen tutkimuslaitos, Opas no. 7, 37 p. (in Finnish).

Saltikoff, B., 1984, Boulder tracing and the Mineral Indication Data Bank in Finland: Prospecting in areas of glaciated terrain: IMM, London, p. 179-191.

Söderholm, B., Sotka, P., and Eloranta, E., 1980, Opas teollisuusmineraalitiedoston käyttäjälle (Guide book for users of the Industrial Mineral Data Bank) Suomen Luonnonvarain Tutkimussäätiö, teollisuusmineraaliprojekti, 39 p. (in Finnish).

Tarvainen, T., 1986, Teollisuusmineraalitietokannan käyttäjän opas (Industrial Minerals Database, User's Guide): Geologian tutkimuskeskus, Opas no. 18, 30 p. (in Finnish).

Tarvainen, T., 1987a, Malmitietokannan käyttäjän opas (Ore Deposits Database, User's Guide): Geologian tutkimuskeskus, Opas no. 19, 49 p. (in Finnish, English summary).

Tarvainen, T., 1987b, Malmiviitetietokannan käyttäjän opas (Mineral Indication Database, User's Guide): Geologian tutkimuskeskus, Opas no. 20, 25 p.(in Finnish, English summary).

Tiainen, J., and Aumo, R., 1988, Raporttitietokanta - käyttöohje/Report Database - User's Guide: Geologian tutkimuskeskus, 18 p. (in Finnish).

Tontti, M., Koistinen, E., and Lehtonen, M.K.A., 1979, Kotalahden nikkelivyöhykkeen monimuuttuja-analyysi. Summary: Multivariate analysis of the Kotalahti Nickel Belt: Geol. Survey Finland, Rept. of Invest. No.36, 44 p.

Tontti, M., Koistinen, E., and Seppänen, H., 1981, Vihannin Zn-Cumalmivyöhykkeen geomatemaattinen arviointi. Summary: Geomathematical evaluation of the Vihanti Zn-Cu ore zone: Geol. Survey Finland, Rept. of Invest. No. 54, 64 p.

RELATED STATISTICAL TECHNIQUES

Regression Analysis of Geochemical Data With Observation Below Detection Limits

C. F. Chung
Geological Survey of Canada, Ottawa

ABSTRACT

In regression analysis of geochemical data, incompleteness of observations of the dependent variable in a statistical model has caused much difficulty. Geochemical data sets with observations below detection limits are typical examples of incomplete data. A new regression technique based upon the maximum likelihood method is presented herein and shown to be a better procedure than the traditional least-squares method. As an illustration this procedure is applied first to a simulated data set and then to a geochemical data set consisting of gold, tungsten, sulphur, and arsenic values in rock samples from the Meguma Terrane in Nova Scotia, Canada.

INTRODUCTION

Suppose that we wish to determine reliable geochemical indicators or pathfinder elements for gold potential in a study region. We can attempt to establish a functional relationship between gold and other geochemical elements by postulating a linear model such as:

$$\text{gold} = \beta_0 + \beta_1 \text{ sulphur} + \beta_2 \text{ arsenic} + \varepsilon, \qquad (1)$$

where (β_0, β_1, β_2) are unknown regression coefficients and e denotes a random component. For such model, we first estimate (β_0, β_1, β_2) based on the observations of the random samples from the study region, then we look at the statistical significance of the estimators, and finally we determine whether sulphur or arsenic can be used as pathfinder elements for the gold exploration.

When the gold contents of the all samples collected for the study are above the detection limit of the measuring instrument, the usual regression techniques based on the least-squares (LS) method are employed to estimate (β_0, β_1, β_2). However, if the gold contents of some of the samples are below the detection limit, then regression estimators are not obtainable. We are proposing here a new technique which is based on the maximum likelihood method to estimate ($ß_0$, $ß_1$, $ß_2$).

Let Y be a random variable with the distribution function F(y)=P{Y≤y}. Suppose that we have N independent observations $Y_1, \ldots, Y_N$ on Y. Then using the usual standard statistical analysis, inferences on Y can be made based on the N observations. However, when some of the N observations are censored (i.e., here values that are less than the detection limit), even simple statistics, such as the sample mean or the variance, can not be estimated without strong assumptions and complex iterative procedures (Chung, 1988).

In this exposition, we discuss how to estimate unknown regression coefficients ($ß_0$, $ß_1, \ldots, ß_p$) in the following linear model:

$$Y = ß_0 + ß_1 X_1 + \ldots + ß_p X_p + \varepsilon, \qquad (2)$$

where X1, ... , Xp denote independent variables and ε represents a random term in the model. Obviously, when the observations of Y are censored, the usual LS method cannot be used to obtain the estimates for ($ß_0, ß_1, \ldots, ß_p$).

In geoscience applications, a usually employed ad-hoc method of estimating ($ß_0, ß_1, \ldots, ß_p$) is the LS method applied to the data set which is modified by substituting some "sensible" values such as 60% of the detection limit for the censored observations. When the number of censored observations is comparably small, the modified estimates may provide geologically reasonable information. However, if the number of censored data is relatively large, then the LS regression coefficients may be uninterpretable.

The maximum likelihood (ML) method proposed here is an extension of the LS method under the normality assumption. It is well known that the LS estimators are identical to the ML estimators in the linear model postulated in Equation (2) under the assumption that ε has the normal distribution (cf. Draper and Smith, 1966). As shown later in this paper, the ML estimates are obtainable even if the observations for Y are censored. The ML estimators have several desirable statistical properties if the normality assumption for ε (and hence Y) is reasonable. In other words, an assumption of normality is the price we are paying to obtain a theoretically sound estimator for ($ß_0, ß_1, \ldots, ß_p$) and σ.

MULTIPLE CENSORING

Consider N independent observations $Y_1, \ldots, Y_N$. Suppose that we do not have the last g observations, but we know that $Y_{N-g+j} < \alpha$ for j=1, ... ,g, where α is a constant, Y_i for i=1, ... , N-g, are observed values, and $Y_i(>\alpha)$ (i.e. instead of $Y_1, \ldots, Y_N$, we have $Y_1, \ldots, Y_{N-g}, <\alpha, \ldots, <\alpha$). Such a data set is termed *single-censored* and the last g observations are termed *left-censored*. A left-censored geochemical data set with observations below a single detection limit A is shown in Table 1 for gold data where the detection limit is 1 part per billion (ppb).

Table 1. Gold, tungsten, arsenic, and sulphur data of 94 samples from lithology "Slates", Meguma Terrane, Nova Scotia, Canada.

Au(ppb)	in	W(ppm)	in	arsenic*	sulphur**	Au(ppb)	in	W(ppm)	in	arsenic*	sulphur**
9.0	0	10.0	0	4.787	-5.319	710.0	0	17.0	0	7.783	-3.042
45.0	0	7.0	0	6.016	-4.895	580.0	0	7.0	0	8.497	-2.262
1.0	-1	7.0	0	4.431	-6.771	42.0	0	16.0	0	8.366	-2.780
8.0	0	11.0	0	4.868	-6.834	34.0	0	12.0	0	6.551	-3.039
67.0	0	1.0	-1	4.595	-4.124	95.0	0	20.0	0	7.901	-2.907
18.0	0	25.0	0	4.700	-3.919	290.0	0	26.0	0	8.102	-2.657
16.0	0	9.0	0	5.347	-4.214	120.0	0	15.0	0	6.721	-2.942
39.0	0	7.0	0	4.700	-3.834	110.0	0	12.0	0	8.366	-2.574
3.0	0	11.0	0	4.605	-3.636	1700.0	0	1.0	-1	7.696	-1.589
11.0	0	11.0	0	4.511	-4.903	180.0	0	13.0	0	5.966	-2.502
1.0	-1	1.0	-1	3.258	-6.478	290.0	0	10.0	0	7.003	-2.145
1.0	-1	1.0	-1	3.401	-6.718	28.0	0	8.0	0	6.016	-1.962
28.0	0	5.0	0	2.833	-6.097	1100.0	0	10.0	0	6.877	-.644
1.0	-1	7.0	0	3.258	-3.701	230.0	0	9.0	0	4.605	-2.848
1.0	-1	5.0	0	2.639	-3.707	98.0	0	11.0	0	4.787	-2.942
1.0	-1	4.0	0	2.773	-3.940	1.0	-1	7.0	0	4.787	-2.980
2.0	0	5.0	0	3.091	-4.285	2.0	0	6.0	0	3.045	-2.781
1.0	-1	3.0	0	1.386	-4.991	160.0	0	16.0	-1	10.491	-1.012
1.0	-1	3.0	0	1.609	-4.577	10.0	0	13.0	0	9.852	-1.404
1.0	-1	1.0	-1	5.394	-5.413	7.0	0	5.0	0	5.670	-2.642
1.0	-1	1.0	0	.693	-4.832	1.0	-1	5.0	0	4.394	-1.663
1.0	-1	6.0	0	3.638	-5.573	5.0	0	3.0	0	3.638	-1.542
1.0	0	5.0	0	5.521	-5.123	12.0	0	5.0	0	3.091	-2.236
6.0	0	8.0	0	4.787	-4.660	1.0	0	14.0	0	8.613	-2.282
67.0	0	7.0	0	5.136	-4.293	1.0	-1	4.0	0	7.824	-2.419
1.0	0	7.0	0	1.609	-5.540	4.0	0	4.0	0	3.951	-1.595
1.0	-1	11.0	0	1.946	-5.387	4.0	0	3.0	0	3.638	-2.831
50.0	0	7.0	0	4.043	-5.125	740.0	0	3.0	0	3.784	-2.074
1.0	-1	8.0	0	1.792	-4.662	4.0	0	4.0	0	6.087	-3.724
1.0	-1	1.0	-1	7.003	-3.955	1.0	-1	6.0	0	3.219	-2.428
1.0	-1	9.0	0	6.016	-2.535	9.0	0	1.0	-1	6.908	-2.120
11.0	0	2.0	-1	7.696	-.944	9.0	0	1.0	-1	4.143	-2.764
4.0	0	5.0	0	7.824	-.854	1.0	0	1.0	-1	3.466	-4.421
4.0	0	1.0	-1	6.131	-1.210	7.0	0	10.0	0	3.219	-4.864
10.0	0	9.0	0	7.824	-2.743	2.0	0	1.0	-1	4.700	-2.838
42.0	0	4.0	0	7.650	-.472	2.0	0	1.0	-1	4.007	-3.128
480.0	0	18.0	0	9.000	-2.220	51.0	0	1.0	-1	5.858	-1.326
110.0	0	12.0	0	8.216	-1.871	110.0	0	7.0	0	5.799	-.897
2.0	0	5.0	0	5.075	-2.576	35.0	0	1.0	-1	7.496	-3.377
48.0	0	2.0	-1	10.519	-1.340	1.0	-1	8.0	0	5.394	-4.534
26.0	0	6.0	0	8.343	-2.515	5.0	0	8.0	0	5.136	-3.695
110.0	0	15.0	0	9.852	-1.511	10.0	0	10.0	0	10.645	-1.753
21.0	0	2.0	0	6.310	-2.904	11.0	0	12.0	0	9.127	-2.247
2500.0	0	16.0	-1	9.306	-1.159	12.0	0	11.0	-1	9.083	-2.131
3300.0	0	14.0	-1	9.798	-.675	10.0	0	14.0	-1	9.547	-1.630
71.0	0	4.0	-1	5.858	-2.024	2500.0	0	4.0	0	7.244	-1.548
87.0	0	20.0	0	7.244	-4.202	25.0	0	12.0	0	5.011	-3.806

in   :indicator variable (See the text for the detail).
arsenic*   :logarithm transformation was applied to the original observation(ppm).
sulphur**   :log-ratio transformation was applied to the original observation(%).

For a data set with multiple censoring, instead of $Y_1, \ldots, Y_N$, we have $Y_1, \ldots, Y_n, A_1, \ldots, A_h, B_1, \ldots, B_g$, where $Y_{n+1} < A_1, \ldots, Y_{n+h} < A_h, Y_{n+h+1} > B_1, \ldots, Y_{n+h+g}(=Y_N) > B_g$, and $A_1, \ldots, A_h$ and $B_1, \ldots, B_g$, are known constants. The first n samples have complete observations, the next h samples have left-censored observations, and the last g samples have right-censored observations. The tungsten data in Table 1 are an example of multiple left-censored data.

Chung (1988) discussed how to obtain the ML estimators for mean and variance of Y assuming that all the samples including the censored data come from a normal distribution.

REGRESSION ANALYSIS OF MULTIPLE CENSORED DATA

As in usual regression analysis, consider N observations on a response (dependent) variable Y and p controlling (independent) variables $X_1, X_2, \ldots, X_p$:

$$Y_1 ; X_{11}, X_{12}, \ldots, X_{1p}$$
$$\vdots$$
$$Y_N ; X_{N1}, X_{N2}, \ldots, X_{Np}$$
(3)

For the linear model

$$Y = \beta_0 + \beta_1 X_1 + \ldots + \beta_p X_p + \varepsilon, \tag{4}$$

we obtain the ordinary LS estimators $\beta_0, \beta_1, \ldots, \beta_p$ from the N observations. These LS estimators also are the ML estimators assuming that:

(A1) $E(Y_i) = \beta_0 + \beta_1 X_{i1} + \ldots + \beta_p X_{ip}$ and $Var(Y_i) = \sigma^2$;
(A2) Y_i has a normal distribution function;
(A3) $Cov(Y_i, Y_j) = 0$ for $i = j$.

Thus the usual statistical tests such as the goodness-of-fit and influence observations, are applied to these LS estimators (cf. Draper and Smith, 1966).

As multiple censored observations for the dependent variable Y, we have $Y_1, \ldots, Y_n, A_1, \ldots, A_h, B_1, \ldots, B_g$, instead of $Y_1, \ldots, Y_N$ where $Y_{n+1} < A_1, \ldots, Y_{n+h} < A_h, Y_{n+h+1} > B_1, \ldots, Y_{n+h+g}(=Y_N) > B_g$, and $A_1, \ldots, A_h$ and $B_1, \ldots, B_g$, are known constants. The first n samples have complete observations, the next h samples have left-censored observations and the last g samples have right-censored observations. The N observations on p independent variables are as usual. Hence the N observations are:

$Y_1 ; X_{11}, X_{12}, \ldots, X_{1p}$

$\cdot \quad \cdot \quad \cdot \quad \cdot \quad \cdot$
$\cdot \quad \cdot \quad \cdot \quad \cdot \quad \cdot$
$\cdot \quad \cdot \quad \cdot \quad \cdot \quad \cdot$

$Y_n ; X_{n1}, X_{n2}, \ldots, X_{np}$
$A_1 ; X_{(n+1)1}, X_{(n+1)2}, \ldots, X_{(n+1)p}$

$\cdot \quad \cdot \quad \cdot \quad \cdot \quad \cdot$
$\cdot \quad \cdot \quad \cdot \quad \cdot \quad \cdot \qquad (5)$
$\cdot \quad \cdot \quad \cdot \quad \cdot \quad \cdot$

$A_h ; X_{(n+h)1}, X_{(n+h)2}, \ldots, X_{(n+h)p}$
$B_1 ; X_{(n+h+1)1}, X_{(n+h+1)2}, \ldots, X_{(n+h+1)p}$

$\cdot \quad \cdot \quad \cdot \quad \cdot \quad \cdot$
$\cdot \quad \cdot \quad \cdot \quad \cdot \quad \cdot$
$\cdot \quad \cdot \quad \cdot \quad \cdot \quad \cdot$

$B_g ; X_{(n+h+g)1}, X_{(n+h+g)2}, \ldots, X_{(n+h+g)p}$,

instead of those in Equation (3).

Because $Y_{n+1}, \ldots, Y_N$ are unknown, it is not possible to obtain LS estimators of ($\beta_0, \beta_1, \ldots, \beta_p$) in Equation (4) from the data in (5). However, under the assumptions (A1), (A2), and (A3), the ML estimators of ($\beta_0, \beta_1, \ldots, \beta_p$) in (4) can be obtained as follows. The likelihood function is:

$$\prod_{i=1}^{n} \phi(Y_i : \mu_i, \sigma) \prod_{j=1}^{h} \Phi(A_j : \mu_{n+j}, \sigma) \prod_{k=1}^{g} (1 - \Phi(B_k : \mu_{n+h+k}, \sigma)),$$

where $\phi(.)$ and $\Phi(.)$ are the normal density and distribution functions respectively, that is

$$\phi(x:\mu,\sigma) = \frac{1}{\sigma\sqrt{2\pi}} e^{-\frac{1}{2}\left(\frac{x-\mu}{\sigma}\right)^2} \quad \text{and} \quad \Phi(y:\mu,\sigma) = \int_{-\infty}^{y} \phi(x:\mu,\sigma) \, dx,$$

$$\mu_t = \beta_0 + \beta_1 X_{t1} + \cdots + \beta_p X_{tp} \quad \text{for } t = 1, \ldots, N. \qquad (6)$$

Hence the log-likelihood function $L(\beta,\sigma)$ is written as

$$L(\beta,\sigma) = \sum_{i=1}^{n} \log \phi(Y_i:\mu_i,\sigma) + \sum_{j=1}^{h} \log \Phi(A_j:\mu_{n+j},\sigma)$$

$$+ \sum_{k=1}^{g} \log (1 - \Phi(B_k:\mu_{n+h+k},\sigma)). \qquad (7)$$

The unknown parameters ($\beta_0, \beta_1, \ldots, \beta_p$) and σ are estimated such that the log-likelihood function L in Equation (7) is maximized and these estimators are termed the ML estimators. We note that the analytic solutions of such estimators are not obtainable, and an iterative procedure is employed to obtain the estimators.

There are several iterative algorithms for obtaining the ML estimators maximizing $L(\beta,\sigma)$. Three widely used techniques are the scoring method (Rao, 1973), the EM-algorithm (Dempster, Laird, and Rubin, 1977), and the conjugate gradients method (Stoer and Bulirsch, 1980). The scoring method to obtain the ML estimators from Equation (7) is illustrated in Appendix A. A documentation of the FORTRAN 77 computer program REGRES written for the algorithm in Appendix A is in preparation for publication in *Computer & Geosciences* together with test data.

EXAMPLE

(i) *Simulated data.* In the unit square ($[0,1] \times [0,1] \in R^2$), 100 points are selected randomly. At each point (u,v), an observed value for a random variable $Y_{u,v}$ is simulated, based on the following model:

$$\log_e Y_u, v = f(u,v) + \varepsilon , \qquad (8)$$

where $f(u,v) = 4 + 3u - 2v - 1.5 u^2 - 3 u v + 2 v^2$, and ε has the standard normal distribution function with mean 0 and variance 1. The simulated value is recorded as "20" if the value is less than 20 or as the actual simulated value if the latter is greater than or equal to 20. In addition, an indicator variable is defined at each point such that it takes either value "-1", if the simulated value is less than 20 (and thus censored), or "0", if that is greater than or equal to 20 (not censored). The 100 value simulated data set with indicator variable and the u-v coordinates in the unit square are shown in Table 2.

For the simulated data set in Table 2, consider the quadratic trend-surface model:

$$Z u,v = \beta_0 + \beta_1 u + \beta_2 v + \beta_3 u^2 + \beta_4 u v + \beta_5 v^2 + \varepsilon , \qquad (9)$$

where ($\beta_0, \beta_1, \ldots, \beta_5$) are unknown parameters, and E has a normal distribution function with mean 0 and unknown variance σ^2. Then we obviously expect to obtain 1 and (4, 3, -2, -1.5, -3, 2) as the estimates of σ^2 and ($\beta_0, \beta_1, \ldots, \beta_5$), respectively.

If we were to have all simulated observations on $Y_{u,v}$ instead of censored observations below the 20, we would first transform the observed $Y_{u,v}$ into the logarithm values and then apply the trend-surface analysis to the transformed data. The least-squares estimates for ($\beta_0, \beta_1, \ldots, \beta_5$) and σ^2 of the trend-surface analysis would be the minimum unbiased linear estimators and would be close to the true values (4, 3, -2, -1.5, -3, 2) and 1.

However 22 of the 100 simulated values are censored at 20 and the only information available for the censored 22 samples is that the observed values are less than 20. In order to apply the trend-surface analysis to this censored data set, all 22 censored observations are assigned first a fixed value, such as $\log_e 10$ ($\log_e(0.5 \times 20)$) or $\log_e 12$ ($\log_e(0.6 \times 20)$). To obtain the least-squares estimates for ($\beta_0, \beta_1, \ldots, \beta_5$) and σ^2, the trend-surface analysis is applied to the log-transformed data. Eleven such estimates for eleven different substituted values for the 22 censored observations are shown in

Table 2. Simulated data of 100 random points for Example (i). See text for detail.

s.data*	in**	u-axis	v-axis	s.data*	in**	u-axis	v-axis
20.0	-1	.4909	.6221	53.5	0	.1332	.2099
20.0	-1	.9752	.9155	55.8	0	.4715	.5454
20.0	-1	.9808	.3010	61.8	0	.2956	.5957
31.4	0	.8018	.2631	73.8	0	.0854	.2945
20.0	-1	.7994	.9643	80.2	0	.2765	.6928
20.0	-1	.7728	.9866	87.0	0	.4479	.7615
20.1	0	.5562	.7558	114.3	0	.6370	.5595
23.1	0	.5645	.9251	191.1	0	.4590	.4225
47.0	0	.9250	.3549	215.5	0	.0764	.1537
59.2	0	.4595	.3448	305.9	0	.5295	.5042
128.0	0	.6569	.1398	20.0	-1	.9011	.8303
41.2	0	.8744	.5656	20.0	-1	.7039	.8819
135.1	0	.4776	.1858	20.0	-1	.2033	.6597
92.0	0	.4517	.4401	20.0	-1	.3785	.8069
177.7	0	.2412	.0553	20.8	0	.4922	.5856
122.8	0	.6785	.4546	45.1	0	.3252	.1466
203.5	0	.7276	.3012	34.7	0	.3830	.3658
503.3	0	.5714	.0762	74.9	0	.5750	.1899
312.8	0	.3196	.2164	125.5	0	.6218	.0752
532.9	0	.4563	.2289	31.4	0	.6919	.6958
20.0	-1	.8387	.8368	26.8	0	.7957	.8092
20.0	-1	.5761	.5813	96.6	0	.1624	.0886
20.0	-1	.3761	.3696	55.4	0	.0190	.7053
20.0	-1	.1894	.5108	72.8	0	.4175	.8490
20.0	-1	.6764	.6387	46.3	0	.7947	.9427
48.3	0	.6251	.2343	174.4	0	.4790	.3120
76.0	0	.4302	.0808	88.7	0	.9109	.6442
20.0	-1	.7560	.9247	183.4	0	.9010	.4310
156.1	0	.6735	.0499	198.2	0	.2348	.6171
67.4	0	.5218	.3098	299.0	0	.7434	.5262
33.8	0	.6342	.8668	20.0	-1	.0765	.2441
56.4	0	.4210	.7898	20.0	-1	.2257	.5281
183.9	0	.7293	.1587	20.0	-1	.0415	.4673
73.1	0	.1284	.4934	36.5	0	.6836	.2180
56.1	0	.8165	.7438	40.0	0	.9790	.2361
115.5	0	.3159	.8795	27.1	0	.9869	.3588
134.6	0	.4328	.8061	20.0	-1	.6267	.7601
158.6	0	.4685	.8812	28.3	0	.4342	.6760
812.3	0	.6478	.0867	112.9	0	.4128	.0214
717.6	0	.8988	.3000	39.6	0	.1984	.5135
20.0	-1	.7402	.1538	98.8	0	.4510	.1751
20.0	-1	.0317	.9373	87.8	0	.0990	.0478
28.4	0	.4775	.1621	46.6	0	.9834	.5252
20.0	-1	.6675	.7059	78.4	0	.3032	.5626
20.1	0	.0785	.8917	111.4	0	.0841	.9983
20.7	0	.2191	.5880	280.1	0	.9116	.1702
21.3	0	.4503	.7361	199.2	0	.6835	.3706
21.4	0	.7312	.6644	153.7	0	.0158	.2286
35.4	0	.2566	.5042	185.3	0	.1892	.6706
27.7	0	.6398	.9630	277.7	0	.0044	.8304

s.data* : simulated data ; in** : indicator variable.

Table 3. Knowing the true values, we observe that the estimates corresponding to $\log_e(0.8 \times 20)$ are close to the true values. Whatever estimates could be selected from the eleven estimates, it would be ad hoc in nature, because there is no firm theoretical reason to select any one particular value for the censored observations.

The ML estimates described in this paper for ($\beta_0, \beta_1, \ldots, \beta_5$) and σ^2 also are shown in Table 3. For the initial estimates, any of the latter eight sets of estimates (not the first three sets) in Table 3 can be used. Of course the same ML estimates are obtained regardless of which one of the eight sets of estimates is used initially. For the ML estimates, substitution is not required for the censored observations. As expected in this simulated data set, when comparing the estimates with the true values, the ML estimates do reflect the unknown truth well.

Table 3. ML estimates and LS estimates for coefficients of quadratic trend surface and variance from the data in Table 2. It also shows true coefficients from which data in Table 2 were generated. First column under "Substituted value" illustrates substituted values for observations shown as "20" with indicator variable "-1".

	Substituted value	con. β_0	u β_1	v β_2	u^2 β_3	uv β_4	v^2 β_5	Variance σ^2
Model	na	4.0	3.0	-2.0	-1.5	-3.0	2.0	1.00
ML estimates	na	4.1	3.2	-1.3	-1.7	-3.7	1.3	1.03
LS estimates	0	3.4	5.5	-1.6	-3.2	-5.4	1.6	1.75
	$\log_e 2$	3.5	5.2	-1.7	-3.0	-5.2	1.6	1.65
	$\log_e 4$	3.6	4.9	-1.7	-2.8	-4.9	1.6	1.55
	$\log_e 6$	3.7	4.5	-1.7	-2.6	-4.6	1.6	1.45
	$\log_e 8$	3.8	4.2	-1.7	-2.3	-4.3	1.6	1.35
	$\log_e 10$	3.9	3.8	-1.8	-2.1	-4.0	1.6	1.26
	$\log_e 12$	4.0	3.5	-1.8	-1.9	-3.7	1.6	1.16
	$\log_e 14$	4.1	3.1	-1.8	-1.7	-3.4	1.6	1.08
	$\log_e 16$	4.2	2.8	-1.8	-1.4	-3.1	1.6	1.00
	$\log_e 18$	4.3	2.4	-1.9	-1.2	-2.8	1.6	0.92
	$\log_e 20$	4.4	2.1	-1.9	-1.0	-2.5	1.6	0.86

na : not applicable.

(ii) *Gold, tungsten, sulphur, and arsenic data in the Meguma Terrane, Nova Scotia, Canada* (Kerswill, 1987). Geological and analytical data for gold, tungsten, sulphur, and arsenic on 272 bedrock samples from gold-bearing districts in the Eastern Meguma Terrane of Nova Scotia were used in this application. Samples were classified as "Wackes"(wackes lacking veins), "Slates"(slates lacking veins), and "Veins"(either wackes or slates containing quartz veins). 99, 94, and 79 of the 272 samples were assigned to the Wackes, Slates, and Veins, respectively. The detection limit for gold was 1 ppb, and 28 (5 Wackes, 20 Slates, and 3 Veins) of 272 samples contain gold less than 1 ppb. The observed values of 94 Slates samples are shown in Table 1. Note that the detection limit for each sample is different.

All subsequent statistical analyses were applied to log-transformed data for gold, tungsten, and arsenic and log-ratio transformed data for sulphur (Aitchison, 1986). The technical reason for these transformations goes beyond the scope of this paper (e.g., gold implies that $\log_e$(gold) was used).

To help determining whether sulphur and arsenic are related to gold, the following linear model is postulated:

$$\text{gold} = \beta_0 + \beta_1 \text{ sulphur} + \beta_2 \text{ arsenic} + \varepsilon,$$

where (β_0, β_1, β_2) are unknown regression coefficients and e denotes a random component. (β_0, β_1, β_2) might be expected to differ according to rock type. The ML estimates for (β_0, β_1, β_2) are shown in Table 4(i) for each rock type. Table 4(i) also contains the LS estimates on the data substituting a value 0.6 ppm (60% x 1 ppm) for the observations below the detection limit. The LS and ML estimates are different for Slates but similar for Wackes and Veins. This reflects the greater number of observations below the detection limit in the Slates (20), whereas few observations in the Wackes (5) and Veins (3) are below the detection limit. The results confirm that interrelationships among gold, sulphur, and arsenic are different in each of the three rock types.

The next application is to determine how sulphur and arsenic are related to tungsten. Tungsten is known to be associated with arsenic and to some extent with sulphur in the study region. The postulated model is:

$$\text{tungsten} = \beta_0 + \beta_1 \text{ sulphur} + \beta_2 \text{ arsenic} + \varepsilon.$$

In order to apply the LS method, a value equivalent to $\log_e$(0.6 x detection limit) was substituted where tungsten data were below the detection limit for each sample. Table 4(ii) contains the LS estimates and the ML estimates for each rock type separately. Examination of the results shows a significant difference between LS and ML estimates for all three rock types. This suggests that the ML method is particularly appropriate in handling multicensored data. Furthermore, tungsten does not seem to be well correlated with either arsenic or sulphur. Work is in progress to further geological knowledge on the relationships between gold, tungsten, sulphur, and arsenic.

In this example, the ML approach to the estimation of regression coefficients in multicensored data sets is an attractive and stable alternative to more traditional LS method. Furthermore, the ML method has produced geologically useful results.

Table 4. ML estimates and LS estimates for regression coefficients of linear models. For lithology "Slate", estimates are based on data given in Table 3.

(i) gold = $\beta_0 + \beta_1$ arsenic + β_2 sulphur

Lithology	Estimates	Constant β_0	Arsenic β_1	Sulphur β_2	Variance s
Wacke	LS	-0.866	0.499	-0.158	1.660
	ML	-1.136	0.520	-0.195	1.729
Slate	LS	1.379	0.397	0.328	1.836
	ML	0.816	0.496	0.437	2.188
Veins	LS	3.569	0.243	0.501	2.090
	ML	3.505	0.249	0.506	2.162

(ii) tungsten = $\beta_0 + \beta_1$ arsenic + β_2 sulphur

Lithology	Estimates	Constant β_0	Arsenic β_1	Sulphur β_2	Variance s
Wacke	LS	0.615	-0.007	-0.224	1.047
	ML	-0.337	-0.053	-0.449	1.744
Slate	LS	0.492	0.134	-0.120	0.869
	ML	0.414	0.132	-0.122	1.049
Veins	LS	0.618	0.120	-0.052	1.092
	ML	0.431	0.078	-0.115	1.543

ACKNOWLEDGMENT

I wish to thank Mr. J.A. Kerswill of the Geological Survey of Canada for providing not only the data set containing gold, tungsten, sulpur, and arsenic in the Meguma Terrane, Nova Scotia, but also useful comments on the earlier draft of this paper.

REFERENCES

Aitchison, J., 1986, The statistical analysis of compositional data: Chapman and Hall, London, 416p.

Chung, C.F., 1988, Statistical analysis of truncated data in geosciences: Sciences de la Terre-Serie Informatique Geologique (Nancy) v.27, p.157-180.

Dempster, A.P., Laird, N.M., and Rubin, D.B., 1977, Maximum likelihood from incomplete data via the EM algorithm: Jour. Roy. Stat. Soc., Ser. B, v.39, p.1-22.

Draper, N., and Smith, H., 1966, Applied regression analysis: John Wiley & Sons, New York, 407 p.

Kerswill, J.A., 1987, Mineralogy and chemistry of metawackes and slates as a guide to gold in Eastern Meguma Terrane of Nova Scotia: Mines and Minerals Branch Rept. Activities 1987, P. A., v. 87-5, p.209-211.

Rao, C.R., 1975, Linear statistical inference and its applications (2nd ed.): John Wiley & Sons, New York, 625 p.

Stoer, J., and Bulirsch, R., 1980, Introduction to numerical analysis: Springer-Verlag, New York, 609 p.

APPENDIX A. SCORING METHOD

As discussed in the section on regression, the log-likelihood function L(ß,σ) is written as:

$$L(\beta,\sigma) = \sum_{i=1}^{n} \log \phi(Y_i:\mu_i,\sigma) + \sum_{j=1}^{h} \log \Phi(A_j:\mu_{n+j},\sigma)$$

$$+ \sum_{k=1}^{g} \log (1 - \Phi(B_k:\mu_{n+h+k},\sigma)) .$$
(A.1)

The ML estimates $\beta_0, \beta_1, \ldots, \beta_p$ and σ are obtained such that the log-likelihood function L in (A.1) is maximized.

The scoring method (Rao, 1973) based on the Taylor series expansion is an iterative procedure as follows:

$$\begin{pmatrix} _{i+1}\beta_0 \\ _{i+1}\beta_1 \\ \vdots \\ _{i+1}\beta_p \\ _{i+1}\sigma \end{pmatrix} = \begin{pmatrix} _{i}\beta_0 \\ _{i}\beta_1 \\ \vdots \\ _{i}\beta_p \\ _{i}\sigma \end{pmatrix} - \begin{pmatrix} \frac{\partial^2 \log L}{\partial \beta_0^2} & \frac{\partial^2 \log L}{\partial \beta_1 \partial \beta_0} & \cdots & \frac{\partial^2 \log L}{\partial \beta_p \partial \beta_0} & \frac{\partial^2 \log L}{\partial \sigma \partial \beta_0} \\ \frac{\partial^2 \log L}{\partial \beta_0 \partial \beta_1} & \frac{\partial^2 \log L}{\partial \beta_1^2} & \cdots & \frac{\partial^2 \log L}{\partial \beta_p \partial \beta_1} & \frac{\partial^2 \log L}{\partial \sigma \partial \beta_1} \\ \vdots & \vdots & \cdots & \vdots & \vdots \\ \frac{\partial^2 \log L}{\partial \beta_0 \partial \beta_p} & \frac{\partial^2 \log L}{\partial \beta_1 \partial \beta_p} & \cdots & \frac{\partial^2 \log L}{\partial \beta_p^2} & \frac{\partial^2 \log L}{\partial \sigma \partial \beta_p} \\ \frac{\partial^2 \log L}{\partial \beta_0 \partial \sigma} & \frac{\partial^2 \log L}{\partial \beta_1 \partial \sigma} & \cdots & \frac{\partial^2 \log L}{\partial \beta_p \partial \sigma} & \frac{\partial^2 \log L}{\partial \sigma^2} \end{pmatrix}^{-1} \begin{pmatrix} \frac{\partial \log L}{\partial \beta_0} \\ \frac{\partial \log L}{\partial \beta_1} \\ \vdots \\ \frac{\partial \log L}{\partial \beta_p} \\ \frac{\partial \log L}{\partial \sigma} \end{pmatrix}_{\substack{\beta_0 = _i\beta_0 \\ \beta_1 = _i\beta_1 \\ \vdots \\ \beta_p = _i\beta_p \\ \sigma = _i\sigma}}$$
A.2)

where $(_1\beta_0, _1\beta_1, \ldots, _1\beta_p)$ and $_1\sigma$ are initial estimates for $(\beta_0, \beta_1, \ldots, \beta_p)$ and σ, and

$$\frac{\partial \log L}{\partial \beta_s} = \sigma^{-1}\left[\sum_{i=1}^{n} y_i X_{is} - \sum_{j=1}^{h} \alpha_j X_{(j+n)s} + \sum_{k=1}^{g} \tau_k X_{(k+n+h)s}\right]$$

$$\frac{\partial \log L}{\partial \sigma} = \sigma^{-1}\left[\sum_{i=1}^{n} y_i^2 - \sum_{j=1}^{h} \alpha_j a_j + \sum_{k=1}^{g} \tau_k b_k - n\right]$$

$$\frac{\partial^2 \log L}{\partial \beta_r \partial \beta_s} = -\sigma^{-2}\left[\sum_{i=1}^{n} X_{ir}X_{is} + \sum_{j=1}^{h}(a_j+\alpha_j)\alpha_j X_{(j+n)r}X_{(j+n)s}\right.$$
$$\left. - \sum_{k=1}^{g}(b_k-\tau_k)\tau_k X_{(k+n+h)r}X_{(k+n+h)s}\right]$$

$$\frac{\partial^2 \log L}{\partial \sigma \partial \beta_s} = -\sigma^{-2}\left[2\sum_{i=1}^{n} y_i X_{is} + \sum_{j=1}^{h}((a_j+\alpha_j)a_j - 1)\alpha_j X_{(j+n)s}\right.$$
$$\left. - \sum_{k=1}^{g}((b_k-\tau_k)b_k - 1)\tau_k X_{(k+n+h)s}\right]$$

$$\frac{\partial^2 \log L}{\partial \sigma^2} = -\sigma^{-2}\left[3\sum_{i=1}^{n} y_i^2 + \sum_{j=1}^{h}((a_j+\alpha_j)a_j-2)\alpha_j a_j - \sum_{k=1}^{g}((b_k-\tau_k)b_k-2)\tau_k b_k\right]$$

$$y_i = \frac{Y_i - \mu_i}{\sigma}, \quad a_j = \frac{A_j - \mu_{j+n}}{\sigma}, \quad b_k = \frac{B_k - \mu_{k+n+h}}{\sigma},$$

$$\alpha_j = \frac{\phi(a_j)}{\Phi(a_j)}, \quad \tau_k = \frac{\phi(a_k)}{1 - \Phi(a_k)}, \quad \text{and}$$

$$\mu_t = \beta_0 + \beta_1 X_{t1} + \cdots + \beta_p X_{tp}, \quad t = 1, \cdots, N.$$

The iteration in (A.2) is continued until the differences of two successive estimates are less than a specified value.

Trend Analysis on a Personal Computer: Problems and Solutions

J. E. Robinson
Syracuse University, Syracuse

ABSTRACT

The proliferation of inexpensive personal computers and user-friendly software has made it easy for geologists to contour and to apply trend-analysis techniques to a broad assortment of geologic data. Many of the current mapping programs designed for personal computers permit both polynomial and Fourier series analysis. Both methods are useful when correctly applied, however the wrong application can lead to serious distortions in the output thus creating apparent anomalies that do not exist. Problems may arise from the selection of contouring method, how the gridding or triangulation algorithm is applied to the data as well as from the trend-analysis method. The original data set may be modified before it is suitable for analysis. Polynomial trend-analysis data sets may need segmentation for valid results whereas Fourier analysis usually requires the set to be expanded in size. Processing problems are not restricted to personal computers and even occur with the most costly main frame software systems. However the more rigid requirements for small computer programs tend to restrict user selection in order to achieve processing speed and simplicity, generally at the expense of accuracy. Fortunately relatively simple tests permit the user to evaluate potential pitfalls in the application of contouring and trend-analysis programs and to determine avoidance strategy.

INTRODUCTION

There are a large number of contouring and trend-analysis programs available to the geologist who has access to a personal computer and who wishes to produce and analyze geologic data that are suitable for contouring as maps. These programs range from public domain software that the geologist can use without charge to costly commercialized systems (Leonard, 1986). They may be relatively simple in that the user must set all the parameters or they may be user friendly to the extent that the user has little control of the program. However in all situations the geologist should be aware of how the program utilizes the input data and how it produces the results. There are instances where even the best of programs can generate erroneous contours and false anomalies. Fortunately a few simple tests can provide the computer user with an overview of how the specific program handles the data. Similarly, simple corrections or changes to the input information may improve markedly the computer output maps.

The basic test is to have the program contour and analyze a known surface. The test surface must be simple because contouring of any complex surface that is defined only by discrete samples is subjective and open to interpretation. A uniform dipping plane is a good initial test surface. For other tests, a single anomalous sample can be located within the surface. The trivial situation of a flat surface is not suitable as many excellent contouring programs have trouble displaying an absolutely flat surface and this would not illustrate suitably the main purpose of the test. However a uniform dipping surface can test many of the contouring algorithms and also is useful for illustrating the biased results from the direct application of the Fourier series analysis included in many trend-analysis programs. A uniform dipping surface with a single anomalous sample is even better for displaying the contouring algorithms and can be used to test the polynomial trend-surface analysis program.

The contouring package used in the illustrations is termed "COGSSURF" (Guth, 1987). It is a good general purpose contouring package with user selection of either gridding or triangulation methods and is capable of both polynomial trend-surface analysis and Fourier series analysis. Quick look, printer plots are used for the illustrations.

CONTOURING

Contouring programs are designed to determine a best estimate of the values between control points to produce a global surface that can be defined at any location. Accuracy limitations are described by sampling theory (e.g. Shannon, 1949). There are two general contouring techniques used in most of the personal computer contouring programs. These are known as Gridding and Triangulation. The gridding method utilizes the original randomly located data points and a series of low-order local surfaces to generate a new and uniformly spaced grid of values that then are contoured by interpolation between grid points. Triangulation joins the original data points with a network of lines to form a series of triangles then interpolates the contours along the sides of the triangles. Smoothing procedures are used in both techniques to produce final maps. Both contouring techniques are useful with the triangulation method having some advantage with sparse or widely separated data points. Neither technique is effective for extrapolation beyond the data limits and it is usually the quality of the final contour smoothing algorithm that determines the acceptance of the map. Because contouring produces a global surface, a uniform grid of values also can be generated from a triangulated map should it be required for subsequent processing.

The first series of tests illustrate the effect of contouring a uniform dipping surface that has a single anomalous value in the center. Figure 1 displays a map of the triangulated contours of the test surface and its orthogonal image. Values are at the apex of the diagonal lines outlining the triangles. Figure 2 is the same input data but pregridded with a small grid interval. The rather erratic surface is caused by the gridding program extending the grid by including computed grid values as input rather than relying solely on the original data. The anomalies are most pronounced in the open areas between original values. Changing the grid size also affects the contours. Figure 3 illustrates the same surface with the grid interval doubled. The map is smoother but there are anomalies in areas that should display only uniform gradients. Figure 4 displays the grid interval doubled once again and finally made larger to an interval approximately equal to the original sample spacing. Both maps show less overall deviation than in the previous figures, however only the map in which the computed grid nodes fall on or close to the input data is a reasonable presentation of the real surface.

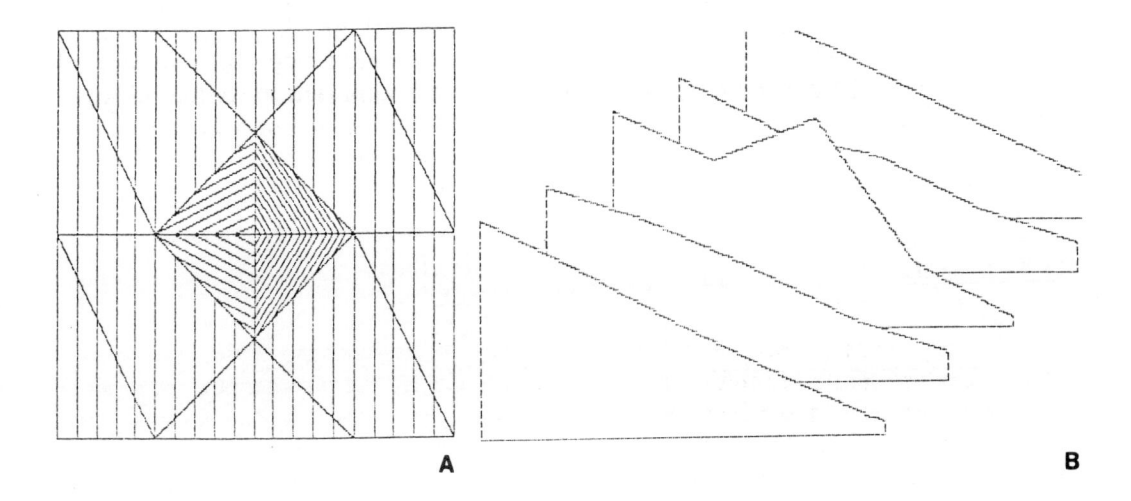

Figure 1. A, Map of uniform dipping plane with single anomaly in center. Contours were produced by triangulation with diagonal lines defining triangle edges. B, Is orthogonal view of test feature.

In this example where the input data are separated widely and has relatively uniform spacing, the triangulation method gives good results. The gridding technique requires more preprocessing analysis and only produces acceptable results when the computed grid nodes are in close proximity to the input sample points. There is distortion when the computed grid interval is less than the original sample spacing and again in the form of a smooth map when the computed grid is large compared to the sample spacing. The most accurate contours by the gridding method occur when the computed grid values fall as closely as possible to the original values. Where

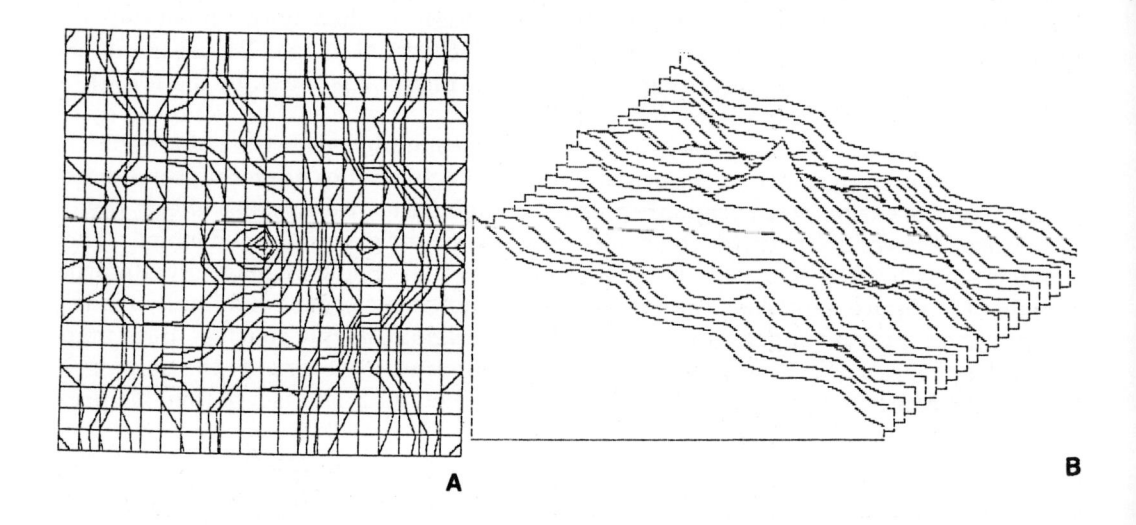

Figure 2. A, Is dipping plane and anomaly contoured with gridding algorithm and small grid interval. Grid is apparent on contoured map. B, Is orthogonal view of same data; numerous distortions are apparent in contoured surface.

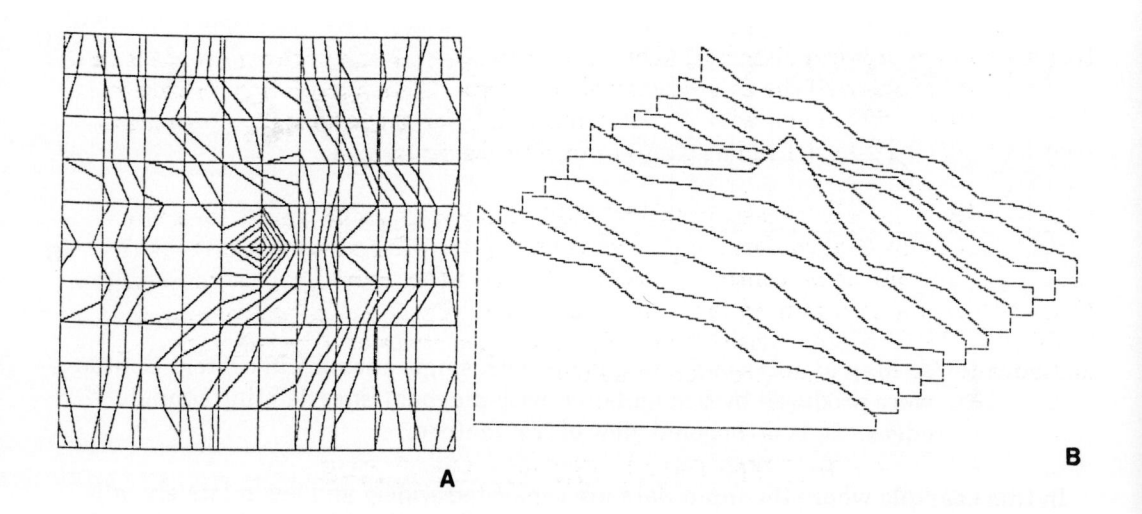

Figure 3. A, Is test example gridded with larger grid spacing. B, is orthogonal view; distortions are fewer and of less magnitude but are present.

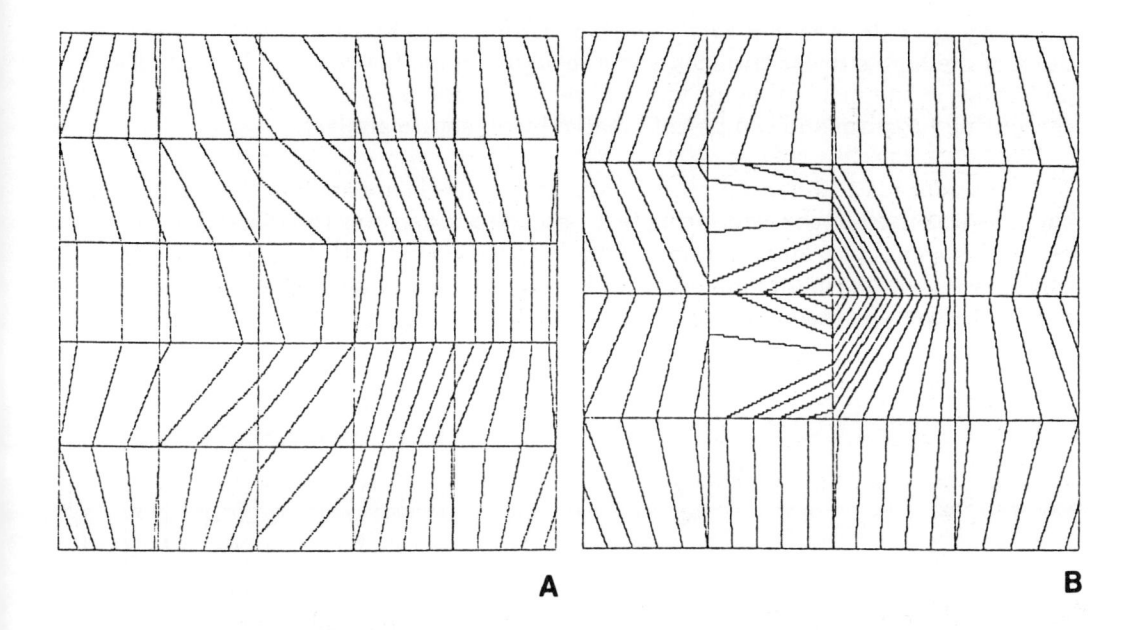

Figure 4. A, Test data contoured with grid interval again enlarged. Grid nodes are offset from input values and map appears smoother without major central anomaly. B, Are data with a larger grid interval but with node that falls on central anomaly.

there is a fairly uniform distribution of random samples, a rule of thumb suggests a grid interval of one-half the average sample interval. The triangulation technique does well with uniform and widely spaced input data however this advantage disappears when the data are random or is grouped in clusters.

Occasionally, it may be necessary to add additional input control values to produce satisfactory maps that conform to theoretical mapping rules and are aestheticly pleasing. The additional control values usually are added only to correct problems that arise during preliminary contouring runs. The values are estimated according to the most likely contour position and placed in open areas or along edges to create a smooth map without uncontrolled anomalies.

POLYNOMIAL TREND ANALYSIS

Polynomial trend-analysis usually is considered to be a procedure in which a low-order polynomial surface, the trend, is computed for the data set. This trend surface then is subtracted from the original data to produce a residual surface which displays local anomalies. Although both trend and residual surfaces may be contoured and interpreted, the main interest is usually in the anomalies displayed in the map of the residuals. Because these anomalies may be considered to be prospects, it is essential that they are valid and accurately located. Unfortunately, particularly in the situation of the higher order surfaces, some residual anomalies may not be real. False anomalies in the residual maps may be created by the polynomial behavior or by the contouring program and can be anticipated by an astute analyst.

The input data is a uniform dipping plane with a single anomaly in the center, the same as was used in the contouring test. Figure 5 illustrates the first-order surface as an orthogonal plot and a contoured map of the residual. The surface is protrayed correctly as a uniform dipping plane and there is a clearly defined single residual anomaly. There is rarely any problem with a first-order trend analysis.

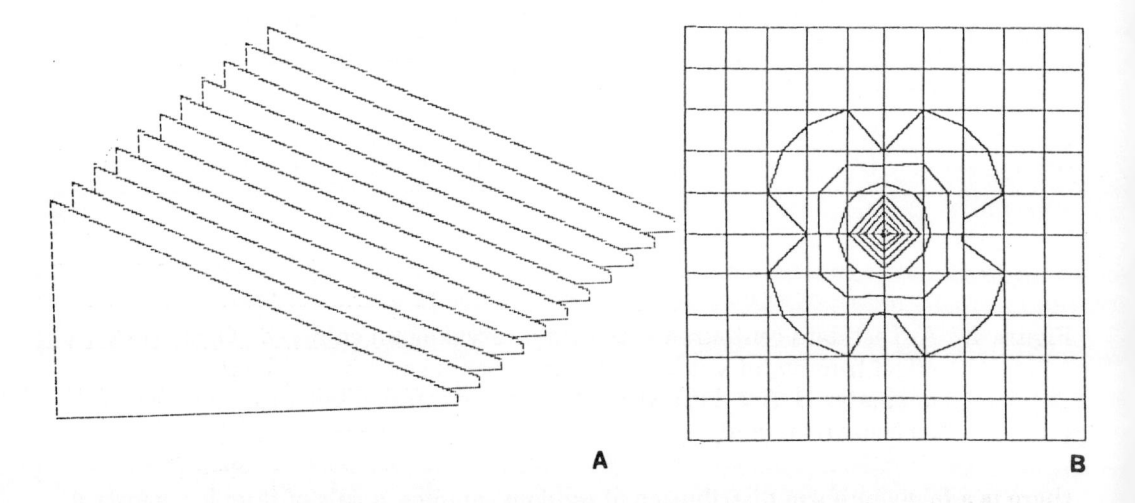

A   B

Figure 5. A, Is orthogonal view of first-order computed surface of dipping plane and anomaly used for contouring test. B, Is residual contoured map with grid superimposed; anomaly is little large but generally reasonable.

Figure 6 illustrates the second-order trend and residual. The map of this surface has to be smooth and if the calculations are done correctly then variations must be the result of the contouring. Fortunately contouring variations take place after the surface has been calculated so should not affect the residual map. However there are anomalies in the residual that are the result of both the surface curvature and of the contouring. Only the large central anomaly is valid.

Figure 7 displays the third-order surface and the residual contoured by both triangulation and gridding. Triangulation shows the central anomaly but also suggests it extends to the edges of the map. The anomaly has been enlarged. Conversely, grid contouring has reduced the size of the anomaly but created minor secondary anomalies about the edges. Neither map is entirely accurate.

Figure 8 illustrates the fourth-order surface and residual maps. Whereas the lower order surfaces were well behaved, the fourth-order surface with its greater freedom for oscillation creates obvious false anomalies around the margins of both the trian-

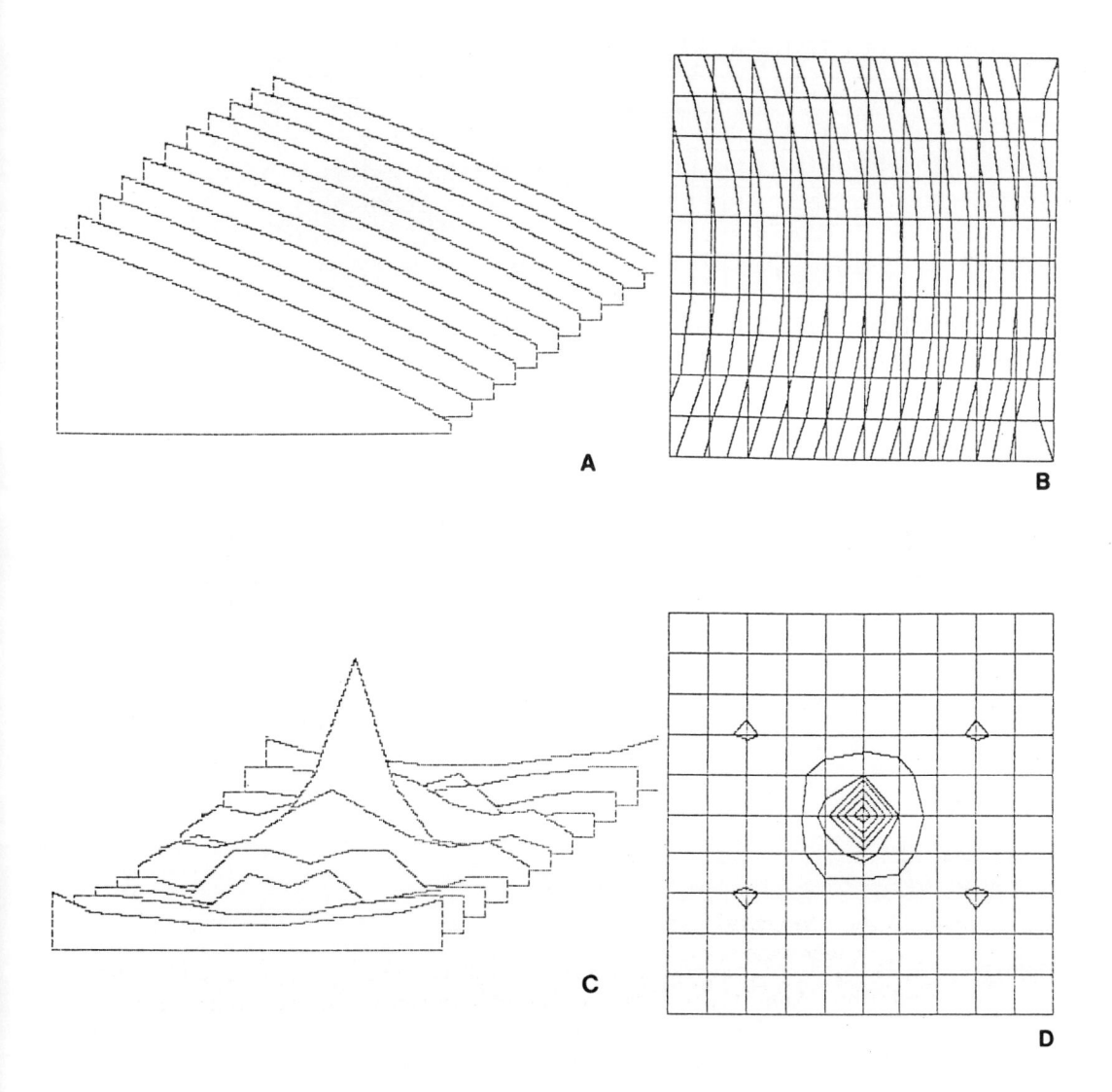

Figure 6. A, Is orthogonal view, and B, map and grid of second-order surface. C, Is orthogonal view, and D, map of residual. Surface is satisfactory except for minor variations along right-hand edge. Residual displays low amplitude false anomalies distinct from real central feature.

gulated and the gridded residual maps. As the spurious anomalies approach the size and amplitude of the real one it would be difficult to select the real anomalies without some preknowledge. In this particular situation, the fourth-order trend-analysis is not valid.

The test examples are simple surfaces with open areas designed to display problems created in the processing. Polynomial surfaces follow mathematical concepts that do not always fit in with geologic expectations. Higher order surfaces will oscillate and unless restricted by the data or by the program will try to create anomalies in areas

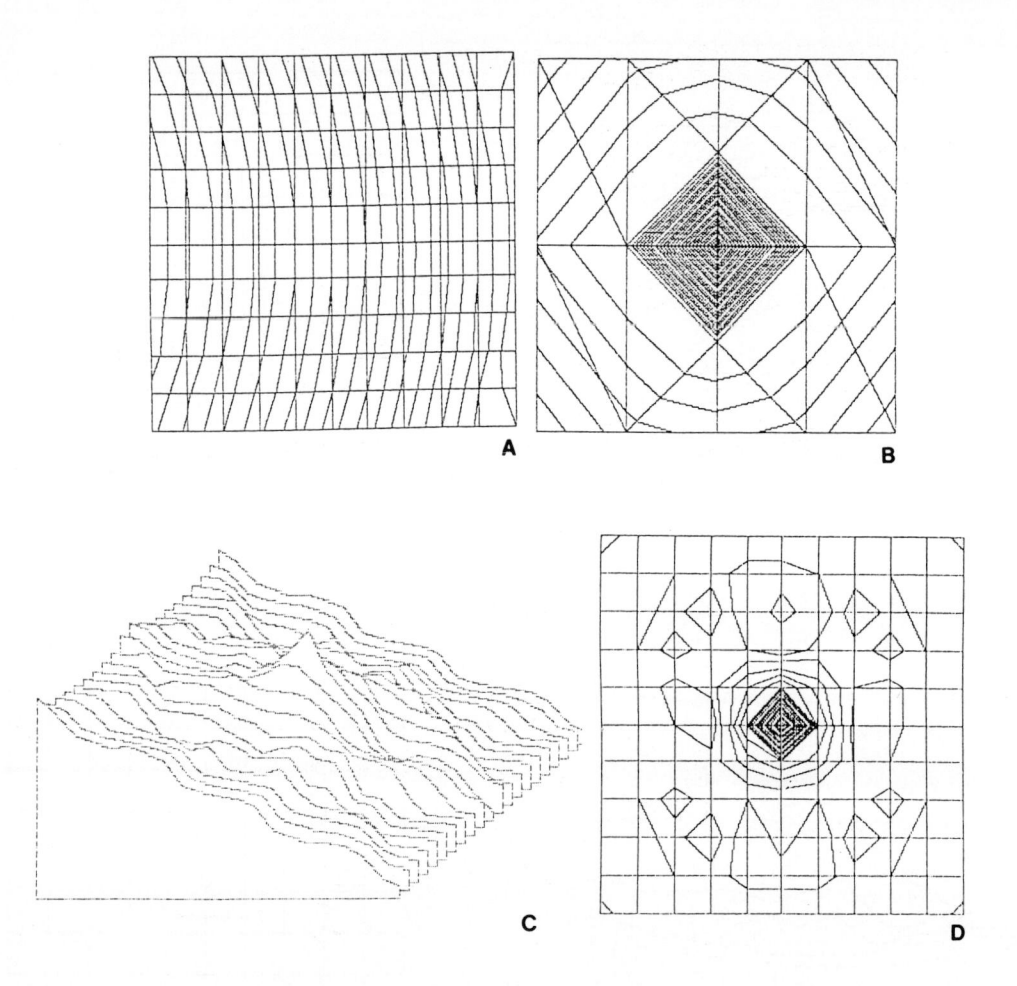

Figure 7. A, Is third-order surface, and B, third-order residual contoured by triangulation. Anomaly is displayed sharply in center and is expanded towards edges of map. C, and D, are orthogonal image and map produced by gridding; central anomaly appears reduced in size however that are false secondary anomalies distributed about sides of map.

of sparse data. There are differences in programs and different reactions to data sets, problems tend to be minimal with sets of closely spaced relatively uniform samples. Fortunately it is easy to test trend-analysis programs to determine the areas where extra care is required. There are few problems with the lower order trends but is is best to check all residual anomalies against the original data display to ensure validity.

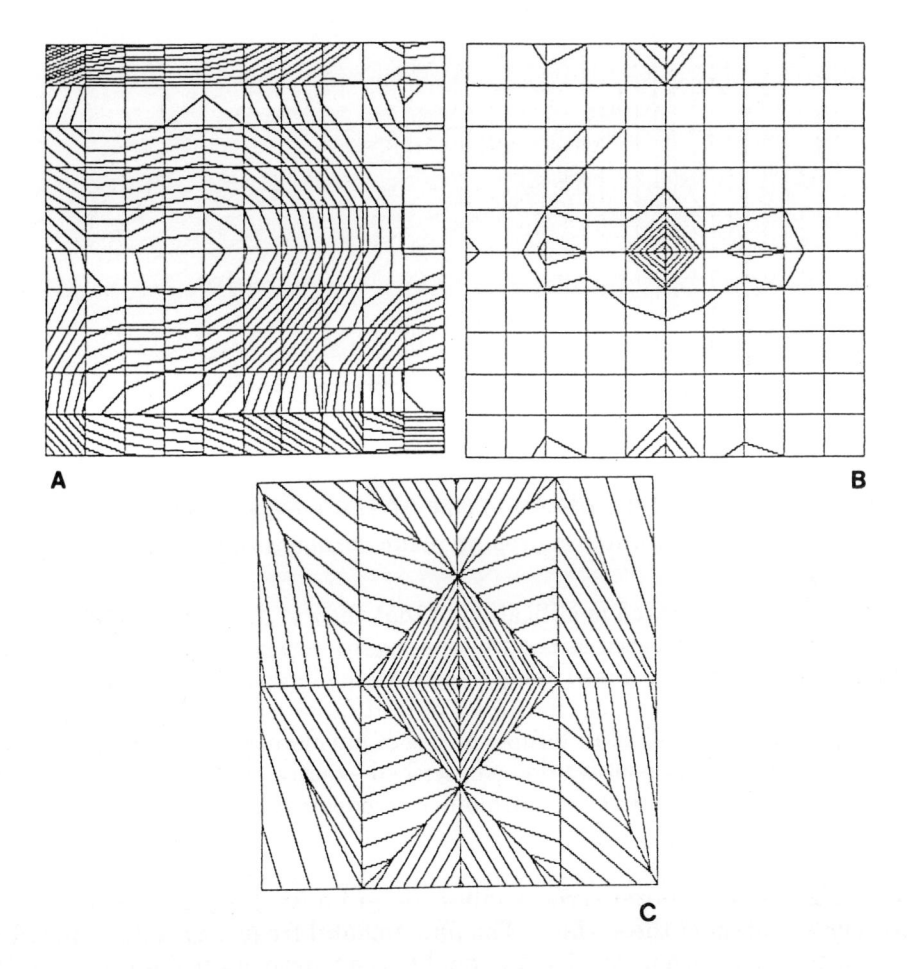

Figure 8. A, Is fourth-order surface. It does not represent known surface and is too high order polynomial for this type of data. B, Is gridded residual and displays false anomalies, as does C, residual produced by triangulation. Neither surface nor residuals are useful for data such as is in test example.

FOURIER SERIES ANALYSIS

Many personal computer mapping programs include an option allowing the user to calculate Fourier series harmonics in order to display cyclical components contained in the data set. This type of analysis can be useful, however unless the original data is screened and adjusted carefully, the displayed cycles relate more to the dimensions of the data set than to any contained features. This effect can be shown by computing the Fourier components for a simple uniform dipping surface (Fig. 9). Fourier theory states that any periodic function of time or distance can be described completely by the sum of a series of sinusiods consisting of a DC component, a fundamental frequency and integer harmonics of that fundamental. Map analysis

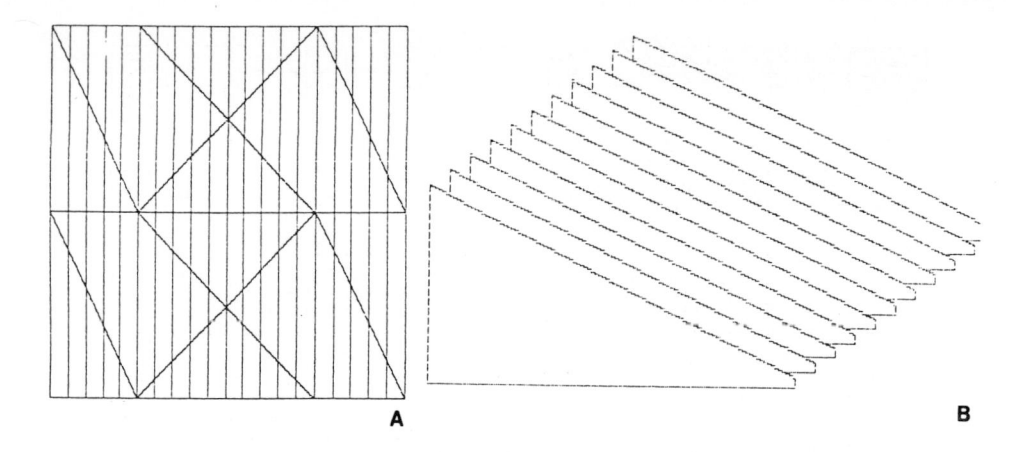

Figure 9. A, Is map, and B, orthogonal view of test surface used in Fourier analysis. Surface is uniform dipping plane without anomalies.

programs tend to be a biased modification of that theory. In practice the average value of the data is subtracted as the DC component. Then, the fundamental frequency is assumed to have a wavelength equal to the length of the data set. Harmonics then are fixed as integer multiples of that fundamental frequency and not by internal data components. If an infinite series of harmonics were computed and summed, the original data set would be described exactly. However the programs usually calculate only the first few harmonics erroneously suggesting cyclic features in the data.

In actual practice the Fourier series analysis program first calculates and removes the average elevation of the surface. The fundamental frequency is determined as the best fit sinusiod with a wavelength equal to the length of the data set (Fig. 10A). Much of the amplitude of the regional frequencies with wavelengths longer than the map is accounted for by this fundamental frequency. When the fundamental has been subtracted, a harmonic with half the previous wavelength is computed (Fig. 10B). Because the computed components are cyclic, the remainders, with one possible exception, also will be cyclic so that all subsequent harmonics will have a measurable amplitude. Fig. 10 (C and D) indicates that both the fifth- and the ninth-order harmonics still have significant relief although the original surface was absolutely plane.

Obviously a direct application of a Fourier Series analysis is not practical. However there are ways in which the procedure can be used to provide a valid description of and contained cyclic components. Regional trends as displayed in mapped data can be considered as the result of fundamental frequencies and harmonics with wavelengths longer than the dimensions of the map. One method of improving the analysis is to enlarge the data set making the map two or three times larger by additions of data about the margins. If these additional data are samples whose value is the average value of the main data, then the effect is to increase the wavelength of the fundamental. Also because the number of harmonics computed by the program usually is a function of the dimensions of the data set, increasing the size of the map

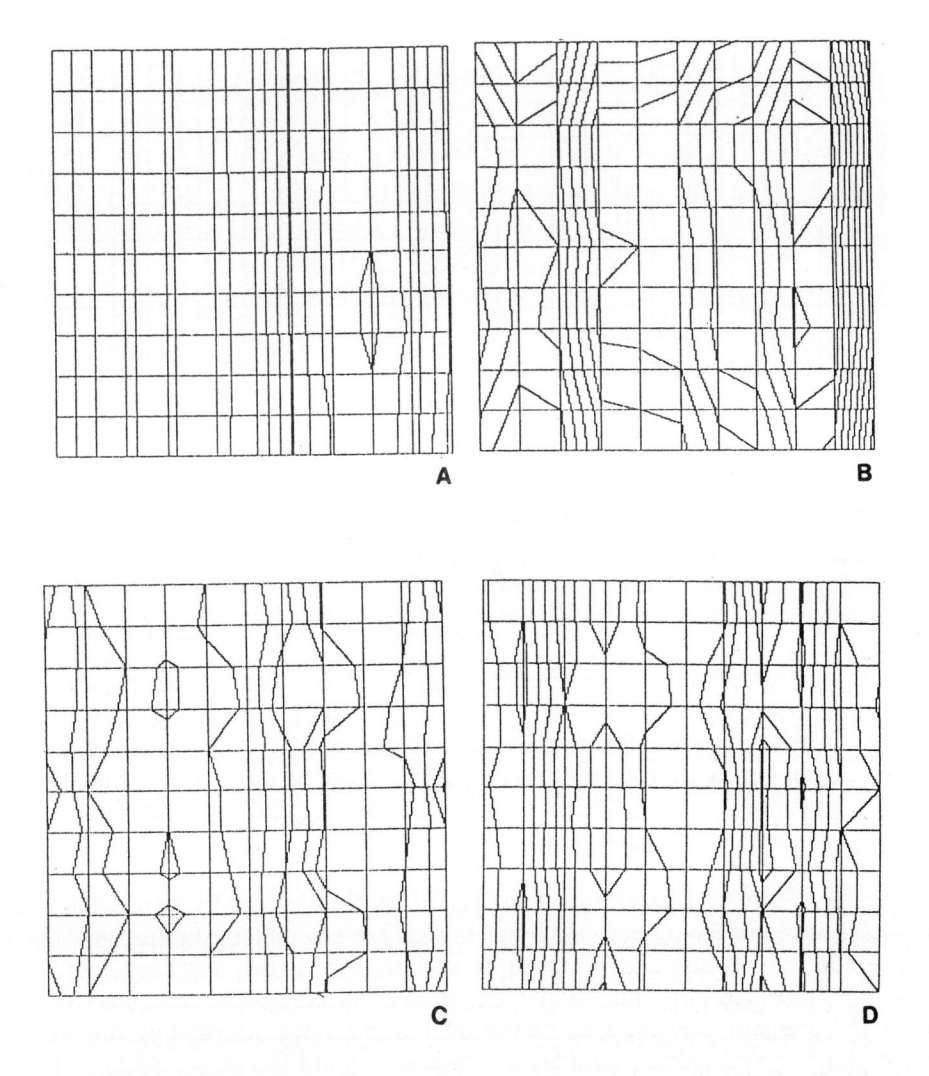

Figure 10. A, Fundamental frequency with wavelength equal to dimensions of map. B, Is first harmonic with wavelength equal to one-half map dimension. C, Is fifth harmonic, and D, is ninth harmonic. Amplitude of higher harmonics are decreasing, however they are accounting for residual amplitudes remaining from oscillations of low frequencies. None of displayed harmonics are real frequencies.

allows additional and more closely spaced harmonics to be computed for the true data set. The result is a more realistic description of cyclic features that actually exist in the data. Increasing the map size improves the validity of a Fourier analysis however, unless the new size is picked so that the slope of the central portion of the fundamental approximates that of the regional trend in the data set there will be problems (Fig. 11).

An even more effective procedure for using a Fourier series analysis to determine the presence of valid cyclic components within a map area is to make two modifications to the data before the analysis. If a first-order polynomial surface is computed and

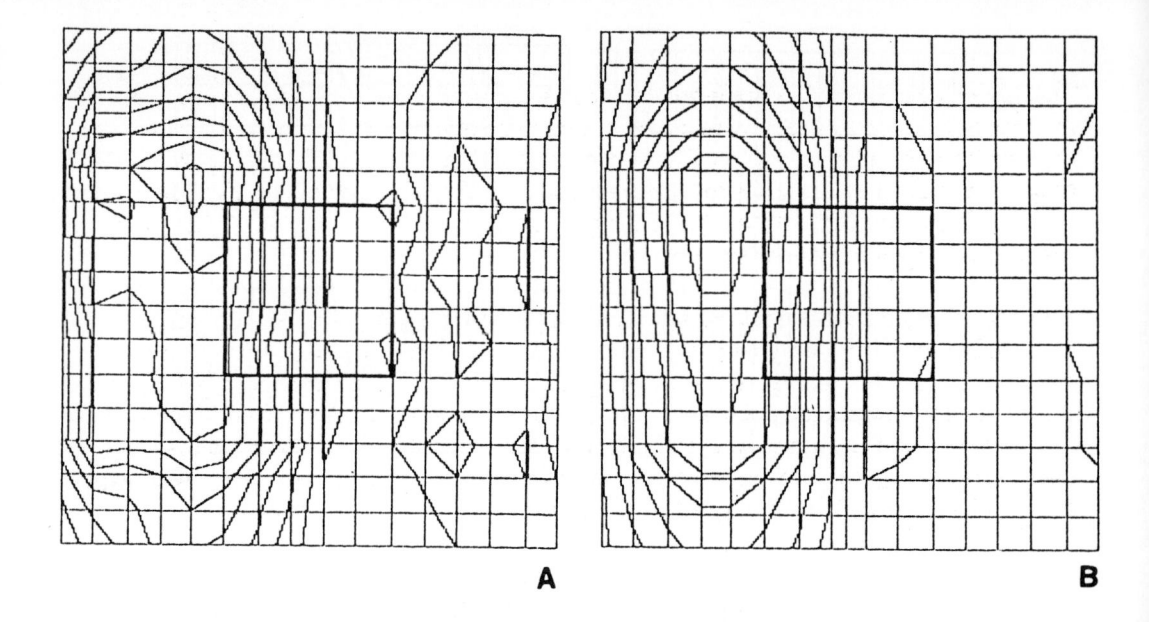

Figure 11. A, Fundamental frequency, and B, fifth-order harmonic computed for map enlarged from test example enclosed in center square. Approximation is improved for analysis of simple test map but for other data could have been better if first-order trend surface first was removed. However if first-order surface had been removed only trivial flat map would have remained in test example.

subtracted from the original data set the results will be markedly improved. This eliminates those low-frequency components that have wavelengths that are longer than the map dimensions and which cannot be described with any degree of accuracy by a simple Fourier series. Removal of the first-order trend also removes the average value or DC component of the map so that the map can be enlarged by the simple addition of a few zero valued samples in the area beyond the map margins. Enlarging the map is useful because it permits the calculation of a wider range of harmonics with wavelengths that are not dependent on the original map dimensions.

CONCLUSIONS

Contouring programs differ from one another in their approach and solution to the contouring dilemmas. They also differ in those areas where they have problems. However contouring tests, especially those that do not at first use a smoothing option, usually will give an indication of how the program handles the data and whether there is a potential problem. The test data then can be contoured a second time using the smoothing option to check on changes to the style and configuration of the contour display. Where some programs have been known to cross contours during smoothing, others approach the best contouring standards. The tests are worthwhile.

There also are a variety of polynomial type trend-analysis programs. Some are modified or restricted to minimize generation of unwarranted anomalies. However a high-order polynomials will oscillate, consequently, unless a different mathematical approach is used in the program, there may be problems with fourth or higher order trends. Tests can indicate the need for care and can be devised to show the program response to specific situations.

All data for Fourier series analysis should be examined and in most situations modified before any analysis is attempted. The only exception is in the limited situation where there are no low-frequency components with wavelengths longer than the map dimensions contained in the data. Such a situation would be indicated if the computed fundamental frequency had zero amplitude. The simple tests are useful and are important because they permit the user, even a user with little experience, to be aware of the strengths and weaknesses of geologic software.

REFERENCES

Leonard, J.E., 1986, CEED II, Special CEED II review: Geobyte, v. 1, no. 4, p. 14-20.

Guth, P.A., 1987, COGSSURF, Contour interactive contouring package: Peter A. Guth, 105-A Chaucer Rd., Mt. Laurel New Jersey 08054.

Shannon, C.E., 1949, Communication in the pressence of noise, IRE Proc., v. 37, 10 p.

Index

Accumulation production rate model, 262
ADVISOR, 362
aeromagnetics, 105
ALGOL, 189
alpha exponent, 260
analysis of landforms, 228
analytical approximation, 342
anomaly, 166, 169
Arabo-Nubian Massif, 276
Archean crust, 40
arsenic, 428
Asia, 400
association analysis, 91
association map, 175, 176, 177, 178
Atkin's notation, 390
Australia, 264, 265, 266
Austria, 98

Baltic Shield, 30
BASIC, 151
basic ultrabasic rocks, 179
basin analysis, 326
batholithic granite intrusions, 191
Bayesian statistics, 5, 26, 89, 115
bedrock lithology, 397
bibliographical file (BIB), 134
biogenic-derived constituents, 275
black box-model, 368
block kriging, 288, 292
BMDP software package, 78
Bohemian Massif, 75
Boolean representation, 171
Bouguer anomalies, 206
Brasil, 156, 371, 373, 375
BSE images, 307

California, 54, 401
Canada, 423, 428
Canadian Shield, 2
Cartographic data, 125, 126
categorical variables, 172
characteristic analysis, 92, 93, 169, 180
China, 371, 373, 375
CIPW norm, 133
C language, 142
classification, 24, 26, 216
CLUSTAN, 89, 92
cluster analysis, 88, 333
CODASYL-type databases, 414
conditional simulation, 243, 250
coefficient of resemblance, 43
COGSSURF, 436
coherency analysis, 67, 73
comparison map, 49
computer languages
    ALGOL, 189
    BASIC, 151
    C, 142
    FORTRAN, 67, 122, 415
computer programs
    ADVISOR, 362
    BMDP, 78
    CLUSTAN, 89,92
    COGSSURF, 436
    FIESTA, 216
    FINDER, 115
    1ST CLASS, 360
    GEONIX, 132, 137
    GMI-PACK, 216
    INTERCRAST, 189
    ISOPERS, 325
    MAPCOMP, 44

MARICA, 214
MINEVAL, 274
NCHARAN, 109
PALEO, 325
PEREC, 344
POROS, 304
PROGNOS, 150
PROSPECTOR, 352
RECLAS, 325
SCANDING, 121
SEAPUP, 253, 255
SPANS, 19
THE CONSULTANT, 353
The Deciding Factor, 352
THE EDITOR, 353
connectivity, 374, 399
contingency table, 152
contouring programs, 436
critical genetic factor, 56, 57
crude oil reserves, 347
cybernetics, 120
Czechoslovakia, 148

Darcy equation, 302
data
   acquisition, 136, 396
   cartographic, 125, 126
   compression, 25
   integration, 109, 113, 116
   missing, 16, 19, 46
   multivariate, 24
   processing, 272
   quantitative, 101
   remote sensing, 216
   simulated, 426
   SPOT, 225, 230
database
   CODASYL - types, 414
   geological, 411
   global, 368, 395, 407
   management functions, 137, 414, 418
   mineral indication, 411, 417
   relational, 414
   utilization, 416
decompaction, 322
Delphi, 86
dendrogram, 333
depositional settings, 349

diagenetic histories, 348, 349
digital image-processing, 196, 207
digital terrain model (DTM), 44, 101, 225
drilling model, 256
drilling profiles, 333

Eastern Alps, 101
economic accumulation size model, 262
empirical discriminant analysis, 25
environmental problems, 291
estimation variance, 282
expert systems, 222, 348
expert systems "shells", 352
exploration
   data, 324
   database system, 418
   model, 260
   target, 204
exploratory integration studies, 111
Europe, 371, 373

Factor analysis, 25
facts and rules entry, 355
false anomalies, 436
fault systems, 243
favorability indices, 87, 89, 93
field parameters evaluation, 165
FIESTA, 216
FINDER, 115
Finland, 23, 187, 195, 295, 207, 410
1ST CLASS, 360
fluid flow, 316, 348
FORTRAN, 67, 122, 415
Fourier series analysis, 57, 443, 444
France, 189
frequency distributions, 201
fundamental frequency, 445, 444
fuzzy-c-means-clustering, 88

GDR, 326
geochemical
   anomaly, 171, 223
   data, 110, 168, 207
   exploration, 220
   fields, 59, 109
   file (GCHEM), 135
   indicators, 421

INDEX

program, 160
signature, 8, 17
geochemistry, 24, 106
Geofields, 54, 70
geographic information systems (GIS), 1, 19, 101, 102, 114, 214, 215
geologic maps, 397
geological databank, 411
geological diversity, 383, 388, 390
geological predicting, 121
geology fields, 58
GEONIX, 132, 137, 143
geophysical imagery, 201, 203
geostatistical analysis, 274, 281
GIS, see geographic information systems
global databank, 368, 395, 407
GMI-PACK, 216
gold, 60, 72, 428
    deposits, 2, 6, 76
    mineralization, 75
    potential, 77, 421
    shear zones, 220
goodness-of-fit, 18
graphic functions, 140
gravity fields, 66
Greece, 84
gridding, 436

Harmonics, 445
hetrogeneities, 348, 349
hydrocarbons, 245, 318
hydrographic pattern, 144
hydrothermal
    alteration, 58
    ore deposits, 186

IBM PC, 360
image processing, 196, 198, 207, 304, 306
    also see digital image processing
Industrial Mineral Data Bank, 411
informatics, 120, 121
integration studies, 109
intelligent classification, 219, 221
INTERCRAST, 188, 189
interpolation methods, 240
intrinsic sample (IS) theory, 54
isodensity contouring, 141

ISOPERS, 325
Israel, 274

Kansas, 46
knowledge base, 353
kriging, 240, 244, 248, 251, 282
    also see block kriging, 288, 292
krynine classification, 369

LANDSAT, 58, 197, 198, 208
Lapland, 29, 203
latin-square approximation method, 343
lead time model, 262
least-squares (LS) method, 421
lineaments, 101
linear features, 235
lithological description, 369
Lodeve Basin, 225

Magnetic fields, 68
Mahalanobis distance, 24
man-machine approaches, 121, 126
MAPCOMP, 44
MARICA, 214
Massif Central, 220
mathematical functions, 138
mathematical models, 322
matrix methods, 340
maximum likelihood method, 422
Meguma Terrane, 2, 6, 423, 428
metal mineralizations, 207
metallic mineral resources, 159
metallogenetic provinces, 30
metallogenic units, 148, 149
Middle East, 371, 373
minerals
    deposits, 84
    exploration, 1, 201
    inventory file, 99
    occurrences, 159
    resources, 368, 383
mineral-bearing sequence, 275
mineral-deposits predicting, 121
mineral indication database, 411, 417
mineral resources
    assessment, 84, 100, 411
    diversity, 388

mineralization, 66, 171
mineralogical descriptors, 412
MINEVAL, 274
missing data, 16, 19, 46
models, 27, 179, 257, 316, 327
    accumulation production rate, 262
    bauxite, 179
    black box, 368
    digital terrain, 44, 101, 225
    drilling, 256
    economic accumulation size, 262
    formulation, 171
    granitoid rocks, 179
    lead time, 262
    limitations, 267
    mathematical, 138
    mineralization, 171
    paleospace, 322
    paleostructural, 334
    prediction, 148, 149
    space, 328, 329, 330
    volcano-sedimentary, 179
Monte Carlo simulation, 86, 340, 341
multiple censored data, 424
multiple censoring, 422
multiple linear regression, 78
multivariate analysis, 24, 77, 88

Natural resource investigations, 125
NCHARAN, 109
Negev, 276
Nevada, 54, 401
Nordkalott project, 23
North Sea Sector, 305, 312, 313
North Sea Well Study, 314, 315
Norway, 23
notched box plots, 385, 386
Nova Scotia, 2, 6, 423, 428
numerical methods, 322

Objective approach, 90
oil accumulation, 261
oil exploration, 239
oil-in-place expectation curves, 345
oil reservoirs, 302
oil shale, 276
onshore petroleum traps, 257

ore-bearing bedrock, 28
orebody, 274
Ore Deposit Data Bank, 410, 411
ore favorability index, 84
ore-localizing conditions, 186

PALEO, 325
paleospace modeling, 322
paleostructural modeling, 334
panning prospecting, 82
PASCAL, 189
pathfinder elements, 421
Pechelbronn Field, 240
PEREC, 344
permeability, 302, 307, 311
permeability barriers, 349
permeability/porosity crossplot, 303
personal computers, 325
petrographic file (EXPLIQ), 135
petroleum prospects, 340
petrological description, 412
phosphate, 276, 290
polynomial trend analysis,
    see trend analysis
pore geometry, 306, 307, 308
pore geometry classification, 311
pore networks, 304
POROS, 304
porosity, 302, 322
porphyry deposit, 190
Precambrian areas, 158
prediction models, 148, 149
principal components analysis (PCA),
    208
probabilistic estimation, 53
producing probability map, 247
PROGNOS, 150
prognostic assessment, 148
PROSPECTOR, 352

Q-analysis, 371, 374, 381, 399
Quaternary lithology, 103
quantitative data, 101, 297
query language, 415

Raster mode processing, 218, 220
RECLAS, 325

INDEX

regionalized variable, 282
Region - SCANDING, 121
regression analysis, 424
relational database, 414
reliability of conclusion, 359
remote-sensing data, 216
reserve estimates, 281
reservoir geometry, 348
reservoir lithologies, 316
residual maps, 440
resources per unit area, 259, 261
Rhein Graben, 240
rock index systems, 413
Rosenbusch-Zirkel classification, 369, 397
Rotem Oil-Shale deposit, 279, 285, 286

Sampling, 161
satellite imagery, 201, 202, 207
SCANDING,
　see region SCANDING
scheme of dissimilarity, 50
SEAPUP, 253, 255
second derivative, 169
sedimentary structures, 354, 360
siliciclastic depositional systems, 350
silver, 60, 72
simulated data, 426
simulated perspective views, 235
South Yorqe'am Phosphate Deposit, 283
space model, 328, 329, 330
spatial analysis, 43
spatial relation, 111
spatial variability, 290
spatial variability patterns, 297
SPANS, 19
SPOT data, 225, 230
spreadsheet approach, 361
standardization, 43, 44
statistical methods, 160
statistical pattern integration, 3
structure fields, 64
subjective approach, 85, 86
sulphur, 428
Sweden, 23

Target area, 167
terrain model,
　see ditigal terrain model (DTM)
Tethys Ocean, 275
Tertiary intrusives, 71
The CONSULTANT, 353
"The Deciding Factor", 352
The EDITOR, 353
thematic map, 111
time-petrographic index, 398
time-petrographic numbers, 369
trend surfaces, 240, 435, 439
triangulation, 436
tungsten, 428
Turkey, 84

Uncertainty, 19
uranium mine, 225
USA, 189, 371, 373, 401
USSR, 121, 128, 187, 189, 397, 400, 401
utilization databanks, 416

Variability patterns, 297
variogram, 168, 244, 246, 251, 282, 284, 288
vector mode processing, 219, 220
vein mineralization, 190
Venn diagrams, 3, 4
volcano-sedimentary model, 179

Washington, 401
wavelength, 444, 445
weights, 45
WTMC, weighted and targeted multi-variate criterion, 56

Zohar phosphate deposit, 293, 294, 296